Stress Response

METHODS IN MOLECULAR BIOLOGY™

John M. Walker, SERIES EDITOR

99. **Stress Response:** *Methods and Protocols*, edited by *Stephen M. Keyse, 2000*

98. **Forensic DNA Profiling Protocols,** edited by *Patrick J. Lincoln and James M. Thomson, 1998*

97. **Molecular Embryology:** *Methods and Protocols*, edited by *Paul T. Sharpe and Ivor Mason, 1999*

96. **Adhesion Protein Protocols,** edited by *Elisabetta Dejana and Monica Corada, 1999*

95. **DNA Topoisomerases Protocols:** *II. Enzymology and Drugs,* edited by *Mary-Ann Bjornsti and Neil Osheroff, 2000*

94. **DNA Topoisomerases Protocols:** *I. DNA Topology and Enzymes,* edited by *Mary-Ann Bjornsti and Neil Osheroff, 1999*

93. **Protein Phosphatase Protocols,** edited by *John W. Ludlow, 1998*

92. **PCR in Bioanalysis,** edited by *Stephen J. Meltzer, 1998*

91. **Flow Cytometry Protocols,** edited by *Mark J. Jaroszeski, Richard Heller, and Richard Gilbert, 1998*

90. **Drug–DNA Interaction Protocols,** edited by *Keith R. Fox, 1998*

89. **Retinoid Protocols,** edited by *Christopher Redfern, 1998*

88. **Protein Targeting Protocols,** edited by *Roger A. Clegg, 1998*

87. **Combinatorial Peptide Library Protocols,** edited by *Shmuel Cabilly, 1998*

86. **RNA Isolation and Characterization Protocols,** edited by *Ralph Rapley and David L. Manning, 1998*

85. **Differential Display Methods and Protocols,** edited by *Peng Liang and Arthur B. Pardee, 1997*

84. **Transmembrane Signaling Protocols,** edited by *Dafna Bar-Sagi, 1998*

83. **Receptor Signal Transduction Protocols,** edited by *R. A. John Challiss, 1997*

82. **Arabidopsis Protocols,** edited by *José M Martinez-Zapater and Julio Salinas, 1998*

81. **Plant Virology Protocols:** *From Virus Isolation to Transgenic Resistance,* edited by *Gary D. Foster and Sally Taylor, 1998*

80. **Immunochemical Protocols (2nd. ed.),** edited by *John Pound, 1998*

79. **Polyamine Protocols,** edited by *David M. L. Morgan, 1998*

78. **Antibacterial Peptide Protocols,** edited by *William M. Shafer, 1997*

77. **Protein Synthesis:** *Methods and Protocols,* edited by *Robin Martin, 1998*

76. **Glycoanalysis Protocols (2nd. ed.),** edited by *Elizabeth F. Hounsell, 1998*

75. **Basic Cell Culture Protocols (2nd. ed.),** edited by *Jeffrey W. Pollard and John M. Walker, 1997*

74. **Ribozyme Protocols,** edited by *Philip C. Turner, 1997*

73. **Neuropeptide Protocols,** edited by *G. Brent Irvine and Carvell H. Williams, 1997*

72. **Neurotransmitter Methods,** edited by *Richard C. Rayne, 1997*

71. **PRINS and *In Situ* PCR Protocols,** edited by *John R. Gosden, 1996*

70. **Sequence Data Analysis Guidebook,** edited by *Simon R. Swindell, 1997*

69. **cDNA Library Protocols,** edited by *Ian G. Cowell and Caroline A. Austin, 1997*

68. **Gene Isolation and Mapping Protocols,** edited by *Jacqueline Boultwood, 1997*

67. **PCR Cloning Protocols:** *From Molecular Cloning to Genetic Engineering,* edited by *Bruce A. White, 1997*

66. **Epitope Mapping Protocols,** edited by *Glenn E. Morris, 1996*

65. **PCR Sequencing Protocols,** edited by *Ralph Rapley, 1996*

64. **Protein Sequencing Protocols,** edited by *Bryan J. Smith, 1997*

63. **Recombinant Protein Protocols:** *Detection and Isolation,* edited by *Rocky S. Tuan, 1997*

62. **Recombinant Gene Expression Protocols,** edited by *Rocky S. Tuan, 1997*

61. **Protein and Peptide Analysis by Mass Spectrometry,** edited by *John R. Chapman, 1996*

60. **Protein NMR Techniques,** edited by *David G. Reid, 1997*

59. **Protein Purification Protocols,** edited by *Shawn Doonan, 1996*

58. **Basic DNA and RNA Protocols,** edited by *Adrian J. Harwood, 1996*

57. **In Vitro Mutagenesis Protocols,** edited by *Michael K. Trower, 1996*

56. **Crystallographic Methods and Protocols,** edited by *Christopher Jones, Barbara Mulloy, and Mark R. Sanderson, 1996*

55. **Plant Cell Electroporation and Electrofusion Protocols,** edited by *Jac A. Nickoloff, 1995*

54. **YAC Protocols,** edited by *David Markie, 1996*

53. **Yeast Protocols:** *Methods in Cell and Molecular Biology,* edited by *Ivor H. Evans, 1996*

52. **Capillary Electrophoresis Guidebook:** *Principles, Operation, and Applications,* edited by *Kevin D. Altria, 1996*

51. **Antibody Engineering Protocols,** edited by *Sudhir Paul, 1995*

50. **Species Diagnostics Protocols:** *PCR and Other Nucleic Acid Methods,* edited by *Justin P. Clapp, 1996*

49. **Plant Gene Transfer and Expression Protocols,** edited by *Heddwyn Jones, 1995*

48. **Animal Cell Electroporation and Electrofusion Protocols,** edited by *Jac A. Nickoloff, 1995*

47. **Electroporation Protocols for Microorganisms,** edited by *Jac A. Nickoloff, 1995*

46. **Diagnostic Bacteriology Protocols,** edited by *Jenny Howard and David M. Whitcombe, 1995*

45. **Monoclonal Antibody Protocols,** edited by *William C. Davis, 1995*

44. **Agrobacterium Protocols,** edited by *Kevan M. A. Gartland and Michael R. Davey, 1995*

43. **In Vitro Toxicity Testing Protocols,** edited by *Sheila O'Hare and Chris K. Atterwill, 1995*

42. **ELISA:** *Theory and Practice,* by *John R. Crowther, 1995*

41. **Signal Transduction Protocols,** edited by *David A. Kendall and Stephen J. Hill, 1995*

40. **Protein Stability and Folding:** *Theory and Practice,* edited by *Bret A. Shirley, 1995*

39. **Baculovirus Expression Protocols,** edited by *Christopher D. Richardson, 1995*

METHODS IN MOLECULAR BIOLOGY™

Vol 99

Stress Response

Methods and Protocols

Edited by

Stephen M. Keyse

ICRF Laboratories, University of Dundee, UK

Humana Press ✳ Totowa, New Jersey

Cover design by Patricia F. Cleary.

Cover photo: Figure 2, from Chapter 17, Analysis of the Mammalian Heat Shock Response: *Inducible Gene Expression and Heat Shock Factor Activity*, by A. Mathew, Y. Shi, C. Jolly, and R. I. Morimoto.

For additional copies, pricing for bulk purchases, and/or information about other Humana titles, contact Humana at the above address or at any of the following numbers: Tel: 973-256-1699; Fax: 973-256-8341; E-mail: humana@humanapr.com, or visit our Website at www.humanapress.com

Library of Congress Cataloging in Publication Data

Stress response : methods and protocols / edited by Stephen M. Keyse.
       p.cm. -- (Methods in molecular biology ; v. 99)
   Includes bibliographical references and index.
   ISBN 0-89603-611-1 (alk. paper)
   1. Stress (Physiology)--Laboratory manuals. 2. Molecular biology--Laboratory
manuals. 3. Cellular signal transduction--Laboratory manuals. 4. Heat shock
proteins--Laboratory manuals. 5. Gene expression--Laboratory manuals. I. Keyse,
Stephen M. II. Methods in molecular biology (Clifton, N.J.) ; v. 99

QP82.2.S8 S886 2000
572--dc21

                                                                                              99-042177

Preface

Mammalian cells have evolved a complex multicomponent machinery that enables them to sense and respond to a wide variety of potentially toxic agents present in their environment. These stress responses are often associated with an increased cellular capacity to tolerate normally lethal levels of an insult. The realization that the mammalian stress response may be intimately linked with many human diseases, including rheumatoid arthritis, ischemia, fever, infection, and cancer, has led to an explosion of interest in this research area.

Stress Response: Methods and Protocols brings together a diverse array of practical methodologies that may be employed to address various aspects of the response of mammalian cells to environmental stress. The protocols are carefully described by authors who have both devised and successfully employed them, and they represent a mixture not only of well-established techniques, but also new technologies at the leading edge of research. The areas covered include the detection and assay of stress-induced damage, the activation of signal transduction pathways, stress-inducible gene expression, and stress protein function. Although no volume of this size can be comprehensive and the topics covered reflect a personal choice, it is hoped that it will prove of substantial interest and use to a wide range of research workers in the field.

As with all the volumes in the *Methods in Molecular Biology* series, the aim is to present protocols that carry sufficient background information and experimental detail to be used as a primary reference at the laboratory bench. To this end the reader is provided with step-by-step guidance for each method, including details of reagents, equipment, and other requirements. However, it should be remembered that many of the procedures described are highly involved and should not be approached without considerable preparation and preliminary validation. In this way it is hoped that they will be used both directly and as the basis for modification and adaptation to specific experimental systems and objectives.

Finally, I would like to thank the series editor, John Walker, for the invitation to compile this volume, my colleagues within ICRF and the research community in Dundee for their continued support and to express my gratitude to all of the authors involved in this project for tolerating my numerous reminders and for the uniformly high quality of their contributions.

Stephen M. Keyse

To Margaret

Contents

Preface v

Contributors xiii

PART I. DETECTION AND ASSAY OF STRESS-INDUCED DAMAGE

1 Identifying and Counting Protein Modifications Triggered
   by Nitrosative Stress
   Prabodh K. Sehajpal and Harry M. Lander 3

2 Determination of Carbonyl Groups in Oxidized Proteins
   Rodney L. Levine, Nancy Wehr, Joy A. Williams,
   Earl R. Stadtman, and Emily Shacter 15

3 Quantitation of 4-Hydroxynonenal Protein Adducts
   Koji Uchida and Earl R. Stadtman 25

4 Detection of Oxidative Stress in Lymphocytes Using
   Dichlorodihydrofluorescein Diacetate
   Cecile M. Krejsa and Gary L. Schieven 35

5 The Measurement of Protein Degradation in Response
   to Oxidative Stress
   Thomas Reinheckel, Tilman Grune,
   and Kelvin J. A. Davies 49

PART II. THE ACTIVATION OF SIGNAL TRANSDUCTION BY CELLULAR STRESS

6 Analysis of the Role of the AMP-Activated Protein Kinase
   in the Response to Cellular Stress
   D. Grahame Hardie, Ian P. Salt, and Stephen P. Davies 63

7 Detection and Activation of Stress-Responsive Tyrosine Kinases
   Gary L. Schieven 75

8 Detection of DNA-Dependent Protein Kinase in Extracts
   from Human and Rodent Cells
   Yamini Achari and Susan P. Lees-Miller 85

9 Expression and Assay of Recombinant ATM
  Yael Ziv, Sharon Banin, Dae-Sik Lim, Christine E. Canman,
  Michael B. Kastan, and Yosef Shiloh 99

10 Detection and Purification of a Multiprotein Kinase Complex
  from Mammalian Cells: *IKK Signalsome*
  Frank Mercurio, David B. Young, and Anthony M. Manning 109

11 Methods to Assay Stress-Activated Protein Kinases
  Ana Cuenda 127

12 Monitoring the Activation of Stress-Activated Protein Kinases
  Using GAL4 Fusion Transactivators
  Chao-Feng Zheng and Li Xu 145

13 Use of Kinase Inhibitors to Dissect Signaling Pathways
  Ana Cuenda and Dario R. Alessi 161

14 The Development and Use of Phospho-Specific Antibodies
  to Study Protein Phosphorylation
  Jeremy P. Blaydes, Borek Vojtesek, Graham B. Bloomberg,
  and Ted R. Hupp 177

15 Peptide Assay of Protein Kinases and Use of Variant Peptides
  to Determine Recognition Motifs
  D. Grahame Hardie 191

PART III. THE ANALYSIS OF STRESS-INDUCED GENE EXPRESSION

16 Assaying NF-κB and AP-1 DNA-Binding
  and Transcriptional Activity
  Judith M. Mueller and Heike L. Pahl 205

17 Analysis of the Mammalian Heat Shock Response:
  Inducible Gene Expression and Heat Shock Factor Activity
  Anu Mathew, Yanhong Shi, Caroline Jolly,
  and Richard I. Morimoto 217

18 Approaches to Define the Involvement of Reactive Oxygen
  Species and Iron in Ultraviolet-A Inducible Gene Expression
  Charareh Pourzand, Olivier Reelfs, and Rex M. Tyrrell 257

19 The Human Immunodeficiency Virus LTR-Promoter Region
  as a Reporter of Stress-Induced Gene Expression
  Michael W. Bate, Sushma R. Jassal, and David W. Brighty 277

20 SAGE: *The Serial Analysis of Gene Expression*
  Jill Powell 297

21 Analysis of Differential Gene Expression Using the SABRE
     Enrichment Protocol
     *Daniel J. Lavery, Phillippe Fonjallaz, Fabienne Fleury-Olela,
     and Ueli Schibler* 321
22 UVB-Regulated Gene Expression in Human Keratinocytes:
     Analysis by Differential Display
     *Harry Frank Abts, Thomas Welss, Kai Breuhahn,
     and Thomas Ruzicka* 347

PART IV. ANALYSIS OF STRESS PROTEIN FUNCTION

23 Heme Oxygenase Activity: *Current Methods and Applications*
     Stefan W. Ryter, Egil Kvam, and Rex M. Tyrrell 369
24 Analysis of Molecular Chaperone Activities Using In Vitro
     and In Vivo Approaches
     *Brian C. Freeman, Annamieke Michels, Jaewhan Song,
     Harm H. Kampinga, and Richard I. Morimoto* 393
25 Analysis of Chaperone Properties of Small Hsp's
     *Monika Ehrnsperger, Matthias Gaestel,
     and Johannes Buchner* 421
26 Analysis of Small Hsp Phosphorylation
     Rainer Benndorf, Katrin Engel, and Matthias Gaestel 431
27 Analysis of Multisite Phosphorylation of the p53
     Tumor-Suppressor Protein by Tryptic Phosphopeptide Mapping
     David W. Meek and Diane M. Milne 447
28 The Development of Physiological Models to Study Stress
     Protein Responses
     Ted R. Hupp 465
Index 485

Contributors

HARRY FRANK ABTS • *Department of Dermatology, Heinrich-Heine-University, Duesseldorf, Germany*

YAMINI ACHARI • *Department of Biological Sciences, University of Calgary, Calgary, Alberta, Canada*

DARIO R. ALESSI • *MRC Protein Phosphorylation Unit, University of Dundee, Dundee, UK*

SHARON BANIN • *Sackler School of Medicine, Tel Aviv University, Tel Aviv, Israel*

MICHAEL W. BATE • *Biomedical Research Centre, University of Dundee, Ninewells Hospital and Medical School, Dundee, UK*

RAINER BENNDORF • *Department of Anatomy and Cell Biology, University of Michigan Medical School, Ann Arbor, MI*

JEREMY P. BLAYDES • *Department of Molecular and Cellular Pathology, Ninewells Hospital and Medical School, University of Dundee, Dundee, UK*

GRAHAM B. BLOOMBERG • *The Department of Biochemistry, University of Bristol, Bristol, UK*

KAI BREUHAHN • *Department of Dermatology, Heinrich-Heine-University, Duesseldorf, Germany*

DAVID W. BRIGHTY • *Biomedical Research Centre, Ninewells Hospital and Medical School, University of Dundee, Dundee, UK*

JOHANNES BUCHNER • *Institut für Organische Chemie und Biochemie, Technische Universität München, Garching, Germany*

CHRISTINE E. CANMAN • *St. Jude's Children's Research Hospital, Memphis, TN*

ANA CUENDA • *MRC Protein Phosphorylation Unit, University of Dundee, Dundee, UK*

KELVIN J. A. DAVIES • *Ethel Percy Andrus Gerontology Center, University of Southern California, Los Angeles, CA*

STEPHEN P. DAVIES • *Department of Biochemistry, University of Dundee, Dundee, UK*

MONIKA EHRNSPERGER • *Institut für Biophysik und Physikalische Biochemie, Universität Regensburg, Regensburg, Germany*

KATRIN ENGEL • *Max-Delbrück-Centrum für Molekulare Medizin, Berlin, Germany*

FABIENNE FLEURY-OLELA • *Département de Biologie Moléculaire, University of Geneva, Geneva, Switzerland*

PHILIPPE FONJALLAZ • *Département de Biologie Moléculaire, University of Geneva, Geneva, Switzerland*

BRIAN C. FREEMAN • *Department of Cellular and Molecular Pharmacology, University of California at San Francisco, San Francisco, CA*

MATTHIAS GAESTEL • *Innovationskolleg Zellspezialisierung, Institut für Pharmazeutische Biologie, Martin-Luther-University Halle-Wittenberg, Halle, Germany*

TILMAN GRUNE • *Clinics of Physical Medicine and Rehabilitation, Humboldt University, Berlin, Germany*

D. GRAHAME HARDIE • *Department of Biochemistry, University of Dundee, UK*

TED R. HUPP • *Department of Molecular and Cellular Pathology, Ninewells Hospital and Medical School, University of Dundee, Dundee, UK*

SUSHMA R. JASSAL • *Biomedical Research Centre, Ninewells Hospital and Medical School, University of Dundee, Dundee, UK*

CAROLINE JOLLY • *Rice Institute for Biomedical Research, Northwestern University, Evanston, IL*

HARM H. KAMPINGA • *Rice Institute for Biomedical Research, Northwestern University, Evanston, IL*

MICHAEL B. KASTAN • *St. Jude's Children's Research Hospital, Memphis, TN*

STEPHEN M. KEYSE • *Imperial Cancer Research Fund Molecular Pharmacology Unit, Biomedical Research Centre, Ninewells Hospital and Medical School, University of Dundee, Dundee, UK*

CECILE M. KREJSA • *Department of Environmental Health, University of Washington, Seattle, WA*

EGIL KVAM • *Department of Pharmacy and Pharmacology, University of Bath, Bath, UK*

HARRY M. LANDER • *Department of Biochemistry, Weill Medical College of Cornell University, New York, NY*

DANIEL J. LAVERY • *Glaxo Wellcome Experimental Research, Institut de Biologie Cellulaire et de Morphologie, University of Lausanne, Lausanne, Switzerland*

SUSAN P. LEES-MILLER • *Department of Biological Sciences, University of Calgary, Calgary, Alberta, Canada*

RODNEY L. LEVINE • *Laboratory of Biochemistry, National Heart, Lung, and Blood Institute, National Institutes of Health, Bethesda, MD*

DAE-SIK LIM • *St. Jude's Children's Research Hospital, Memphis, TN*

ANTHONY M. MANNING • *Signal Pharmaceuticals Inc, San Diego, CA*

ANU MATHEW • *Rice Institute for Biomedical Research, Northwestern University, Evanston, IL*

DAVID W. MEEK • *Biomedical Research Centre, Ninewells Hospital and Medical School, University of Dundee, Dundee, UK*

FRANK MERCURIO • *Signal Pharmaceuticals Inc, San Diego, CA*

ANNAMIEKE MICHELS • *Faculty of Medical Sciences, University of Groningen, Groningen, The Netherlands*

DIANE M. MILNE • *Biomedical Research Centre, Ninewells Hospital and Medical School, University of Dundee, Dundee, UK*

RICHARD I. MORIMOTO • *Rice Institute for Biomedical Research, Northwestern University, Evanston, IL*

JUDITH M. MUELLER • *Department of Obstetrics and Gynecology, University Hospital Freiburg, University of Freiburg, Freiburg, Germany*

HEIKE L. PAHL • *Head of Experimental Anesthesiology Section, University Hospital Freiburg, University of Freiburg, Freiburg, Germany*

CHARAREH POURZAND • *Department of Pharmacy and Pharmacology, University of Bath, Bath, UK*

JILL POWELL • *The Richard Dimbleby Department of Cancer Research, Imperial Cancer Research Fund Laboratory, St. Thomas' Hospital, London, UK*

OLIVIER REELFS • *Department of Pharmacy and Pharmacology, University of Bath, Bath, UK*

THOMAS REINHECKEL • *Department of Experimental Surgery, University of Magdeburg, Germany*

THOMAS RUZICKA • *Department of Dermatology, Heinrich-Heine-University, Duesseldorf, Germany*

STEFAN W. RYTER • *Division of Endocrinology, Metabolism, and Molecular Medicine, Department of Internal Medicine, Southern Illinois University School of Medicine, Springfield, IL*

IAN P. SALT • *Division of Biochemistry, Institute of Biomedical and Life Sciences, University of Glasgow, Glasgow, UK*

UELI SCHIBLER • *Département de Biologie Moléculaire, University of Geneva, Geneva, Switzerland*

GARY L. SCHIEVEN • *Immunology, Inflammation and Pulmonary Drug Discovery, Bristol–Myers Squibb Pharmaceutical Research Institute, Princeton, NJ*

PRABODH K. SEHAJPAL • *Department of Biochemistry, Weill Medical College of Cornell University, New York, NY; Department of Molecular Biology and Biochemistry, Guru Nanak Dev University, Amritsar, India*

EMILY SHACTER • *Division of Hematologic Products, Center for Biologics Evaluation and Research, Food and Drug Administration, Rockville, MD*

YANHONG SHI • *Rice Institute for Biomedical Research, Northwestern University, Evanston, IL*

YOSEF SHILOH • *Sackler School of Medicine, Tel Aviv University, Tel Aviv, Israel*

JAEWHAN SONG • *Rice Institute for Biomedical Research, Northwestern University, Evanston, IL*

EARL R. STADTMAN • *Laboratory of Biochemistry, National Heart, Lung, and Blood Institute, National Institutes of Health, Bethesda, MD*

REX M. TYRRELL • *Research Professor, Department of Pharmacy and Pharmacology, University of Bath, Bath, UK*

KOJI UCHIDA • *Laboratory of Food and Biodynamics, Nagoya University Graduate School of Bioagricultural Sciences, Nagoya, Japan*

BOREK VOJTESEK • *Masaryk Memorial Cancer Institute, Brno, Czech Republic*

NANCY WEHR • *Laboratory of Biochemistry, National Heart, Lung, and Blood Institute, National Institutes of Health, Bethesda, MD*

THOMAS WELSS • *Department of Dermatology, Heinrich-Heine-University, Duesseldorf, Germany*

JOY A. WILLIAMS • *Division of Hematologic Products, Center for Biologics Evaluation and Research, Food and Drug Administration, Rockville, MD*

LI XU • *Stratagene Cloning Systems, Inc, La Jolla, CA*

DAVID B. YOUNG • *Signal Pharmaceuticals Inc, San Diego, CA*

CHAO-FENG ZHENG • *Stratagene Cloning Systems, Inc, La Jolla, CA*

YAEL ZIV • *Sackler School of Medicine, Tel Aviv University, Tel Aviv, Israel*

I

DETECTION AND ASSAY
OF STRESS-INDUCED DAMAGE

1

Identifying and Counting Protein Modifications Triggered by Nitrosative Stress

Prabodh K. Sehajpal and Harry M. Lander

1. Introduction

Nitric oxide (NO) is a small, labile molecule that plays a key role as an intercellular and intracellular messenger. Interest in the mechanisms by which NO exerts its effects has increased tremendously along with the vast experimental evidence implicating NO in a variety of physiological and pathophysiological conditions. Here, we describe a method to evaluate the structural consequences of a NO-protein interaction.

1.1. Protein Modifications Triggered by Nitrosative Stress

Nitric oxide, one of the 10 smallest molecules found in nature, is a paramagnetic radical, thus making it very reactive. It is derived through the oxidation of one of the terminal guanidino-nitrogen atoms of L-arginine *(1)* by the enzyme nitric oxide synthase (NOS), which exists in three isoforms. Neuronal nitric oxide synthase (nNOS) (type I) *(2)* and endothelial nitric oxide synthase (eNOS) (type III) *(3)* were initially cloned from neuronal and endothelial cells, are calcium-calmodulin dependent and are expressed constitutively. Inducible nitric oxide synthase (iNOS) (type II) *(4)*, which was first identified in macrophages, is calcium independent and inducible. Many tissues have now been shown to express these isoforms. The regulation of these enzymes is complex and requires five cofactors: flavin adenine dinucleotide (FAD), flavin mononucleotide (FMN), heme, calmodulin and tetrahydrobiopterin and three cosubstrates, L-arginine, NADPH and O_2 *(5)*. NO has an unpaired electron on the nitrogen atom, making it highly reactive. Its reaction with redox modulators yields many reactive species such as NO^+, $ONOO^-$, and NOx^-. Hence the term "reactive nitrogen species" (RNS) is used in this chapter and refers to

From: *Methods in Molecular Biology, vol. 99: Stress Response: Methods and Protocols*
Edited by: S. M. Keyse © Humana Press Inc., Totowa, NJ

those species whose origin is the free radical •NO, but whose final chemical nature depends on its interaction with local redox modulators and the redox milieu of the cell (*6*).

Detailed structural analysis of proteins is essential for understanding their biological functions. The interaction of NO with proteins has been documented to play a critical role in diverse biological functions such as the regulation of blood pressure, neurotransmission, host defense and gene expression (*5,6*). Known targets of NO on proteins are cysteine and tyrosine residues and transition metals (*7*). The nitrosative reactions occurring at these sites have been shown to modify significantly the target protein's structure. This can lead to enzymatic activation, inhibition or even altered function (*8,9*). Much work has been carried out in characterizing the nature of NO–heme interactions, but less is known regarding NO–cysteine interactions. Nitrosothiol (RSNO) bonds are believed to occur via a substitution by NO or other RNS on free sulfhydryl groups (*7*). Nitrosothiol formation is a likely outcome of NO–cysteine interaction at an acidic pH or in an acidic microenvironment provided by adjacent amino acids (*7*). The function of several proteins has been postulated to be modulated by NO through formation of RSNO. The poor understanding of the role of NO or RNS in redox regulation is compounded by the lack of adequate techniques to identify the reactive species causing the change and to characterize the chemical intermediates and final structure.

1.2. Studying Proteins with Mass Spectrometry

Since the mid-1960s, mass spectrometry (MS) has been successfully used for exact mass measurement and determination of elemental composition of organic compounds. The only reason why this technique could not be applied to protein analysis was the inability to ionize the proteins into the gaseous phase. This process is a prerequisite for mass spectrometry, and many attempts were made to refine the technology. With the development of new ionization methods for protein molecules, such as laser desorption or electrospray ionization (ESI), a new area has opened that allows one to monitor and study subtle changes in protein structure. Commercial availability of these instruments has greatly increased the access to MS technology in the biochemical analysis of proteins and will soon be routine.

1.2.1. Matrix-Assisted Laser Desorption Ionization-Mass Spectrometry

A commonly used technique, matrix-assisted laser desorption ionization-mass spectrometry (MALDI–MS), utilizes a laser to ionize proteins into the gas phase. In order to ionize proteins, the laser energy must be transferred to the protein. This is done by incorporating a matrix, such as ferulic acid or sinapinic acid, in the protein solution and allowing this to dry on the surface of a probe, yielding tiny crystals. The laser then excites the matrix, which then ionizes the proteins. Resolution in MALDI–MS is dependent on good co-crystalization of sample and matrix. In MALDI–MS, the energy level can be regulated by varying the peak width and intensity of the laser. Energy is typically given in spurts of 100 ns to prevent any pyrolysis and fragmentation of proteins. MALDI is commonly used in combination with time-of-flight (TOF) mass analyzers. Because the TOF analysis is not limited by an upper mass range, it is a good choice for MALDI, which can produce high m/z ions *(10)*. MALDI–MS analysis is useful in analyzing heterogenous samples. The technique is generally less reliable in correctly predicting the mass of large proteins, as the resolution of the spectra decreases in the high mass range *(11)*. MALDI-MS has been used successfully in mapping linear epitopes of proteins *(12)*, identification of disulfide bonds *(13)* and locating cysteine groups *(14)*.

Despite the utility of using MALDI–MS in protein analysis, its use in studying delicate NO–protein interaction is limited. Although MALDI–MS can be used to identify stable end products of NO–protein interactions such as tyrosine nitration *(15)*, MALDI–MS cannot identify the more labile interactions such as the formation of S–NO and S–Fe complexes *(16)*. Because the energy required to ionize the proteins using MALDI-MS is very high, the labile NO–protein interactions are destroyed by its use. Therefore, we will focus on the gentle technique of electrospray ionization–mass spectrometry (ESI–MS).

1.2.2. Electrospray Ionization–Mass Spectrometry

Electrospray ionization–mass spectrometry (ESI–MS) analyzes proteins in an aqueous phase. A basic understanding of the ESI–MS technique is critical for obtaining meaningful information from an analysis. Most ionization sources create ions with a single charge, but ESI is unique in that it creates multiply charged ions. The electrospray is generated by applying a high electrical field to an aqueous or aqueous/organic solvent delivered by a capillary tube. The electric field transfers the charge to the liquid surface, and the spray then permits the formation of small charged droplets at the tip of the capillary. Dry gas, heat or both are used to evaporate the solvent around the charged droplets leading to a decrease in the size of the droplets and increase in the surface charge density. A curtain of nitrogen gas is then used to exclude large droplets from entering the mass analyzer and also to decluster the ions. An ESI source is typically interfaced with a quadrupole mass analyzer (**Fig. 1**).

Mass analyzers identify the ions based on their mass/charge ratio and not on their mass. One advantage of producing multiply charged ions by ESI is that a mass analyzer of limited mass/charge range can be used to study large proteins with an accuracy of 0.01% *(17)*.

counterflow gas

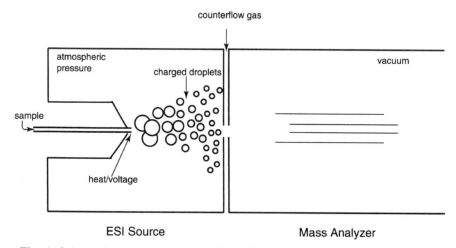

ESI Source   Mass Analyzer

Fig. 1. Schematic representation of ESI–MS. An aqueous sample is pumped into the ESI source where heat and/or voltage potential is applied to it. This nebulization leads to charged droplets of protein and solvent. As solvent is stripped off the protein, the charged droplets get smaller and smaller until a protein devoid of any solvent is analyzed in the mass analyzer. This process takes approx 1 ms.

The gentleness of an ESI source allows labile complexes to remain intact for detection by the mass analyzer, and hence ESI–MS is an ideal technique to determine protein–NO interactions in a rapid and sensitive manner. Mirza et al. *(18)*, using ESI–MS, monitored and identified specific modifications of a native peptide as a result of exposure to NO. Lander et al. *(19)* revealed the structural basis of the interaction between NO and the ras oncogene product p21. To identify the exact site of *S*-nitrosylation, they chemically digested Ras with cyanogen bromide (CNBr) and studied the fragments with ESI–MS. These studies resulted in identifying Cys 118 as the molecular target of NO on Ras. Ferranti et al. *(20)* used ESI–MS to characterize S-nitrosoproteins and quantitate *S*-nitrosothiols in blood. Matthew et al. *(21)* used ESI-MS to localize NO to Cys 62 on the p50 subunit of the transcriptional factor NF-kB.

Studies like these exemplify the profound structural insight that ESI–MS can provide and should serve as a signal that the time is ripe to use this technique. We now discuss how to identify the exact chemical modifications on proteins induced by NO and how to quantitate them.

2. Materials

1. A peptide (KNNLKECGLY, Mass = 1181.4 Da) corresponding to the C-terminus of the G-protein $G\alpha_{i1}$ was commercially synthesized.
2. The Ras protein was purified as described previously *(22)*.

3. NO gas cylinder (Matheson, East Rutherford, NJ) (*see* **Note 1**).
4. Nitrogen gas cylinder (Matheson).
5. 20 mM ammonium bicarbonate, pH 8.0 (*see* **Note 2**).
6. 1 mM N-ethylmaleimide (NEM). (*see* **Note 3**).
7. Cyanogen bromide (CNBr) (*see* **Note 1**).
8. Water/methanol/acetic acid (1:1:0.05, v/v/v, pH 3.0).
9. Water/methanol (1:1, v/v, pH 7.8).

3. Methods
3.1. Preparation of NO Solution

1. Sparge a solution of 20 mM ammonium bicarbonate solution, pH 8.0, in a rubber-stoppered polypropylene tube (*see* **Note 4**) for 15 min with nitrogen and then with NO gas. This results in a saturated solution of NO (1.25 mM). This solution may contain higher oxides of NO that are not quantified (*see* **Notes 1** and **2**).

3.2. Preparation of Samples for ESI–MS

3.2.1. In Vitro Samples

1. Native peptide: Mix 100 μM synthetic peptide in 5 μL of 20 mM ammonium bicarbonate, pH 7.8, with 20 μL of water/methanol/acetic acid (1:1:0.05, v/v/v, pH 3.0) and spray directly for ESI–MS analysis (*see* **Note 5**).
2. Peptide treated with NO: Add 100 μM NO to the peptide and immediately mix with water/methanol/acetic acid (1:1:0.05, v/v/v, pH 3.0) and analyze. In a separate experiment perform the same reaction at pH 7.8 and mix with water/methanol (1:1, v/v, pH 7.8) before analysis.
3. Treatment with N-ethylmaleimide (NEM): Add 1 mM NEM (*see* **Note 3**) to the peptide and incubate at 22°C for 5 min. Mix with water/methanol/acetic acid (1:1:0.05, v/v/v, pH 3.0) and analyze. In a separate experiment, treat the peptide with 1 mM NEM for 5 min at 22°C and then add 100 μM NO . Incubate for a further 5 min at 22°C and then analyze by ESI–MS (*see* **Note 6**).
4. Kinetics of the NO–peptide interaction: Add 100 μM NO to peptide in several tubes and incubate the mixtures for 2, 7, 24, and 70 h at 22°C (*see* **Note 7**). In a separate experiment add 1 mM NEM to peptide and incubate for 5 min at 22°C. 100 μM NO is then added, and the sample incubated at 22°C for 70 h. In all of these experiments, the incubation mixture is removed after the requisite time and then mixed with a combination of water/methanol/acetic acid (1:1:0.05, v/v/v, pH 3.0) and analyzed immediately (*see* **Notes 4–7**).
5. Chemical digestion: Add one small crystal of CNBr to 100 pmol of Ras protein in 20 μL of 0.1 M HCl in a 0.5-mL polypropylene tube (*see* **Note 1**). Digest for 10 min at room temperature. Following digestion, the sample is directly electrosprayed and analyzed (*see* **Note 8**).
6. Enzymatic digestion: Proteins can be digested by enzymes to yield known fragments. These fragments are then analyzed to yield information on the site of modification by NO (*see* **Note 9**).

3.3. Instrument Conditions for ESI–MS

Electrospray ionization-mass spectra are obtained on a Finnigan-MAT TSQ-700 triple quadrupole instrument. Because every commercial mass spectrometer is different in design, the parameters described here will be different for other mass spectrometers and these should be calibrated accordingly. Peptide and protein samples are electrosprayed from acidified (acetic acid) 50% methanolic solutions, pH 3.0. Concentrations of the analyte electrospray solution are in the range of 10–20 μM. The measurements of pH are made with a PHM 95 pH meter (Radiometer, Copenhagen) calibrated in aqueous solutions. No corrections are applied for the pH measurements of solutions containing methanol. The analyte solutions are infused into the mass spectrometer source using a Harvard syringe pump (model 24000-001) at a rate of 3 μL/min through a 100-μM (inner diameter) fused-silica capillary. The positive ion spectra obtained are an average of 16 scans and are acquired at a rate of 3 s/scan. The ion signals are recorded by a Finnigan ICIS data system operated on a DEC station 5000/120 system. The reconstructed molecular mass profiles are obtained by using a deconvolution algorithm (FinniganMat). Unless otherwise indicated, all data are acquired with a capillary transfer tube temperature at 125°C and 80 V potential difference between the capillary transfer tube and the tube lens of the mass spectrometer (*see* **Notes 10–12**).

3.4. Data Interpretation

Although every commercially available mass spectrometer provides a software package to calculate the molecular mass of multiply charged ions from the m/z ratios by using various deconvolution algorithms, it is always prudent to check the masses generated by the software with those manually calculated based on the m/z ratios. For the positive ion spectra of proteins, the calculation of molecular mass is based on two assumptions: (1) the adjacent peaks in a given spectrum differ by only one charge and (2) the charge is due to a cation attachment (usually a proton and rarely a mixture of proton and Na^+) to the molecular ion. With these assumptions, it follows that the observed mass/charge ratios for each member of the distribution of multiply protonated molecular ions are related by a series of simple linear simultaneous equations where mass (m) and charge (z) are unknown. Any two ions are sufficient to determine m and z. In a given spectra, the m/z values are obtained from the centroid of the peak consistent with the fact that isotopic contributions for these large multiply-charged ions are not resolved. A detailed account of the determination of molecular mass and the principles behind it have been reported *(23)*. In brief, based on the above presumptions, the relationship between a multiply charged ion at a given m/z (called p_1) with a charge z_1 and the mass (Mr) is:

$$p_1z_1 = Mr + Ma\, z_1 = Mr + 1.0079z_1 \tag{1}$$

Here, it is presumed that the charge-carrying species (Ma) is a proton and p_1 is the m/z of a chosen peak. The molecular mass of a second multiply-protonated ion at m/z (p_2, where $p_2 > p_1$), that is j peaks away from p_1 (e.g., $j_1 = 1$ for two adjacent peaks) is given by

$$p_2(z_1 - j) = Mr + 1.0079(z_1 - j) \tag{2}$$

Eqs. 1 and **2** can be solved for the charge p_1:

$$z_1 = j(p_2 - 1.0079)/(p_2 - p_1) \tag{3}$$

The molecular mass can be obtained by taking z_1 as the nearest integer value and, thus, the charge of each peak in the multiple charge distribution can be determined.

This approach can be used to determine the mass of the charged adduct species:

$$p_1 = (Mr/z_1) + M_a$$

As an example, **Fig. 2** shows a typical multiply-charged spectrum for horse myoglobin. Let us take two unknown charged ions, p_1 and p_2, with m/z ratios of 997.97 and 1060.27 respectively. The charge on the p_1 ion can be calculated by applying **Eq. 3**, presuming the charge-carrying adduct is a proton:

$$
\begin{aligned}
z_1 \quad &= j(p_2 - 1.0079)\,/\,(p_2 - p_1) \tag{3}\\
&= j(1060.27 - 1.0079)\,/\,(\,1060.27 - 997.97)\\
&= j(1059.26)/62.30\\
&= 1(1059.26)/62.30,\ \text{as } j = 1 \text{ for two adjacent multiply-charged ions}\\
&= 17
\end{aligned}
$$

The molecular mass of the p_1 ion can be calculated by placing the value of z_1 in **Eq. 1**:

$$
\begin{aligned}
p_1z_1 \quad &= \ Mr + 1.0079z_1\\
(997.97)(17) &= \ Mr + (1.0079)(17)\\
16{,}965.49 &= \ Mr + 17.1343\\
16{,}948.36 &= \ \text{mass (calculated)}
\end{aligned}
$$

The predicted molecular mass of horse myoglobin is 16,951.50 Da. The m/z spectrum in **Fig. 2** was deconvoluted via a computer algorithm, and the calculated mass of the protein was 16,951.00 Da. The manually calculated mass has an error of 0.018%, which indicates that the parameters selected for computer-

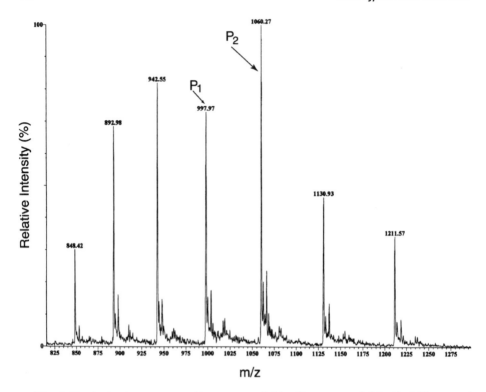

Fig. 2. Typical *m/z* spectrum of horse myoglobin. Horse myoglobin (1 µ*M* in water/ acetic acid, 1:0.05, v/v) was sprayed at 80°C and 60 V.

assisted algorithms are correct. For cases in which the mass of the protein is not known, a manually calculated mass serves as a good indicator of the mass of the protein. It is important to mention that wrongly selected deconvolution parameters could lead to erroneous mass calculations.

4. Notes

1. All the reactions with NO gas and CNBr should be performed in a fume hood.
2. Water used in all experiments should be ultrapure.
3. Ideally, the NEM solution should be made just prior to use. However, the frozen solution is stable for several days.
4. Samples should be stored in good polypropylene tubes. Glass can introduce contaminants such as sodium and potassium into the solution.
5. Once a sample is in solution, it should be analyzed as soon as possible to avoid decomposition and to minimize losses through adsorption.
6. Analysis of the test peptide, which contains one cysteine residue, yielded a mass of 1181.4 Da. Treatment with NO gas for 5 min at pH 3.0 led to the formation of a new species with a mass of 1210.5 Da. The observed mass difference of 29 Da

Table 1.
Theoretical Mass Shift Predicted in Various NO–Protein Interactions

Type of modification	Predicted mass shift (Da)
R–S–NO	+29
R–S–S–R	–2
R–SO$_3$H	+49
R–Tyr–NO$_2$	+45

agrees precisely with the molecular mass of NO (30 Da) minus the mass of the substituted proton (1 Da), thus suggesting very strongly the occurrence of a substitution reaction. This can be confirmed by treating the peptide with NEM (125 Da), an irreversible thiol-binding reagent. This led to the detection of a new species with a mass of 1306.5 Da, which is equal to the combined mass of peptide and NEM. Subsequent treatment of peptide + NEM with NO gas does not yield any new species of higher or lower mass. This indicates that NEM blocked the reaction of NO with the peptide and that NO formed an RSNO with the peptide. Treatment of peptide with NO at pH 7.8 leads to nitrosothiol formation, but to a much lesser extent. The major end product at pH 7.8 is the peptide dimer with a mass of 2362.8 Da.

7. Time is another variable that influences NO-dependent modification on the peptide. The peptide is allowed to react with NO for a variable period of time and then analyzed with ESI–MS. After 5 min of exposure to NO, nitrosothiol is the major product at pH 3.0. (mass = 1210.4 Da). After 2 h, and especially at 7 h, two new species arise with masses of 1230.4 and 1275.4 Da, respectively. One species has an increased mass of 49 Da, likely corresponding to a sulfonic acid derivative (R–SO$_3$H). The second species has an increased mass of 94 Da, most likely corresponding to a peptide with Tyr-NO$_2$ (45 Da) and sulfonic acid. After 24 h, very little starting peptide or nitrosothiols remains and the majority of the peptide had either sulfonic acid (mass = 1230.4 Da) or both Tyr–NO$_2$ and sulfonic acid (mass = 1275.4 Da). Formation of sulfonic acid is confirmed by pretreatment of the test peptide with NEM. Treatment of the peptide–NEM complex (mass = 1307 Da) for 70 h with NO yielded both the starting complex (1307 Da) and a peptide-NEM complex with a Tyr–NO$_2$. Neither sulfonic acid nor the nitration with sulfonic acid derivatization is seen (expected mass = 1356 and 1401 Da, respectively). *See* **Table 1**.

8. Chemical or enzymatic digestion of proteins is often used to identify potential targets of nitrosative reactions. Lander et al. *(24)* demonstrated that a single *S*-nitrosation event on Ras corresponded with enhanced guanine nucleotide exchange. To identify the site of *S*-nitrosation, Ras is digested with CNBr, yielding three major fragments, each with one cysteine residue. ESI–MS analysis of chemically digested Ras identifies three fragments with masses of 7203, 4540, and 6225 Da, respectively. In a separate experiment, Ras treated with NO gas and

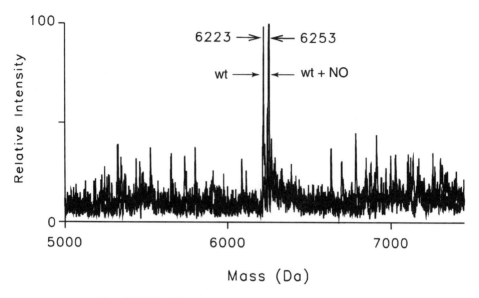

Fig. 3. ESI–MS analysis of Ras digested with CNBr.

subsequently digested with CNBr identified fragment 3 with a new molecular mass of 6253 Da. An increment of 29 Da in the mass is clearly indicative of S-nitrosylation at Cys 118 residue in this fragment (**Fig. 3**). This was later confirmed *(25)* by creating a Ras mutant, where cysteine 118 was mutated to serine (Ras C118S). ESI–MS analysis of NO-treated Ras C118S failed to identify an *S*-nitrosated CNBr fragment with a mass of 6253 Da.

9. Proteins can be digested with various enzymes to yield known fragments. The choice of the enzyme is dictated by the experimental design. Protocols for different enzymes, their specificity, pH optima and other relevant information for use in mass spectrometery can be accessed from the website http:\\prowl.rockefeller.edu.

10. The energy used to ionize the peptide and to decluster ions is critical for detection of the NO–peptide interaction. Too much energy that can be delivered either by increasing the declustering voltage or by increasing the temperature of the capillary transfer tube can lead to the destruction of labile *S*-NO bonds. Mirza et al. *(18)* found that nitrosated peptide could be easily detected at temperatures of 125°C and 150°C (mass = 1181.4 Da). At 175°C, approx 50% of the nitrosothiol bonds were broken (mass = 1210.4 Da), leading to the re-emergence of the native peptide (1181.4 Da), and at 200°C, nitrosothiols were undetectable.

11. Low settings of these parameters will allow solvent molecules to remain attached to the ionized protein molecules. This can lead to poor resolution and high noise levels in the spectra. Hence, it is very important to adjust carefully the settings of these two parameters to impart adequate energy to obtain a good spectra.

12. The choice of the solvent mixture is an important consideration for the ESI–MS analysis. Methanol is commonly included to assist in droplet formation. How-

ever, when working with heme protein, it has been observed that a higher volume of methanol in the solvent mixture may remove heme from the protein molecule and can also influence the intramolecular structure of the protein *(26)*. A mixture of 10% methanol and 5 m*M* ammonium bicarbonate has been reported to be a good solvent for the study of NO-modified heme interactions in MS experiments *(27)*.

References

1. Palmer, R. M. J., Rees, D. D., Ashton, D. S., and Moncada, S. (1988) L-arginine is the physiological precursor for the formation of nitric oxide in endothelium dependent relaxation. *Biochem. Biophys. Res. Commun.* **153,** 1251–1256.
2. Bredt, D. S. and Synder, S. H. (1994) Nitric oxide: a physiologic messenger molecule. *Annu. Rev. Biochem.* **63,** 175–195.
3. Lamas, S., Marsden, P. A., Li, G. K., Tempst, P., and Michel, T. (1992) Endothelial nitric oxide synthase: molecular cloning and characterization of a distinct constitutive enzyme isoform. *Proc. Natl. Acad. Sci. USA* **89,** 6348–6352.
4. Xie, Q.-W., Cho, H. J., Calaycay, J., Mumford, R. A., Swiderek, K. M., Lee, T. D., Ding, A., Troso, T., and Nathan, C. (1992) Cloning and characterization of inducible nitric oxide synthase from mouse macrophages. *Science* **256,** 225–228.
5. Nathan, C. and Xie, Q. W. (1994) Nitric oxide synthases: roles, tolls and controls. *Cell* **78,** 915–918.
6. Lander, H. M. (1997) An essential role for free radical derived species in signal transduction. *FASEB J.* **11,** 118–124.
7. Stamler, J. S. (1994) Redox signaling: nitrosylation and related target interactions of nitric oxide. *Cell* **78,** 931–936.
8. Estrada, C., Gomez, C., Martin-Nieto, J., De Frutos, T., Jimenez, A., and Villalobo, A. (1997) Nitric oxide reversibly inhibits the epidermal growth factor receptor tyrosine kinase. *Biochem. J.* **326,** 369–376.
9. Gardner, P. R., Costantino, G., Szabo, C., and Salzman, A. L. (1997) Nitric oxide sensitivity of the acotinases. *J. Biol. Chem.* **272,** 25,071–25,076.
10. Hillenkamp, F. and Karas, M. (1990) Mass spectrometry of peptides and proteins by matrix assisted ultraviolet laser desorption/ionization. *Methods Enzymol.* **193,** 281–295.
11. Rudiger, A., Rudiger, M., Weber, K., and Schomburg, D. (1995) Characterization of post translational modifications of brain tubulin by matrix assisted -laser desorption/ionization mass spectrometery: direct one-step analysis of a limited subtilisin digest. *Anal. Chem.* **224,** 532–537.
12. Zhao, Y. and Chait, B. T. (1994) Protein epitope mapping by mass spectrometry. *Anal. Chem.* **66,** 3723–3726.
13. Gehrig, P. M. and Biemann, K. (1996) Assignment of disulfide bonds in napin, a seed storage protein from Brassica napus using matrix assisted laser desorption/ionization mass spectrometery. *Pept. Res.* **9,** 308–314.
14. Wu, J., Gage, D. A., and Watson, A. (1996) A strategy to locate cysteine residues in protein by specific chemical cleavage following matrix assisted laser desorption /ionization time of flight mass spectrometry. *Anal. Biochem.* **235,** 161–174.

15. Plough, M., Rahbek-Nielsen, H., Ellis, V., Roestorff, P., and Dano, K. (1995) Chemical modification of the urokinase-type plasminogen activator and its receptor using tetranitromethane. Evidence for the involvement of specific tyrosine residues in both molecules during receptor-ligand interaction. *Biochemistry* **34,** 12,524–12,534.

16. Osipov, A. N., Gorbunov, N. V., Day, B. W., Elasyed, N. M., and Kagan, V. E. (1996) Electronspin resonance and mass spectral analysis of interaction of ferryl hemoglobin and ferryl myoglobin with NO. *Methods Enzymol.* **268,** 193–203.

17. Loo, J. A., Udseth, H. R., and Smith R. D. (1989) Peptide and protein analysis by electrospray ionization mass spectrometry and capillary electrophoresis. *Anal. Biochem.* **179,** 404–412.

18. Mirza, U. A., Chait, B. T., and Lander, H. M. (1995) Monitoring reactions of nitric oxide with peptides and proteins by electrospray ionization-mass spectrometry. *J. Biol. Chem.* **270,** 17,185–17,188.

19. Lander, H. M., Milbank, A. J., Tauras, J. M., Hajjar, D. P., Hempstead, B. L., Schwartz, G. D., Kraemer, R. T., Mirza, U. A., Chait, B. T., Campbell-Burk, S., and Quilliam, L. A. (1996) Redox regulation of cell signalling. *Nature* **381,** 38–381.

20. Ferranti, P., Malorni, A., Mammone, G., Sannolo, N., and Marino, G. (1997) Characterisation of *S*-nitrosohaemoglobin by mass spectrometry. *FEBS Lett.* **400,** 19–24.

21. Matthew, J. R., Botting, C. H., Panico, M., Morris, H. R., and Hay, R. T. (1996) Inhibition of NF-kB DNA binding by nitric oxide. *Nucleic Acids Res.* **24,** 2236–2243.

22. Campbell-Burk, S. L. and Carpenter, J. W. (1995) Refolding and purification of Ras proteins. *Methods Enzymol.* **250,** 3–13.

23. Mann, M., Meng, C. K., and Fenn, J. B. (1989) Interpreting mass of multiply charged ions. *Anal. Chem.* **61,** 1702–1708.

24. Lander, H. M., Ogiste, J. S., Pearce, S. F. A., Levi, R., and Novogrodsky, A. (1995) Nitric oxide-stimulated guanine nucleotide exchange on p21[ras]. *J. Biol. Chem.* **270,** 7017–7020.

25. Lander, H. M., Hajjar, D. P., Hempstead, B. L., Mirza, U. A., Chait, B. T., Campbell, S., and Quilliam, L. A. (1997) A molecular redox switch on p21[ras]. *J. Biol. Chem.* **272,** 4323–4326.

26. Katta, V. and Chait, B.T. (1991) Conformational changes in proteins probed by hydrogen exchange electrospray-ionization mass spectrometery. *Rapid Commun. Mass Spectrom.* **5,** 214–217.

27. Upmacis, R. K., Hajjar, D. P., Chait, B. T., and Mirza, U. A. (1997) Direct observation of nitrosylated heme in myoglobin and hemoglobin by electrospray ionization mass spectrometery. *J. Am. Chem. Soc.* **119,** 10,424–10,429.

Determination of Carbonyl Groups in Oxidized Proteins

Rodney L. Levine, Nancy Wehr, Joy A. Williams, Earl R. Stadtman, and Emily Shacter*

1. Introduction

There now exists a bewildering array of biological processes in which free radicals have been implicated *(1)*, and we assume that enzymes and structural proteins may be attacked whenever free radicals are generated. As a consequence, oxidative modification of proteins may occur in a variety of physiologic and pathologic processes. Although the distinction is sometimes arbitrary, these modifications may be primary or secondary. Primary modifications occur in metal-catalyzed oxidation, in radiation-mediated oxidation, and in oxidation by ozone or oxides of nitrogen. Secondary modifications occur when proteins are modified by molecules generated by oxidation of other molecules. One important example is the covalent modification of proteins by hydroxynonenal produced by oxidation of lipids *(2, see also*, Chapter 3 of this volume).

Carbonyl groups (aldehydes and ketones) may be introduced into proteins by any of these reactions, and the appearance of such carbonyl groups is taken as presumptive evidence of oxidative modification. The word *presumptive* is important because the appearance of carbonyl groups is certainly not specific for oxidative modification. For example, glycation of proteins may add carbonyl groups onto amino acid residues. Despite this caveat, the assay of carbonyl groups in proteins provides a convenient technique for detecting and quantifying oxidative modification of proteins.

Methods for determination of carbonyl content were discussed in an earlier article *(3)*. The methodology and discussion in that article remain useful and

*The authors of this chapter are government employees who performed their work in this capacity. The contents of this chapter are not copyrighted.

From: *Methods in Molecular Biology, vol. 99: Stress Response: Methods and Protocols*
Edited by: S. M. Keyse

can be read in conjunction with this chapter, which presents newer methods based on the reaction of carbonyl groups with 2,4-dinitrophenylhydrazine to form a 2,4-dinitrophenylhydrazone. These assays provide substantial improvements in both sensitivity and specificity. They employ high-performance liquid chromatography (HPLC) gel filtration or electrophoresis for removal of excess reagent and introduce Western blotting for sensitive and specific detection of the 2,4-dinitrophenyl group *(4)*. A similar technique was developed independently by Keller and colleagues *(5)*. This contribution has been slightly modified from **refs.** *6* and *7*.

1.1. Reaction with 2,4-Dinitrophenylhydrazine in 6 M Guanidine Followed by HPLC

2,4-Dinitrophenylhydrazine is a classical carbonyl reagent *(8)*, and it has emerged as the most commonly used reagent in the assay of oxidatively modified proteins. However, quantitative derivatization requires that a large excess of the reagent be present and that the reagent must be removed to allow spectrophotometric determination of the protein-bound hydrazone. Removal is required because the 2,4-dinitrophenylhydrazine reagent has significant absorbance at 370 nm so that residual reagent can cause an artifactual increase in the apparent carbonyl content of the sample. Recent investigations suggest that this may occasionally be a problem with the filter paper method, especially with samples containing low amounts of protein *(3)*. Previous articles described several methods for removal of reagent, including extraction by ethanol/ethyl acetate, reverse-phase chromatography, and gel filtration *(3,9)*. After extraction, the hydrazone can be determined spectrophotometrically by its absorbance at 370 nm.

High-pressure liquid chromatographs are widely available, and gel filtration by HPLC has proven to be a convenient and efficient technique for removal of excess reagent. Derivatized proteins are also separated by molecular weight, allowing a more specific analysis of carbonyl content. HPLC spectrophotometric detectors are also far more sensitive than stand-alone spectrophotometers; therefore, much less sample is required for quantitation of carbonyl content. The HPLC method has sufficient sensitivity for analysis of cells from a single tissue culture dish (~500,000 cells) or for analysis of the small amounts of tissue available from biopsy and autopsy samples. Most HPLC detectors provide chromatograms at two or more wavelengths, allowing carbonyl content to be followed at 370 nm and protein content at around 276 nm, obviating the need for a separate protein assay. If a diode-array detector is available, then one also obtains full spectra of the peaks. These can be useful in checking for contaminants that might artifactually affect either carbonyl or protein determination. As noted earlier, nucleic acid contamination could potentially interfere with the assay *(3)*; such contamination would be suspected if the protein spectrum were skewed from the 276-nm protein peak toward the 260-nm nucleic acid peak. Also, any chromophore which absorbs at 370 nm would interfere with quantitation of the hydrazone. Although total background can be determined on a blank sample not treated with 2,4-dinitrophenylhydrazine, availability of spectra may assist in identifying the chromophore and in devising methods for minimizing the background. Examples of such chromophores include heme from contaminating hemoglobin and retinoids from tissues such as liver (L. Szweda and R. L. Levine, unpublished observations). One can also check for artifactual elevations of the carbonyl content by pretreating the sample with sodium borohydride, which will reduce the carbonyl groups and render them unreactive to 2,4-dinitrophenylhydrazine *(3)*. If the borohydride-treated sample has an apparent carbonyl content above background, one should strongly suspect that it is artifactual.

Derivatizations with 2,4-dinitrophenylhydrazine are classically performed in solutions of strong acids such as 2 *M* HCl. There are two disadvantages of using of 2 *M* HCl in preparing samples for HPLC analysis. First, very few HPLC systems or columns can tolerate the HCl. Second, many proteins are insoluble in HCl and must be solubilized before injection into the HPLC system. These problems are dealt with by derivatization in 6 *M* guanidine at pH 2.5. The guanidine effectively solubilizes most proteins, whereas pH 2.5 is compatible with most HPLC systems (*see* **Note 1**).

The rate of reaction of proteins studied thus far is much faster than suggested earlier *(3)*, being essentially complete within 5 min. The time-course may vary for different proteins and should be checked if this is a concern. Moreover, samples should not be allowed to stand in the derivatization solution longer than about 15 min because a slow reaction leads to the introduction of the 2,4-dinitrophenyl moiety in a nonhydrazone linkage. When either glutamine synthetase or bovine serum albumin were incubated for 120 min, the apparent carbonyl content was artifactually increased by approx. 0.2 mol/subunit. The nature of the reaction is not yet understood, but it can be avoided simply by injecting the sample onto the HPLC at a fixed reaction time, such as 10 min. Autosamplers sold by different manufacturers are capable of performing the derivatization and then injecting the sample at fixed reaction times.

1.2. Reaction with 2,4-Dinitrophenylhydrazine in Sodium Dodecyl Sulfate Followed by HPLC or Western Blotting

The 6 *M* guanidine buffer specified in **Subheading 1.1.** can take its toll on the HPLC (*see* **Note 1**). Moreover, the guanidine must be removed if one wants to analyze the derivatized proteins by sodium dodecyl sulfate (SDS)-polyacrylamide gel electrophoresis. To obviate these two problems we developed a method to derivatize proteins in SDS instead of guanidine *(10)*.

We have less experience with the SDS system for HPLC than with the guanidine system, and pilot experiments should be performed to determine whether the SDS or guanidine systems may be preferred for particular samples. When checked with several samples of oxidized glutamine synthetase, the analytical results were the same with both methods. There are at least two advantages of the SDS system over the guanidine system: Backpressure is substantially lower with the SDS system, and the problem of salt corrosion is essentially eliminated. One disadvantage of SDS is that the resolution of proteins from each other and from the reagent may not be as good as in guanidine, presumably due to the relatively large size of the SDS micelles *(11)*. We had previously found that separation of reagent from proteins was actually better in the SDS system than in guanidine *(6)*. This is not true for Zorbax columns available currently, evidently due to a change in manufacturing procedure (*see* **Note 2**).

2. Materials

2.1. Reaction with 2,4-Dinitrophenylhydrazine in 6 M Guanidine

1. Buffer for gel filtration and for derivatization blank: 6.0 *M* guanidine HCl, 0.5 *M* potassium phosphate, pH 2.5
2. Derivatization solution: 10 m*M* 2,4-dinitrophenylhydrazine, 6.0 *M* guanidine HCl, 0.5 *M* potassium phosphate, pH 2.5 (*see* **Note 3**).
3. Column: HPLC gel filtration column such as TosoHaas QC-pak TSK 200, cat. no. 16215 (Montgomeryville, PA) (*see* **Note 2**), run at up to 0.5 mL/min.

2.2. Reaction with 2,4-Dinitrophenylhydrazine in SDS (see Note 4)

1. Buffer for gel filtration: 200 m*M* sodium phosphate, pH 6.5, 1% SDS
2. Derivatization solution: 20 m*M* 2,4-dinitrophenylhydrazine in 10% trifluoroacetic acid (v/v).
3. Derivatization blank solution: 10% trifluoroacetic acid (v/v).
4. Neutralization solution: 2 *M* TRIS (free base, not the HCl salt), 30% glycerol (*see* **Note 5**).
5. 12% SDS (warming the water will speed dissolution of the solid).
6. Column: HPLC gel filtration column such as TosoHaas QC-pak TSK 200, cat. no. 16215 (*see* **Note 2**), run at up to 0.6 mL/min.

2.3. Western Blotting

1. Antibodies to the 2,4-dinitrophenyl moiety. These are available from several suppliers as monoclonal and polyclonal antibodies. We have most experience with the rabbit polyclonal from Dako (V0401, Carpenteria, CA). Typically a 1:1000 dilution is used, but it is best to check the optimal dilution first.
2. Labeled secondary antibody to the species chosen for the antibody to the 2,4-dinitrophenyl moiety. When the first antibody is rabbit, we have used the goat anti-rabbit IgG conjugated to alkaline phosphatase (part of the Western-Light

chemiluminescent kit WL10RC from Tropix, Bedford, MA), typically at a 1:5000 dilution (*see* **Note 6**).

3. Methods

3.1. Reaction with 2,4-Dinitrophenylhydrazine in 6 M Guanidine

3.1.1. Preparation of Buffer for Gel Filtration and for Derivatization Blank (1 L)

1. Dissolve 573 g guanidine HCl in water. Use only about 200 mL water initially to avoid overdilution after the guanidine dissolves.
2. Bring the volume to about 850 mL, warming as needed to dissolve the guanidine.
3. Add 33.3 mL concentrated phosphoric acid (85%) slowly with stirring.
4. Bring the pH to 2.5 with 10 M KOH.
5. Adjust the volume to 1000 mL and filter through a 0.45-μm filter.

3.1.2. Preparation of Derivatization Solution (100 mL)

1. Dissolve the solid 2,4-dinitrophenylhydrazine in 3.33 ml concentrated phosphoric acid (85%). The 100 mL derivatization solution requires 198 mg 2,4-dinitrophenylhydrazine. You must take into account the actual content of 2,4-dinitrophenylhydrazine in your supply, which is typically 60–70%, with the remainder being water.
2. Dissolve 57.3 g guanidine HCl in water to give about 80 mL volume. This will require adding only about 25 mL water.
3. With stirring, add the 2,4-dinitrophenylhydrazine solution dropwise to the guanidine solution.
4. Use 10 M KOH, added dropwise, to bring the solution to pH 2.5.
5. Adjust the volume to 100 mL.

3.1.3. Derivatization (see **Note 7**)

1. Prepare the sample as desired. For example, concentrate by precipitation with trichloroacetic acid or ammonium sulfate.
2. Split the sample in half if a derivatization blank is also being prepared, as generally recommended.
3. Add 3 volumes of derivatization solution and mix to dissolve the sample.
4. If a blank is being prepared, treat it with the buffer without 2,4-dinitrophenylhydrazine.
5. Allow to stand 10 min at room temperature.
6. An in-line filter will remove particulates, but it is possible to lengthen the time between filter changes by centrifuging the sample for 3 min in a tabletop microcentrifuge (11,000g).
7. Inject the sample and follow the chromatograms at 276 nm and 370 nm (*see* **Note 8**). Inject the next sample when the reagent has washed through and the baseline has restabilized.
8. Calculate the results as described in **Subheading 3.4.**

3.2. Reaction with 2,4-Dinitrophenylhydrazine in SDS

3.2.1. Preparation of Derivatization Solution (100 mL)

1. Take into account the actual content of 2,4-dinitrophenylhydrazine in the reagent, as most manufacturers supply 2,4-dinitrophenylhydrazine with at least 30% water content. For 100 mL stock, dissolve the 396 mg of 2,4-dinitrophenylhydrazine in 10 mL of pure trifluoroacetic acid (15.4 g); then dilute to 100 mL with water. Precipitates that appear during storage can be removed by centrifugation.

3.2.2. Derivatization (see **Note 7**)

1. Prepare the sample as desired. Samples with high protein concentrations (> 10 mg/mL) usually require dilution to assure solubility after the addition of SDS and acid. Potassium dodecyl sulfate is less soluble than the sodium salt so that the sample should not contain more than about 50 mM potassium.
2. Split the sample in half if a derivatization blank is also being prepared, as is recommended. The volumes mentioned in **steps 3–5** always refer to the volume of the sample at this point, i.e., before the addition of derivatizing reagents.
3. Add 1 volume of 12% SDS with mixing. Do this even if the sample contains some SDS, because it is important to have at least 6% SDS before adding the trifluoroacetic acid.
4. Add 2 volumes of the 2,4-dinitrophenylhydrazine solution and mix.
5. If a blank is desired, treat it with 2 volumes of 10% trifluoroacetic acid alone.
6. Allow to stand 10 min at room temperature.
7. Bring the solution approximately to neutrality by adding 2 M TRIS/30% glycerol (*see* **Note 5**).
8. For HPLC analysis, inject all or part of the sample and follow the chromatograms at 276 nm and 360 nm (*see* **Note 8**). This is the same as for the guanidine system, except that the hydrazones are monitored at 360 nm in SDS instead of the 370 nm used in guanidine.
9. Calculate the results as described in **Subheading 3.4.**

3.3. Immunodetection of Protein-bound 2,4-Dinitrophenylhydrazones (Western Blotting; see **Note 9**)

1. Follow **steps 1–7** of **Subheading 3.2.2.** for derivatization in SDS.
2. Promptly load the samples and perform SDS gel electrophoresis followed by transfer to nitrocellulose using standard techniques (*see* **Note 10**).
3. Immunodetect the labeled proteins by Western blotting with chemiluminescent detection using standard techniques (*see* **Notes 11** and **12**).

3.4. Calculations

Carbonyl content may be expressed either as nmol carbonyl/mg protein or as mol carbonyl/mol protein. In either case, integrate the chromatograms to obtain the area of the protein peaks at 276 and 370 nm. Stop integration before the reagent begins to elute, preferably at a specific molecular weight determined by calibration of the column. We typically exclude material eluting below 10,000 Da.

If a blank was run, subtract its area at 370 nm from that of the sample. For the molar absorbtivity of the hydrazone, use $\varepsilon_{M_{370nm}} = 22,000$ and $\varepsilon_{M_{276nm}} = 9460$ (43% of that at 370 nm). If the molar absorbtivity of the protein is known, use it. If not, use 50,000, which is a good estimate of the molar absorbtivity of a protein of average amino acid composition *(12)* and a molecular weight of 50,000. Individual proteins can deviate substantially from this average value. For example, glutamine synthetase from *Escherichia coli* has $\varepsilon_{M_{276nm}} = 33,000$. To determine the mol of carbonyl per mole of protein,

$$\text{Mol carbonyl/Mol protein} = [(\varepsilon_{\text{protein}_{276}})(\text{Area}_{370})]/[(22,000)(\text{Area}_{276} - 0.43\text{Area}_{370})]$$

Because the result is a ratio, this calculation is independent of the amount of the sample injected. However, when desired, one can determine the actual mass in a given peak. For a typical injection, one will have nanogram quantities of protein (pmols) that will contain pmols of carbonyl. The general equation for determination of the amount of material in a peak is:

$$\text{mol} = [(\text{Area})(\text{Flow})]/[(\varepsilon_M)(\text{Path length})]$$

Note that the path lengths of HPLC detectors are sometimes not 1.0 cm. In the case of the Hewlet Packard diode-array detector, the path length is 0.6 cm and the area units are mAU-sec. Switching to micromolar absorbtivity and pmols, the equation becomes

$$\text{pmols} = [(\text{Area})(\text{Flow})]/[(\varepsilon_{\mu M})(36)]$$

Given a flow rate of 2 mL/min and $\varepsilon_{M_{370nm}} = 22,000$, the calculation for carbonyl is simply

$$\text{pmols carbonyl} = (2.53)(\text{Area}_{370nm})$$

For protein determination with $\varepsilon_{M_{276nm}} = 50,000$ the equation would be,

$$\text{pmols protein} = [(1.11)(\text{Area}_{276nm} - 0.43\text{Area}_{370nm})]$$

with the area at 267 nm being corrected for any contribution from the hydrazone, as noted earlier. For a mol wt of 50,000 then,

$$\text{ng protein} = [(55.5)(\text{Area}_{276nm} - 0.43\text{Area}_{370nm})]$$

4. Notes

1. The HPLC gel filtration method described in this chapter has been used for a large number of analyses in our laboratory. It is especially convenient if an autosampler is available, obviating the need to manually derivatize and inject

samples. However, 6 M guanidine HCl is not kind to pump seals, injector rotors, and other components of the HPLC system. Even small leaks will deposit substantial amounts of guanidine, which may cause corrosion. We make it a practice to flush with water and to inspect the system at the end of each day's use, taking care to correct any small leaks that may be noted. Pump seals should be changed at the earliest sign of wear. With most solvent systems, a delay in changing the seals is of no consequence, but with 6 M guanidine one risks creating a very unpleasant scene in the pump.

    The guanidine solution is relatively viscous, leading to rather high backpressures, and systems with long runs of microbore tubing will see even higher pressures. Only HPLC systems can generate the required pumping pressure, so this method cannot be used with low-pressure pumping systems such as Pharmacia's FPLC.

2. Be certain to validate the performance of the selected column. As mentioned in **Note 1**, some columns cannot tolerate the backpressure generated in HPLC systems, and others do not sufficiently separate the excess reagent from the peptides and proteins of interest. We previously recommended Zorbax columns (now distributed by Agilent) *(6)*, but, unfortunately, a manufacturing change now renders them unsuitable with either the guanidine or SDS systems.

3. Solutions of 2,4-dinitrophenylhydrazine develop precipitates during storage, and these can be removed by centrifugation before use. Stock solutions are usable for months.

4. The procedure for derivatization in SDS is very similar to that for guanidine, but it has been designed to facilitate the preparation of the sample for both HPLC gel filtration and Western blotting, if desired.

5. Glycerol is included to facilitate analysis by SDS gel electrophoresis. It may be omitted if only the HPLC analysis is being conducted.

6. If the first antibody is a mouse monoclonal IgE (Sigma D-8406, St. Louis, MO), we have used biotinylated rat anti-mouse IgE from Southern Biotechnology (1130-08, Birmingham, AL), typically at a 1:4000 dilution. A biotin–avidin amplification system provides increased sensitivity and works well with purified proteins. However, false-positive bands are often observed in cruder mixtures that contain biotin-binding proteins. In our hands, it is not possible to reproducibly prevent this false positive by blocking, so we simply use a second antibody conjugated to horseradish peroxidase and avoid the biotin–avidin problems.

7. Derivatization is conveniently carried out in 0.4- or 1.5-mL plastic tubes with a screw-top or a snap-top, or in standard autosampler vials if derivatization is being automated on the autosampler. Sensitivity is comparable to the tritiated borohydride method, so that one can use 10 µg of protein containing 1 mol carbonyl/mol protein or 100 µg of protein containing 0.10 mol carbonyl/mol protein. As with any gel filtration method, larger sample volumes will cause peak broadening. If one wishes to estimate the molecular weights of the labeled proteins, then the sample volume before derivatization should be 75 µL or less for most columns. Note that with larger volumes, it will take longer for the later-eluting reagent to

come off the column, requiring that the injection of the next sample be delayed by a few minutes to permit the absorbance to return to baseline.

8. A 10-nm bandwidth is appropriate for detectors with adjustable bandwidths. If the detector can only follow one wavelength, use 370 nm to detect the hydrazone and determine protein content by a separate assay. With the TosoHaas QC-Pak column in a Hewlett Packard 1050 HPLC, the proteins will begin to elute after 7 min, whereas the reagent will elute after 12.5 min. These times will be different for other columns and other HPLCs.

9. A kit containing the reagents for Western blotting is available from Intergen as their Oxyblot kit, catalog S7150-KIT (Gaithersburg, MD).

10. After derivatization and neutralization with the 2 M TRIS/30% glycerol, the samples may be loaded directly onto the gel and electrophoresed as usual because the sample is now in a solution similar to that used for electrophoresis by the method of Laemmli *(10)*. Note that samples are not heated before analysis because heating is generally not required for reduction of disulfide bonds, and we do not know the stability of protein-bound hydrazones to heating. When desired, β-mercaptoethanol can be added to the sample solution after neutralization. Inclusion of β-mercaptoethanol generally intensifies the bands in the chemiluminescent detection system so that it is best to either include or omit the thiol from all samples that are to be directly compared.

11. Either colorimetric or chemiluminescent detection protocols may be used. With chemiluminescence, the lower limit of detection is about 30 ng of oxidized glutamine synthetase containing 0.5 carbonyl groups per subunit (i.e., about 0.3 pmol carbonyl). If chemiluminescence is chosen, it is convenient to make exposures of varying lengths. Times of 0.5–5 min have worked well when using the alkaline phosphatase kit from Tropix. Shorter times, 2–30 s, were used with Amersham's ECL kit (RPN 2109, Arlington Heights, IL) which employs a horseradish peroxidase linked secondary antibody.

12. The excess reagent from the sample does not interfere with gel electrophoresis or with Western blotting. This lack of interference may result from poor accessibility of the antibody to reagent on nitrocellulose, a suggestion which follows from the observation that the monoclonal antibody (Sigma D-8406) did not detect the reagent spotted directly onto the nitrocellulose. As with all Western blot techniques, this method is not quantitative, but serial dilutions of the sample should provide an estimate of the amount of labeled material. In addition, it is important to include standard proteins of known carbonyl content in each gel. These facilitate the selection of development times, which allow the distinction between carbonyl-positive and carbonyl-negative proteins. As noted in **Subheading 1.**, specificity for derivatized carbonyl groups can also be checked by testing samples that have been pretreated with sodium borohydride to reduce carbonyl groups and Schiff bases *(3)*. The technique also works with isoelectric focusing gels and have been used with two-dimensional gels *(13)*. Use of these antibodies for immunoaffinity purification and for immunocytochemical localization of oxidized proteins is also feasible *(14,15)*.

References

1. Halliwell, B. and Gutteridge, J. M. (1990) Role of free radicals and catalytic metal ions in human disease: An overview. *Methods Enzymol.* **186,** 1–85.
2. Levine, R. L., Oliver, C. N., Fulks, R. M., and Stadtman, E. R. (1981) Turnover of bacterial glutamine synthetase: Oxidative inactivation precedes proteolysis. *Proc. Natl. Acad. Sci. USA* **78,** 2120-2124.
3. Levine, R. L., Garland, D., Oliver, C. N., Amici, A., Climent, I., Lenz, A. G., Ahn, B. W., Shaltiel, S., and Stadtman, E. R. (1990) Determination of carbonyl content in oxidatively modified proteins. *Methods Enzymol.* **186,** 464–478.
4. Shacter, E., Williams, J. A., Lim, M., and Levine, R. L. (1994) Differential susceptibility of plasma proteins to oxidative modification. Examination by Western blot immunoassay. *Free Rad. Biol. Med.* **17,** 429–437.
5. Keller, R. J., Halmes, N. C., Hinson, J. A., and Pumford, N. R. (1993) Immunochemical detection of oxidized proteins. *Chem. Res. Toxicol.* **6,** 430–433.
6. Levine, R. L., Williams, J. A., Stadtman, E. R., and Shacter, E. (1994) Carbonyl assays for determination of oxidatively modified proteins. *Methods Enzymol.* **233,** 346–357.
7. Shacter, E., Williams, J. A., Stadtman, E. R., and Levine, R. L. (1996) Determination of carbonyl groups in oxidized proteins, in *Free Radicals: A Practical Approach* (Punchard, N. A. and Kelly, F. J., eds.), IRL, Oxford University Press, Oxford, pp. 159–170.
8. Jones, L. A., Holmes, J. C., and Seligman, R. B. (1956) Spectrophotometric studies of some 2,4-dinitrophenyhydrazones. *Anal. Chem.* **28,** 191–198.
9. Levine, R. L. (1984) Mixed-function oxidation of histidine residues. *Methods Enzymol.* **107,** 370–376.
10. Laemmli, U. K. (1970) Cleavage of structural proteins during the assembly of the head bacteriophage T4. *Nature* **227,** 680–685.
11. Helenius, A., McCaslin, D. R., Fries, E., and Tanford, C. (1979) Properties of detergents. *Methods Enzymol.* **56,** 734–749.
12. *Protein Identification Resource, Release 26. 0.* (1990) National Biomedical Research Foundation Washington, D. C.,
13. Iwai, K., Drake, S. K., Wehr, N. B., Weissman, A. M., LaVaute, T., Minato, N., Klausner, R. D., Levine, R. L., and Rouault, T. A. (1998) Iron-dependent oxidation, ubiquitination, and degradation of iron regulatory protein 2: implications for degradation of oxidized proteins. *Proc. Natl. Acad. Sci. USA* **95,** 4924–4928.
14. Pompella, A. & Comporti, M. (1991) The use of 3-hydroxy-2-naphthoic acid hydrazide and Fast Blue B for the histochemical detection of lipid peroxidation in animal tissues—a microphotometric study. *Histochemistry* **95,** 255–262.
15. Smith, M. A., Perry, G., Richey, P. L., Sayre, L. M., Anderson, V. E., Beal, M. F., and Kowall, N. (1996) Oxidative damage in Alzheimer's. *Nature* **382,** 120–121.

3

Quantitation of 4-Hydroxynonenal Protein Adducts

Koji Uchida and Earl R. Stadtman*

1. Introduction

Lipid peroxidation has been associated with important pathophysiological events in a variety of diseases, drug toxicities, and traumatic or ischemic injuries. It has been postulated that free radicals and aldehydes generated during this process may be responsible for these effects because of their ability to damage cellular membrane, protein, and DNA. Recent studies have shown that the cytotoxicity of products of lipid peroxidation is due in part to the formation of α,β-unsaturated aldehydes, especially 4-hydroxy-2-alkenals *(1,2)*. 4-Hydroxy-2-alkenals elicit a variety of cytopathological effects, including inactivation of enzymes *(1)*, lysis of erythrocytes *(2)*, chemotactic activity toward neutrophils *(3)*, and inhibition of protein and DNA synthesis *(4)*. Among the aldehydes formed, 4-hydroxynonenal is the major product of lipid peroxidation and has been suggested to play a major role in liver toxicity associated with lipid peroxidation *(2,5–7)*.

It is generally accepted that 4-hydroxy-2-alkenals exert these effects because of their facile reactivity toward molecules with sulfhydryl groups *(8–10)*. The α,β double bond of 4-hydroxy-2- alkenals reacts with sulfhydryl groups to form thioether adducts via a Michael-type addition. Whereas it was generally accepted that the aldehyde moiety of 4-hydroxynonenal and other 2-alkenals reacts with primary amino groups to form α,β-unsaturated aldimines *(11,12)*. It is now evident that the ε-amino group of lysine residues in proteins may also undergo an Michael addition reactions with the α,β double bond of 4-hydroxynonenal to form secondary amines possessing an aldehyde group *(13)*. Furthermore, the imidazole groups of histidine residues in proteins also undergo Michael addition type of reactions *(14)*. Accordingly, it is not surprising that the modification of low-density lipoprotein (LDL) *(15,16)* and several other

*The authors of this chapter are government employees who performed their work in this capacity. The contents of this chapter are not copyrighted.

From: *Methods in Molecular Biology, vol. 99: Stress Response: Methods and Protocols*
Edited by: S. M. Keyse

proteins *(17)* by 4-hydroxynonenal is associated with a significant loss of lysine and histidine residues. It is thus important to establish the procedures to detect products derived from covalent attachment of 4-hydroxynonenal to amino acid side chains in order to understand the mechanism of a large number of biological effects induced by 4-hydroxynonenal.

1.1. Quantitation of 4-Hydroxynonenal Protein Thioether Adducts

Taking advantage of the fact that Raney nickel catalyzes the cleavage of thioether bonds *(18–21)*, a procedure has been developed *(22)* by which lipid-derived protein carbonyl derivatives can be distinguished from protein carbonyl derivatives generated by other reactions, including metal-catalyzed oxidation and glycation reactions *(23)*. The strategy used is illustrated in **Fig. 1**. In this procedure, the aldehyde moiety of the 4-hydroxynonenal thioether adduct is reduced to the corresponding [^{3}H]-labeled dihydroxyalkane by treatment with tritiated borohydride ($NaB[^3H]H_4$). The [^{3}H]-labeled dihydroxyalkane is then released from the protein by cleavage (desulfurization) of the thioether bond by means of Raney nickel catalysis. The free [^{3}H]-labeled dihydroxyalkane is then separated from the reaction mixture by extraction with an organic solvent and is quantitated by radioactivity measurement. This method can be used to measure the amount of 4-hydroxynonenal that is conjugated to proteins by a thioether linkage only. It cannot be used to measure that fraction of 4-hydroxynonenal that is bound to proteins via both a thioether linkage and also by a Schiff base linkage, as would occur if the aldehyde moiety of the primary Michael addition product (thioether) undergoes a secondary reaction with the free amino group of lysine residues in the same or a different protein subunit. Upon reduction with $NaB[^3H]H_4$, the Schiff base would be converted to a stable [^{3}H]-labeled secondary amine. Subsequent treatment of the complex with Raney nickel will lead to cleavage of the thioether linkage, but the derivatized 4-hydroxynonenal would still be tethered to the protein via the secondary amine linkage and would, therefore, not be extractable by organic solvents.

2. Materials
2.1. Reagents and Equipment

1. *N*-Acetylcysteine and glutathione (Sigma, St. Louis, MO).
2. 4-Hydroxynonenal: A stock solution of *trans*-4-hydroxy-2-nonenal is prepared by the acid treatment (1 m*M* HCl) of 4-hydroxynonenal diethylacetal and stored at –20°C. Concentration of 4-hydroxynonenal stock solution is determined from the molar extinction coefficient of 4-hydroxynonenal at 224 nm (ε = 13,750). The purity of the stock solution should be checked by high-performance liquid chromatography (HPLC) prior to the experiment.
3. Acetonitrile: 0–100% (v/v) in 0.05% trifluoroacetic acid.
4. EDTA: 10 m*M* and 100 m*M*.

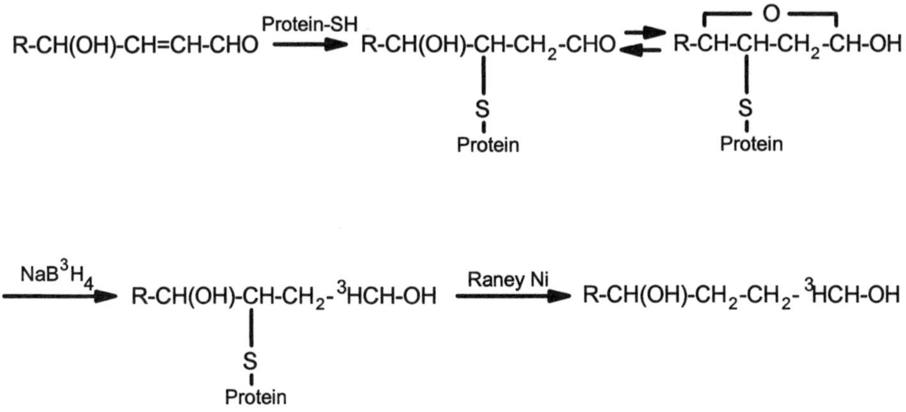

R-CH(OH)-CH=CH-CHO $\xrightarrow{\text{Protein-SH}}$ R-CH(OH)-CH-CH$_2$-CHO $\rightleftarrows$ R-CH-CH-CH$_2$-CH-OH

(with S—Protein branches)

$\xrightarrow{\text{NaB}^3\text{H}_4}$ R-CH(OH)-CH-CH$_2$-^{3}HCH-OH $\xrightarrow{\text{Raney Ni}}$ R-CH(OH)-CH$_2$-CH$_2$-^{3}HCH-OH

(with S—Protein branch)

Fig. 1. Experimental scheme to determine 4-hydroxynonenal covalently attached to protein sulfhydryl groups by means of thioether linkage. R:CH$_3$(CH$_2$)$_4$$^-$.

5. NaOH: 100 mM and 1 M.
6. HCl: 1 M and 6 M.
7. Methanol.
8. Sodium phosphate buffer 50 mM: pH 7.2 and pH 8.0.
9. Tricholoroacetic acid 20% (w/v).
10. NaB[^{3}H]H$_4$, 100 mM, in NaOH, specific activity about 100 mCi/mmol: Add required volume of 1 M NaBH$_4$ and 0.1 N NaOH to the NaB[^{3}H]H$_4$ and store at –20°C.
11. Guanidine HCl, 8 M, with 13 mM EDTA and 133 mM Tris, adjusted to pH 7.2 with HCl.
12. Guanidine hydrochloride 6 M.
13. The Raney nickel-activated catalyst (Sigma): Rinse thoroughly with water and ethanol prior to use (*see* **Note 1**).
14. Chloroform/methanol 9:1.
15. Sodium sulfate (dehydrated).
16. *N*-Acetylhistidine and *N*-acetyllysine (Sigma).
17. 0.05% Trifluoroacetic acid (solvent A) as a 100–0% gradient in acetonitrile (solvent B).
18. HPLC: Reverse-phase HPLC is performed on a Hewlett Packard Model 1090 chromatograph equipped with a Hewlett Packard model 1040A diode-array ultraviolet (UV) detector.
19. *o*-Phthaldialdehyde (OPA, Sigma): Dissolve 3.09 g boric acid in 40 mL HPLC-grade water and adjust the pH to 10.4 with KOH pellets. Dissolve 134 mg OPA in 0.75 mL HPLC-grade methanol in a Sarstedt-type tube and then add to the borate. Add 0.21 mL β-mercaptoethanol, water, and KOH to bring the pH back to 10.4 and the volume to 50 mL. Pipet into a 1.5-mL Sarstedt tube fitted with an O-ring and a cap and store at –20°C.
20. Brij-35: Make 10 mL by mixing 250 mL 2 M NaCl, 167 µL 30% Brij-35, and water (*see* **Note 2**).

3. Methods

3.1. Preparation of [³H]-Labeled 4-Hydroxynonenal Adducts of N-Acetylcysteine and Glutathione

1. To prepare standard samples of 4-hydroxynonenal-cysteine adduct, 2 mM N-acetylcysteine or glutathione is incubated with 2 mM 4-hydroxynonenal in 1 mL of 50 mM sodium phosphate buffer (pH 7.2) at 37°C for 2 h. Formation of 4-hydroxynonenal-N-acetylcysteine and 4-hydroxynonenal-glutathione adducts can be followed by reverse-phase HPLC on an Apex Octadecyl 5U column. The adducts are eluted at 1 mL/min with a linear gradient of 0–100% (v/v) acetonitrile in 0.05% trifluoroacetic acid for 25 min. The elution profiles are monitored by the absorbance at 210 nm. Under these chromatographic conditions, the 4-hydroxynonenal-N-acetylcysteine and 4-hydroxynonenal-glutathione adducts are eluted at 10.7 and 11.0 min, respectively *(22)*.
2. Take an aliquot (400 µL) of the reaction mixture and add 10 mM EDTA (40 µL)/ 1 M NaOH (40 µL)/100 mM NaB[³H]H$_4$ (40 µL) in a 1.5-mL Sarstedt tube fitted with an O-ring and a cap. After incubation for 1 h at 37°C, terminate the reaction by adding 200 µL of 1 N HCl, allow to stand 5 min in the hood, and then load the mixture on a Sep-Pak C-18 cartridge. Elute with 5 mL water followed by 5 mL of methanol. Concentrate the methanol fraction in a vacuum centrifuge (Savant), redissolve in 200 µL of methanol, and apply to reverse-phase-HPLC to purify the [³H]-labeled adducts.
3. Collect the eluted fractions in 1-mL aliquots and monitor the amount of [³H] in each fraction by scintillation counting. Collect the fractions which contain the [³H]-labeled products, concentrate, and then redissolve the adducts in 50 mM sodium phosphate buffer (pH 7.2). This solution should be stored at –20°C.

3.2. Preparation of [³H]-Labeled 4-Hydroxynonenal-Modified Proteins

1. Incubate 1 mg protein with 2 mM 4-hydroxynonenal in 1 mL of 50 mM sodium phosphate buffer (pH 7.2) to prepare 4-hydroxynonenal-modified proteins.
2. After incubation for 2 h at 37°C, take an aliquot (400 µL) of 4-hydroxynonenal-modified protein and add an equal volume of 20% trichloroacetic acid (w/v, final concentration).
3. Centrifuge the tubes in a tabletop microcentrifuge at 11,000g for 3 min and discard the supernatant.
4. Dissolve the precipitate with 400 µL of 8 M guanidine HCl/13 mM EDTA/133 mM Tris (pH 7.2) and treat with 0.1 M EDTA (40 µL)/1 N NaOH (40 µL)/0.1 M NaB[³H]H$_4$ (40 µL) in a 1.5-mL Sarstedt tube fitted with an O-ring and a cap.
5. After incubation for 1 h at 37°C, add 1 N HCl (100 µL) to terminate the reaction and then apply to the PD-10 column (Sephadex G-25), equilibrated in 6 M guanidine HCl, in order to separate protein-bound counts from free radioactivity.
6. Collect every 500 µL and determine protein recovery spectrophotometrically and radioactivity by liquid scintillation counting.

3.3. Raney Nickel Desulfurization

1. Take an aliquot (50 µL) of [^{3}H]-labeled 4-hydroxynonenal-*N*-acetylcysteine adduct, 4-hydroxynonenal-glutathione adduct, or 4-hydroxynonenal-modified proteins prepared as earlier and mix with 400 mg of Raney nickel and 300 µL of 8 *M* guanidine HCl/13 m*M* EDTA/133 m*M* Tris (pH 7.2) in a 1.5-mL Sarstedt tube fitted with an O-ring and a cap.
2. After incubation for 15 h at 55°C, released products are extracted twice with 500 µL each of chloroform/methanol (9:1). Add the chloroform/methanol solution to the mixture, vortex vigorously, and centrifuge the tubes in a tabletop microcentrifuge at 11,000*g* for 3 min.
3. After centrifugation, collect chloroform (lower) layer and add dehydrated sodium sulfate (50–100 mg) to the chloroform solution.
4. After vortexing, take an aliquot (50 µL) of chloroform extract, mix with 5 mL of scintillation liquid, and count the radioactivity to measure the amount of released product (*see* **Notes 3** and **4**).

3.4. Quantitation of 4-Hydroxynonenal-Histidine and 4-Hydroxynonenal-Lysine Adducts by HPLC

The oxidation of low-density lipoprotein (LDL) is accompanied by a loss of histidine and lysine residues *(24,25)* and its conversion to a form that is taken up by macrophages, giving rise to foam cells that have been implicated in atherogenesis *(25)*. A role of 4-hydroxynonenal in LDL oxidation and its possible involvement in atherogenesis is underscored by the observations: (1) 4-hydroxynonenal is formed during the oxidation of LDL *(12)*; (2) treatment of unoxidized LDL with 4-hydroxynonenal leads to the modification of lysine and histidine residues *(15,16)* and to the conversion of LDL to a form that is taken up by the scavenger receptor on macrophages *(26)*. The development of procedures for the detection and quantitation of the 4-hydroxynonenal-lysine and histidine adducts in proteins is therefore fundamental to assessment of the role such adducts have in the cytotoxicity of lipid peroxidation and atherogenesis.

3.4.1. Sample Preparation

1. *N*-Acetylhistidine and *N*-acetyllysine are used to prepare standard samples of 4-hydroxynonenal-histidine and 4-hydroxynonenal-lysine adduct, respectively. Treat 50 mg of *N*-acetylhistidine (or *N*-acetyllysine) with 5–10 m*M* 4-hydroxynonenal in 2 mL of 50 m*M* sodium phosphate buffer (pH 7.2) for 20 h at 37°C.
2. The products are isolated by reverse-phase HPLC, using a TSK-GEL ODS-80 TM column (0.46 × 25 cm), and a linear gradient of 0.05% trifluoroacetic acid in water (solvent A)–acetonitrile (solvent B) (time = 0, 100% A; 20 min, 0% A), at a flow rate of 1 mL/min. Under these chromatographic conditions, 4-hydroxynonenal adducts of *N*-acetylhistidine and *N*-acetyllysine are eluted at 11.2 min and 10.8 min, respectively.
3. 4-Hydroxynonenal-modified proteins are prepared according to the procedures described earlier.

3.5. Acid Hydrolysis of 4-Hydroxynonenal-Modified Samples

1. An aliquot (0.1 mL) containing 0.1 mg of 4-hydroxynonenal-modified proteins or 1–10 nmol of purified 4-hydroxynonenal-*N*-acetylhistidine or 4-hydroxy-nonenal-*N*-acetyllysine adduct is placed in a hydrolysis vial.
2. Add 10 µL each of 10 m*M* EDTA, 1 N NaOH, and 0.1 *M* NaBH$_4$, and incubate for 1 h at 37°C (*see* **Note 5**).
3. Add 30 µL of 1 *N* HCl to terminate the reaction and concentrate in a vacuum centrifuge.
4. Add 200 µL of 6 *N* HCl and flush the headspace with nitrogen for 60 s, and then cap using a screw cap fitted with a 12-mm Teflon/silicone liner with the Teflon facing the HCl.
5. Place the vial in a benchtop heater at 110°C and incubate for 20 h.
6. Concentrate the mixture in a vacuum centrifuge and redissolve in the desired volume of 50 m*M* sodium phosphate buffer (pH 8.0) containing 1 m*M* EDTA.

3.6. Amino Acid Analysis

3.6.1. Derivatization

1. Pipette 10 µL *o*-phthaldialdehyde onto the bottom of a Sarstedt tube.
2. Tilt the tube and place 10 µL of the sample on the side of the tube and mix with a vortex and start the timer.
3. At 0.25 min, add 180 µL Brij-35.
4. At 1 min, inject 100 µL of the above mixture and start the gradient.

3.6.2. HPLC Analysis

Reverse-phase HPLC is performed on a Hewlett Packard Model 1090 chromatograph equipped with a Hewlett Packard model 1046A programmable fluorescence detector (excite at 340 nm and follow emission at 450 nm). The column we used to use is a 15 cm C$_{18}$ from Jones Chromatography (5 µ size). Control the column temperature at 30°C. The flow rate is 2.0 mL/min. The following gradient program is used. All gradients are linear (A: 100 m*M* NaCl, B: water, C: CH$_3$OH)

Time (min)	Gradient
0.0	A = 50.0%, B = 50.0%
0.3	A = 37.5%, B = 37.5%, C = 25.0%
7.0	A = 37.5%, B = 37.5%, C = 25.0%
11.0	A = 30.0%, B = 30.0%, C = 40.0%
12.0	A = 25.0%, B = 25.0%, C = 50.0%
16.0	A = 22.5%, B = 22.5%, C = 55.0%
18.0	A = 12.5%, B = 12.5%, C = 75.0%
20.0	A = 12.5%, B = 12.5%, C = 75.0%
21.0	A = 50.0%, B = 50.0%

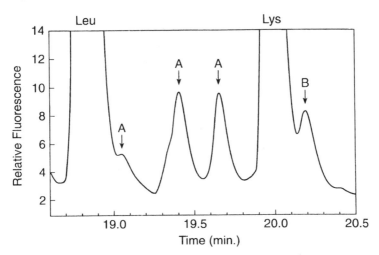

Fig. 2. Separation of the 4-hydroxynonenal adducts by HPLC. The enzyme (1 mg/ mL) was incubated with 2 mM 4-hydroxynonenal in 50 m*M* sodium phosphate buffer (pH 7.2) for 2 h at 37°C. After reduction with NaBH$_4$, reaction mixtures were hydrolyzed and the amino acid composition was analyzed by HPLC following derivatization with OPA. The arrows in section A indicate the peaks corresponding to the 4-hydroxynonenal-histidine adducts. The arrow in B indicates the 4-hydroxynonenal-lysine peak. The peaks marked Leu and Lys correspond to the OPA derivatives of leucine and lysine, respectively.

In the HPLC system used, *o*-phthaldialdehyde derivatives of the reduced forms of 4-hydroxynonenal-histidine and 4-hydroxynonenal-lysine adduct can be separated from all other normal amino acids. As shown in Fig. 2, the histidine derivatives appear as three separate peaks (isomers?) which elute between leucine and lysine at 19.05, 19.4, and 19.7 min; whereas the lysine adduct elutes just after lysine at 20.2 min. By integration of the areas under these peaks and comparison with the areas of amino acid standards, the amounts of histidine and lysine 4-hydroxynonenal adducts can be quantitated.

3.7. Concluding Remarks

We found that the major product formed in the reaction of 4-hydroxynonenal with α-*N*-acetyllysine or poly-L-lysine is the Michael addition product; that is, a secondary amine produced by addition of the ε-amino group of lysine to the double bond (C$_3$) of 4-hydroxynonenal *(13)*. For a number of 4-hydroxynonenal-modified proteins tested (glyceraldehyde phosphate dehydrogenase, insulin, bovine serum albumin, low-density lipoproteins), the number of 4-hydroxynonenal-histidine residues detected by these procedures was almost exactly equal to the number of histidine residues that disappeared.

However, the number of lysine-hydroxynonenal and cysteine–hydroxy-nonenal adducts that can be detected by HPLC assay of the *o*-phthaldialdehyde derivatives and the Raney nickel treatment, respectively, is often considerably lower than the number of these residues that disappeared upon 4-hydroxy-nonenal treatment. For example, reaction of 4-hydroxynonenal with glyceral-dehyde-3-phosphate dehydrogenase under the conditions described here led to the modification of 5 histidine, 3.5 lysine, and 2.5 cysteine residues *(17)*. By means of the Raney nickel procedure, only 17% of the modified cysteine could be attributed to a simple Michael addition reaction, whereas 90% and 28%, respectively, of the histidine and lysine residues were present as simple Michael addition products, as determined by HPLC of the *O*-phthaldehyde derivatives of NaBH$_4$-treated acid-hydrolyzed samples. It was proposed that the poor recovery of lysine and cysteine residues might be due to secondary reactions in which the aldehyde groups of some primary Michael addition products react with proximal lysine residues to form Schiff-base crosslinks, which would be stabilized by reduction with NaBH$_4$ *(17)*. This possibility is supported by (1) the observation that the number of cysteine plus histidine residues that could not be accounted for as Michael addition products is equal to the number of lysine residues that could not be accounted for and (2) by the appearance of protein conjugates that upon SDS gel electrophoresis exhibited molecular weights about two times that of the native subunit *(17)*.

4. Notes

1. Avoid complete drying becuase raney nickel is inflammable.
2. Make this solution up on the day of use, as the Brij-35 is not stable.
3. Upon treatment with Raney nickel, the [^{3}H]-labeled product can be recovered in 80–90% yield from both 4-hydroxynonenal-*N*-acetylcysteine and 4-hydroxy-nonenal-glutathione adducts in a solvent 10% methanol/chloroform-extractable form *(22)*. Before Raney nickel treatment, only 1% and 25% of the radioactivity in the *N*-acetylcysteine and glutathione adducts, respectively, were extracted by the solvent. It was considered that the small amount (25%) of solvent-extractable radioactivity present in the 4-hydroxynonenal-glutathione derivative before Raney nickel treatment is due to the contamination of the adduct preparation with radiolabeled reduced forms of free 4-hydroxynonenal.
4. When glyceraldehyde-3-phosphate dehydrogenase which contains 4 sulfhydryl groups per subunit, was modified with 4-hydroxynonenal, 3.2 of these sulfhydryl groups are lost; however, 0.54 mol/mol of the labeled adduct was released in a solvent-extractable form upon treatment with Raney nickel. Therefore, only 17% of the modified cysteine residues is present as a simple 4-hydroxynonenal-thioether adduct. Upon HPLC analysis, [^{3}H]-labeled product released after treat-ment with Raney nickel was indistinguishable from the products obtained following Raney nickel treatment of the 4-hydroxynonenal-*N*-acetylcysteine and

4-hydroxynonenal-glutathione adducts. Thus, the procedure using [^{3}H]-labeling followed by Raney nickel treatment enables one to determine bonafide 4-hydroxynonenal-protein thioether adducts and attests to the fact that the reaction of 4-hydroxynonenal with cysteine residues of proteins may lead to derivatives other than simple thioether adducts (*see* **Subheading 3.7.**).

5. Treatment of the 4-hydroxynonenal-histidine and 4-hydroxynonenal-lysine adducts with NaBH$_4$ stabilizes the histidyl-4-hydroxynonenal linkage. If the reduction step is omitted, acid hydrolysis of the 4-hydroxynonenal adducts leads to quantitative release of the histidyl and lysyl moieties as free amino acids. These observations are consistent with the Michael addition mechanism, provided that reversal of the addition reaction is catalyzed by strong acid. Thus, the reduction of the aldehyde moiety would preclude reversibility and, thereby, lead to formation of an acid-stable derivatives.

Acknowledgment

We thank Dr. H. Esterbauer (University of Graz) for his generous gift of 4-hydroxynonenal diethylacetal. We also thank Dr. R. L. Levine and B. S. Berlett (National Institutes of Health) for their advice on amino acid analysis.

References

1. Schauenstein, E. and Esterbauer, H. (1979) Formation and properties of reactive aldehydes, in *Submolecular Biology of Cancer*, Ciba Foundation Series 67, Excerpta Medica/Elsevier, Amsterdam, pp. 225–244.
2. Benedetti, A., Comporti, M., and Esterbauer, H. (1980) Identification of 4-hydroxynonenal as a cytotoxic product originating from the peroxidation of liver microsomal lipids. *Biochem. Biophys. Acta* **620,** 281–296.
3. Cuzio, M., Esterbauer, H., Mauro, C. D., Cecchini, G., and Dianzani, M. U. (1986) Chemotactic activity of the lipid peroxidation product 4-hydroxynonenal and homologous hydroxyalkenals. *Biol. Chem. Hoppe-Seyler* **367,** 321–329.
4. Esterbauer, H., Zollner, H., and Lang, J. (1985) Metabolism of the lipid peroxidation product 4-hydroxynonenal by isolated hepatocytes and by liver cytosolic fractions. *Biochem. J.* **228,** 363–373.
5. Benedetti, A., Esterbauer, H., Ferrali, M., Fulceri, R., and Comporti, M. (1982) Evidence for aldehydes bound to liver microsomal protein following CCl$_4$ or BrCCl$_3$ poisoning. *Biochem. Biophys. Acta* **711,** 345–356.
6. Esterbauer, H., Cheeseman, K. H., Dianzini, M. U., Poli, G., and Slater, T. F. (1982) Separation and characterization of the aldehydic products of lipid peroxidation stimulated by ADP-Fe^{2+} in rat liver microsomes. *Biochem. J.* **208,** 129–140.
7. Benedetti, A., Pompella, A., Fulceri, R., Romani, A., and Comporti, M. (1986) Detection of 4-hydroxynonenal and other lipid peroxidation products in the liver of bromobenzene-poisoned mice. *Biochim. Biophys. Acta* **876,** 658–666.
8. Schauenstein, E., Taufer, M., Esterbauer, H., Kylianek, A., and Seelich, T. (1971) The reaction of protein-SH-groups with 4-hydroxy-pentenal. *Monatsh. Chem.* **102,** 571.

9. Esterbauer, H., Zollner, H., and Scholz, N. (1975) Reaction of glutathione with conjugated carbonyls. *Z. Naturforsch.* **30c,** 466–473.

10. Esterbauer, H., Ertl, A., and Scholz, N. (1976) The reaction of cysteine with α,β-unsaturated aldehydes. *Tetrahedron* **32,** 285.

11. Suyama, K., Tachibana, A., and Adachi, S. (1979) Kinetic studies on the reaction of α,β- unsaturated aldehydes with the amino group of glycine. *Agric. Biol. Chem.* **43,** 9.

12. Esterbauer, H., Schaur, R. J., and Zollner, H. (1991) Chemistry and biochemistry of 4-hydroxynonenal, malondialdehyde, and related aldehydes. *Free Radical Biol. Med.* **11,** 81–128.

13. Szweda, L. I., Uchida, K., Tsai, L., and Stadtman, E. R. (1993) Inactivation of glucose-6- phosphate dehydrogenase by 4-hydroxy-2-nonenal: Selective modification of an active site lysine. *J. Biol. Chem.* **268,** 3342–3347.

14. Uchida, K. and Stadtman, E. R. (1992) Modification of histidine residues in proteins by reaction with 4-hydroxynonenal. *Proc. Natl. Acad. Sci. USA* **89,** 4544–4548.

15. Jürgens, G., Lang, J., and Esterbauer, H. (1986) Modification of human low-density lipoprotein by the lipid peroxidation product 4-hydroxynonenal. *Biochim. Biophys. Acta* **875,** 103–114.

16. Uchida, K. and Stadtman, E. R. (1992) 4-hydroxynonenal-histidine adducts generated in lipoproteins. *FASEB J.* **6,** A371.

17. Uchida, K. and Stadtman, E. R. (1993) Covalent attachment of 4-hydroxynonenal to glyceraldehyde-3-phosphate dehydrogenase. *J. Biol. Chem.* **268,** 6388–6393.

18. Schaffer, M. H. and Stark, G. R. (1976) Ring cleavage of 2-iminothiazolidine-4-carboxylates by catalytic reduction. A potential method for unblocking peptides formed by specific chemical cleavage at half-cystine residues. *Biochem. Biophys. Res. Commun.* **71,** 1040–1047.

19. Farnsworth, C. C., Wolda, S. L., Gelb, M. H., and Glomset, J. A. (1989) Human lamin B contains a farnesylated cysteine residue. *J. Biol. Chem.* **264,** 20,422–20,429.

20. Rilling, M. C., Breunger, E., Epstein, W. W., and Crain, P. F. (1990) Prenylated proteins: the structure of the isoprenoid group. *Science* **247,** 318–320.

21. Farnsworth, C. C., Gelb, M. H., and Glomset, J. A. (1990) Identification of geranylgeranyl- modified proteins in HeLa cells. *Science* **247,** 320–322.

22. Uchida, K. and Stadtman, E. R. (1992) Selective cleavage of thioether linkage in proteins modified with 4-hydroxynonenal. *Proc. Natl. Acad. Sci. USA* **89,** 5611–5615.

23. Stadtman, E. R. (1991) Covalent modification reactions are marking steps in protein turnover. *Biochemistry* **29,** 6323–6331.

24. Fong, L. G., Parthasarathy, S., Witztum, J. L., and Steinberg, D. (1987) Nonenzymatic oxidative cleavage of peptide bonds in apoprotein B-100. *J. Lipid Res.* **32,** 1466–1477.

25. Steinberg, D., Parthasarathy, S., Carew, T. E., Koo, J. C., and Witztum, J. L. (1989) Beyond cholesterol modifications of low-density lipoprotein that increase athrogenicity. *New Eng. J. Med.* **320,** 915–924.

26. Hoff, H. F., O'Neil, J., Chisolm III, G. M., Cole, T. B., Quehenberger, O., Esterbauer, H., and Jürgen, G. (1989) Modification of low density lipoprotein with 4-hydroxynonenal induces uptake by macrophages. *Arteriosclerosis* **9,** 538–549.

4

Detection of Oxidative Stress in Lymphocytes Using Dichlorodihydrofluorescein Diacetate

Cecile M. Krejsa and Gary L. Schieven

1. Introduction

1.1. Generation of Intracellular Oxidants

Normal oxidative metabolism leads to the generation of reactive oxygen species (ROS), in particular, superoxide anion (O_2-), and its dismutation product hydrogen peroxide (H_2O_2), which may escape from the electron-transport chain *(1)*. In addition, oxidative stress may be generated through the action of specialized enzymes, e.g., NADPH oxidase, nitric oxide synthase, cycloxygenase, and lipoxygenase, which produce H_2O_2, O_2-, NO, and lipid hydroperoxides [ROOH] *(2,3)*. Toxicants can also induce oxidative stress by a variety of mechanisms. Compounds that uncouple or block the electron-transport chain lead to increased leakage of ROS from mitochondria to the cytosol. Redox cycling of metals can result in H_2O_2, hydroxyl radical ($\cdot$OH), and thionyl radical (RS$\cdot$) production, and can deplete cellular thiol pools *(4)*. In addition, toxicants may directly inhibit antioxidant enzymes or compromise cellular reducing capacity by depletion of NAD(P)H, and glutathione (GSH) through the cytochrome P450 oxidoreductase and glutathione-*S*-transferase detoxification pathways, and through the action of the GSH peroxidase/reductase cycle *(5)*. In cells with diminished natural antioxidant capacity, normal metabolic sources of ROS may overwhelm the redox balance and push cells into a condition of oxidative stress.

Many methods for assessment of intracellular oxidative stress have been described, including quantitation of low-molecular-weight thiols such as GSH, assays of other cellular reducing equivalents, and detection of damage caused by ROS, such as lipid peroxides (*see also* Chapter 3, this volume) , reactive aldehydes, protein carbonyls (*see also* Chapter 2, this volume), and DNA strand

From: *Methods in Molecular Biology, vol. 99: Stress Response: Methods and Protocols*
Edited by: S. M. Keyse © Humana Press Inc., Totowa, NJ

breaks *(6,7)*. Although these methods are extremely useful for determining the effects of increased ROS within the cell, they do not provide a direct measurement of the reactive oxygen being generated. In some cases it is useful to know the extent of initial ROS production; for this, tracers which compete with cellular antioxidant systems for ROS *in situ* may be utilized.

The method described in this chapter uses one such tracer, dichlorodihydrofluorescein (H_2DCF) to measure the development of intracellular oxidation within lymphocytes. This allows the study of ROS generation and cellular resistance to oxidative stress in live cells by fluorimetric plate reader or flow cytometry.

1.2. H₂DCF-DA: Theoretical Considerations

Dichlorodihydrofluorescein–diacetate (H_2DCF-DA) has been used as a fluorimetric detector for hydrogen peroxide for over 30 yr *(8)*. Recent interest in the role of ROS in normal cellular function and cellular toxicity has stimulated the utilization of this compound and its derivatives for intracellular visualization of oxidative stress *(9–11)*. The acetyl groups on H_2DCF–DA are cleaved by intracellular esterases to yield the less stable product H_2DCF (also known as dichlorofluorescin), which has minimal fluorescence *(12)*. One-electron oxidation of H_2DCF yields the highly fluorescent product, dichlorofluorescein (DCF), which has excitation and emission spectra conducive to detection by most fluorescence plate readers and flow cytometers *(13)*. The reaction H_2O_2 + $2H_2DCF \longrightarrow 2H_2O + 2DCF$ is catalyzed by intracellular peroxidases; in addition, redox-active metals such as iron may also catalyze the oxidation of H_2DCF to DCF *(14)*. Although the original use of H_2DCF–DA was to quantitate H_2O_2, the compound has been reported to be oxidized by a wide variety of ROS, including H_2O_2, O_2-,·OH, ROOH, and NO *(7,14–18)*. These properties of H_2DCF–DA make it useful as a general detector of cellular oxidative stress.

The compound H_2DCF–DA is cell permeable and can be loaded into lymphocytes at 37°C. Through the action of esterases, H_2DCF–DA is deacetylated to yield H_2DCF, which is moderately well retained inside the cell *(see **Note 1**)*. However, transport of H_2DCF–DA and H_2DCF from the cell has been reported and must be considered as a possible confounder *(16)*. This is especially true when redox-active compounds are present in the medium. In addition, the oxidized, highly fluorescent product DCF may be transported from the cell over time, leading to a diminished signal in long-term oxidation studies. Through careful assay design, these effects may be minimized, so it is wise to consider the possible fate of the nonfluorescent (reduced) and fluorescent (oxidized) tracer in each particular system. For instance, if long-term generation of ROS is to be assayed, a plate assay may provide the most complete data, by collecting both intracellular DCF and DCF transported to the culture medium over time. In contrast, if a redox-active compound is used in the culture medium to treat cells, the flow-cytometric assay may be more appropriate. Flow cytometry allows the quantitation of fluorescence associated with individual cells, removing the potentially confounding media effects arising extracellular oxidation of H_2DCF. Another option is to perform a plate assay that incorporates a washing step to remove the contribution of extracellularly generated DCF fluorescence. These considerations of experimental design and alternatives to the basic protocols are discussed below.

1.3. Applications of H₂DCF-DA in Immunology

Assessment of the nature of the respiratory burst in phagocytic cells was the first important application of H_2DCF-DA to cellular systems, and it was shown that granulocytes from patients with chronic granulomatous disease were deficient in H_2O_2 generation following IgG stimulation *(19)*. Studies using H_2DCF–DA in myeloid cells have continued for over two decades, as the nature of the ROS produced by the respiratory burst has been explored *(15,18)*. The role of oxidative stress from chronic inflammation in the development of many diseases has led to investigations of ROS stimulated by inflammatory mediators in target cells as well as cells of lymphoid or myeloid origin, using H_2DCF-DA in biochemical, flow-cytometric, and microscopic assays to measure and localize ROS production *(14,20–24)*. DCF fluorescence has also been utilized to demonstrate a role for ROS in signaling pathways leading to T-lymphocyte activation and proliferation *(25,26)*.

In addition, DCF fluorescence, in combination with other measures of oxidative stress, has been used to investigate the role of ROS in apoptosis of thymocytes, B-cells, and adult T-cells *(27–30)*. Bcl-2, an antiapoptotic protein associated with B-cell lymphoma, was shown to confer resistance to oxidative stress generated during apoptosis, as measured by DCF fluorescence *(31)*. We recently utilized H_2DCF–DA as an intracellular probe for ROS generation stress in lymphocytes treated with kinase and phosphatase inhibitors, as part of a broader study of the effects of oxidative stress on protein tyrosine phosphatases *(32)*. DCF fluorescence has also proved useful for toxicity studies, as many classes of chemical toxicants act in part by inducing changes in intracellular redox status, either directly or through the activation of inflammatory mechanisms *(5,33)*.

2. Materials

2.1. Microtitre Plate Assay

2.1.1. Equipment

1. Tissue culture facilities, including appropriate incubators, pipettors, and so forth.
2. Fluorescence plate reader equipped with filters to allow excitation at 485 nm and emission at 530 nm.

3. Swinging bucket centrifuge equipped with 96-well plate holder (only necessary if samples are to be washed prior to assay).
4. Multichannel pipettor suitable for sample delivery to 96-well plates.

2.1.2. Supplies and Reagents.

1. Flat-bottom 96-well plates; one for each time-point to be assayed.
2. RPMI-1640 medium prepared with no phenol red, *or* phosphate buffered saline (PBS) (*see* **Notes 2–4**).
3. N-2-Hydroxyethylpiperazine-N'-2-ethanesulfonic acid (HEPES) buffer, 1 M, sterile filtered.
4. Assay medium: RPMI-1640 (no phenol red), buffered with 10 mM HEPES, *or* PBS with 10 mM HEPES. (*See* **Notes 2–7** for details about selection of appropriate assay media.)
5. Dichlorodihydrofluorescein diacetate stock solution: 5 mM H_2DCF-DA in DMSO. (Store protected from light at 4°C. This stock solution is stable for several months to a year if stored properly. The solution should be colorless; if a yellowish color develops, discard and remake the stock solution.) H_2DCF–DA is available from Molecular Probes (Eugene, OR).
6. Lymphocytes to be tested: Generally, 1×10^5 cells per well is sufficient for a good signal.

2.2. Flow Cytometry Assay

2.2.1. Equipment

1. Tissue culture facilities, including appropriate incubators, pipettors, and so forth.
2. Flow cytometer equipped with lasers and filters to allow excitation at 488 nm and emission at 530 nm (standard fluorescein spectra).
3. Refrigerated, swinging bucket centrifuge (equipped with a plate holder if samples are to be incubated in multiwell plates).

2.2.2. Supplies and Reagents

1. 24-, 48-, or 96-well tissue culture plates, or enough small tissue culture flasks for each condition to be assayed.
2. RPMI-1640 medium.
3. N-2-Hydroxyethylpiperazine-N'-2-ethanesulfonic acid (HEPES) buffer, 1 M, sterile filtered.
4. Assay medium: RPMI-1640, buffered with 10 mM HEPES (*see* **Note 3**).
5. Dichlorodihydrofluorescein diacetate stock solution: 5 mM H_2DCF–DA in dimethyl sulfoxide (DMSO). (Store protected from light at 4°C. This stock solution is stable for several months to a year if stored properly. The solution should be colorless; if a yellowish color develops, discard and remake the stock solution.)
6. Lymphocytes to be tested: Generally, the fluorescence of up to 1×10^4 cells is assayed per sample run; *see* **Subheading 3.2.1.** for a discussion of estimating total cell requirements for flow cytometry.

3. Methods

3.1. Microtitre Plate Assay

3.1.1. Pretreatment of Cells

Lymphocytes are treated according to experimental design; this may include pretreatment of cells with specific compounds which are expected to enhance or reduce production of ROS following the experimental treatments. In general, pre-treatment of cells may be performed in normal tissue culture flasks, containing the medium in which the cells normally grow (e.g., RPMI-1640 containing 10% fetal bovine serum [FBS] and antibiotics). This reduces the stress to cells associated with changing media conditions prior to the assay. In addition, it is recommended that the cells be resuspended in assay medium immediately prior to plating into the microtiter plate. To accomplish this may require that plates be set up in advance of cell pretreatment, depending on the length of the pretreatment timecourse. For very short pretreatments, it may be convenient to incorporate the pre-treatment protocol when plating the cells into the microtiter plate.

1. Estimate the number of lymphocytes needed per pretreatment group. A 96-well plate assay requires 1×10^5 cells/well; in general, the assay is performed in triplicate. The outside wells of the plate may be eliminated from the assay due to reader variation, in which case each plate has sufficient wells for 60 treatments, or 30 treatments if cell-free treatment controls are run (*see* **Notes 4** and **5**). As repeated scans in the plate reader have a tendency to decrease lymphocyte viability, it is recommended that a separate plate be run for each time-point tested.
2. Pretreat lymphocytes as required by experimental design. Following pretreatment, wash the cells once by centrifuging to pellet, remove pretreatment media, then resuspend cells at 2×10^6 cells/mL in assay medium (RPMI-1640 containing no phenol red, supplemented with 10 mM HEPES buffer). For experiments that do not require pretreatment, spin the cells out of their growing medium and resuspend at 2×10^6 cells/mL in assay medium.

3.1.2. Plate Setup

The treatment conditions are plated in triplicate in each microtitre plate such that the assay plates are ready for the addition of cells at the time the cells are washed from their pre-treatment media. This may require that the plates be prepared in advance of lymphocyte pre-treatment. Each experimental design will differ, so individual experience with the microtitre plate assay will determine the best way to accomplish the preparation of the cells and layout of the plate conditions. For most applications, a multichannel pipettor should be used to decrease the plating time. All reagent and plate preparations should be made using sterile technique, preferably in a tissue culture hood.

1. Add 100 µL of assay medium containing the treatment conditions at two times the final dosage to triplicate wells. Final assay volume will be 200 µL/well. If using cell-free treatment controls, add the assay medium with treatment conditions to six wells.
2. Add 50 µL of assay medium containing 16 mM H$_2$DCF–DA to each well.
3. Add 50 µL of lymphocyte suspension (1×10^5 cells) to each well. Add 50 µL of assay medium to cell-free treatment controls, if applicable.
4. Place plates into the cell culture incubator for appropriate time periods.

3.1.3. Data Aquisition and Analysis

Dichlorofluorescein (DCF) fluorescence is quickly read on a fluorescence plate reader, using a typical fluorescein filter set with excitation at 485 nm and emission at 530 nm. Most fluorescence plate readers contain programmable or preset plate dimension settings to adjust for slight variations in the size of 96-well microtiter plates from different sources. Select the appropriate settings or program the plate reader according to manufacturer's directions. In general, the ultraviolet (UV) light source must be activated in advance of reading plates, to allow for warm-up and stabilization of the light intensity. Follow the manufacturers guidelines for powering up the plate reader prior to data aquisition.

It is recommended that a time-point be made immediately following plating ($T = 0$) to assess the background fluorescence of the H$_2$DCF–DA preparation and to identify possible confounding or interfering factors in the medium (*see* **Notes 3–5**).

1. Power up the fluorescence plate reader at an appropriate interval prior to the anticipated time of the $T = 0$ sample. Define the plate dimension settings and select a filter set appropriate for excitation and detection of fluorescein fluorescence.
2. Read the DCF fluorescence on each plate and print the plate data or save in digital form.
3. Average the readings from the triplicates for each experimental condition. Calculate the percent coefficient of variation (% CV) for each of the averages (*see* **Note 8**).
4. If required, subtract the fluorescence of the cell-free treatment controls from that of the lymphocytes treated under the same experimental conditions to estimate intracellular DCF fluorescence. Calculate the mean fluorescence and % CV on each triplicate as described earlier, and after confirming that the tripicates are consistent, subtract the fluorescence average of the the cell-free treatment control triplicates from the average for cell fluorescence.

3.2. Flow Cytometry Assay

3.2.1. Pretreatment of Cells

Lymphocytes are treated according to experimental design; this may include pretreatment of cells with specific compounds that are expected to enhance or reduce production of ROS following the experimental treatments. In general, pretreatment of cells may be performed in normal tissue culture flasks, containing the medium in which the cells normally grow (e.g., RPMI-1640 containing 10% FBS and antibiotics). This reduces the stress to cells associated with changing media conditions prior to the assay.

1. Estimate the number of lymphocytes needed per pretreatment group. A flow cytometry run may be set to acquire 1×10^4 gated events per sample. This data aquisition occurs after the appropriate setting of gates, so the number of cells required per sample will be a function of the percentage of cells falling within the gates, in addition to the cells used during the setting of the gates. Thus, it is wise to overestimate the expected number of cells needed for any treatment group. Unless the cells are very precious or rare, we recommend allowing 1×10^5 cells/ sample for ease of handling. Multiply the estimated number of lymphocytes needed for each flow cytometry run by the number of conditions and time-points to be tested.
2. Pretreat lymphocytes as required by experimental design. Following pretreatment, wash the cells once by centrifuging to pellet them. Remove pretreatment media and resuspend the cell pellet to 2×10^6 cells/mL in assay medium (RPMI-1640, supplemented with 10 mM HEPES buffer). For experiments that do not require pretreatment, spin the cells out of their growth medium and resuspend the cells in assay medium.

3.2.2. Set-up of Assay Conditions and Incubation.

1. Prepare assay medium (RPMI-1640 with 10 mM HEPES) containing appropriate treatment conditions at twice the desired final dose.
2. Add 1 volume of cells, 1 volume of assay medium containing 8 µM H$_2$DCF-DA, and 2 volumes of assay medium containinng treatments at two times the final dosage for each condition to be assayed. It may be convenient to set up the assay conditions in 24-, 48-, or 96-well plates, depending on the number of cells required and the length of the incubation period. For 96-well plates, use 50 µL of cell suspension, 50 µL H$_2$DCF-DA solution, and 100 µL of 2X treatment medium. For larger well sizes or tissue culture flasks, increase the volumes accordingly.
3. For each treatment, prepare an unstained sample by adding the appropriate volume of assay medium without H$_2$DCF–DA; this will be used for setting gates and for establishing baseline fluorescence of the cell population.
4. Incubate samples at 37°C for the period dictated by experimental design.
5. Following sample incubation, collect the cells at the bottom of the multiwell plates by centrifugation and aspirate the supernatants. Resuspend the cells with ice-cold RPMI-1640 containing 10 mM HEPES buffer and hold the plates on ice, protected from light, until analysis by flow cytometry.

3.2.3. Data Aquisition and Analysis

It is best to proceed with flow-cytometric analysis as soon as possible after treatment and staining of the samples. The flow cytometer should be set with standard fluorescein emission and excitation filters, and in general a logarithmic scale is used for depicting the histograms of fluorescence intensity versus event counts.

1. Use the unstained control cells, and unstained treatment cells if applicable, to adjust gates for capture of appropriate cell populations. With certain treatments, this may involve the exclusion of dead or dying cells, selection of single cells versus cell clusters, or, in some cases, the selection of a cell population based on counterstaining for surface markers or other characteristics.
2. Use H_2DCF–DA-loaded control cells and cells from a treatment group that is expected to be high in DCF fluorescence to adjust the fluorescence intensity associated with the fluorescein fluorescence detector. A wide range of intensities is desirable, with the control samples having low fluorescence, allowing differences between treatments to be detectable and on scale. Dye-loaded control samples will typically be about an order of magnitude higher in fluorescence than unstained control lymphocytes.
3. Run the samples through the flow cytometer and collect data according to the specifications of the operating system. If numerous samples are to be processed, it is recommended that a control sample be run at the beginning and again at the end of the flow cytometry assay to assess the stability of the DCF signal over time.

3.3. Alternative Protocols

3.3.1. "Snapshot" Assay of DCF Fluorescence

The fluorimetric tracer H_2DCF–DA has been used extensively for the study of intracellular ROS generation and is conducive to many variations of the treatment protocol. The protocols described earlier have focused on experimental designs in which the tracer is present in the lymphocytes during treatment, in effect competing with normal cellular ROS scavenging mechanisms to provide a record of the generation of ROS generation over the treatment time-course. However, DCF fluorescence may also be utilized to assay for ROS in a "snapshot" fashion, and this type of protocol may be advisable in instances where many dissimilar treatments are to be compared. The protocol for detection of DCF fluorescence is essentially the same as those described above, except that treatments are performed in normal growth medium and cells are handled as described in **Subheadings 3.1.1. and 3.2.1.** Following treatment, the cells are incubated with assay medium containing H_2DCF–DA, usually for 30–60 min at 37°C. If the DCF fluorescence is to be read by a plate reader, the cells are added to plates in triplicate, as described earlier. For flow-cytometric detection, the cells may be loaded with H_2DCF–DA in any convenient tissue culture dish or flask.

1. Treat the cells as required by experimental design and wash from treatment media as detailed in **Subheading 3.1.1.** Resuspend in assay medium to a density of 2×10^6 cells/mL.

2a. *(Plate reader)* Prepare triplicate wells by plating 50 μL of assay medium (no phenol red) containing 8 μM H_2DCF–DA into each well. Add 50 μL of treated lymphocytes in assay medium (1×10^5 cells total) to each of the triplicate wells. Incubate for 30 min at 37°C and read as described in **Subheading 3.1.3.**

2b. *(Flow cytometer)* After washing cells from treatment media, resuspend them directly in 2 μM H_2DCF–DA in assay medium. (If working with numerous samples, resuspend the pellets in assay medium, then add an equal volume of 4 μM H_2DCF–DA to each sample.) Mix and incubate for 30 min at 37°C. Following incubation, pellet the cells by centrifugation, aspirate the supernatants, and resuspend in ice-cold RPMI-1640 containing 10 m*M* HEPES. Place the samples on ice and quantify fluorescence by flow cytometry as described in **Subheading 3.2.3.**

3.3.2. Preloading with H_2DCF–DA

For some types of studies, such as detection of ROS in association with cell signaling events, it may be necessary to preload the lymphocytes with H_2DCF–DA prior to the experimental treatment. In this case, the cells should be incubated for 15–30 min with H_2DCF–DA in a 37°C incubator to allow for conversion of the stain to H_2DCF by intracellular esterases. The length of loading time will need to be determined empirically, as the rate of deacetylation may differ between cell types, as will the oxidation of the H_2DCF to fluorescent DCF via normal metabolic processes. An optimal loading time will provide the greatest increase in DCF fluorescence following the experimental treatment versus the slow increase in DCF fluorescence over time in untreated cells. The magnitude of the oxidative stress rapidly induced by the experimental treatment will, in part, determine the amount of loading required for a good signal.

1. Load cells with H_2DCF-DA at 4 μM for plate assay or 2 μM for flow cytometry assay in RPMI-1640 containing 10 m*M* HEPES. Incubate cells at 37°C for 15–30 min, depending on the cell type and stimulation to be used.

2. Centrifuge cells to pellet them and aspirate the supernatants. Resuspend the cells in fresh medium (without H_2DCF–DA) at a density of 2×10^6/mL (for plate assay) or as the experimental design dictates (for flow cytometry).

3a. *(Plate assay)* Add 50 μL cell suspension to wells containing 50 μL of assay medium with treatments at twice the desired final concentration. Incubate at 37°C for time-points as dictated by experimental design, then read DCF fluorescence on the plate reader as described in **Subheadng 3.1.3.**

3b. *(Flow cytometry)* Add an appropriate number of cells to the assay medium containing experimental treatments, or resuspend cells, following wash (**step 2**) directly into the various treatment conditions. Incubate at 37°C for time-points as dictated by experimental design, then read DCF fluorescence by flow cytometry as described in **Subheading 3.2.3.**

4. Notes

1. Newer derivatives of DCF are reported to have better intracellular retention following deacetylation than H_2DCF-DA. These include a carboxy derivative that carries a negative charge at physiological pH, and a chloromethyl derivative that is thought to attach to cellular thiols following deacetylation *(7,12)*. These compounds are available from Molecular Probes.

2. Media containing phenol red are not compatable with the plate-reader assay, as the presence of the pH indicator dye greatly attenuates the DCF signal. Medium without phenol red is available in liquid and powder form (Life Technologies, Gaithersburg, MD). The medium should be reconstituted according to manufacturer's directions no more than 1–2 mo prior to assay, and stored at 4°C.

3. The inclusion of fetal bovine serum in the assay medium may interfere with the delivery of a pro-oxidant treatment to the cells and assessment of the generation and export of ROS from the cells, due to the action of serum factors such as catalase. In addition, serum esterases can interfere with the loading of H_2DCF–DA into cells. Therefore, FBS has not been included in the assay medium in the protocols described above.

4. The oxidation of H_2DCF to DCF can be catalyzed by agents capable of electron-transfer functions (e.g., Fe^{2+} in solution) *(14)*. RPMI-1640 medium contains factors that enable the oxidation of H_2DCF–DA, even in cell-free systems. This creates problems for working with H_2DCF–DA, especially in situations such as the plate-reader assay where cells are treated with a chemical pro-oxidant in the same medium that contains the tracer dye. To assess the extent of extracellular H_2DCF oxidation, cell-free media controls may be run for each treatment condition. If the time frame of the experiment permits the use of PBS without affecting cell viability, then PBS buffered with 10 m*M* HEPES, which causes little extracellular H_2DCF oxidation, should be used as the assay medium.

5. H_2O_2 can directly oxidize H_2DCF-DA. Although this effect is much greater in RPMI-1640 medium, it is also significant in PBS assay medium. If working with pro-oxidants that are likely to produce H_2O_2 or ·OH radicals, be certain to include cell-free treatment controls to assess the extent of extracellular H_2DCF oxidation (*see also* **Note 7**).

6. Media conditioning may result in the accumulation of H_2DCF in the assay medium over time, through intracellular deacetylation and subsequent export *(16)*. As H_2DCF is more rapidly oxidized than H_2DCF-DA, this "conditioning" effect may amplify the problems listed in **Notes 4** and **5**, namely the extracellular formation of DCF by oxidants present in the medium. Although in some cases it is safe to assume that the well fluorescence *en toto* is an adequate representation of intracellular oxidation, there are instances where extracellular formation of DCF may occur independently of the cellular redox status. For these cases, a washing step is recommended for the plate assay (*see* **Note 7**); alternatively, the flow-cytometric assay, which measures only intracellular DCF fluorescence, should be used (also *see* **Note 1**).

7. In the event that there is significant extracellular oxidation in cell-free treatment controls of the plate assay, especially if the incubation times are long enough to cause significant "conditioning" of the medium with deacetylated H_2DCF (*see* **Note 6**), it is advisable to perform a washing step immediately prior to reading the plate fluorescence. This is easily accomplished by centrifuging the plates in a swinging bucket centrifuge, aspirating the assay medium, and replacing with 200 μL PBS containing 10 mM HEPES per well. (The wash may be repeated to remove all traces of extracellular DCF.) The plate is read immediately on the fluorescence plate reader, as described in **Subheading 3.1.3.**

8. The % CV is the standard deviation of the samples divided by the mean, times 100% *(34)*; this value allows a rapid assessment of triplicate variation independent of the magnitude of the sample fluorescence reading. If the % CV is very high (>10%), review the original plate data. Pipetting error or other sample-handling problems may lead to high % CV values. In some instances an obvious outlier may be discarded as nonrepresentative of the mean, in other cases, the cause of high experimental variation is unclear and the plate assay will need to be rerun.

Acknowledgments

This work was supported by NIH grant ES04696 and by Bristol-Myers Squibb. C.M.K. was supported by NIEHS Environmental Pathology/Toxicology Training Grant T32 ES07032. The authors wish to thank Dr. Terrance J. Kavanagh for helpful suggestions.

References

1. Richter, P., Gogvadze, V., Laffranchi, R., Shalapbach, R., Schweizer, M., Suter, M., Walter, P., and Yaffee, M. (1995) Oxidants in mitochondria: from physiology to diseases. *Biochem. Biophys. Acta* **1271,** 67–74.
2. Gille, G. and Sigler, K. (1995) Oxidative stress and living cells. *Folia Microbiol. Praha.* **40(2),** 131–152.
3. Cohen, G. (1994) Enzymatic/nonenzymatic sources of oxyradicals and regulation of antioxidant defenses. *Ann. N.Y. Acad. Sci.* **738,** 8–14.
4. Stohs, S. and Bagchi, D. (1995) Oxidative mechanisms in the toxicity of metal ions. *Free Radical Biol. Med.* **18(2),** 321–336.
5. Seis, H. and de-Groot, H. (1992) Role of reactive oxygen species in cell toxicity. *Toxicol. Lett.* **64-65** Spec. No. pp. 547–551.
6. Tyson, C. and Frazier, J. (eds.) (1994) *Methods in Toxicology, Vol 1B: In Vitro Toxicity Indicators.* .Academic Press, Orlando, FL.
7. Poot, M., Kavanagh, T., June, C., and Rabinovitch, P. (1995) Assessment of cell physiology by flow cytometry, in *Weir's Handbook of Experimental Immunology* 5th ed. (Weir, D.,Blackwell, C., Herzenberg, L., and Herzenberg, L., eds.), Oxford Science Inc., Oxford, UK, pp. 53.1–53.11.
8. Keston, A. and Brandt, R. (1965) The fluorometric analysis of ultramicro quantities of hydrogen peroxide. *Anal. Biochem.* **11,** 1–5.

9. Paul, B. and Sbarra, A. (1968) The role of the phagocyte in host-parasite interactions. XIII. The direct quantitative estimation of H_2O_2 in phagocytizing cells. *Biochem. Biophys. Acta.* **156,** 168–178.

10. Bass, D., Parce, J., Dachatelet, L., Szejda, P., Seeds, M., and Thomas, M. (1983) Flow cytometric studies of oxidative product formation by neutrophils: A graded response to membrane stimulation. *J. Immunol.* **130,** 1910–1917.

11. Burrow, S. and Valet, G. (1987) Flow cytometric characterization of stimulation, free radical formation, peroxidase activity and phagocytosis of human granulocytes with 2,7-dichlorofluorescein (DCF). *Eur. J. Cell. Biol.* **43,** 128.

12. Haugland, R. (1996) Assaying oxidative activity in live cells and tissue, in *Handbook of Fluorescent Probes and Research Chemicals* (Spence, M., ed.), Molecular Probes, Inc., Eugene, OR, pp. 491,492.

13. Black, M. and Brandt, R. (1974) Spectrofluorometric analysis of hydrogen peroxide. *Anal. Biochem.* **58,** 246-254.

14. LeBel, C., Ischiropoulos, H., and Bondy, S. (1992) Evaluation of the probe 2',7'-dichlorofluorescin as an indicator of reactive oxygen species formation and oxidative stress. *Chem. Res. Toxicol.* **5,** 227–231.

15. Rothe, G. and Valet, G. (1990) Flow cytometric analysis of respiratory burst activity in phagocytes with hydroethidine and 2',7'-dichlorofluorescin. *J. Leukocyte Biol.* **47,** 440–448.

16. Royall, J. and Ischiropoulos, H. (1993) Evaluation of 2',7'-dichlorofluorescin and dihydrorhodamine 123 as fluorescent probes for intracellular H_2O_2 in cultured endothelial cells. *Arch. Biochem. Biophys.* **302(2),** 348–355.

17. Cathcart, R., Schwiers, E., and Ames, B. (1983) Detection of picomole levels of hydroperoxides using a fluorescent dichlorofluorescein assay. *Anal. Biochem.* **134,** 111–116.

18. Rao, K., Padmanabhan, P., Kilby, D., Cohen, H., Currie, M., and Weinberg, J. (1992) Flow cytometric analysis of nitric oxide production in human neutrophils using dichlorofluorescein diacetate in the presence of a calmodulin inhibitor. *J. Leukocyte Biol.* **51,** 496–500.

19. Homan-Muller, J., Weening, R.S., and Roos, D. (1975) Production of hydrogen peroxide by phagocytizing human granulocytes. *J. Lab. Clin. Med.* **85(2),** 198–207.

20. Horio, F., Fukuda, M., Katoh, H., Petruzzelli, M., Yano, N., Ritterhaus, C., Bonner-Weir, S., and Hattori, M. (1994) Reactive oxygen intermediates in autoimmune islet cell destruction of the NOD mouse induced by peritoneal exudate cells (rich in macrophages) but not in T cells. *Diabetologia* **37(1),** 22–31.

21. Birdsall, H. (1991) Induction of ICAM-1 on human neural cells and mechanisms of neutrophil-mediated injury. *Am. J. Pathol.* **139(6),** 1341–1350.

22. Fukumura, D., Kurose, I., Miura, S., Tsuchiya, M., and Ishii, H. (1995) Oxidative stress in gastric mucosal injury: role of platelet-activating factor-activated granulocytes. *J. Gasterentol.* **30(5),** 565–571.

23. Minamiya, Y., Abo, S., Kitamura, M., Izumi, K., Kimura, Y., Tozawa, K., and Saito, S. (1995) Endotoxin-induced hydrogen peroxide production in intact pulmonary circulation of rat. *Am. J. Respir. Crit. Care Med.* **152(1),** 348–354.

24. al-Mehdi, A., Shuman, H., and Fisher, A. (1994) Fluorescence microtopography of oxidative stress in lung ischemia-reperfusion. *Lab. Invest.* **70(4),** 579–587.
25. Rouhi, N., Levallois, C., Favier, F., and Mani, J. (1989) Cyclooxygenase and lipoxygenase inhibitors act differently on oxidative product formation by immune mononuclear cells: a flow cytometric investigation. *Int. J. Immunopharmacol.* **11(6),** 681–686.
26. Jeitner, T., Kneale, C., Christopherson, R., and Hunt, N. (1994) Thiol-bearing compounds selectively inhibit protein kinase C-dependent oxidative events and proliferation in human T cells. *Biochem. Biophys. Acta* **1223(1),** 15–22.
27. Wang, J., Jerrells, T., and Spitzer, J. (1996) Decreased production of reactive oxygen intermediates is an early event during in vitro apoptosis of rat thymocytes. *Free Radical Biol. Med.* **20(4),** 533–542.
28. Fernandez, A., Kiefer, J., Fosdick, L., and McConkey, D. (1995) Oxygen radical production and thiol depletion are required for Ca^{2+}-mediated endogenous endonuclease activation in apoptotic thymocytes. *J. Immunol.* **155,** 5133–5139.
29. Toledano, B., Bastien, Y., Noya, F., Baruchel, S., and Mazer, B. (1997) Platelet activating factor abrogates apoptosis in a human B lymphoblastoid cell line. *J. Immunol.* **158(8),** 3705–3715.
30. Kohno, T., Yamada, Y., Hata, T., Mori, H., Yamamura, M., Tomonaga, M., Urata, Y., Goto, S., and Kondo T. (1996) Relation of oxidative stress and glutathione synthesis to CD95(Fas/APO-1) -mediated apoptosis of adult T cell leukemia cells. *J. Immunol.* **156,** 4711–4728.
31. Hockenbery, D., Oltval, Z., Yin, X-M., Milliman, C., and Korsmeyer, S. (1993) Bcl-2 functions in an antioxidant pathway to prevent apoptosis. *Cell* **75,** 241–251.
32. Krejsa, C., Nadler, S., Esselstyn, J., Kavanagh, T., Ledbetter, J., and Schieven, G. (1997) Role of oxidative stress in the action of vanadium phosphotyrosine phosphatase inhibitors. *J. Biol. Chem.* **272,** 11,541–11,549.
33. Burchiel, S., Kerkvliet, N., Geberick, G., Lawrence, D., and Ladics, G. (1997) Assessment of immunotoxicity by multiparameter flow cytometry. *Fundam. Appl. Toxicol.* **38(1),** 38–54.
34. Woolson, R. (1987) *Statistical Methods for the Analysis of Biomedical Data,* Wiley, New York, pp. 23,24.

5

The Measurement of Protein Degradation in Response to Oxidative Stress

Thomas Reinheckel, Tilman Grune, and Kelvin J. A. Davies

1. Introduction

Molecular oxygen is an excellent acceptor of electrons and is therefore employed by nature for a wide variety of highly important biochemical reactions. The chemical reactivity of oxygen, however, also leads to the formation of oxygen radicals as by-products of metabolism. These radicals result in a condition that is commonly referred to as oxidative stress, in which the formation of reactive oxygen species represents a functional challenge, or even endangers survival of a cell or organism. Oxidative stress can arise from increased production of reactive oxygen species, from exposure to an oxidant, or from decreased function of antioxidant systems or oxidant repair systems. Important examples for the involvement of reactive oxygen species in pathological conditions are inflammatory diseases, ischemia/reperfusion injuries, and many neurodegenerative disorders.

It has been shown in numerous studies that proteins are susceptible to reactive oxygen species attack. The side chains of amino acids are the primary sites of free-radical damage. Well-characterized examples are the oxidation of cysteine (SH) to cystine (S–S), oxidation of methionine to methionine sulfoxide, the formation of 3,4-dihydroxyphenylalanine (DOPA) from tyrosine, and leucine oxidation, which results in various hydroxyleucines (1–3). Because oxidative stress is characterized by a complex pattern of radical reactions, affecting virtually all cellular constituents, secondary protein modification by products of lipid peroxidation (i.e., by 4-hydroxynonenal; see Chapter 3, this volume) or glycoxidation also occurs in vivo (4,5).

The alterations of primary protein structure described are highly likely to cause changes in higher-order structures. As a consequence, the protein unfolds, over-

From: *Methods in Molecular Biology, vol. 99: Stress Response: Methods and Protocols*
Edited by: S. M. Keyse © Humana Press Inc., Totowa, NJ

all hydrophobicity increases as hydrophobic amino acid residues become exposed, and, eventually, protein function is impaired *(6)*. The increase in overall hydrophobicity may also lead to noncovalent protein aggregation based on hydrophobic interactions. High doses of oxidants can cause covalent intermolecular crosslinks, as observed for the formations of intermolecular disulfide bridges or dityrosine crosslinks. Another mechanism of protein damage is the fragmentation of proteins. This term refers to the direct disintegration of protein (main-chain scission and side-chain scission) by mechanisms other than peptide bond hydrolysis *(6)*.

The accumulation of oxidatively damaged proteins that lack enzymatic activity or other functional properties could lead to a pool of useless cellular debris. More serious complications, like metabolic disturbances or immunological problems, can be imagined. Thus, the removal of oxidized proteins or protein fragments by proteases and the subsequent reutilization of the released amino acids for protein synthesis or energy metabolism seems to "make sense" in order to maintain cellular integrity. For this concept, the term "proteolysis as a secondary antioxidant defense" was coined *(1,7)*. It was shown by several groups that oxidized proteins are preferentially degraded by proteases (i.e., trypsin, chymotrypsin, 20S-proteasome) as compared with their nonoxidized forms (**Fig. 1**). The full picture is, however, more complex (for review *see* **ref. 8**). There is an optimal reactive oxygen species dose to increase the proteolytic susceptibility of a protein. Further exposure to reactive oxygen species actually leads to decreased degradation (**Fig. 1**). This is most likely the result of protein aggregation and crosslinking, which limit the access of proteases to their specific cleavage sites. Increased intracellular proteolysis was, however, found after exposing cell cultures to mild oxidative stress (**Fig. 2**; *9–11*). The 20S form of the proteasome was identified as the protease responsible for most of the breakdown of cytosolic, oxidatively modified proteins *(9–11)*. Interestingly, no upregulation of the cytosolic proteasomal system was found in this work. Thus, the increase in proteolysis induced by oxidative stress mainly appears to be a consequence of enhanced substrate proteolytic susceptibility resulting from oxidative protein modification.

As described earlier, the relations among proteases, proteins and oxidative stress are fairly complex. The following questions may be of interest when approaching this field:

1. How susceptible is a given oxidized protein (i.e., an isolated protein of interest) to degradation by the cellular proteolytic system or a purified protease?
2. Is there an increased protein turnover after exposure of an experimental system (i.e., cell culture) to oxidants or to a stress where reactive oxygen species-formation is expected?
3. If an increased degradation of cellular proteins is detected, is the enhanced proteolysis the result of upregulation of the proteolytic system(s) or the result of preferential degradation of oxidatively modified proteins?

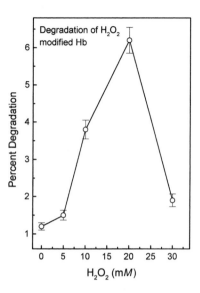

Fig. 1. Degradation of H_2O_2 modified hemoglobin by human K562 cell lysates. [³H]-hemoglobin was oxidized and subsequently used for the proteolysis assay as described in **Subheading 3.1.2.** Data for this figure were drawn from **ref. 10** with copyright permission.

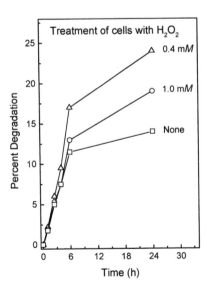

Fig. 2. Degradation of short-lived proteins in oxidatively stressed, human K562 hematopoietic cells. Newly synthesized cell proteins were metabolically labeled with [³⁵S]-methionine/cysteine for 2 h and exposed for 30 min to H_2O_2 as described in **Subheading 3.2.2.** Data for this figure were drawn from **ref. 10** with copyright permission.

The selection of methods given in this chapter aims to describe the basic tools for answering these questions. Here, we focus on the assessment of proteolytic activity as the ultimate event of physiological importance. We have attempted to provide methods with relatively "high output," because usually many conditions have to be tested in order to characterize the proteolytic response to an oxidative stress. The regulation of proteolytic systems at the levels of mRNA and protein content are beyond the scope of this chapter. In principle, the fate of proteins during oxidative stress can also be elucidated by two-dimentional-electrophoresis or by immunoblotting. When using Western blotting, care should be always taken to check if the decreased immunostaining is the result of protein degradation or the oxidation of amino acids that are essential for binding of the antibody. The electrophoresis-based techniques are, however, omitted in this chapter, as these methods are routine in many laboratories dealing with stress responses.

Probably the most convenient way of measuring protease activities in cell lysates is the use of highly sensitive fluorogenic peptides that are usually specific for a particular protease, or for a class of proteases. These peptides consist of a small number of defined amino acids linked to a fluorogenic group (e.g., 7-amino-4-methylcurmarine; MCA). After cleavage of the peptide the fluorogenic group is released and can be quantified by fluorescence spectroscopy. The specificity of the measurements can often be further enhanced by use of specific protease inhibitors. Because the proteasome appears to be strongly involved in the degradation of many/most oxidatively modified proteins a method for the measurement of the "chymotrypsinlike" activity of the proteasome using the fluorogenic peptide "succinyl-Leucine-Leucine-Valine-Tyrosine-MCA " (suc-LLVY-MCA) as substrate is described in **Subheading 2.1.1.**

Short peptides are obviously not the appropriate substrates for assessing the activity of cellular proteases toward oxidized proteins. Therefore, a method for oxidation of [^{3}H]-labeled proteins and their subsequent use as substrates in a proteolysis assay is described. The assay is based on the inability of trichloroacetic acid to precipitate peptides of less than about 5 kDa in size. Thus, after breakdown of the radioactively labeled substrate–protein, increases in acid-soluble radioactivity represent a measure of protein degradation. This assay can be applied for the estimation of the proteolytic activity of cell lysates as well as of purified proteases toward oxidized proteins.

Because the estimation of proteolysis based on nonradioactive quantification of amino acids is of very limited use in functional cells (*see* **Note 1**), the measurement of cellular protein degradation relies on metabolic radiolabeling of intracellular proteins. Proteolysis experiments using this approach (also known as the "pulse-chase" technique) consist of the following principal steps:

1. Labeling of proteins in intact cells occurs by incorporation of exogenously added radioactive amino acids into proteins during protein biosynthesis ("pulse"), followed by washout of the unincorporated label and supplementation of the medium with an excess of nonradioactive amino acids ("cold chase"). The duration of the "pulse" determines the fraction of proteins labeled (*see* **Note 2**).
2. Induction of (oxidative) stress.
3. Estimation of proteolysis at defined time points by precipitation of intact proteins with trichloroacetic acid (TCA) and measurement of acid-soluble counts.

In contrast to intact cells, nonradioactive methods are needed for the estimation of stress-induced protein breakdown in organelles or cellular fractions, unless organelles are prepared after metabolic labeling. Using isolated mitochondria as an example, the derivatization of TCA soluble primary amines, which represent amino acids or small peptides from degraded mitochondrial proteins, with fluorescamine and their subsequent quantification by fluorescence spectroscopy is described.

2. Materials

2.1. Proteolytic Activity of Cell Lysates Toward Exogenously Added Oxidized and Nonoxidized Substrates

2.1.1. Degradation of suc-Leu-Leu-Val-Tyr-MCA (suc-LLVY-MCA) in Cell Lysates

1. General materials: Cell culture equipment, fluorescence spectrometer, shaking water bath at 25°C and 37°C, liquid nitrogen or dry ice, assay for protein quantification.
2. Sterile phosphate-buffered saline (PBS).
3. Lysis buffer: 0.25 M sucrose, 25 mM HEPES, pH 7.8, 10 mM MgCl$_2$, 1 mM EDTA, 1 mM dithiothreitol (DTT); store at −20°C; add DTT immediately before use.
4. Proteolysis buffer: 50 mM Tris, pH 7.8, 20 mM KCl, 5 mM MgOAc, 0.5 mM DTT; store at 4°C; add DTT before use.
5. Suc-LLVY-MCA-stock: 4 mM succinyl-Leu-Leu-Val-Tyr-MCA (BACHEM Ltd., Saffron Walden, UK) in dimethyl sulfoxide (DMSO), store in 1-mL aliquots at −20°C.
6. "Free" MCA: 1 mM MCA (BACHEM) in DMSO, store in 0.5-mL aliquots at −20°C.
7. Lactacystin-stock: 1 mM lactacystin (Biomol Inc., Plymouth Meeting, PA) in sterile distilled water; store aliquots at −20°C; stable for at least 6 mo.
8. Ethanol (absolute); cool to −20°C before use.
9. Boric acid 0.125 M, pH 9.0; store at room temperature.

2.1.2. Degradation of Oxidized and Nonoxidized Proteins in Cell Lysates

1. General materials: Cell culture equipment, equipment to work with radiochemicals, β-counter, shaking water bath at 25°C and 37°C, liquid nitrogen or dry ice, assay for protein quantification, dialysis tubes with a molecular size cut-off at about 10,000 Da).

2. [^{3}H]-labeled protein (*see* **Note 3**) at a concentration >2 mg/mL.
3. H_2O_2 stock of known concentration (*see* **Note 4**).
4. Oxidation buffer: 20 mM phosphate buffer, pH 7.4, store at 4°C.
5. Dialysis buffer: 5 mM phosphate buffer, pH 7.4, 10 mM KCl , prepare at least 5 L; store at 4°C.
6. Proteolysis buffer: 50 mM Tris, pH 7.8, 20 mM KCl, 5 mM MgOAc, 0.5 mM DTT; store at 4°C, add DTT before use.
7. 3% BSA (w/v) in proteolysis buffer.
8. TCA: 20% (w/v) in water, keep on ice before use.
9. Scintillation cocktail.

2.2. Measurement of Protein Degradation in Organelles and Intact Cells

2.2.1. Nonradioactive Assessment of Mitochondrial Protein Breakdown by Fluorescamine Reactivity

1. General materials: Mitochondria (or another cellular organelle of interest), fluorescence spectrometer.
2. TCA: 10% (w/v) in water, keep on ice before use.
3. Neutralizer: 1 M HEPES, pH 7.8, store at room temperature.
4. Fluorescamine solution: 0.3 mg/mL in acetone; prepare immediately before use.
5. Glycine stock: 50 mM in PBS; store in aliquots at –20°C.

2.2.2. Degradation of Metabolically Labeled Cellular Proteins After Exposure of Intact Cells to Oxidative Stress

1. General material: Cell culture equipment, equipment to work with radiochemicals, β-counter, centrifuge (should be able to handle multiwell plates, *see* **Note 5**).
2. Sterile phosphate buffered saline (PBS).
3. Complete cell growth medium.
4. Methionine/cysteine-free cell growth medium (i.e., minimal essential Eagle's medium).
5. "Cold" methionine / cysteine (cell culture grade).
6. [^{35}S] methionine/cysteine (i.e., Tran[^{35}S]-label, ICN Biomedicals, Basingstoke, UK; *see* **Note 6**).
7. TCA: 20% (w/v) in water, keep on ice before use.
8. Scintillation cocktail.

3. Methods

3.1. Proteolytic Activity of Cell Lysates Toward Exogenously Added Oxidized and Nonoxidized Substrates

3.1.1. Degradation of suc-Leu-Leu-Val-Tyr-MCA (suc-LLVY-MCA) in Cell Lysates

The artificial peptide suc-LLVY-MCA is readily degraded by the "chymotrypsin-like" activity of the proteasome (*see* **Note 7**). The use of lactacystin, a proteasome inhibitor, helps to ensure the specificity of the assay.

1. Perform the experiment and harvest tissue culture cells. Count the cells and pellet between 10^5 and 10^7 of them by centrifugation at 800g for 10 min at 4°C. Wash the cell pellet three times with PBS.
2. Resuspend the cells in 200 μL lysis buffer and rapidly freeze/thaw the samples three times. Use liquid nitrogen or dry ice and a water bath at 25°C (*see* **Note 8**).
3. Centrifuge the samples at 14,000g for 10 min at 4°C. Transfer the supernatant in a new cup. Take an aliquot for protein determination (*see* **Note 9**). Store the samples on ice.
4. Dilute the samples with proteolysis buffer to 50 μg protein/mL. Keep the samples on ice.
5. Add 1 volume of the suc-LLVY-MCA stock to 9 volumes proteolysis buffer. If desired, add 2, 10, 20, 40, and 60 μL from the lactacystin stock during dilution of the peptide stock to yield aliquots of 1 mL (*see* **Note 10**).
6. Prepare standards with increasing amounts of "free" MCA in proteolysis buffer (i.e., 0.4 nmol/mL, 0.8 nmol/mL, and so forth).
7. Use 2-mL plastic cups (*see* **Note 11**). Take 100 μL of the diluted samples and add 100 μL of the diluted substrate. If desired, use for some samples 100 μL of the lactacystin containing substrate dilutions. Prepare blanks and standards without sample (proteolysis buffer only). Incubate in a shaking water bath at 37°C for 1 h (*see* **Note 12**).
8. Add 200 μL ice-cold absolute ethanol in order to stop the reaction. Add 1.6 mL sodium borate (0.125 M, pH 9.0). Measure the degradation of suc-LLVY-MCA by fluorometry at 365-nm excitation and 460-nm emission wavelengths.
9. Subtract the blanks from all values. Calculate the amount of substrate degraded using a calibration curve. Check the specificity of the method by evaluation of the lactacystin-inhibited samples.

3.1.2. Degradation of Oxidized and Nonoxidized Proteins in Cell Lysates

Proteins should be assessed for oxidation-induced proteolytic susceptibility immediately after exposure to oxidants. Freezing also affects the stability of oxidized proteins and may therefore modulate the effects of oxidation. Thus, the preparation of oxidized substrate proteins and the subsequent proteolysis assay is described in a single 2-d protocol. See **Fig. 1** for a typical result of the experiment described.

First Day Procedures

1. Dilute a [³H]-labeled extensively dialyzed protein (i.e., hemoglobin; *see* **Note 3**) with oxidation buffer to 0.66 mg protein/mL. Prepare five aliquots.
2. Dilute a H_2O_2 stock (*see* **Note 4**) with oxidation buffer to H_2O_2 concentrations of 10, 20, 40, and 60 mM (*see* **Note 13**). For preparation of a nonoxidized controls, omit the oxidant in one aliquot.
3. Add 1 volume of the H_2O_2 dilutions to 1 volume [³H]-protein (*see* **Note 14**). For the nonoxidized control, use the oxidation buffer without H_2O_2. Leave for 2 h at room temperature. Vortex briefly every 30 min.
4. Transfer the protein solutions in pre-wetted dialysis tubes with a molecular-weight cutoff of 10,000 Da (*see* **Note 15**). Dialyze against 2.5 L dialysis buffer at 4°C. Exchange the buffer after 2.5 h and continue the dialysis at 4°C overnight.

Second Day Procedures

5. Prepare cell lysates (as in **steps 1–3** in **Subheading 3.1.1.**).
6. Dilute the samples with proteolysis buffer to 0.5 mg protein/mL. Keep the samples on ice. Dilute 1 volume of the nonoxidized and oxidized [^{3}H]-protein substrates with 4 volumes of proteolysis buffer.
7. Use 2-mL plastic tubes for the assay and start the reactions as follows:
   a. Samples: 0.1 mL diluted lysate + 0.1 mL diluted [^{3}H]-protein substrate (use the differentially oxidized samples and the nonoxidized control).
   b. Blanks and totals for each [^{3}H]-protein substrate: 0.1 mL diluted [^{3}H]-protein + 0.1 mL 3% BSA.
8. Incubate in a shaking water bath at 37°C for 2 h.
9. End the reactions rapidly by adding 50 µL of 3% BSA as a precipitation carrier to each tube, followed by the immediate addition of 1.5 mL ice-cold TCA (12.5%) to samples and blanks. Add 1.5 mL distilled water to the totals. Vortex and keep all vials on ice for 30 min.
10. Centrifuge at 3000g for 15 min at 4°C.
11. Carefully transfer 0.5 mL of the supernatants to scintillation tubes. Add 4.5 mL scintillation cocktail and count each sample using an appropriate [^{3}H] filter.
12. Subtract the acid-soluble blank counts from the sample counts and from the total counts for each oxidized and nonoxidized [^{3}H]-protein substrate. Now calculate as follows:

Percent degradation = (Acid-soluble sample counts/Total counts) × 100

3.2. Measurement of Protein Degradation in Organelles and Intact Cells

3.2.1. Nonradioactive Assessment of Mitochondrial Protein Breakdown by Fluorescamine Reactivity (see **Note 1**)

1. Prepare mitochondria (or another cellular fraction) by standard procedures. The method described below, was used for mitochondrial preparations containing 8 mg protein/mL incubation medium (*see* **Note 16**).
2. Take 50-µL aliquots at the desired time-points of your experiment and add 450 µL ice-cold TCA (10%).
3. Allow the samples to precipitate 30 min on ice and centrifuge at 3000g for 10 min at 4°C.
4. Carefully transfer 250 µL of the supernatant in a 5-mL plastic tube and add 1.25 mL of 1 M HEPES (pH 7.8). Check the pH of each sample (*see* **Note 17**).
5. Construct a standard curve using various concentrations of glycine, in the range from 0.05 to 10 mM. Treat the standards as described in **steps 2** and **3** of this section.
6. Slowly add 0.5 mL of the fluorescamine solution while vortexing each tube, and keep samples in the dark.
7. Measure the samples and standards by fluorometry at an excitation wavelength of 390 nm and an emission wavelength of 475 nm, at constant intervals after the addition of fluorescamine (*see* **Note 18**).

8. Calculate the content of free primary amines using the glycine calibration curve. Increased free primary amines indicate enhanced degradation of mitochondrial proteins.

3.2.2. Degradation of Metabolically Labeled Cellular Proteins After Exposure of Intact Cells to Oxidative Stress

See **Fig. 2** for a typical result of the experiment described.

1. Grow cells in appropriate medium to about 70% of their maximal density (*see* **Note 5**).
2. Remove the growth medium, wash three times with PBS by centrifugation at 800*g* for 5 min and add methionine-free minimal essential Eagle's medium.
3. Add [^{35}S]-methionine/cysteine (i.e., Tran[^{35}S]-label) to reach an activity of 50 μCi/mL medium (*see* **Note 6**) and incubate the cells at 37°C for 2 h (*see* **Note 2**).
4. Wash the cells three times with PBS by centrifugation at 800*g* for 5 min. Collect all of the washout for subsequent estimation of label incorporation (*see* **Note 19**).
5. Treat the remaining cells with the desired oxidant (i.e., H_2O_2) in PBS for 30 min.
6. Wash the cells twice with PBS and collect all of the washout for subsequent estimation of label incorporation (*see* **Note 19**).
7. Add appropriate growth medium supplemented with 10 m*M* methionine/cysteine (*see* **Note 20**).
8. Treat about six samples 1:1 (v/v) with ice-cold TCA (20%) to estimate background radioactivity. Keep on ice for 30 min and centrifuge at 3000*g* for 10 min at 4°C (*see* **Note 21**).
9. Add an equal volume of ice-cold TCA (20%) to the samples at the desired time-points. Incubate on ice for 30 min and centrifuge at 3000*g* for 10 min at 4°C.
10. Carefully transfer aliquots from the supernatants of samples and background, dilute about 10-fold with scintillation cocktail, and count the acid-soluble radioactivity. Count aliquots of the washout as well.
11. Calculate the radioactivity incorporated into cellular proteins as the difference between amount of label used for the "pulse" and the content of label in the washout. Subtract the background from the sample counts. Now calculate as follows:

Percent degradation = (Acid soluble sample counts/Incorporated counts) × 100

4. Notes

1. For proteolysis measurements in functionally intact cells the use of techniques based on the nonradioactive quantification of amino acids is limited. Amino acid reincorporation due to ongoing protein synthesis as well as *de novo* synthesis of amino acids may lead to underestimation or overestimation of the actual intracellular proteolysis. In exceptional cases, like the detection of alanine in red blood cells that are neither able to synthesize proteins nor this amino acid, nonradioactive methods can be used for estimation of cellular protein degradation (*12,13*). In general, we feel that the use of the fluorescamine assay described here should

be restricted to isolated organelles, cellular fractions, and activity assessment of purified proteases.

2. The design of the labeling procedure will largely influence the fractions of labeled cellular proteins. In general, the duration of labeling should be dictated by the half-life of the protein (or protein fraction) of interest. Short-time labeling of up to 2 h will result in radioactivity incorporation in "short-lived" proteins or proteins that are synthesized in large quantities (*see* also **Fig. 2**). Longer incorporation times will lead to the detection of "long-lived" proteins. This protein fraction can be further defined if a 16-h labeling "pulse" is, for example, followed by a "cold chase" of 2 h (for degradation of "short-lived" proteins) before the start of the actual experiment.

3. Many commercially available protein preparations contain substances such as EDTA, salicylate, or ammonium sulfate for preservation. All of these additives have significant effects on radical/oxidant reactions, and most of them also affect proteases and proteolysis. To remove these contaminations, extensive dialysis is required before use. Purified proteins can be radiolabeled for oxidative stress studies by reductive methylation procedures. $[^3H]$-labeling or $[^{14}C]$-labeling with formaldehyde (as the source of the isotope) as described by Rice and Means *(14)* and Jentoft and Dearborn *(15)* has been used by us successfully for a wide variety of proteins. For the study of radical damage to proteins, one should avoid labeling with strong radiation sources (i.e., $[^{125}I]$ or $[^{131}I]$). The high specific activity of such labels will itself lead to radical formation and, therefore, to oxidative modification of the protein substrate. For example, it was shown as early as 1957 by Yallow and Berson *(16)* that the in vivo proteolytic susceptibility of serum albumin is enhanced by $[^{131}I]$-labeling. The proteolysis assay described represents a tracer method with the appropriate controls for each oxidized and nonoxidized substrate. Therefore moderate levels of label incorporation are sufficient.

4. The molar concentrations of H_2O_2 in the commercially available stocks (usually about 30%) should be checked routinely by a 1:1000 dilution in water and subsequent spectrophotometry. The absorption coefficient at 240 nm is $39.4/cm^{-1}M^{-1}$.

5. Multiwell plates (i.e., 24 wells, each well about $1.9\ cm^2$) will be sufficient for many cell cultures. Make sure that a centrifuge is equipped to handle these plates at $3000g$.

6. Not all proteins contain methionine. The use of a $[^{35}S]$-methionine/cysteine-cocktail (i.e., Tran$[^{35}S]$-label) ensures labeling of all proteins synthesized during the "pulse." In general, all amino acids available in radioactively labeled form can be used for metabolic labeling. For instance, we have recently used $[^3H]$leucine and $[^{14}C]$leucine as alternatives to $[^{35}S]$ isotopes.

7. Other proteolytic activities of the proteasome and its respective peptide substrates include the trypsinlike activity (N-Boc-Leu-Ser-Thr-Arg-MCA) and the peptidylglutamyl-peptide hydrolase activity (N-Cbz-Leu-Leu-Glu-NA).

8. The cell lysis is critical for the activity of the proteasome. The freeze/thaw cycles should be performed as fast and reproducibly as possible for all samples. The technique described is adapted from a protocol for purification of the 20*S* and

26S forms of the proteasome *(17)*. Alternatively, cells can be lysed by hypotonic suspension in 1 mM DTT in distilled water at 4°C for 1 h.

9. The method for protein quantification should detect proteins in the microgram-range. Commercially available assays (i.e., from Bio-Rad, Hercules, CA) are sufficient.

10. A lactacystin concentration of 5 µM is sufficient for most cases (*see* **Note 12**). However, the inhibitor should be tested with any new cell line.

11. The assay is adjusted to 2.0-mL fluorescence cuvets. If smaller cuvets are available all volumes can be scaled down accordingly.

12. Final concentrations: suc-LLVY-MCA 200 µM; lowest MCA-standard 0.2 µM, lactacystin 1–30 µM, sample protein 25 µg/mL.

13. The final H_2O_2 concentrations are 0, 5, 10, 20, and 30 mM. These H_2O_2 concentrations are a good starting point but have to be adjusted for each protein of interest. Other oxidants such as hypochlorite and peroxynitrite are also suitable for protein oxidation. More complex treatments such as with Fe^{2+}/ascorbate or Fe^{2+}/H_2O_2 work as well but may exhibit lower reproducibility.

14. It is essential to add 1 volume of oxidant to 1 volume of the protein solution. The addition of a small volume of highly concentrated oxidant will immediately lead to denaturation of proteins around the pipet-tip. Total volumes of 2–20 mL work best.

15. The 10-kDa cutoff is needed not only for the removal of the oxidant but also for the removal of protein fragments of less than 5 kDa. These fragments do not precipitate with TCA and would therefore cause high backgrounds in the proteolysis assay.

16. The assay is based on the reactivity of fluorescamine with primary amines, such as the α-amino groups of all amino acids, and the ε-amino groups of lysine residues. Because Tris is itself a primary amine, care should be taken to avoid Tris-based buffers (and other primary amines) during the preparation, the experiments, and the assay. Buffers containing HEPES represent a convenient alternative.

17. The fluorescence intensity of fluorescamine is strongly influenced by pH and is highest at pH 9.0. Make sure that all samples, standards, and blanks are at a constant pH (preferably pH 9.0 for maximum sensitivity). If necessary, adjust with KOH.

18. Because fluorescence of fluorescamine is not stable, samples must be measured at constant intervals after the addition of fluorescamine. Intervals of 5–10 min usually work best.

19. A convenient way to estimate the incorporation of the radiolabel into cell proteins is the calculation of the difference between the amount of label used and the content of label in the washout. Independent of cell type and the duration of the "pulse" (*see* **Note 2**), 10–50% of the label should be incorporated. This should be confirmed for every cell line and labeling procedure by precipitation of cell proteins with 10% trichloroacetic acid, subsequent washing with 100% acetone (at –20°C), resolubilization of the resulting protein powder (e.g., with 6 M guanidine hydrochloride), and determination of the protein label by scintillation counting.

20. The excess "cold" methionine/cysteine is added to prevent reincorporation of proteolysis-derived, labeled amino acids into newly synthesized proteins.

21. This step defines the start of proteolysis measurement.

References

1. Davies, K. J. A., Delsignore, M. E., and Lin, S. W. (1987) Protein damage and degradation by oxygen radicals. II. Modification of amino acids. *J. Biol. Chem.* **262,** 9902–9907.
2. Stadtman, E. R. (1993) Oxidation of free amino acids and amino acid residues in proteins by radiolysis and by metal-catalyzed reactions. *Annu. Rev. Biochem.* **62,** 797–821.
3. Dean, R. T., Fu, S., Stocker, R., and Davies, M. J. (1997) Biochemistry and pathology of radical-mediated protein oxidation. *Biochem. J.* **324,** 1–18.
4. Uchida, K. and Stadtman, E. R. (1992) Modification of histidine residues in proteins by reaction with 4-hydroxynonenal. *Proc. Natl. Acad. Sci. USA* **89,** 4544–4548.
5. Lee, Y. and Shacter, E. (1995) Role of carbohydrates in oxidative modification of fibrinogen and other plasma proteins. *Arch. Biochem. Biophys.* **321,** 175–181.
6. Davies, K. J. A. and Delsingnore, M. E. (1987) Protein Damage and Degradation by Oxygen Radicals III. Modification of secondary and tertiary structure. *J. Biol. Chem.* **262,** 9902–9907.
7. Davies, K. J. A., Lin, S. W., and Pacifici, R. E. (1987) Protein Damage and Degradation by Oxygen Radicals-IV. Degradation of denatured protein. *J. Biol. Chem.* **262,** 9914–9920
8. Grune, T., Reinheckel, T., and Davies, K. J. A. (1997) Degradation of oxidized proteins in mammalian cells. *FASEB J.* **11,** 526–534.
9. Grune, T., Reinheckel, T., Joshi, M., and Davies, K. J. A. (1995) Proteolysis in cultured liver epithelial cells during oxidative stress. *J. Biol. Chem.* **270,** 2344–2351.
10. Grune, T., Reinheckel, T., and Davies, K. J. A. (1996) Degradation of oxidized proteins in human hematopoietic cells by proteasome. *J. Biol. Chem.* **271,** 15,504–15,509.
11. Grune, T., Blasig, I. E., Sitte, N., Roloff, B., Haseloff, R., and Davies, K. J. A. (1998) Peroxynitrite increases the degradation of aconitase and other cellular proteins by proteasome. *J. Biol. Chem.* **273,** 10,857–10,862.
12. Davies, K. J. A. and Goldberg, A. L. (1987) Oxygen radicals stimulate intracellular proteolysis and lipid peroxidation by independent mechanisms in erythrocytes. *J. Biol. Chem.* **262,** 8220–8226.
13. Davies, K. J. A. and Goldberg, A. L. (1987) Proteins damaged by oxygen radicals are rapidly degraded in extracts of red blood cells. *J. Biol. Chem.* **262,** 8227–8234.
14. Rice, R. and Means, G. E. (1971) Radioactive labeling of proteins *in vitro*. *J. Biol. Chem.* **246,** 831,832.
15. Jentoft, N. and Dearborn, D. G. (1979) Labeling of proteins by reductive methylation using sodium cyanoborohydride. *J. Biol. Chem.* **254,** 4359–4365.
16. Yallow, R. S. and Berson, S. A. (1957) Chemical and biological alterations induced by irradiation of [^{131}I]-labelled serum albumin. *J. Clin. Invest.* **36,** 44–50.
17. Hough, R., Pratt, G., and Rechsteiner, M. (1987) Purification of two high molecular weight proteases from rabbit reticulocyte lysate. *J. Biol. Chem.* **262,** 8303–8313.

II

THE ACTIVATION OF SIGNAL
TRANSDUCTION BY CELLULAR STRESS

6

Analysis of the Role of the AMP-Activated Protein Kinase in the Response to Cellular Stress

D. Grahame Hardie, Ian P. Salt, and Stephen P. Davies

1. Introduction

The AMP-activated protein kinase (AMPK) is the central component of a protein kinase cascade that is activated by cellular stresses causing ATP depletion and has been referred to as a "fuel gauge" or "metabolic sensor" of the eukaryotic cell *(1,2)*. The kinase is activated by phosphorylation by an upstream protein kinase termed AMP-activated protein kinase kinase (AMPKK) *(3)*. Elevation of 5'-AMP activates the cascade by a complex mechanism involving binding of the nucleotide to both the upstream kinase (AMP-activated protein kinase kinase, AMPKK) and the downstream kinase, AMPK (*see* **Subheading 1.2.**). These effects of AMP are also antagonized by high concentrations (m*M*) of ATP. The AMP:ATP ratio in the cell varies approximately as the square of the ADP:ATP ratio, due to the action of adenylate kinase which maintains its reaction (2ADP ´ ATP + AMP) close to equilibrium at all times. Therefore, any cellular stress that affects the ability of the cell to maintain a high ATP:ADP ratio (normally approx 10:1 in an unstressed cell) leads to activation of the AMPK cascade. Cellular stresses can do this either by inhibiting ATP production or by increasing ATP consumption, and stresses shown to cause AMPK activation include heat shock *(4)*, various mitochondrial inhibitors such as arsenite, antimycin A, dinitrophenol, and azide *(4,5)*, ischemia/hypoxia in heart muscle *(6)*, and exercise in skeletal muscle *(7)*. ATP can also be depleted, and AMPK activated, by incubation of cells with high concentrations of certain sugars which trap cellular phosphate, such as fructose *(8)* and 2-deoxyglucose *(9)*. Detachment of cultured cells from their substrate by trypsinization has also been reported to increase cellular AMP:ATP and to inhibit lipid synthesis, consistent with the activation of AMPK *(10)*. Down-

From: *Methods in Molecular Biology, vol. 99: Stress Response: Methods and Protocols*
Edited by: S. M. Keyse © Humana Press Inc., Totowa, NJ

stream targets for the system include biosynthetic pathways that are inhibited, thus conserving ATP, and catabolic pathways tht are activated, thus generating more ATP *(1,2)*. Although most of the currently known targets for the system are metabolic enzymes, the yeast homolog of AMPK (i.e., the SNF1 complex) regulates gene expression *(2)*. At least one isoform of AMPK is partly localized to the nucleus (*see* **Subheading 1.1.**), and it seems very likely that the mammalian system will also turn out to regulate gene expression.

1.1. Subunit Structure of AMP-Activated Protein Kinase

The mammalian kinase is a heterotrimer comprising α-, β-, and γ-subunits, of which the former is the catalytic subunit. Each of these subunits exists in at least two isoforms (α_1, α_2, β_1, β_2, γ_1, γ_2, etc.) encoded by distinct genes *(11–18)*. The α_1- and β_1-subunits are widely expressed, whereas α_2 is expressed at high levels in liver, skeletal muscle and cardiac muscle, and β_2 in skeletal and cardiac muscle. Subunit-specific antibodies have recently been developed, which allow assays of complexes containing specific isoforms to be made in immunoprecipitates *(14)*. This is of interest because complexes containing the α_1-and α_2-subunits differ both in their subcellular localization and in their dependence on AMP. Complexes containing the α_2-subunit are partly localized in the nucleus, and are activated five- to sixfold by AMP, whereas complexes containing α_1 are not present in the nucleus and are only activated 1.5- to twofold by AMP *(19)*.

1.2. Regulation of AMP-Activated Protein Kinase

The AMP-activated protein kinase is present in all cell types examined to date. Although there are tissue-specific differences in expression of different subunit isoforms (*see* **Subheading 1.1.**), at present there is no evidence that the expression of the protein subunits is acutely regulated. However, the kinase activity of existing protein is exquisitely regulated. Methods to measure the kinase activity of the complex in response to different stresses are therefore important, as are methods to artificially manipulate the kinase activity in intact cells. These form the main topic of this chapter.

Elevation of AMP (coupled with depression of ATP) activates the system by no less than four mechanisms *(3,20,21)*: (1) Binding of AMP causes allosteric activation of the downstream kinase, AMPK; (2) binding of AMP to dephosphorylated AMPK causes it to become a much better substrate for the upstream kinase, AMPKK; (3) binding of AMP to phosphorylated AMPK causes it to become a much worse substrate for protein phosphatases, especially protein phosphatase-2C; (4) the upstream kinase, AMPKK, is also allosterically activated by AMP. The effects of mechanism 1 do not survive homogenization and purification and so cannot be readily measured in cell extracts, but mechanisms 2–4 all result in increased phosphorylation (primarily at Thr-172 in the α-subunit *[3]*) and can be measured under appropriate conditions in cell-free extracts prepared from the cells.

1.3. Assay of AMP-Activated Protein Kinase in Cell Extracts

At least two important conditions must be met before one can be confident that the activity measured in extracts reflects the true activity in the intact cells.

1. The AMPK is extremely sensitive to any form of stress, so the challenge is to find a method for harvesting and homogenizing the cells that does not itself result in artifactual activation of the kinase. Rapid cooling is the key factor. In intact tissues, freeze clamping is essential for preventing AMPK activation: dissecting out the tissue at ambient temperature, no matter how rapid, tends to cause activation, probably due to hypoxia *(22)*. In isolated cells in suspension, centrifuging and resuspending the cells activates the kinase, and the best methods are either to add concentrated homogenization medium directly to the cell suspension and freeze the entire mixture in liquid nitrogen prior to homogenization *(4)*, or to dilute the cells into a large volume of ice-cold medium prior to centrifugation *(23,24)*. For cells attached to a culture dish, the best method is to pour off the medium and immediately add a small volume of ice-cold lysis buffer containing nonionic detergent. This method is described in this chapter (**Subheading 3.2.**).
2. Because activation of AMPK is caused by phosphorylation, the homogenization medium must prevent both phosphorylation and dephosphorylation occurring subsequent to homogenization. Addition of EDTA (5 mM, to bind cellular Mg^{2+} and prevent phosphorylation) and phosphatase inhibitors (e.g., 50 mM NaF plus 5 mM Na pyrophosphate) to the homogenization medium are essential. Any purification carried out prior to assay must also be carried out rapidly and at as low a temperature as possible.

The peptide assays described in this chapter were originally shown to be rather specific for AMPK in extracts of rat liver *(25)*, but they may not be completely specific in all cell types. The peptides are also phosphorylated by Ca^{2+}/calmodulin-dependent protein kinase I *(26)*, although this is routinely overcome by the inclusion of EGTA in the homogenization and assay buffers. Nevertheless, in some cultured cells, up to 50% of the activity detected with the peptide substrates (even in stressed cells where the kinase is activated) can be the result of other unidentified protein kinases. In our experience these other kinases are not activated by stress treatments, but they result in a high apparent "basal" activity. The assays can be made more specific by partially purifying AMPK by polyethylene glycol precipitation (**Subheading 3.3.**), and completely specific by immunoprecipitation (**Subheading 3.4.**), prior to assay.

1.4. Artificial Manipulation of AMPK Activity in Intact Cells

Identification of novel cellular targets and processes regulated by AMPK would be greatly aided by the development of methods for activating and inhibiting the kinase in intact cells. There are currently no specific pharmacological inhibitors of the system; molecular biological approaches such as the use of knockouts, dominant negative mutants, ribozymes or antisense technology are under development but are not yet routinely available. There is, however, a pharmacological method for *activation* of the kinase cascade in intact cells that has already been quite widely used. 5-Aminoimidazole-4-carboxamide riboside (AICA riboside) is taken up into cells and is phosphorylated by adenosine kinase to the monophosphorylated form (AICA ribotide or ZMP). Although ZMP is a natural intermediate in purine nucleotide synthesis, in many (but not all) cells, the formation of ZMP from extracellular AICA riboside is much more rapid than its subsequent metabolism, so ZMP accumulates. ZMP mimics the effects of AMP both on allosteric activation of AMPK *(24,27)*, and on phosphorylation by the upstream kinase *(24)*. A distinct advantage of this method is that incubation of suitable cells with AICA riboside leads to accumulation of ZMP and activation of AMPK, without affecting the cellular content of AMP, ADP, or ATP. It is therefore a much more specific method for activating AMPK than giving some form of cellular stress that elevates AMP and depletes ATP, where there may be many secondary consequences of changes in the levels of these nucleotides. Incubation of cells or tissues with AICA riboside has, for example, been used to demonstrate that AMPK activation causes inhibition of fatty acid and sterol synthesis in hepatocytes *(24,27)*, inhibition of lipolysis in adipocytes *(24)*, activation of fatty acid oxidation in hepatocytes and perfused muscle *(28,29)*, and activation of glucose transport in perfused muscle *(29)*.

2. Materials

The laboratory should be equipped for general biochemical techniques, including facilities for the handling of radioisotopes and liquid scintillation counting.

2.1. Assay of AMP-Activated Protein Kinase

1. HEPES-Brij buffer: 50 mM Na HEPES, pH 7.4, 1 mM dithiothreitol (DTT), 0.02% Brij-35: can be kept for a few days at 4°C, otherwise store at –20°C.
2. 100 mM unlabeled ATP: Dissolve slowly with stirring in HEPES-Brij buffer, keeping the pH just above 7.0 with NaOH solution. ATP solutions must be neutralized *before* addition of MgCl$_2$, otherwise an insoluble MgATP complex will precipitate during neutralization.

3. [γ-^{32}P]ATP: we use the approx 30-Ci/mmol formulation for protein kinase assays (Amersham, cat. no. PB10132). A laboratory stock solution is prepared in the following manner: 1 mCi (100 μL) is added to 890 μL of water and 10 μL of 100 mM unlabeled ATP to give a solution (1 mCi/mL, 1 mM) that can be stored at –20°C. This stock solution is then diluted to the desired specific radioactivity with 1 mM unlabeled ATP with 5 μL/mL of 5 M MgCl$_2$ added (final Mg^{2+} concentration 25 mM). The exact concentration of this working solution is determined spectrophotometrically (A$_{260\ nm}$ of 1 mM solution = 15).
4. 1 mM AMP (dissolved in HEPES-Brij buffer) can be kept for a few days at 4°C, otherwise store in aliquots at –20°C.
5. *SAMS* peptide (HMRSAMSGLHLVKRR) or *AMARA* peptide (AMARA ASAAALARRR) (*see* **Note 1**), 1 mM in HEPES-Brij buffer. Peptides can be kept for a few days at 4°C, otherwise store in aliquots at –20°C. Peptides made by a peptide synthesis service should be high-performance liquid chromatography (HPLC) purified and the concentration determined by amino acid analysis: oxidation of the methionines during prolonged storage at 4°C can affect their ability to act as substrates.
6. P81 phosphocellulose paper (Whatman): cut into 1 cm^2 squares.
7. 1% v/v phosphoric acid
8. Optiscint "Hisafe" scintillation cocktail (Wallac).

2.2. Rapid Lysis of Cultured Cells for Kinase Assay

1. Krebs-HEPES buffer: 20 mM Na HEPES , pH 7.4, 118 mM NaCl, 3.5 mM KCl, 1.3 mM CaCl$_2$, 1.2 mM MgSO$_4$, 1.2 mM KH$_2$PO$_4$, 10 mM glucose, 0.1% (w/v) bovine serum albumin (BSA): prepare fresh buffer prior to use.
2. Lysis buffer: 50 mM Tris-HCl, pH 7.4 at 4°C, 50 mM NaF, 5 mM Na pyrophosphate, 1 mM EDTA, 1 mM EGTA: prepare as 10X stock. Before use, add (final concentrations) 250 mM mannitol, 1% (v/v) Triton X-100, 1 mM DTT, and the proteinase inhibitors 1 mM benzamidine, 0.1 mM phenylmethane sulfonyl fluoride (PMSF), and 5 μg/mL soybean trypsin inhibitor (SBTI).
3. Lysate assay buffer: 62.5 mM Na HEPES, pH 7.0, 62.5 mM NaCl, 62.5 mM NaF, 6.25 mM Na pyrophosphate, 1.25 mM EDTA, 1.25 mM EGTA, 1 mM DTT, 1 mM benzamidine, 1 mM PMSF, 5 μg/mL SBTI. The DTT, benzamidine, PMSF, and SBTI are added just before use. The remainder can be stored at –20°C for several months.

2.3. Polyethylene Glycol Precipitation Prior to Assay

Polyethylene glycol (PEG) 6000 (Merck/BDH). Prepare 25% (w/v) stock immediately prior to use.

2.4. Immunoprecipitation Prior to Assay

1. IP buffer: 50 mM Tris-HCl, pH 7.4 at 4°C, 150 mM NaCl, 50 mM NaF, 5 mM Na pyrophosphate, 1 mM EDTA, 1 mM EGTA. Prepare 1X stock. Before use add 1 mM DTT, 0.1 mM benzamidine, 0.1 mM PMSF, 5 μg/mL soybean trypsin inhibitor (final concentrations).

2. Protein G-Sepharose (Pharmacia Biotech)
3. Anti-AMPK α subunit antibodies: these are affinity purified sheep anti-AMPK antibodies as described in Woods et al. *(14)*. They may soon become commercially available through Upstate Biotechnology Inc.

2.5. Activation of AMP-Activated Protein Kinase in Intact Cells with AICA Riboside

AICA riboside (Sigma): dissolve in Krebs-HEPES buffer immediately before use.

3. Methods

3. 1. Assay of AMP-Activated Protein Kinase

This is based on the assay originally described by Davies et al. *(25)*. The method now outlined is a modified method that is suitable for purified AMPK preparations. Demonstrating activation by AMP is difficult, except when using highly purified enzyme (*see* **Note 2**), and is not usually worth trying in crude cell extracts.

1. Reaction mixtures are prepared on ice containing the following:

5 µL	1 m*M* [γ-^{32}P]ATP, 25 m*M* MgCl$_2$ (specific activity 250–500 cpm/pmol)
5 µL	1 m*M* AMP in HEPES–Brij buffer
5 µL	1 m*M* SAMS or AMARA peptide in HEPES–Brij buffer
5 µL	HEPES-Brij buffer

   Blank reactions are also performed containing HEPES-Brij buffer in place of peptide.
2. Reactions are initiated by the addition of AMPK (5 µL, 1–5 U/mL). In skilled hands, assays can be started and stopped at 15-s intervals.
3. Incubate at 30°C for 10 min, remove 15 µL and spot onto a P81 paper square (number the squares with a hard pencil before use). After the liquid has soaked in 1–2 s), the square is dropped into a beaker containing 1% (v/v) phosphoric acid.
4. After all incubations have been stopped, the paper squares are stirred gently on a magnetic stirrer for 2–3 min. The phosphoric acid (*caution:* radioactive!) is poured off and a second phosphoric acid wash is performed. The papers are rinsed briefly in water before soaking in acetone. They are then laid out on a paper towel and allowed to dry.
5. Dried filters are counted after immersing in 5 ml of Optiscint "Hisafe" scintillation cocktail.
6. The AMPK activity is expressed in units of nanomoles of phosphate incorporated into substrate peptide per minute. To determine the specific radioactivity of the ATP, spot 5 µL onto a paper square, dry it, and count in scintillation fluid. Keep this vial and recount it every time you count some assays using the same ATP. The counts obtained correspond to 5 nmol of ATP, and radioactive decay is corrected for automatically. Using this method of counting, the counting efficiency is constant and need not be determined.

7. If you use a nonaqueous scintillation cocktail such as Optiscint "Hisafe," the vial of scintillation fluid can usually be reused after removing the paper square. Do not reuse the vial that you used to count the ATP.

3.2. Rapid Lysis of Cultured Cells for Kinase Assay

This method is designed for cultured cells attached to the dish. Different methods are required for cells in suspension or intact tissue samples (*see* **Subheading 1.3.**). Depending on the cell type, it may be difficult to measure the kinase accurately in a crude lysate (*see* **Note 3**). If this is the case, it can be partially purified prior to assay by PEG precipitation or immunoprecipitation as described in **Subheading 3.3.** and **3.4.**, respectively.

1. Cells are seeded in 10-cm culture dishes and grown until nearly confluent. The medium is removed and the cells are washed with 3×5 mL of Krebs-HEPES buffer at 37°C. Cells are then preincubated in 5 ml Krebs-HEPES buffer at 37°C for 30 min.
2. The medium is removed and replaced with a further 5 mL Krebs-HEPES buffer containing any substances or treatments to be tested, and the cells incubated for appropriate times.
3. The bulk medium is poured off, residual medium removed with a Pasteur pipet, the dish placed on ice, and 0.5–1.0 ml of ice-cold lysis buffer immediately added.
4. The cells are scraped from the dish with a cell scraper and the lysate transferred to a microcentrifuge tube. The lysate is kept on ice throughout this and subsequent procedures. The lysate is centrifuged ($18,000g$, 4°C, 3 min).
5. The supernatant can be assayed immediately, or it can be snap-frozen in liquid N_2 and stored at –80°C for at least 2 wk prior to assay.
6. Carry out assays as in **Subheading 3.1.**, except that the HEPES-Brij buffer is replaced by lysate assay buffer (*see* **Note 4**).

3.3. Polyethylene Glycol Precipitation Prior to Assay

Polyethylene glycol (PEG) precipitation typically purifies AMPK by three- to eightfold from the crude lysate, removing inhibitors and concentrating the enzyme prior to assay (*see* **Note 3**).

1. One volume of PEG 6000 (25% [w/v]) is added to 9 volumes of the supernatant (final concentration 2.5%). The mixture is centrifuged ($18,000g$, 4°C, 3 min) and the supernatant removed to another microfuge tube.
2. One volume of PEG 6000 (25% [w/v]) is added to 9 volumes of the 2.5% PEG supernatant (final concentration 4.75%, *see* **Note 5**). The mixture is centrifuged as before and the supernatant carefully removed.
3. The resultant pellet is resuspended in assay buffer (typically 50–100 µL) and assayed directly as in **Subheading 3.2.**, or it can be snap-frozen in liquid N_2 and stored at –80°C for up to 1 mo.

3.4. Immunoprecipitation Prior to Assay

Cell lysates can also be assayed in immunoprecipitates using the antibodies described in Woods et al. *(14)*. Complexes containing the α_1 or α_2 isoforms of the catalytic subunit may be precipitated specifically using anti-α_1 or anti-α_2 antibodies. Alternatively, total activity may be assayed using a pan-α antibody, a mixture of anti-α_1 and anti-α_2 antibodies, or a suitable anti-β or anti-γ subunit antibody.

1. Wash 40 µL (packed volume) of Protein G-Sepharose (*see* **Note 6**) with 5×1 mL of IP buffer, and resuspend in 120 µL IP buffer. This should be enough for 30 assays.
2. Add 40 µg of sheep anti-AMPK antibody to the Protein G-Sepharose slurry and mix on a roller mixer for 45 min at 4°C.
3. Centrifuge the slurry (18,000g, 1 min, 4°C) and wash the pellet five times with 1 mL ice-cold IP buffer. Resuspend the final pellet with another 120 µL IP buffer and divide the slurry into 5-µL aliquots.
4. To each 5-µL aliquot add 50–100 µL of cell lysate and mix for 2 h at 4°C on a roller mixer.
5. Centrifuge the mixture (18,000g, 1 min, 4°C) and wash the pellet with 5×1 mL of ice-cold IP buffer containing 1 M NaCl (to remove nonspecifically bound protein). Wash the pellet with 3×1 mL of lysate assay buffer and resuspend in 30 µL HEPES-Brij buffer prior to assay. It is best to assay immunoprecipitates immediately, but they can be stored at –20°C for a few days.
6. Assay the immunoprecipitates using the method described in **Subheading 3.1.**, except that the reactions are performed on a orbital shaker (*see* **Note 7**).

3.5. Activation of AMP-Activated Protein Kinase in Intact Cells with AICA Riboside

1. Cells are preincubated in Krebs-HEPES buffer for 30 min at 37°C as described in **Subheading 3.2.**
2. The medium is removed and replaced with a further 5 mL Krebs-HEPES buffer, 100 µL of the AICA riboside solution is added to the medium and the cells incubated for various times at 37°C (*see* **Note 8**).
3. The cellular process under study is then examined. To assay AMPK activity the medium is removed and the cells rapidly lysed as in **Subheading 3.2.**

4. Notes

1. The AMPK assay was originally developed using the *SAMS* peptide. With mammalian AMPK, the V_{max} is about 50% higher using the *AMARA* peptide *(26)*, so it can be advantageous to use this, especially when assaying in crude cell extracts where the activity is low. For reasons that remain unclear, the kinase is more dependent on AMP using *SAMS* rather than *AMARA* as the substrate, which is why the use of *SAMS* has not been completely replaced. The yeast homologue,

SNF1, is less active with *AMARA* (because of a high K_m) so *SAMS* is used routinely in assays of yeast extracts. With the higher plant homologues, activities are considerably higher using *AMARA (26)*.

2. Activation by AMP is difficult to demonstrate in crude extracts for two reasons: (1) there may be contamination with endogenous AMP; (2) there is almost always contamination with adenylate kinase, which generates AMP from ADP formed by the kinase and other ATPases. Problem (1) can be overcome by treating extracts with snake venom 5'-nucleotidase (from *Crotalus adamanteus*, Sigma). If agarose-bound 5'-nucleotidase is used, this can be removed by centrifugation prior to assay. Problem (2) is more difficult but can sometimes be overcome by including 5'-nucleotidase in the assay itself. However the best way to demonstrate AMP-dependence is to assay the enzyme in well washed immunoprecipitates.

3. With crude lysates of some cells, the signal to noise ratio is low (i.e., the counts obtained in the presence of substrate peptide, and those obtained in the blank without peptide, may not be very different). An additional problem is that the assays may not be linear with the amount of protein added, because the extract contains inhibitors (possibly alternative substrates *[25]*) that have to be diluted out. Partially purifying the kinase by PEG precipitation, or purifying it by immunoprecipitation, helps the first problem because the kinase is concentrated, and the second because inhibitors are removed. These methods, particularly immunoprecipitation also remove other kinases which might phosphorylate the substrate peptide.

4. When assaying the kinase in crude cell lysates, we replace the HEPES-Brij buffer by the lysate assay buffer for all components except the MgATP solution. The lysate assay buffer contains protein phosphatase inhibitors to prevent dephosphorylation of the kinase during the assay, EGTA to inhibit Ca^{2+}-dependent protein kinases, and proteinase inhibitors.

5. In our experience this 2.5–4.75% PEG cut precipitates the kinase almost quantitatively (>80%) from most cell lysates, but this should be confirmed with any new cell type under study. Commercial PEG preparations vary in chain length, so also check the optimal PEG concentrations if a different source of PEG is used.

6. When pipeting any solution containing Sepharose beads, use pipet tips with the last 5-mm cut-off to avoid trapping the beads in the tip. This includes the spotting of reactions onto P81 paper.

7. When assaying kinases coupled to antibodies that are immobilized on protein G-Sepharose, it is important to keep the Sepharose beads suspended throughout the assay. We use a small orbital shaker (IKA-VIBRAX-VXR, Janke & Kunkel) fitted with a holder that takes microcentrifuge tubes (type VX 2E) and mounted in a small benchtop air incubator (Stuart Scientific Incubator S.I.60). A conventional orbital incubator could be used, but it is not convenient for inserting and removing tubes rapidly. Because of problems in pipeting Sepharose beads, the reproducibility of these assays is slightly lower than assays where everything is in the liquid phase, and we routinely perform them in triplicate.

8. Typically, 500 μM to 1 mM AICA riboside for 10 to 30 min gives maximal stimulation of AMPK, but with any new cell type the effect of riboside concentration

and incubation time on kinase activity should be studied. In some cells, e.g., cardiomyocytes *(30)*, incubation with AICA riboside does not give rise to accumulation of ZMP or kinase activation, so so a lack of effect on some cellular process may be inconclusive unless you show that the kinase is being activated. ZMP and other cellular nucleotides can be monitored by HPLC *(24)*, but the most reliable method to check whether AICA riboside is effective in activating AMPK in a new cell type is to assay the kinase directly.

Acknowledgments

Studies in this laboratory were supported by a Programme Grant from the Wellcome Trust.

References

1. Hardie, D. G. and Carling, D. (1997) The AMP-activated protein kinase: fuel gauge of the mammalian cell? *Eur. J. Biochem.* **246,** 259–273.
2. Hardie, D. G., Carling, D., and Carlson, M. (1998) The AMP-activated/SNF1 protein kinase subfamily: metabolic sensors of the eukaryotic cell? *Ann. Rev. Biochem.* **67,** 821–855.
3. Hawley, S. A., Davison, M., Woods, A., Davies, S. P., Beri, R. K., Carling, D., and Hardie, D. G. (1996) Characterization of the AMP-activated protein kinase kinase from rat liver, and identification of threonine-172 as the major site at which it phosphorylates and activates AMP-activated protein kinase. *J. Biol. Chem.* **271,** 27,879–27,887.
4. Corton, J. M., Gillespie, J. G., and Hardie, D. G. (1994) Role of the AMP-activated protein kinase in the cellular stress response. *Current Biol.* **4,** 315–324.
5. Witters, L. A., Nordlund, A. C., and Marshall, L. (1991) Regulation of intracellular acetyl-CoA carboxylase by ATP depletors mimics the action of the 5'-AMP-activated protein kinase. *Biochem. Biophys. Res. Comm.* **181,** 1486–1492.
6. Kudo, N., Barr, A. J., Barr, R. L., Desai, S., and Lopaschuk, G. D. (1995) High rates of fatty acid oxidation during reperfusion of ischemic hearts are associated with a decrease in malonyl-CoA levels due to an increase in 5'-AMP-activated protein kinase inhibition of acetyl-CoA carboxylase. *J. Biol. Chem.* **270,** 17,513–17,520.
7. Winder, W. W., and Hardie, D. G. (1996) Inactivation of acetyl-CoA carboxylase and activation of AMP-activated protein kinase in muscle during exercise. *Am. J. Physiol.* **270,** E299–E304.
8. Moore, F., Weekes, J., and Hardie, D. G. (1991) AMP triggers phosphorylation as well as direct allosteric activation of rat liver AMP-activated protein kinase. A sensitive mechanism to protect the cell against ATP depletion. *Eur. J. Biochem.* **199,** 691–697.
9. Sato, R., Goldstein, J. L., and Brown, M. S. (1993) Replacement of Serine-871 of hamster 3-hydroxy-3-methylglutaryl CoA reductase prevents phosphorylation by AMP-activated protein kinase and blocks inhibition of sterol synthesis induced by ATP depletion. *Proc. Natl. Acad. Sci. USA* **90,** 9261–9265.

10. Page, K. and Lange, Y. (1997) Cell adhesion to fibronectin regulates membrane lipid biosynthesis through 5'-AMP-activated protein kinase. *J. Biol. Chem.* **272,** 19,339–19,342.

11. Carling, D., Aguan, K., Woods, A., Verhoeven, A. J. M., Beri, R. K., Brennan, C. H., Sidebottom, C., Davison, M. D., and Scott, J. (1994) Mammalian AMP-activated protein kinase is homologous to yeast and plant protein kinases involved in the regulation of carbon metabolism. *J. Biol. Chem.* **269,** 11,442–11,448.

12. Gao, G., Widmer, J., Stapleton, D., Teh, T., Cox, T., Kemp, B. E., and Witters, L. A. (1995) Catalytic subunits of the porcine and rat 5'-AMP-activated protein kinase are members of the SNF1 protein kinase family. *Biochim. Biophys. Acta* **1266,** 73–82.

13. Stapleton, D., Mitchelhill, K. I., Gao, G., Widmer, J., Michell, B. J., Teh, T., House, C. M., Fernandez, C. S., Cox, T., Witters, L. A., and Kemp, B. E. (1996) Mammalian AMP-activated protein kinase subfamily. *J. Biol. Chem.* **271,** 611–614.

14. Woods, A., Salt, I., Scott, J., Hardie, D. G., and Carling, D. (1996) The α1 and α2 isoforms of the AMP-activated protein kinase have similar activities in rat liver but exhibit differences in substrate specificity *in vitro. FEBS Lett.* **397,** 347–351.

15. Woods, A., Cheung, P. C. F., Smith, F. C., Davison, M. D., Scott, J., Beri, R. K., and Carling, D. (1996) Characterization of AMP-activated protein kinase β and γ subunits: assembly of the heterotrimeric complex *in vitro. J. Biol. Chem.* **271,** 10,282–10,290.

16. Gao, G., Fernandez, S., Stapleton, D., Auster, A. S., Widmer, J., Dyck, J. R. B., Kemp, B. E., and Witters, L. A. (1996) Non-catalytic β- and γ-subunit isoforms of the 5'-AMP-activated protein kinase. *J. Biol. Chem.* **271,** 8675–8681.

17. Thornton, C., Snowden, M. A., and Carling, D. (1998) Identification of a novel AMP-activated protein kinase β subunit isoform which is highly expressed in skeletal muscle. *J. Biol. Chem.* **273,** 12,443–12,450.

18. Stapleton, D., Woollatt, E., Mitchelhill, K. I., Nicholl, J. K., Fernandez, C. S., Michell, B. J., Witters, L. A., Power, D. A., Sutherland, G. R., and Kemp, B. E. (1997) AMP-activated protein kinase isoenzyme family: subunit structure and chromosomal location. *FEBS Lett.* **409,** 452–456.

19. Salt, I. P., Celler, J. W., Hawley, S. A., Prescott, A., Woods, A., Carling, D., and Hardie, D. G. (1998) AMP-activated protein kinase - greater AMP dependence, and preferential nuclear localization, of complexes containing the α2 isoform. *Biochem. J.* **334,** 177–187.

20. Carling, D., Clarke, P. R., Zammit, V. A., and Hardie, D. G. (1989) Purification and characterization of the AMP-activated protein kinase. Copurification of acetyl-CoA carboxylase kinase and 3-hydroxy-3-methylglutaryl-CoA reductase kinase activities. *Eur. J. Biochem.* **186,** 129–136.

21. Davies, S. P., Helps, N. R., Cohen, P. T. W., and Hardie, D. G. (1995) 5'-AMP inhibits dephosphorylation, as well as promoting phosphorylation, of the AMP-activated protein kinase. Studies using bacterially expressed human protein phosphatase-2Cα and native bovine protein phosphatase-2A$_C$. *FEBS Lett.* **377,** 421–425.

22. Davies, S. P., Carling, D., Munday, M. R., and Hardie, D. G. (1992) Diurnal rhythm of phosphorylation of rat liver acetyl-CoA carboxylase by the AMP-acti-

vated protein kinase, demonstrated using freeze-clamping. Effects of high fat diets. *Eur. J. Biochem.* **203,** 615–623.

23. Gillespie, J. G., and Hardie, D. G. (1992) Phosphorylation and inactivation of HMG-CoA reductase at the AMP-activated protein kinase site in response to fructose treatment of isolated rat hepatocytes. *FEBS Lett.* **306,** 59–62.

24. Corton, J. M., Gillespie, J. G., Hawley, S. A., and Hardie, D. G. (1995) 5-Aminoimidazole-4-carboxamide ribonucleoside: a specific method for activating AMP-activated protein kinase in intact cells? *Eur. J. Biochem.* **229,** 558–565.

25. Davies, S. P., Carling, D., and Hardie, D. G. (1989) Tissue distribution of the AMP-activated protein kinase, and lack of activation by cyclic AMP-dependent protein kinase, studied using a specific and sensitive peptide assay. *Eur. J. Biochem.* **186,** 123–128.

26. Dale, S., Wilson, W. A., Edelman, A. M., and Hardie, D. G. (1995) Similar substrate recognition motifs for mammalian AMP-activated protein kinase, higher plant HMG-CoA reductase kinase-A, yeast SNF1, and mammalian calmodulin-dependent protein kinase I. *FEBS Lett.* **361,** 191–195.

27. Henin, N., Vincent, M. F., Gruber, H. E., and Van den Berghe, G. (1995) Inhibition of fatty acid and cholesterol synthesis by stimulation of AMP-activated protein kinase. *FASEB J.* **9,** 541–546.

28. Velasco, G., Geelen, M. J. H., and Guzman, M. (1997) Control of hepatic fatty acid oxidation by 5'-AMP-activated protein kinase involves a malonyl-CoA-dependent and a malonyl-CoA-independent mechanism. *Arch. Biochem. Biophys.* **337,** 169–175.

29. Merrill, G. M., Kurth, E., Hardie, D. G., and Winder, W. W. (1997) AICAR decreases malonyl-CoA and increases fatty acid oxidation in skeletal muscle of the rat. *Am. J. Physiol.* **36,** E1107–E1112.

30. Javaux, F., Vincent, M. F., Wagner, D. R., and van den Berghe, G. (1995) Cell-type specificity of inhibition of glycolysis by 5-amino-4-imidazolecarboxamide riboside. Lack of effect in rabbit cardiomyocytes and human erythrocytes, and inhibition in FTO-2B rat hepatoma cells. *Biochem J.* **305,** 913–919.

7

Detection and Activation
of Stress-Responsive Tyrosine Kinases

Gary L. Schieven

1. Introduction

Recent studies have determined that a variety of protein tyrosine kinases can be activated by the exposure of cells to oxidative stress *(1–3)*. The stress may arise from chemical agents such as hydrogen peroxide, as well as from irradiation with ultraviolet or ionizing radiation. Oxidative stress can activate tyrosine phosphorylation signaling pathways normally regulated by cell-surface receptors. Because tyrosine kinases are frequently the proximal signaling enzyme to cell-surface receptors, their activation by stress can lead to activation of signal cascades, including the activation of serine/threonine kinases *(4)*. However, because this activation occurs in the absence of a natural ligand, oxidative stress is capable of activating the receptor signal pathways outside of normal receptor control. This process has been extensively characterized in lymphocytes, where oxidative stress from hydrogen peroxide or ultraviolet (UV) radiation has been found to activate tyrosine kinases associated with the antigen receptor in T- and B-cells, giving rise to signaling patterns similar to those induced by direct antigen-receptor stimulation *(5–8)*.

Ligands can act on receptors to activate two types of tyrosine kinases. Receptor tyrosine kinases are transmembrane proteins that contain an intracellular tyrosine kinase catalytic domain *(9)*. These receptor tyrosine kinases are activated by ligand-induced dimerization *(9)*, with each member of the dimer phosphorylating its partner in trans to achieve kinase activation. The resultant tyrosine phosphorylation acts as a scaffold to recruit additional signaling molecules, which can bind via SH2 (Src homology 2) domains that recognize phosphotyrosine in the context of specific amino acid sequences *(10)*.

From: *Methods in Molecular Biology, vol. 99: Stress Response: Methods and Protocols*
Edited by: S. M. Keyse © Humana Press Inc., Totowa, NJ

The second type of tyrosine kinases that can be activated are the nonreceptor tyrosine kinases. The Src-family kinases are an important example of this type of kinase. Src kinases act as proximal signaling kinases for a variety of receptors in hematopoietic cells, such as the T-cell receptor, the B-cell antigen receptor, and Fcε and Fcγ receptors that bind antibodies *(11,12)*. In these cells, the Src-family kinases such as Lck, Lyn, and Fyn are responsible for activating a second round of tyrosine kinases, such as the Syk-family kinases Syk and ZAP-70, and the Tec-family kinases such as Btk and Itk *(13,14)*. However, Src-family kinases are also activated downstream of growth factor receptor tyrosine kinases such as the EGF receptor and PDGF receptor *(15)*. The activity of the Src family kinases is essential for the mitogenic activity of these receptors *(16)* and is also important for the ability of these receptors to induce changes in gene transcription in response to stress such as UV irradiation *(17,18)*.

In this chapter, two methods of inducing tyrosine kinase activity will be described. The methods are hydrogen peroxide treatment of cells and UV irradiation of cells. In addition, two methods of detecting tyrosine kinase activation will be described. The first method is the detection of increased tyrosine phosphorylation of an immunoprecipitated kinase. Many tyrosine kinases, such as receptor tyrosine kinases and the Syk- and Tec-family kinases, are activated by tyrosine phosphorylation, and thus an increase in tyrosine phosphorylation reflects the activation of these kinases. The second method is the detection of increased enzymatic activity of the kinase. Although this approach can be used for many tyrosine kinases, the Src-family kinases will be used as a specific example. The Src-family kinases consist of Src, Blk, Fgr, Fyn, Hck, Lyk, Lyn, Yes, and Yrk, the last kinase having been detected only in avian cells *(11)*. These kinases have two major sites of tyrosine phosphorylation. Tyrosine phosphorylation of the activation loop of the kinase is activating, whereas phosphorylation of a C-terminal tyrosine inhibits the enzyme by interacting with the kinase's SH2 domain in an intramolecular reaction. Thus, detection of tyrosine phosphorylation of Src-family kinases is not a measure of their activation, as the phosphorylation can either be activating or inhibitory.

2. Materials

2.1. Reagents

1. PBS (phosphate buffered saline).
2. 10X stock hydrogen peroxide: 100 mM hydrogen peroxide in PBS (*see* **Note 1**).
3. Lysis buffer: 50 mM Tris-HCl, pH 8.0, 150 mM NaCl, 1% Nonidet P-40 (NP-40), 0.5% sodium deoxycholate, 1 mM sodium orthovanadate, 1 mM sodium molybdate, 8 μg/mL aprotinin, 5 μg/mL leupeptin, 500 μM phenylmethylsulfonyl fluoride (PMSF) (*see* **Note 2**).
4. Anti-phosphotyrosine antibody (monoclonal or affinity-purified polyclonal).

5. Antibodies to tyrosine kinases of interest.
6. Protein-A–Sepharose (e.g., Repligen).
7. Rabbit antimouse antibodies for precipitation of monoclonal antibodies.
8. 1X Sodium dodecyl sulfate (SDS) sample buffer: Dilute commercial 2X buffer (Novex or Bio-Rad) 1:1 or prepare by combining 4 mL distilled water, 1 mL 0.5 M Tris-HCl, pH 6.8, 1.6 mL 10% SDS, 0.2 mL 0.05% Bromophenol blue. Add β-mercaptoethanol to a final concentration of 5% just before use.
9. Blocking buffer: 5% bovine serum albuminBSA, 1% ovalbumin, 0.05% Tween-20 in PBS (*see* **Note 3**).
10. Washing buffer: 0.05% Tween-20 in PBS.
11. Goat antimouse antibody HRP (horseradish peroxidase) conjugate (e.g., Boehringer Mannheim).
12. ECL (enhanced chemiluminescence) reagents (e.g., Amersham).
13. Rabbit muscle enolase in ammonium sulfate suspension (e.g., Sigma E-0379 [St. Louis, MO] or equivalent).
14. Enolase buffer: 50 mM HEPES, pH 7.2, 1 mM MgCl$_2$, 1 mM dithtithreitol (DTT).
15. Glycerol.
16. 50 mM acetic acid.
17. 0.5 M 3-[N-morpholino]propane sulfonic acid (MOPS), pH 7.0.
18. ATP.
19. [γ-^{32}P] ATP (must be stored and used with shielding to protect from radiation hazard).
20. Protein molecular weight markers for sodium dodecyl sulfate-polyacrylamide gel electrophoresis (SDS-PAGE).
21. Gel-fixing solution: 50% methanol, 10% acetic acid.
22. Kinase gel-wash solution: 10% trichloroacetic acid, 10 mM sodium pyrophosphate.
23. Coomassie Blue staining solution: 0.05% Coomassie Brilliant Blue R-250, 50% methanol, 10% acetic acid, 40% water.
24. Gel-destain solution: 5% methanol, 7% acetic acid, 88% water.
25. Optional high sensitivity destain solution: 10% acetic acid.

2.2. Equipment

1. Tissue culture facility with appropriate incubators, centrifuges, and so forth.
2. Stratolinker UV crosslinker.
3. Cell-scraping tools.
4. 60-mm tissue culture dishes.
5. Boiling water bath or heating block capable of 100°C temperature.
6. SDS-PAGE equipment.
7. Immunoblotting tank transfer electrophoretic unit.
8. Polyvinylidene difluoride (PVDF) membrane made for immunoblotting (e.g., Immobilon from Millipore).
9. X-ray or enhanced chemiluminescence (ECL) film and a film processor.
10. 30°C water bath.
11. Plexiglas shielding suitable for work with [^{32}P].

3. Methods

3.1. Treatment of Cells with Hydrogen Peroxide

1. Grow sufficient cells to use approx 1×10^7 cells/sample. Non-adherent cells such as lymphocytes should be suspended in a volume of 900 µL of media in a microcentrifuge tube. Adherent cells should be grown in tissue culture dishes such as 60-mm dishes.
2. Add 1 vol of 10X stock of hydrogen peroxide to 9 volumes of media containing the cells, for a final concentration of 10 mM hydrogen peroxide. A total volume of 1 mL for suspension cells and 2–3 mL for adherent cells is convenient.
3. Harvest the cells after a 5-min exposure to 10 mM hydrogen peroxide. Suspended cells in a volume of 1 mL are centrifuged for 5 s in a microcentrifuge. The supernatant is then aspirated. For adherent cells in dishes, aspirate the media.
4. Lyse the cells in 1 mL cold lysis buffer. For adherent cells on dishes, the dishes should be kept on ice and scraped with a cell scraper. The lysate is then transferred to a microcentrifuge tube. For cells in suspension, add 1 mL lysis buffer to the cell pellet following centrifugation and pipet vigorously to lyse the cells.
5. Keep the cell lysate on ice at least 10 min, then centrifuge at 10,000g in a microcentrifuge at 4°C. Transfer the supernatant solution to fresh microcentrifuge tubes for immunoprecipitation and keep on ice, or freeze at −70°C for storage.

3.2. Ultraviolet Irradiation of Cells

1. Obtain cells as described in **Subheading 3.1.** Use adherent or suspended cells in a 60-mm tissue culture dish with media to a depth of approx 1–2 mm. Do not exceed this depth or the UV light will be excessively absorbed by the media.
2. Place the uncovered dish in a Stratolinker UV irradiator. Set the power to the desired level, or use the auto crosslink function (*see* **Note 4**).
3. Irradiate the cells using the Stratolinker.
4. For adherent cells, remove the media, and lyse the cells with 1 mL lysis buffer as described in **Subheading 3.1.** For nonadherent cells, transfer the cells to a microcentrifuge tube, centrifuge, remove the supernatant, and lyse the cells as described in **Subheading 3.1.**

3.3. Immunoprecipitation of Tyrosine Kinases

1. All steps should be performed on ice or in a cold room, except for brief centrifugation steps. Add antibody against the kinase of interest (*see* **Note 5**) to the lysate from approx 1×10^7 cells. The amount of antibody to use is dependent on the antibody concentration if a monoclonal antibody, or the titer of the antiserum if polyclonal. In general, 1–10 µL of antiserum is required. Follow the manufacturer's directions for commercial antibodies. Mix the samples gently and allow to incubate on ice or at 4°C for 1–16 h. Longer incubation times may increase the efficiency of immunoprecipitation.

2. Wash Protein-A–Sepharose with PBS. Approximately 50 μL of a 1:1 suspension of Protein A-Sepharose is required per sample. Add sufficient Protein-A–Sepharose to a microcentrifuge tube, centrifuge 30 s at medium speed, and aspirate the supernatant. Add an equal volume of PBS, and repeat, resuspending the washed Protein-A–Sepharose in an equal volume of PBS.

3. If a monoclonal antibody was used to immunoprecipitate the kinase, a secondary antibody step is needed, as mouse antibody may not bind to Protein-A efficiently. Mix 50 μL of rabbit antimouse IgG per milliliter of Protein-A-Sepharose for 30 min on an end-over-end mixer such as a Labquake rotator. Centrifuge 30 s at medium speed in a microcentrifuge, remove the supernatant, and wash with 1 mL lysis buffer. Centrifuge, aspirate the media, and wash again. Suspend the Protein-A–Sepharose in an equal volume of lysis buffer to give a 1: 1 suspension. If rabbit antiserum was used to immunoprecipitate the tyrosine kinase, skip this step and proceed to **step 4**, as a second antibody is not needed.

4. Add approximately 50 μL of the 1:1 suspension of the Protein-A– or Protein-G–Sepharose in PBS to the sample. Mix 1 h on an end-over-end mixer.

5. Wash the Protein-A–Sepharose beads 3–4 times with lysis buffer. For each wash, centrifuge the beads for 30 s at medium speed in a microcentrifuge, aspirate the buffer, add 1 mL lysis buffer, and mix briefly.

3.4. Antiphosphotyrosine Immunoblotting of Immunoprecipitated Tyrosine Kinases

1. Wash beads containing the immunoprecipitate from **step 5** of **Subheading 3.3.** with PBS. Carefully remove all the supernatant and add 50 μL 1x SDS sample buffer. Heat in a boiling water bath or a 100°C heating block for 4–5 min.

2. Allow sample to cool to room temperature, load sample on a 10% SDS-PAGE gel (*see* **Note 6**) and electrophorese.

3. Transfer proteins from the gel to a PVDF membrane using an immunoblotting electrophoretic tank transfer apparatus, following the procedures recommended by the manufacturer.

4. Block the blot by incubating 2–16 h in blocking buffer. All procedures from this point on are performed at room temperature.

5. Probe the filters with a monoclonal antiphosphotyrosine antibody such as 4G10 or PY20 at a concentration of approx 1 μg/mL (or as recommended by the manufacturer) diluted into the blocking buffer. Incubate on a rocker for 2 h.

6. Wash the filters six times for 10 min each in washing buffer on a rocker.

7. Incubate the filters with goat anti-mouse HRP conjugate (for Boehringer Mannheim, use a dilution of 1:30000 in the blocking buffer or as directed by the manufacturer) on a rocker for 1–2 h.

8. Wash the filters 5x for 10 min in washing buffer and then 1x for 10 min in PBS on a rocker.

9. Use ECL reagents to detect antibody binding. Incubate the filters for 30 s in ECL reagent as directed by the manufacturer, then place the filters in a plastic-sheet protector or wrap in Saran Wrap and detect light emission using X-ray or ECL film.

3.5. Immune Complex Tyrosine Kinase Assays

1. For assays of Src-family kinases, prepare the enolase substrate. Transfer 1 mg of rabbit muscle enolase in ammonium sulfate suspension to a microcentrifuge tube and centrifuge for 1 min at high speed at 4°C. Remove the supernatant and resuspend in 100 μL of enolase buffer. Add 100 μL of glycerol and mix gently. This solution can be stored indefinitely at –20°C. Just before beginning the kinase reaction, mix 1 volume of enolase solution with 1 volume of 50 mM acetic acid. Incubate 10 min in a 30°C waterbath. The solution will become turbid (*see* **Note 7**). Add 1 volume of 0.5 M MOPS, pH 7.0. Keep on ice until use.

2. Wash the beads from **step 5** of **Subheading 3.3.** with kinase buffer (20 mM MOPS, pH 7.0, 10 mM MgCl$_2$) one time and resuspend in 25 μL of kinase buffer. The samples should be in screw-top 1.5-mL microcentrifuge tubes such as Sarstedt brand tubes (*see* **Note 8**).

3. Place samples and radioactive substrates behind an angled Plexiglas shield. Set up an appropriate radioactive waste container.

4. Start the enzyme reaction by adding 5 μL of reaction mix containing 3 μg denatured enolase (1.25 μL of the enolase prepared in **step 1**), 25 μCi [γ-^{32}P] ATP, and 5 μM unlabeled ATP). Incubate for 10 min at room temperature.

5. Stop the reaction by adding 30 μL of 2X SDS sample buffer and heat 5 min at 100°C in a boiling water bath or a heat block (*see* **Note 9**).

6. Allow sample to cool to room temperature, load samples with an appropriate set of protein molecular-weight markers on a 10% SDS-PAGE gel and electrophorese.

7. Remove the gel from the electrophoresis unit. Cut the gel slightly above the dye front and discard the lower portion in an appropriate radioactive-waste container, as this portion of the gel will contain [^{32}P]-ATP from the kinase reaction.

8. Fix the gel in gel-fixing solution by washing for 30 min. Wash twice, 30 min each, in 10% trichloroacetic acid, 10 mM sodium pyrophosphate (*see* **Note 10**). These wash solutions will contain [^{32}P] and should be disposed of as radioactive waste. Stain the gel in Coomassie Blue staining solution for 1–2 h. Destain the gel by repeated 30 min washings in destaining solution until the blue protein bands are visible and the background is clear (*see* **Note 11**).

9. Dry the gel and detect the radiolabeled protein by autoradiography or phosphor image analysis.

4. Notes

1. The hydrogen peroxide stock should be prepared fresh each day. Hydrogen peroxide is usually supplied as a 30% solution. This factor and the density of 1.463 g/mL for 100% hydrogen peroxide should be taken into account during calculations for the 10X stock solution.

2. It is important to include phosphotyrosine phosphatase inhibitors such as vanadate and molybdate to prevent cellular phosphatases from degrading the sample.

3. Nonfat dry milk, which is often used in blocking immunoblots, cannot be used for analysis of phosphotyrosine because milk contains substantial amounts of phosphotyrosine.

4. The units used on the Stratolinker are in $mJ/cm^2 \times 100$. These are equivalent to the units of J/m^2 used in most published studies of biological effects of UV irradiation. The auto crosslink option gives a dose of $1200\ J/m^2$, which strongly activates signaling in lymphocytes.

5. The choice of antibody will depend on the identity of the cells being studied. In general, it is best to choose the antibody based on published studies of what particular species of tyrosine kinase is known to be important in the function of the cell type of interest. For example, in B-cells, Syk and Btk are known to be important. To detect potential unknown tyrosine kinases, immune-complex kinase assays of proteins precipitated by antiphosphotyrosine antibodies can be used because almost all tyrosine kinases are tyrosine phosphorylated. This approach was used to detect tyrosine kinase responses activated by ionizing radiation *(19)*. Antiphosphotyrosine immunoblotting of the proteins precipitated by antiphosphotyrosine antibodies is used to detect the proteins phosphorylated in the cell by the activated tyrosine kinases.

6. The percentage of acrylamide SDS-PAGE gel to use is dependent on the size of the kinase of interest. A 10% gel is suitable for most kinases.

7. Acid-denatured enolase is an excellent substrate for Src-family kinases and is readily resolved from these kinases by SDS-PAGE. The enolase must be partially denatured to serve as a good substrate; hence, the solution will appear turbid. The 30°C temperature is critical, as higher temperatures give excessive denaturation. If the enolase precipitates out of solution, it has become excessively denatured and should not be used in the assay.

8. Appropriate safety precautions should be used when working with [^{32}P]. The [^{32}P] should be shielded and all waste should be disposed of properly. Snap-top tubes should not be used to heat [^{32}P] samples because the tops can pop open during heating (**step 5** of the immune complex kinase assay). The [^{32}P] samples must remain securely closed during heating because radioactivity can be carried into the air by escaping water vapor containing dissolved [^{32}P].

9. As an alternative procedure to reduce the amount of [^{32}P] loaded on the gels, stop the reaction by adding 1 mL PBS containing 100 mM EDTA. Wash beads twice with the PBS/EDTA solution, then heat the beads in 50 μL 1X SDS sample buffer. The EDTA stops the kinase reaction by chelating the divalent metal ion (Mg^{2+}) required by kinases for the phosphorylation reaction.

10. Washing with TCA and pyrophosphate removes the [^{32}P]–ATP from the gel.

11. For higher sensitivity, once the gel is partially destained, finish destaining using 10% acetic acid without methanol. The molecular weight markers, the enolase bands and the heavy chain of the immunoprecipitating antibody should be visible.

References

1. Schieven, G. L. and Ledbetter, J. A. (1994) Activation of tyrosine kinase signal pathways by radiation and oxidative stress. *Trends Endocrinol. Metab.* **5,** 383-388.
2. Schieven, G. L.(1998) Activation of lymphocyte signal pathways by oxidative stress: role of tyrosine kinases and phosphatases, in *Oxidative Stress in Cancer,*

AIDS, and Neurodegenerative Diseases (Montagnier, L., Olivier, R., and Pasquier, C., eds.), Marcel Dekker, New York, pp. 35–44.

3. Schieven, G. L. (1997) Tyrosine phosphorylation in oxidative stress, in *Oxidative Stress and Signal Transduction* (Forman, H. J. and Cadenas, E., eds.), Chapman & Hall, New York, pp. 181–199.

4. Uckun, F. M., Schieven, G. L., Tuel-Ahlgren, L. M., Dibirdik, I., Myers, D. E., Ledbetter, J. A., and Song, C. W. (1993) Tyrosine phosphorylation is a mandatory proximal step in radiation-induced activation of the protein kinase C signaling pathway in human B-lymphocyte precursors. *Proc. Natl. Acad. Sci. USA* **90,** 252–256.

5. Schieven, G. L., Kirihara, J. M., Myers, D. E., Ledbetter, J. A., and Uckun, F. M. (1993) Reactive oxygen intermediates activate NF-$_\kappa$B in a tyrosine kinase dependent mechanism and in combination with vanadate activate the p56^{lck} and p59^{fyn} tyrosine kinases in human lymphocytes. *Blood* **82,** 1212–1220.

6. Schieven, G. L., Kirihara, J. M., Burg, D. L., Geahlen, R. L., and Ledbetter, J. A. (1993) p72^{syk} tyrosine kinase is activated by oxidizing conditions which induce lymphocyte tyrosine phosphorylation and Ca^{2+} signals. *J. Biol. Chem.* **268,** 16,688–16,692.

7. Schieven, G. L., Kirihara, J. M., Gilliland, L. K., Uckun, F. M., and Ledbetter, J. A. (1993) Ultraviolet radiation rapidly induces tyrosine phosphorylation and calcium signaling in lymphocytes. *Molec. Biol. Cell* **4,** 523–530.

8. Schieven, G. L., Mittler, R. S., Nadler, S. G., Kirihara, J. M., Bolen, J. B., Kanner, S. B., and Ledbetter, J. A. (1994) ZAP-70 tyrosine kinases, CD45 and T cell receptor involvement in UV and H$_2$O$_2$ induced T cell signal transduction. *J. Biol. Chem.* **269,** 20,718–20,726.

9. Ullrich, A. and Schlessinger, J. (1990) Signal transduction by receptors with tyrosine kinase activity. *Cell* **61,** 203–212.

10. Pawson, T. (1995) Protein modules and signaling networks. *Nature* **373,** 573–579.

11. Chow, L. M. L. and Veillette, A. (1995) The Src and Csk families of tyrosine protein kinases in hemopoietic cells. *Sem. Immunol.* **7,** 207–226.

12. Bolen, J. B., Rowley, R. B., Spana, C., and Tsygankov, A. Y. (1992) The src family of protein kinases in hemopoietic signal transduction. *FASEB J.* **6,** 3403–3409.

13. Weiss, A. and Littman, D. R. (1994) Signal transduction by lymphocyte antigen receptors. *Cell* **76,** 263–274.

14. Li, Z., Wahl, M. I., Eguinoa, A., Stephens, L. R., Hawkins, P. T., and Witte, O. N. (1997) Phosphatidylinositol 3-kinase-gamma activates Bruton's tyrosine kinase in concert with Src family kinases. *Proc. Natl. Acad. Sci. USA* **94,** 13,820–13,825.

15. Kypta, R. M., Goldberg, Y., Ulug, E. T., and Courtneidge, S. A. (1990) Association between the PDGF receptor and members of the src family of tyrosine kinases. *Cell* **62,** 481–492.

16. Twamley-Sein, G. M., Pepperkok, R., Ansorge, W., and Courtneidge, S. A. (1993) The Src family tyrosine kinases are required for platelet-derived growth factor-mediated signal transduction in NIH 3T3 cells. *Proc. Natl. Acad. Sci. USA* **90,** 7696–7700.

17. Devary, Y., Gottlieb, R. A., Smeal, T., and Karin, M. (1992) The mammalian ultraviolet response is triggered by activation of src tyrosine kinases. *Cell* **71,** 1081-1091.

18. Rosette, C. and Karin, M. (1996) Ultraviolet light and osmotic stress: activation of the JNK cascade through multiple growth factor and cytokine receptors. *Science* **274,** 1194–1197.

19. Uckun, F. M., Tuel-Ahlgren, L., Song, C. W., Waddick, K., Myers, D. E., Kirihara, J., Ledbetter, J. A., and Schieven, G. L. (1992) Ionizing radiation stimulates unidentified tyrosine-specific protein kinases in human B-lymphocyte precursors triggering apoptosis and clonogenic cell death. *Proc. Natl. Acad. Sci. USA* **89,** 9005–9009.

8

Detection of DNA-Dependent Protein Kinase in Extracts from Human and Rodent Cells

Yamini Achari and Susan P. Lees-Miller

1. Introduction

The DNA-dependent protein kinase, DNA-PK, is required for DNA double-strand break repair and V(D)J recombination (1–3). DNA-PK is composed of a large catalytic subunit of approx 460 kDa (DNA-PKcs) and a heterodimeric DNA targeting subunit, Ku. Ku is composed of 70-kDa and approx 80-kDa subunits (called Ku70 and Ku80, respectively) and targets DNA-PKcs to ends of double-stranded DNA. DNA-PKcs is related by amino acid sequence to the phosphatidyl inositol kinase (PI-3 kinase) protein family (4), however, in vitro DNA-PK acts as a serine/threonine protein kinase (5–7). The physiological substrates of DNA-PK are not known, however in vitro DNA-PK is known to phosphorylate many proteins including hsp90, SV40 large T antigen, serum response factor, and p53 (reviewed in **ref.** 1). In these substrates, DNA-PK phosphorylates serine (or threonine) that is followed by glutamine (an "SQ motif") (8), although some DNA-PK substrates are phosphorylated at "non-SQ" sites (9). A synthetic peptide, derived from amino acids 11–24 of human p53, that contains an "SQ motif" is a specific substrate of DNA-PK, whereas a similar peptide in which the glutamine is displaced from the serine by one amino acid is not phosphorylated (8). These peptides have been used to assay for DNA-PK activity in extracts from human tissue culture cells (8–10) and normal human cells (11). DNA-PK activity is readily detected in extracts from human and monkey cell lines but not in protein extracts from nonprimate cells that, for reasons that are not well understood, contain 50- to 100-fold less DNA-PK activity (8,10,12). DNA-PK activity in nonprimate cells is detected using the same peptide substrates, but using a more sensitive "DNA-cellulose pull-down" assay in which cellular proteins are prebound to double-stranded (ds)

From: *Methods in Molecular Biology, vol. 99: Stress Response: Methods and Protocols*
Edited by: S. M. Keyse © Humana Press Inc., Totowa, NJ

DNA cellulose resin and assayed on the resin using the dsDNA cellulose itself as the activator *(12,13)*. Here, we describe these methods for the assay and detection of DNA-PK in extracts from human and rodent cells. It is important to note that both of these assays are useful for quantifying the amount of DNA-PK present that can be activated by exogenous DNA; however, at present, no assay is available to measure the amount of DNA-PK that is actually active in a cell at any particular time.

2. Materials

Unless otherwise specified, all reagents were purchased from Sigma or BDH and were of standard grade or higher. Rodent or human tissue culture cell lines should be maintained under appropriate conditions and methods will not be described here. The laboratory should be equipped for standard biochemical techniques and use of radioactive reagents. Work with [^{32}P] should be carried out with appropriate shielding. The radioactive area should be equipped with a water bath set at 30°C, a rocking platform and a microfuge. Leucite microtube boxes are useful for storing or transporting radioactive samples. Access to a liquid scintillation counter is also required. Access to a refrigerated microfuge (not in the radioactive area) is preferred for preparation of protein extracts; however, a microfuge situated in the cold room will suffice.

2.1. Harvesting Tissue Culture Cells for DNA-PK Assays

1. Dithiothreitol (DTT): Make 1 M stock and store in aliquots at –80°C.
2. Phosphate-buffered saline (PBS): Make up 10X Stock solution: 1.4 M NaCl, 25 mM KCl, 25 mM Na$_2$HPO$_4$, 15 mM KH$_2$PO$_4$, pH to 6.9. Dilute 10-fold (to 1X) with dd H$_2$O and adjust pH to 7.4 before use.
3. Low salt buffer (LSB): Make up 50X stock solution: 0.5 M HEPES, 1.25 M KCl, 0.5 M NaCl, 55 mM MgCl$_2$, 5 mM EDTA, pH 7.4. Autoclave and store in aliquots at 4°C. Dilute 50-fold (to 1X) with dd H$_2$O, adjust pH to 7.2 and add DTT to 0.1 mM before use.
4. Phenylmethylsulphonylflouride (PMSF): Make 0.2 M stock in 100% methanol and store at –20°C (*see* **Note 1**).
5. Protease inhibitors: Make 100X stock solution to contain 50 mM PMSF, 0.5 mg/mL aprotinin, 0.5 mg/mL leupeptin made up in 100% methanol. Store at –20°C and add directly to the sample, as PMSF is rapidly inactivated in aqueous buffers (*see* **Note 1**).

2.2. Making High Salt Extracts for DNA-PK Assays

1. DTT: see **Subheading 2.1., item 1**.
2. MgCl$_2$: Make 1 M stock, autoclave, and store at 4°C.
3. High-salt extraction buffer (HSEB): 5 M NaCl, 100 mM MgCl$_2$, 10 mM DTT. Store in 0.1-mL aliquots at –20°C.
4. NaCl: Make 5 M stock, filter, and store at room temperature.

5. 0.5 *M* salt wash buffer: Make from stock solutions: 1 X LSB containing 0.5 *M* NaCl, 10 m*M* MgCl$_2$, 1 m*M* DTT. Store in 1-mL aliquots –20°C.
6. PMSF: see **Subheading 2.1., item 4.**

2.3. Solution Assay for DNA-PK Using High-Salt Extracts from Human Cells

1. DTT: see **Subheading 2.1., item 1.**
2. EDTA, pH 8.0: Make 0.5 *M* stock, pH 8.0 with NaOH, autoclave, and store at room temperature.
3. EGTA, pH 7.2: Make 0.2 *M* stock, pH 7.2 with NaOH, store in aliquots –20°C.
4. HEPES, pH 7.5: Make 1 *M* stock, pH to 7.5 with KOH, autoclave, and store at 4°C.
5. KCl: Make 2 *M* stock, autoclave, and store at room temperature.
6. MgCl$_2$: see **Subheading 2.2., item 2.**
7. NaCl: see **Subheading 2.2., item 4.**
8. Tris-HCl, pH 8.0: Make 1 *M* stock, pH to 8.0 with HCl, autoclave, and store at room temperature.
9. TE: Make from stock solutions: 10 m*M* Tris-HCl, 1 m*M* EDTA, pH 8.0.
10. HEPES, KCl buffer: Make from stock solutions: 50 m*M* HEPES pH 7.5, 50 m*M* KCl. Adjust final to pH 7.5.
11. Synthetic peptides: Synthetic peptides should be synthesized with a free amino terminus and an amide at the carboxyl terminus and should be purified by high-performance liquid chromatography (HPLC) before purchase. We routinely use the following pairs of peptides:

    Peptide A: EPPLSQEAFADLWKK; Peptide B: EPPLSEQAFADLWKK, or
    Peptide C: PESQEAFADLWKK; Peptide D: PESEQAFADLWKK

    Peptide A is derived from the amino terminus of human p53 and is a good substrate for DNA-PK in vitro. Peptide B is not phosphorylated by DNA-PK and serves as a negative control *(8)*. Peptide C is derived from peptide A and is about twice as effective a substrate as peptide A *(9)*. Peptide D is the negative control for peptide C *(9)*. Upon receipt, all peptides should be white a fluffy powder with no odor. If there is any trace of organic odor, resuspend the peptides in sterile distilled water and lyophilize to dryness to drive off any traces of organics.
    To make stock solutions of the peptides: Resuspend the peptides in sterile distilled water to approximately 5 mM by weight. Vortex and allow the peptides to redissolve for about 30 min then spin at 10,000*g* for 10 min at 4°C to remove any insoluble material. Remove a small aliquot and send for amino acid analysis to determine the exact concentration of the peptide. The amount required will depend on the method of amino acid analysis used at your facility. Dilute the peptides to 2.5 m*M* in sterile distilled water, aliquot, and store at –20°C.
12. Acetic acid washing solution: Make 2 L of 15% acetic acid. Use to wash phosphocellulose squares.
13. Phosphocellulose P81 paper (Whatman, cat. no. 3698-875): Cut into 2 x 2-cm squares.
14. Acetic acid stop solution: Make 30% acetic acid, 5 m*M* ATP; store in 1-mL aliquots at –20°C.

15. Unlabeled ATP: Weigh approx 0.1 g ATP (Boehringer Mannheim) and dissolve in 2 mL of sterile distilled water. Remove a small aliquot, dilute in pH 7.0 buffer and measure the absorbance at 259 nm. Determine the concentration of ATP using the molar extinction coefficient $\varepsilon = 15.4 \times 10^3$. Dilute ATP to 25 m$M$ and store in small aliquots at –20°C.

16. Radioactively labeled ATP: [γ-^{32}P]ATP can be purchased from any standard supplier (NEN Life Science Products, Amersham Life Science, etc.). The specific activity should be 3000 Ci/mmol. In our experience, "stabilized" [γ-^{32}P]ATP gives more reproducible results when comparing DNA-PK assays from day to day. We therefore routinely use Redivue (Amersham Life Science) and Easytide (NEN Life Science Products) [γ-^{32}P]ATP. The radioactive ATP is essentially carrier free and must be diluted to the correct specific activity before use in assays. Make a working stock of 2.5 mM unlabeled ATP containing 0.5 µCi [γ-^{32}P]ATP/ µL. Use 2 µL of this working stock per 20 µL assay to give a final concentration of 0.25 mM ATP containing approx 1 µCi [γ-^{32}P]ATP in the assay. The K_m for ATP is approx 25 µM *(6)*.

17. Calf thymus DNA: Double-stranded DNA oligonucleotides of greater than 20 base pairs (bp) will activate DNA-PK; however, we routinely use sonicated calf thymus DNA as the activator. Cut calf thymus DNA (Aldrich Chemical Co.) into small pieces and place in a tared 15 mL conical tube and weigh. Add TE buffer pH 8.0 to give approximately 5 mg/mL and allow DNA to rehydrate overnight at 4°C. Next day, pack the tube in wet ice and sonicate until the DNA is no longer viscous. A sample of the sonicated DNA run on an agarose gel should show a smear of fragments of several kilobasepairs in length. Dilute the DNA to 1 mg/mL in TE using absorbance of 1 at 260 nm as equivalent to 50 µg DNA. Aliquot the DNA and store at –20°C (*see* **Note 2**).

18. Magnesium, EDTA, DTT buffer (MED): Make 10X stock from stock solutions: 100 mM MgCl$_2$, 2 mM EDTA, 10 mM DTT. Store in 0.5-mL aliquots at –20°C.

2.4. DNA-Cellulose "Pull-Down" Assay for DNA-PK Using High-Salt Extracts from Rodent Cells

Note: Solutions 1–15 are exactly as described in **Subheading 2.3.**

16. Radioactively labeled ATP: *see* **Subheading 2.3.**, **items 15** and **16** for making unlabeled 25 mM ATP stock and purchasing [γ-^{32}P]ATP. Make a working stock solution of 500 µM unlabeled ATP containing 5 µCi/µL [γ-^{32}P]ATP. Use 2 µL of the working stock per 20 µL reaction for a final concentration of 50 µM ATP containing 10 µCi [γ-^{32}P]ATP per reaction.

17. Pull-down buffer: Make from stock solutions: 25 mM HEPES, 50 mM KCl, 10 mM MgCl$_2$, 1 mM DTT, 5% glycerol, 0.5 mM EDTA, 0.25 mM EGTA. Adjust final pH to 7.9. Store in aliquots at –20°C.

18. Pull-down buffer containing 500 mM KCl: Make from stock solutions: 25 mM HEPES, 500 mM KCl, 10 mM MgCl$_2$, 1 mM DTT, 5% glycerol, 0.5 mM EDTA, 0.25 mM EGTA. Adjust final pH to 7.9. Store in aliquots at –20°C.

19. HEPES buffer: Make up from stock solutions: 25 mM HEPES, 1 mM DTT, 5% glycerol, 0.5 mM EDTA, 0.25 mM EGTA. Adjust final pH to 7.9.
20. EDTA stop solution: Make from stock solutions: 50 mM EDTA pH 8.0, 10 mM ATP (unlabeled). Store in 1-mL aliquots –20°C.
21. Double-stranded DNA cellulose (Sigma, cat. no. D8515): Place approx 0.1 g dry double-stranded DNA cellulose powder in a 1.5-mL tube. Add 1 mL of pull-down buffer. Mix tube by inversion a few times and allow to sit on ice for 10 min. Spin 30 s 10,000g at 4°C, remove and discard the supernatant. To the DNA cellulose pellet add 1 mL of pull-down buffer containing 500 mM KCl, mix and spin as above. Resuspend the DNA cellulose in 1 mL pull-down buffer, mix, and spin. Repeat wash with another 1 mL pull-down buffer. Resuspend the DNA cellulose in a minimal amount of pull-down buffer (about 200 µL). You should be able to pipette the slurry using a pipet tip with the end cut off (*see* **Note 3**).

2.5. Western Blot of DNA-PK

1. Sodium dodecyl sulfate-polyacrylamide gel electrophoresis (SDS-PAGE): Poly-acrylamide resolving gels should be prepared according to standard procedures to contain 8.5% acrylamide and 0.075% bisacrylamide. The stacking gel and running buffer should be made according to standard procedures.
2. Electroblot transfer buffer: Make 1 L of 50 mM Tris base, 10 mM glycine, 20% methanol, 0.035% SDS. Do not adjust pH. Store at 4°C. Electroblot can be reused up to 10 times.
3. Western blot reagents should be purchase and used according to the manufacturers instructions. X-ray film may also be required for some methods of Western blot detection.
4. Antibodies to DNA-PKcs and Ku are commercially available; for example from Oncogene Research Products, Santa Cruz Biotechnology, Inc., and Serotec Inc.

3. Methods

3.1. Harvesting Tissue Culture Cells for DNA-PK Assay

1. Grow cells under normal specified conditions. We typically start with $2-4 \times 10$ cm^3 plates of cells at approx 70% confluence. For suspension cells, start with approx 1×10^7 cells.
2. Scrape or trypsinize cells, transfer to a 50-mL conical plastic tube, and pellet by centrifugation at 1500g at 4°C for 5 min.
3. Resuspend the cell pellet in 50 mL of ice-cold 1 X PBS using gentle aspiration. If tryspinization was used to harvest the cells, PMSF (0.2 mM) should be added to all washes to inhibit residual trypsin activity.
4. Pellet the cells as in **step 2**.
5. Resuspend the cells in 10 mL of ice-cold PBS (containing 0.2 mM PMSF if required) and transfer to a 15-mL conical plastic tube.
6. Pellet the cells as described in **step 2**.
7. Remove the PBS wash and add 12 mL of ice-cold 1X LSB containing 0.1 mM DTT (plus 0.2 mM PMSF if required). Resuspend the cell pellet by gentle aspiration (*see* **Note 4**).

8. Pellet the cells by centrifugation at 1500*g* at 4°C for 5 min.
9. Remove the LSB wash, resuspend the cells in 1 mL of 1X LSB containing 0.1 m*M* DTT and transfer to a 1.5-mL tube. Centrifuge in a microfuge at 1500*g* for 5 min at 4°C to gently repellet the cells. Remove the LSB and resuspend the cell pellet in 100 μL of 1X LSB containing 0.1 m*M* DTT. If the packed cell volume of cells is small, resuspend in 50 μL.
10. Add protease inhibitor mix to the resuspended cell pellet to give concentrations of 0.5 m*M* PMSF and 5 μg/mL each aprotinin and leupeptin. Add the protease inhibitors directly to the sample (*see* **Note 1**).
11. Incubate the sample on ice for 5 min (to allow cells to expand fully), then immerse the tube into liquid nitrogen or a dry-ice/ethanol bath and allow sample to freeze for at least 5 min (*see* **Note 5**).

3.2. Making High-Salt Extracts for DNA-PK Assay

Cells should be harvested as described in **Subheading 3.1.** DNA-PK should not be assayed in cytoplasmic or nuclear extracts only, as lysis of cells results in two populations of DNA-PK activity—one that is readily released into the cytoplasmic fraction and one that is retained in the nuclear fraction and that can only be released by buffers containing high concentrations of salt (0.4 *M* to 0.5 *M*) in the presence of magnesium *(8,10)*. In the procedure described here, the cells are lysed in hypotonic buffer by a freeze/thaw cycle and chromatin-bound proteins are released by the addition of salt to 0.5 *M* to create a whole-cell extract that contains both populations of DNA-PK *(14)*.

1. Quick-thaw the cell lysate by placing in a 37°C water bath. Do not allow the lysate to warm above freezing. Remove the tube as soon as the contents begins to thaw and place it on ice until it is completely thawed. Additional protease inhibitors can be added at this stage if desired.
2. Mix the contents of the tube by gently pipetting up and down twice and transfer 100 μL of the lysed cell extract to an ice-cold 1.5-mL tube. Any remaining sample can be stored at –80°C for later use.
3. To the 100 μL of lysed cell extract add 11 μL of HSEB (**Subheading 2.2., item 3**) and tap the tube several times to mix the contents. (If the volume of your extract is 50 μL, add 5.5 μL of the HSEB) (*see* **Note 6**).
4. Place extract on ice for 5 min, then centrifuge at 10,000*g* for 3 min at 4°C.
5. **Carefully remove the supernatant with a pipet being very careful not to disturb the pelleted chromatin.** If the pellet is disturbed, respin the samples.
6. Place the supernatant in a fresh ice-cold 1.5-mL tube.
7. To the pellet, add 50 μL of 0.5 *M* salt wash buffer (**Subheading 2.2., item 5**) and tap the tube several times to mix.
8. Centrifuge tube at 10,000*g* for 3 min, as above.
9. Carefully remove the supernatant and pool it with the first. This forms the high-salt total protein extract (*see* **Note 5**).

3.3. Solution Assay for DNA-PK in High-Salt Extracts from Human Cells

1. Quick-thaw an aliquot of the high-salt extract and place the tube on ice as soon as the contents begin to thaw.
2. Determine the protein concentration in the extract using any standard protein concentration assay method. We routinely use the Biorad protein assay using BSA as standard. The protein concentration of the high-salt extract should be between 1 and 4 mg/mL total protein. We have found that DNA-PK is relatively unstable to storage when the total protein concentration in these extracts is below about 0.5 mg/mL.
3. Set up three 1.5-mL plastic centrifuge tubes on ice. If highly accurate measurements are required, presiliconized 1.5-mL tubes can be used.
4. Pipet into each tube: 10 µL 50 mM HEPES, 50 mM KCl, pH 7.5 (**Subheading 2.3., item 10**); 2 µL 10 X MED (**Subheading 2.3., item 18**); 2 µL 0.1 mg/mL sonicated calf thymus DNA (**Subheading 2.3, item 17**).
5. To each tube add either 2 µL water, 2 µL peptide A (or peptide C) or 2 µL peptide B (or peptide D). See **Subheading 2.3.** for descriptions of synthetic peptides.
6. To each tube add 2 µL of high-salt extract prepared as described in **Subheadings 3.1.** and **3.2.** It is important that the final salt concentration in the assay be between 50 and 100 mM as DNA-PK is inhibited at salt concentrations in excess of 120 mM *(8)*. Up to 4 µL of the high-salt extract can be used in the assay as long as the final salt concentration in the assay is adjusted to 100 mM.
7. Start the each reaction by adding 2 µL of ATP containing [γ-^{32}P]ATP (final concentration in the assay of 0.25 mM ATP, 1 µCi [γ-^{32}P]ATP).
8. Tap the tubes to mix contents, cap, and place in a 30°C waterbath.
9. Start each sample at 15-s intervals.
10. After 5 min, remove the first tube and add 20 µL of acetic acid stop solution. Shake the tube to mix and place in a leucite block. Stop each tube at 15-s intervals (*see* **Note 7**).
11. Transfer tubes to a microfuge and spin at full speed for about 30 s.
12. With a 2H pencil, label two 2 × 2 cm squares of phosphocellulose paper for each sample and arrange the squares on a piece of clean Whatman 3M paper.
13. Spot 15 µL of each sample on each of two labeled squares. **Important:** Dispose of the Whatman blotter in the radioactive waste, as it will be contaminated with traces of unincorporated radioactive ATP.
14. Place the paper squares into a Pyrex crystallization dish containing 200–500 mL of 15% acetic acid. There is no need to let the samples dry before placing in the acetic acid solution. There should be enough acetic acid solution to completely cover the squares. Cover the dish with a piece of leucite to reduce your exposure to radioactivity and wash phosphocellulose squares on a standard lab orbital or shaking platform. A Reax3 platform mixer (Caframo) works well.
15. Allow the samples to wash for 5 min then carefully remove the liquid by aspiration. **Important:** (1) Most of the unincorporated [γ-^{32}P]-ATP will be in the first wash. (2) All washes must be disposed of according to the radiation safety rules

of your institution. (3) Be very careful not to touch the phosphocellulose squares with the aspirator, as they may be damaged.

16. Add another 200–500 mL of 15% acetic acid and repeat the wash for another 5 min. Aspirate the liquid as before.

17. Wash squares a total of three or four times, 5 min each. Do not leave the phosphocellulose squares washing for extended periods, as they tend to break down. Similarly, do not treat them roughly, as they can fall apart.

18. Remove the phosphocellulose paper squares and arrange on a *clean* Whatman paper blotter. There is no need to let them dry.

19. Using forceps, place each square in a scintillation vial and add enough water to cover.

20. Cap vials and count by Cerenkov radiation in a scintillation counter set on wide open window.

21. *Calculating enzyme activity:* One unit of activity is defined as the amount of enzyme required to incorporate 1 nmol of phosphate into the substrate peptide (over incorporation into the mock peptide) per minute per mg protein under the conditions of the assay.

    To calculate the specific activity of the stock radioactive ATP: Dilute the 2.5 mM working stock ATP solution 1/100 in water. Remove 2 µL and place directly into a scintillation vial containing water and count by Cerenkov emission. Calculate the specific activity of the stock ATP as cpm per pmole ATP.

    To calculate the enzyme activity: Determine the number of cpm incorporated into the peptide minus the cpm incorporated into the mock peptide. Divide this number by the specific activity of the ATP (calculated in the previous step). Multiply by the dilution factor, divide by the number of minutes of the assay (usually 5), and divide by the number of microliters of protein sample used in the assay. This will give you pmoles of phosphate incorporated into the substrate peptide per minute per microliter of extract (or U/mL of extract). Divide the value for U/mL by the protein concentration (in mg/mL) to get the specific activity of the enzyme in U/mg (nmoles phosphate transferred per minute under the conditions of the assay/mg protein).

Representative results of assays using extracts from HeLa, MO59K, and the radiosensitive DNA-PKcs minus cell line, MO59J *(15)* are shown in **Fig. 1**.

3.4. DNA Cellulose "Pull Down" Assay for DNA-PK in Nonprimate Cells

This assay is based on the published procedure of Finnie et al. *(12)* as modified by Danska et al. *(13)*. Rodent cells contain 50–100 times less DNA-PK protein and DNA-PK activity than human cells, and the solution-based DNA-PK assay described in **Subheading 3.3.** is not sensitive enough to detect DNA-PK in rodent or other nonprimate cells. The "pull-down" assay is based on the binding of DNA-PK to double-stranded (ds) DNA cellulose resin that serves to concentrate the DNA-PK that can then be assayed directly on the resin, using the dsDNA cellulose itself as the activator *(12,13)* (*see* **Note 8**). The "pull-

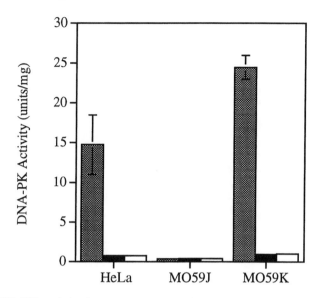

Fig. 1. DNA-PK activity in extracts from human cell lines: Hela S3, MO59J and MO59K cells were grown as described *(15)*. Cells were harvested, high-salt extracts were made, and DNA-PK was assayed as described in **Subheadings 2.1.–2.3.** and **3.1.–3.3.** Each 20-μL reaction contained between 0.1 and 0.2 μg of total protein. Hatched bars represent activity with peptide C, solid black bars represent activity with peptide D, open bars represent activity with no peptide. Assays were performed in triplicate. Similar results were obtained using peptides A and B. In the example shown, background values for mock peptides have not been subtracted.

down" procedure can also be used to concentrate DNA-PK (and the many other proteins that bind to double-stranded DNA cellulose in this assay) for Western blot or to phosphorylate other protein substrates (*see* **Note 9**).

1. Cells are harvested, and high-salt extracts are prepared exactly as described in **Subheadings 3.1.** and **3.2.** For rodent cell lines, start with at least 1×10^7 cells per sample or enough to give 0.5 mg total protein per sample in the high-salt extract (*see* **Note 5**).
2. Remove a volume of high-salt extract containing 0.5 mg of total protein and dilute to 50 mM salt concentration by addition of 9 vol of HEPES buffer (**Subheading 2.4.,** item 19).
3. Add 30 μL of double-stranded DNA cellulose slurry (prepared as described in **Subheading 2.4.**) to the diluted extract and incubate on a rotator at 4°C (e.g., a Labquake Shaker (Barnstead/Thermolyne) located in the cold room) for 15–60 min. Alternatively the samples can be left on ice with occasional mixing.
4. Spin for 1 min 10,000*g*.
5. Remove the supernatant and keep on ice (*see below*).

6. Wash the DNA cellulose pellet in 1 mL pull-down buffer (**Subheading 2.4.**). Spin as above and remove the supernatant.
7. Resuspend the DNA cellulose pellet in 50 µL pull-down buffer.
8. Mix the DNA cellulose slurry well by pipetting up and down a few times. Pipet 16 µL of DNA cellulose slurry into each of three 1.5-mL tubes placed on ice. Carefully expel the DNA cellulose slurry into the bottom of each tube. Cut the end off the pipet tip if the slurry is difficult to pipet.
9. To one tube, add 2 µL water; to another add 2 µL mock synthetic peptide B (or D); to the third add 2 µL substrate peptide A (or C). Synthetic peptides are described in **Subheading 2.3., item 11**.
10. Start each reaction by adding 2 µL of the working stock solution of radioactive ATP (to give a final concentration of 50 μM ATP, containing 10 µCi [γ-^{32}P]ATP per reaction. See **Subheading 2.4.** for ordering and preparing radioactive ATP). Start samples at 15-s intervals.
11. Incubate samples for 10 min at 30°C with occasional agitation.
12. Stop the reaction by the addition of 20 µL of EDTA stop solution (**Subheading 2.4., item 20**).
13. Spin for 1 min at 10,000g.
14. Carefully remove 30 µL of the supernatant, being careful not to disturb the DNA cellulose pellet, and pipet it into a fresh tube containing 30 µL of 30% acetic acid stop solution (**Subheading 2.3., item 14**).
15. Mix and spin samples 30 s in a microfuge.
16. Spot 20 µL of each sample onto each of two 2 × 2 cm phosphocellulose paper squares.
17. Wash the phosphocellulose squares as described in **Subheading 3.3., items 13–20** (*see* **Note 10**).

Representative results of assays in extracts from NIH 3T3 and SCID cells are shown in **Fig. 2**. The assays are reproducible to about 10%; however, we caution that care should be taken when interpreting data in which the counts incorporated are less than 10-fold that of the background incorporation.

3.5. Western Blot of DNA-PK

The large size of the DNA-PKcs polypeptide (approx 460 kDa) makes it difficult to transfer using standard conditions. We therefore use a low percentage of crosslinker in our SDS gels and include SDS in the electroblot transfer buffer in order to maximize efficiency of transfer (*see* **Subheading 2.5.** for details). Ku subunits can also be detected on the same blot.

The SDS polyacrylamide resolving gels should be prepared to contain 8.5% acrylamide and 0.075% bis-acrylamide and run at approx 2–4 V/cm^2 until the Bromophenol blue dye runs off. After electrophoresis, remove the stacking gel and equilibrate the gel in SDS electroblot transfer buffer (**Subheading 2.5.**) for 5 min. For gels of 0.75-mm thickness, transfer in SDS

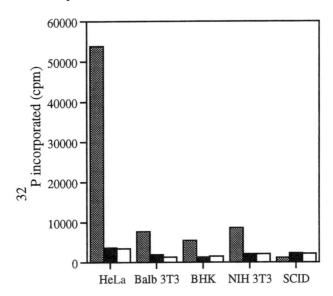

Fig. 2. DNA-PK activity in extracts from rodent cells using the DNA pull-down assay: High salt extracts from Balb/c 3T3, BHK (baby hamster kidney), NIH 3T3 and SCID (SCGR11) cells were made as described in **Subheadings 2.1., 2.2., 2.4., 3.1., 3.2., and 3.4.** For each assay, 500 μg of total protein was bound to 30 μL of DNA cellulose resin slurry. The resin was resuspended in 50 μL pull-down buffer and 16 μL was used in each assay. Also shown are results of a pull-down assay performed using 100 μg of total extract from HeLa cells. Hatched bars represent activity with peptide C; solid black bars represent activity with peptide D; open bars represent activity with no peptide. Similar results were obtained using peptides A and B.

electroblot at 100V (250 mA constant current) for 1 h at room temperature. Under these conditions, DNA-PKcs, Ku70, and Ku80 are transferred to the membrane. Alternatively, DNA-PKcs can be transferred at 15 V for 16 h; however, Ku70 can blow through the membrane under these conditions. Western blot detection should be carried out with suitable reagents according to the manufacturers' instructions. Using most antibodies, between 20 and 100 ng of purified DNA-PKcs can be detected using a 15 well 0.75-mm gel and ECL detection (Amersham), used according to the manufacturer's instructions. DNA-PKcs and Ku can be readily detected using 10–15 μg of total protein from a high-salt extract from human cells (prepared as described earlier). Alternatively, 10^5 human cells can be boiled in SDS sample buffer, the DNA denatured by sonication or passage through a 23-gauge needle, and DNA-PK detected by Western blot as previously stated. In contrast, 200 μg of total protein is required for visualization of DNA-PKcs in high-salt extracts from NIH 3T3 cells.

4. Notes

1. PMSF is highly toxic therefore take adequate safety precautions. In addition, PMSF is rapidly inactivated in aqueous buffers; therefore, add directly to sample. Do not add to buffers ahead of time.

2. Do not dilute the DNA in water or store DNA at concentrations below 0.05 mg/mL, as this can severely compromise its ability to activate DNA-PK. Dilute the 1 mg/mL stock sonicated calf thymus DNA 10-fold in TE pH 8.0 buffer and store in 100-μL aliquots at –20°C. Discard diluted DNA solutions after 10 freeze/thaw cycles.

3. The amount of DNA coupled to the DNA cellulose resin varies from batch to batch. We have obtained lots that vary from 3 to 8 mg DNA/g solid. The optimum amount of DNA cellulose to use will need to be determined for each batch. The DNA cellulose slurry can be stored in pull-down buffer at 4°C for at least 1 wk.

4. It is important to wash the cells well in LSB to remove traces of PBS. The cells will start to swell in the hypotonic LSB buffer and become fragile; therefore, do not leave the cells in LSB for longer than necessary.

5. If using human cells, the sample can be thawed and extracted immediately or samples can be transferred to a –80°C freezer and extracted at a later date. When making high-salt extracts, the sample can be aliquoted and stored at -80°C or used directly for DNA-PK assays or Western blot. We have stored samples from human cells for several months at this stage with little loss of DNA-PK activity. If using rodent or nonprimate cells, the samples should be processed and assayed immediately, as we have found that DNA-PK activity is less stable on storage than in human cells. Do not store extracts at –20°C or 4°C.

6. Do not increase the salt concentration to more than 0.5 *M* in the sample, as this will cause the chromatin to decondense and the protein sample will not be able to be recovered.

7. The assay is generally linear for about 10 min, but this should be determined for your particular assay system.

8. Do not use immobilized single-stranded DNA or denatured DNA to "pull-down" the DNA-PK, as single-stranded DNA does not activate DNA-PK *(6)*.

9. The sensitivity of detection of DNA-PK activity in rodent cells may be increased by using the pull-down procedure and phosphorylating a protein substrate such as RPA or p53, followed by SDS-PAGE autoradiography or phosphorimaging.

10. When performing the assay for the first time, keep the supernatant from the DNA-cellulose "pull-down" and add to it another aliquot of dsDNA cellulose slurry. Repeat the binding and assay procedure in order to determine if all the DNA-PK from the protein extract bound to the DNA cellulose. If not, reduce the amount of protein, or increase the amount of DNA cellulose slurry accordingly, until you have quantitative binding.

References

1. Lees-Miller, S. P. (1996) DNA dependent protein kinase: ten years and no ends in sight. *Biochem. Cell. Biol.* **74,** 503–512.

2. Chu, G. (1997) Double-strand break repair. *J. Biol. Chem.* **272,** 24,097–24,100.
3. Jackson, S. P. (1997) DNA-dependent protein kinase. *Int. J. Biochem. Cell Biol.* **29,** 935–938.
4. Hartley, K. O., Gell, D., Zhang, H., Smith, G. C. M., Divecha, N., Connelly, M. A., Admon, A., Lees-Miller, S. P., Anderson, C. W., and Jackson, S. P. (1995) DNA-dependent protein kinase catalytic subunit: a relative of phosphatidylinositol 3-kinase and the ataxia telangiectasia gene product. *Cell* **82,** 849–856.
5. Carter, T., Vancurová, I., Sun, I., Lou, W., and DeLeon, S. (1990) A DNA-activated protein kinase from HeLa cell nuclei. *Mol. Cell. Biol.* **10,** 6460–6471.
6. Lees-Miller, S .P., Chen, Y.-R., and Anderson, C. W. (1990) Human cells contain a DNA activated protein kinase that phosphorylates simian virus 40 T antigen, mouse p53, and the human Ku autoantigen. *Mol. Cell. Biol.* **10,** 6472–6481.
7. Gottlieb, T. M. and Jackson, S. P. (1993) The DNA-dependent protein kinase: requirement for DNA ends and association with Ku antigen. *Cell* **72,** 131–142.
8. Lees-Miller, S. P., Sakaguchi, K., Ullrich, S., Appella, E., and Anderson, C. W. (1992) The human DNA-activated protein kinase phosphorylates serines 15 and 37 in the aminotransactivation domain of human p53. *Mol. Cell. Biol.* **12,** 5041–5049.
9. Anderson, C. W, Connelly, M. A., Lees-Miller, S. P., Lintott, L. G., Zhang, H., Sipley, J. A., Sakaguchi, K., and Appella, E. (1995) The Human DNA-activated protein kinase, DNA-PK: substrate specificity, in *Methods in Protein Structure Analysis.* (Atassi, M. Z., Appella, E., eds.), Plenum Press, New York, pp. 395–406.
10. Anderson, C. W. and Lees-Miller, S. P. (1992) The nuclear serine/threonine kinase, DNA-PK, *in CRC Critical Reviews in Eukaryotic Gene Expression,* (Stein, G. S., Stein, J. L., and Lian, J. B., eds.), pp. 283–314.
11. Muller, C. and Salles, B. (1997) Regulation of DNA-dependent protein kinase activity in leukemic cells. *Oncogene* **15,** 2343–2348.
12. Finnie, N. J, Gottlieb, T. M., Blunt, T., Jeggo, P. A., and Jackson, S. P. (1995) DNA-dependent protein kinase activity is absent in xrs-6 cells: Implications for site-specific recombination and DNA double-strand break repair. *Proc. Natl. Acad. Sci. USA* **92,** 320–324.
13. Danska, J. S., Holland, D. P., Mariathasan, S., Williams, K. M., and Guidos, C. J. (1996) Biochemical and genetic defects in the DNA-dependent protein kinase in murine *scid* lymphocytes. *Mol. Cell. Biol.* **16,** 5507–5517.
14. Allalunis-Turner, J. M., Lintott, L. G., Barron, G. M., Day, III, R. S., and Lees-Miller, S. P. (1995) Lack of correlation between DNA-dependent protein kinase activity and tumor cell radiosensitivity. *Cancer Res.* **55,** 5200–5202.
15. Lees-Miller, S. P., Godbout, R., Chan, D. W., Weinfeld, M. W., Day III, R. S., Barron, G. M., and Allalunis-Turner, J. (1995) Absence of the p350 subunit of DNA-activated protein kinase from a radiosensitive human cell line. *Science* **267,** 1183–1185.

9

Expression and Assay of Recombinant ATM

Yael Ziv, Sharon Banin, Dae-Sik Lim, Christine E. Canman, Michael B. Kastan, and Yosef Shiloh

1. Introduction

A variety of genetic disorders involve genome instability and abnormal response to DNA damaging agents. Investigation of these disorders has revealed different metabolic pathways responsible for damage repair on one hand, and for signaling the presence of the damage to cellular regulatory systems on the other hand. Ataxia–telangiectasia (A-T) is a typical example of such a disorder. This autosomal recessive disease is characterized by degeneration of the cerebellum, thymus, and gonads, immunodeficiency, premature aging, cancer predisposition, and acute sensitivity to ionizing radiation. A-T cells show hypersensitivity to ionizing radiation and radiomimetic chemicals as well as defects in various signal-transduction pathways induced by these agents, most notably the activation of cell cycle checkpoints *(1–3)*.

The A-T gene, *ATM*, encodes a large protein with a carboxy-terminal domain resembling the catalytic subunit of phosphatidylinositol 3-kinases (PI 3-kinases) *(3)*. This PI 3-kinase-related region is characteristic of a growing family of large proteins identified in yeast, *Drosophila*, and mammals, which are involved in the maintenance of genome stability, telomere length, cell cycle control, or damage responses *(4)*. These proteins are considered protein kinases rather then lipid kinases. Indeed, protein kinase activity has been documented in several members of this family: the DNA-dependent protein kinase (DNA-PK, *see* also Chapter 8, this volume), mTOR (also designated FRAP or RAFT1), ATM, and ATR *(5–9)*.

A central junction of the cascades leading to cell-cycle arrest following DNA damage is regulated by the p53 protein (*10, see also* Chapter 27, this

From: *Methods in Molecular Biology, vol. 99: Stress Response: Methods and Protocols*
Edited by: S. M. Keyse © Humana Press Inc., Totowa, NJ

volume). Following DNA damage, p53 undergoes *de novo* phosphorylation on a Serine residue at position 15 and accumulates in the cells via posttranscriptional mechanisms *(11–13)*. Both Serine-15 phosphorylation and p53 accumulation are significantly delayed in irradiated A-T cells, indicating this response is downstream of ATM *(9,11,12)*. Indeed, ATM was found to phosphorylate Serine-15 of p53 in vitro. This activity is manganese dependent and is inhibited by the fungal metabolite wortmannin, similar to the way it inhibits DNA-PK and mTOR. ATM's kinase activity is enhanced several fold minutes after treatment of cells with ionizing radiation or radiomimetic chemicals *(8,9)*.

Straightforward detection of ATM's catalytic activity can be achieved with immunoprecipitated endogenous ATM *(8,9)*. The use of ectopically expressed ATM enables manipulation, alteration, and therefore functional dissection of the molecule. The initial generation of full-length *ATM* cDNA was challenging because of its large size and inherent instabilty. However, using appropriate hosts and vectors, plasmids containing ATM's 9.2 kilobase (kb) open reading frame can be stably maintained in bacteria *(14,15)*.

Recombinant ATM protein can be produced in human cells using stable or transient expression systems *(8,9,14–17)* and in insect cells using a baculovirus vector *(14–16)*. Stable expression in human cells is based on an episomal expression vector which is maintained in the cells by selective pressure from the drug hygromycin *(14,15,17)*. This expression system is limited to A-T cell lines (lymphoblasts or immortalized fibroblasts), because of reasons not clear yet, expression of recombinant ATM driven by this system seems to be suppressed in cells expressing endogenous ATM (Y. Ziv, unpublished observations). The cellular level of recombinant ATM obtained in this way in A-T fibroblasts is comparable to that of endogenous ATM in various cell lines *(14,17)*. Higher amounts of the protein can be obtained by transient expression of ATM using vectors of the pCDNA3.0 series *(9)*, but this mode of expression requires repeated transfections of the cells.

The kinase activity of recombinant ATM is indistinguishable from that of endogenous ATM *(8,9)*. This chapter will describe the protocols for obtaining active recombinant ATM using stable or transient expression, as well as the assay conditions for ATM's kinase activity.

2. Materials

2.1. Bacterial Strains and Human Cell Lines

1. STBL2 competent cells (Life Technologies, Paisley, UK) (*see* **Note 1**).
2. AT22IJE-T is a fibroblast line from an A-T patient, which was immortalized using an origin defective SV40 genome *(18)*.
3. 293 cells *(19)*.

2.2. Buffers

1. Phosphate-buffered saline (PBS): 137 mM NaCl, 2.7 mM KCl, 4.3 mM Na$_2$HPO$_4$, 1.4 mM KH$_2$PO$_4$.
2. DM lysis buffer: 10 mM Tris-HCl, pH 7.5, 150 mM NaCl, 0.2% (w/v) Dodecyl β-D-Maltoside (Sigma), 5 mM EDTA, 50 mM NaF. (Can be stored up to 1 wk at 4°C). Add protease and phosphatase inhibitors just before use: 2 mM phenylmethylsulfonylflouride (PMSF), 100 µM NaVO$_4$, and 2µM aprotonin.
3. TGN lysis buffer: 50 mM Tris-HCl, pH 7.5, 150 mM NaCl, 1%(v/v) Tween-20, 0.2% (v/v) Nonidet-P40. (Can be stored up to 1 wk at 4°C). Add fresh 1 mM PMSF, 1 mM NaF, 1 mM NaVO$_4$, 10 µg/mL aprotonin, 2 µg/mL pepstatin, and 5 µg/mL leupeptin.
4. Wash buffer: TGN buffer containing 0.5 M LiCl.
5. ATM kinase buffer: 50 mM Hepes, pH 7.5, 150 mM NaCl, 4 mM MnCl$_2$, 6 mM MgCl$_2$, 10% glycerol, 1 mM dithiothreitol (DTT), and 100 µM NaVO$_4$ (prepare fresh).
6. TTBS: 20 mM Tris-HCl, pH 7.5, 150 mM NaCl, 0.1% Tween-20 (prepare a 10X stock).
7. Running buffer: 50 mM Trizma Base, 200 mM glycine, 1% SDS.
8. Transfer buffer: 50 mM Trizma Base, 200 mM glycine, 10% methanol, 0.005% SDS (prepare a 10X stock).
9. Loading buffer: 50 mM Tris-HCl, pH 6.8, 2% SDS, 2% β-mercaptoethanol, 0.2% bromophenol blue, and 10% glycerol (prepare a 10X stock).

2.3. Other Materials

1. Bacterial media: Luria broth (LB) supplemented with 50 µg/mL ampicillin.
2. Tissue culture media: Dulbecco's modified Eagle medium (DMEM) and HAMX F-10 supplemented with 10–20% fetal bovine serum (FBS).
3. Mammalian Transfection Kit (Stratagene, La Jolla, CA).
4. Hygromycin B solution (50 mg/mL, Boehringer Mannheim)
5. Anti-FLAG M2 and M5 antibodies (Eastman-Kodak).
6. Anti-ATM antibodies: Ab-3, a polyclonal antibody raised against a peptide spanning positions 819-844 of ATM (Oncogene Research Products, Cambridge, MA).
7. ATM132, a monoclonal antibody raised against the same peptide as Ab-3 *(8)*.
8. Sheep antimouse IgG (Jackson ImmunoResearch Laboratories, West Grove, PA).
9. Magnetic beads coated with sheep antimouse IgG (Dynabeads M450, Dynal, Oslo, Norway).
10. Magnetic stand (Dynal).
11. Protein A/G sepharose CL-4B (Sigma).
12. Acrylamid:bisacrylamid 37.5:1 40% solution (BioProbe).
13. [γ-^{32}P]ATP (3000 Ci/mM, Amersham).
14. Nitrocellulose MFF membranes (Pharmacia Biotech, San Francisco, CA).
15. Gel-Code: Commassie like staining solution (Pierce, Rockford, Il).
16. Chemiluminescence detection system: Super Signal System (Pierce).
17. Substrates for ATM's kinase activity: the PHAS-I substrate (Stratagene, La Jolla, CA). Any recombinant derivative of the p53 protein that contains the aminoterminus of this protein (*see* **Note 2**).

2.4. ATM Expression Constructs

1. pEBS-YZ5 *(14)* is an ATM full-length cDNA cloned in the episomal expression vector pEBS7 *(20)*. The protein is tagged with an amino-terminal FLAG epitope *(21)*. This construct is used for stable expression of ATM in A-T cells (*see* **Note 3**).
2. pCDNA-FLAG-ATMwt *(9)*, contains the same cDNA as in pEBS-YZ5 (including the FLAG epitope) in the vector pCDNA3.0 (Invitrogen). This construct is used for transient expression of ATM (*see* **Note 3**).
3. pCDNA-FLAG-ATMkd *(9)* is a derivative of pCDNA-FLAG-ATMwt in which 2 critical amino acids within the putative kinase domain were substituted (D2870A and N2875K), rendering the protein inactive ("kinase dead").

3. Methods
3.1. Preparation of Recombinant Plasmids

Amplify the appropriate plasmid in STBL2 bacteria in LB medium with ampicillin. The final volume of the culture depends on the desired amount. Usually, 100 µg of plasmid DNA are obtained from 150 mL culture. Grow the culture overnight at 30°C with constant shaking. Note that these recombinant plasmids exist at low copy numbers (*see* **Note 1**).

3.2. Growing the Cell Lines

The immortalized A-T fibroblast line AT22IJE-T is used for stable expression of ATM using the episomal vector pEBS7 *(14)*. These cells grow in monolayer. Use DMEM culture medium supplemented with 15% FBS and antibiotics. Split the culture at 1:6 ratio when confluent (usually once every 3–4 d). Stable transfectants are grown in the presence of 100 µg/mL hygromycin B. 293 cells are used for transient expression of the ATM protein using the pCDNA3.0 constructs. Use culture medium DMEM containing 10% FBS and split at 1:6 ratio when confluent, usually once every 3–4 d.

3.3. Transfection and Selection of Transfectants
3.3.1. Stable Expression of ATM

1. Plate AT22IJE-T cells in 9-cm dishes (approx 5×10^6 cells/dish) containing 10 mL medium. Make sure the cells are evenly dispersed. Grow overnight.
2. On the next day transfect the cells in each culture with 10µg of pEBS-YZ5 construct using the Mammalian Transfection Kit (Stratagene), according to the manufacturer's instructions. Include one dish of mock transfection and transfect one dish with an empty pEBS7 vector (*see* **Note 4**).
3. The following day wash the plates twice with medium without serum, add fresh complete medium, and continue growing the cultures for another day. Do not pour the medium directly onto the cells, as they are loosely attached to the surface at this stage.

4. On the third day add the selective antibiotic, hygromycin B, directly to each plate to a final concentration of 200 µg/mL.
5. Replace the medium every 4–6 d until the cells in the mock transfection have all died (usually between 7–10 d). When massive cell death is observed, switch the medium to F-10 HAM supplemented with 20% FBS and 200 µg/mL of hygromycin B, in order to maintain the survivor cells' colonies. Continue to grow in this medium until colonies are clearly visible (14–21 d; *see* **Note 5**).
6. Trypsinize colonies, collect the cells, and transfer them into a T25 flask. At this stage switch back to DMEM containing 100 µg/mL hygromycin B.
7. Continue to expand the cultures in DMEM containing 100 µg/mL hygromycin B (*see* **Note 6**). Freeze down aliquots in medium without hygromycin B. Thaw up in medium containing the drug (*see* **Note 6**).

3.3.3. Transient Expression of ATM

1. Plate 2×10^6 293 cells per 9-cm dish.
2. Transfect cells with pCDNA-FLAG-ATMwt or pCDNA-FLAG-ATMkd as in **Subheading 3.3.2.** *(1–3)*, using DMEM medium with 10% FBS throughout the procedure.
3. Forty-eight hours after transfection harvest the cells by scraping and proceed to protein detection and assay.

3.4. Immunoprecipitation and Detection of Recombinant ATM

3.4.1. Cell Harvesting

All steps are carried out on ice.

1. Discard the medium and wash cell monolayer once with ice-cold PBS.
2. Add fresh cold PBS and scrape the cells with a scraper.
3. Collect cells into a 50-mL polypropylene tube. Wash the plate again with PBS and add the wash to the same tube. Fill the tube to the top with cold PBS.
4. Spin in a refrigerated centrifuge at 1000*g* for 10 min.
5. Discard the supernatant and resuspend the cells in 1 mL cold PBS.
6. Transfer to a 1.5-mL tube and spin at 4000*g* in a refrigerated microfuge.
7. Proceed immediately to analysis or remove supernatant and store the cell pellets at –70°C.

3.4.2. Immunoprecipitation (see **Note 7**)

Two alternative protocols for immunoprecipitation of active ATM from cell lysates were developed independently *(8,9)*. Both work equally well and are presented below.

Protocol A

1. Resuspend 5×10^6–10^7 cells in 700 µL of DM lysis buffer. Leave on ice for 30 min and centrifuge at maximal speed in a refrigerated microfuge for 20 min.

Transfer supernatant into a new tube. Keep a 40-μL aliquot for immunoblot analysis of the extracts (*see below*).

2. Add 10 μL of hybridoma spent medium (for antibody ATM132), or 5 μg of anti-FLAG M2 or anti-FLAG M5 antibodies to the cell lysate and rotate for 2 h in the cold room.

3. Prepare the appropriate amount of magnetic beads (50 μL/10^7 cells) by washing Dynabeads coated with sheep antimouse IgG twice with DM lysis buffer and resuspend in DM lysis buffer in the original volume.

4. Add to each tube of the immune complexes 50 μL of washed magnetic beads, and continue rotating for another 2 h.

5. Collect the immune complexes using a magnetic stand (Dynal), and wash the beads three times with 500 μL of DM lysis buffer.

Protocol B

1. Harvest 10^7 cells, resuspend them in 1.5 mL TGN buffer and leave on ice for 30 min.

2. Spin at maximal speed for 15 min in a refrigerated microfuge and transfer supernatant into a new tube.

3. Preclear the samples for 1 h by rotating the mixture in the cold room with 10 μg purified mouse IgG and 40 μL of protein A/G sepharose beads. Pellet the beads and transfer the precleared lysates to a new tube.

4. Add antibodies as described in **Protocol A, step 2**, and rotate for 2 h.

5. Add 40 μL of protein A/G sepharose beads and rotate for 1 h.

6. Wash the beads twice with TGN buffer, and once with the same buffer containing 0.5 *M* LiCl.

3.4.3. Detection of Immunoprecipitated ATM by Immunoblotting

1. Prepare an 8% polyacrylamide-SDS gel (mini-gel size, e.g., 10 × 10.5 cm, Hoefer).

2. Resuspend the immune complexes in 1X loading buffer and boil the samples for 5 min.

3. Load the samples on gel and electrophorese at 40 mA (for each gel) for approx 2 h.

4. Transfer the gel onto a nitrocellulose membrane overnight at 250–300 mA in transfer buffer.

5. Block the membrane for 1.5 h with 3% BSA in TTBS.

6. Change to a fresh blocking solution containing 20 μg/mL anti-FLAG M5 antibody or an anti-ATM antibody and rotate gently for 2–4 h at room temperature.

7. Rinse the blot six times, 5 min each, with 2% low fat milk, and 0.2% BSA in TTBS.

8. Incubate the membrane for 50–60 min with horseradish peroxidase-conjugated antimouse IgG diluted at 1:40000 in blocking solution.

9. Wash the blots five times with TTBS and perform ECL detection according to the manufacturer's instructions. Wrap the membrane immediately with Saran Wrap and expose it to an X-ray film. The recombinant ATM is visualized as a 370 kD band, usually after exposure of 10–60 s.

3.5. ATM Kinase Assay

The in vitro assay of ATM's kinase activity is performed using the immune complexes bound to the beads (magnetic beads or protein A/G sepharose). The reaction is carried out immediately after immunoprecipitation (**Subheading 3.4.2.**). The following steps are performed at room temperature unless otherwise indicated.

1. Wash the bound immune complexes twice with 1 mL kinase buffer and resuspend in 20 µL of fresh kinase buffer.
2. Add cold ATP to final concentration of 20 μM, 10 µCi of [γ-^{32}P]ATP and 200 ng–1 µg substrate. Incubate at 30°C for 3–15 min depending on the aim of the experiment (*see* **Note 8**).
3. Stop the reaction by adding 0.5 vol of 3X loading buffer.
4. Prepare a biphasic polyacrylamide gel, 12.5% at its lower half, and 8% at its upper half (*see* **Note 9**). First, prepare 6 mL of 12.5% polyacrylamide solution for a 10 × 10.5 gel, pour, and let the gel polymerize. Then, pour on top of it 6 mL of 8% gel solution and let it solidify. Finally, prepare the stacking layer: 2.5 mL of 4% gel solution.
5. Boil the samples for 5 min and electrophorese until the blue front of the loading buffer runs off the bottom of the gel.
6. Separate the two halves of the gel. The top part (8%) is used for immunoblot analysis as described in **Subheading 3.4.3.**
7. Wash the bottom part of the gel (12.5%) three times, 5 min each, with ddH$_2$O and stain with the Gel-Code until the substrates become visible. Wash again with water to reduce the blue background. The protein bands should become more pronounced (*see* **Note 10**).
8. Put the gel onto a Whatman 3MM filter paper, cover it with Saran Wrap and dry between several layers of Whatman under vacuum at 80°C for 1.5 h.
9. Expose the dried gel to an X-ray film with an intensifying screen at –80°C, until the phosphorylated products become clearly visible (usually 0.5–2 h).
10. The intensity of the bands can be quantitated using densitometry or phosphor-imager.

4. Notes

1. Plasmids containing the long (9.2 kb) open reading frame of ATM may show considerable instability in standard bacterial hosts under commonly used growth conditions. The STBL2 strain was engineered to stabilize DNA sequences which are prone to rearrangements. Specific recombinogenic functions have been eliminated. Growth at 30°C is recommended in order to minimize the time of culture saturation, since stressful conditions may stimulate processes involving DNA recombination.
2. Since Serine15 of p53 is the only target of ATM *(8,9)*, a p53 derivative must contain this residue in order to serve as an ATM substrate.
3. Transient expression of ATM described above has the advantage of producing relatively high levels of normal or kinase-dead ATM protein; stable transfectants

produce moderate levels of wild type protein and do not express mutant ATM. However, stable ATM expression in A-T cells allows study of the biological effects of the recombinant protein over longer periods of time, such as the effect on various features of the cellular phenotype *(14)*. Another advantage of using recombinant ATM stably expressed in A-T cells is that it can be immunoprecipitated using anti-ATM antibodies (e.g., AT132), which are more efficient for immunoprecipitation than the anti-FLAG antibodies. This is particularly important when the recombinant protein has been manipulated and should be separated from the endogenous protein. (293 cells used for transient expression express a high level of endogenous ATM.)

4. In the stable expression system, mock transfection is essential for evaluating the efficiency of selection. An important negative control is the transfection with an empty vector.

5. HAMS F-10 medium supplemented with 20% FCS supports colony formation of fibroblasts considerably better than DMEM, making its use in the critical stage of transfectant colony formation essential. Colonies usually emerge 7 d after transfection and beginning of selection, but should be allowed to develop before first trypsinization. During colony formation the medium must be changed every 4–5 d, especially in the initial selection period which is characterized by massive cell death.

6. Stable transfectants must be kept under hygromycin selection, or the episomal expression vectors are rapidly lost from the culture. Note that after the colonies become visible the concentration of the selective drug, hygromycin B, is reduced from 200 to 100 µg/mL.

7. There are numerous methods for protein extraction and immunoprecipitation. The two protocols presented here were found to be particularly suitable for the ATM kinase assay.

8. When candidate substrates are tested, or phosphorylation sites mapped, the reaction may proceed for 15–20 min. When the specific activity of ATM is assessed *(8,9)* reaction times of 3–4 min are recommended.

9. The biphasic gel permits convenient quantitation of the ATM protein and its activity using a one gel system. It provides maximum accuracy in assessing ATM's specific activity, since the entire amount of the ATM protein actually present in the kinase reaction mix is visualized.

10. Staining the gel is recommended not only for visualizing the substrate bands, but also for washing out excess radioactive background.

References

1. Shiloh, Y. (1997) Ataxia-telangiectasia and the Nijmegen breakage syndrome: related disorders but genes apart. *Ann. Rev. Genet.* **31,** 635–662.
2. Gatti, R. A. (1998) Ataxia-telangiectasia, in *The Genetic Basis of Human Cancer* (Vogelstein, B. and Kinzler, K. W., eds), McGraw-Hill, New York, pp. 275–300.
3. Rotman, G. and Shiloh, Y. (1998) ATM: From gene to function. *Hum. Mol. Genet.* **7,** 1555–1563.
4. Hoekstra, M. F. (1997) Responses to DNA damage and regulation of cell cycle checkpoints by the ATM protein kinase family. *Curr. Opin. Genet. Dev.* **7,** 170–175.

5. Jeggo, P. A. (1997) DNA-PK: at the cross-roads of biochemistry and genetics. *Mut. Res.* **384,** 1–14.

6. Brunn, G. J., Hudson C. C., Sekulic, A., Williams, J. M., Hosoi, H., Houghton, P. J., Lawrence, J. C., and Abraham, R. T. (1997) Phosphorylation of the translational repressor PHAS-I by the mammalian target of rapamycin. *Science* **277,** 99–101.

7. Burnett, P. E., Barrow, R. K., Cohen, N .A., Snyder, S. H., and Sabatini, D. M. (1998) RAFT1 phosphorylation of the translational regulators p70 S6 kinase and 4E-BP1. *Proc. Natl. Acad. Sci. USA* **95,** 1432–1437.

8. Banin, S., Moyal, L., Shieh, S.-Y., Taya, Y., Anderson, C.W., Chessa, L., Smorodinsky, N.I., Prives, C., Shiloh, Y., and Ziv, Y. (1998) DNA damage-enhanced phosphorylation of p53 by ATM. *Science* **281,** 1674–1677.

9. Canman, C. E., Lim, D.-S., Cimprich, K. A., Taya, Y., Tamai, K., Sakaguchi, K., Appella, E., Kastan, M. B., and Siliciano, J. (1998) ATM is a protein kinase induced by ionizing radiation which phosphorylates p53. *Science* **281,** 1677–1679.

10. Levine, A. J. (1997) p53, the cellular gatekeeper for growth and division. *Cell* **88,** 323–331.

11. Kastan, M. B., Zhan,Q., El-Deiry, W. S., Carrier, F., Jacks, T., Walsh, W. V., Plunkett, B. S., Vogelstein, B., and Fornace Jr. A. J. (1992) A mammalian cell cycle checkpoint pathway utilizing p53 and GADD45 is defective in ataxia-telangiectasia. *Cell* **71,** 589–597.

12. Siliciano, J. D., Canman, C. E., Taya, Y., Sakaguchi, K., Appella, E., and Kastan, M. B. (1997) DNA damage induces phosphorylation of the amino terminus of p53. *Genes Dev.* **11,** 3471–3481.

13. Shieh, S.-Y., Ikeda, M., Taya, Y., and Prives, C. (1997) DNA damage-induced phosphorylation of p53 alleviates inhibition by MDM2. *Cell* **91,** 325–334.

14. Ziv, Y., Bar-Shira, A., Pecker, I., Russell, P., Jorgensen, T. J., Tsarfati, I., and Shiloh, Y. (1997) Recombinant ATM protein complements the cellular A-T phenotype. *Oncogene* **15,** 159–167.

15. Zhang, N., Chen, P., Khanna, K. K., Scott, S., Gatei, M., Kozlov, S., Watters, D., Spring, K., Yen, T., and Lavin, M. (1997) Isolation of full-length ATM cDNA and correction of the ataxia-telangiectasia cellular phenotype. *Proc. Natl. Acad. Sci. USA* **94,** 8021–8026.

16. Scott, S. P., Zhang, N., Khamma, K. K., Khromykh, A., Hobson, K., Watters, D., and Lavin, M. (1998) Cloning and expression of the ataxia-telangiectasia gene in baculovirus. *Biochem. Biophys. Res. Commun.* **245,** 144–148.

17. Shiloh, Y., Bar-Shira, A., Galanty, Y., and Ziv, Y. (1998) Cloning and expression of large mammalian cDNAs: lessons from ATM, in *Genetic Engineering, Principles and Methods* (Anderson, C., Brown, D. D., Day, P., Helsinki, D., and Setlow, J., eds.), Plenum Press, New York, pp. 239–248.

18. Ziv, Y., Etkin, S., Danieli, T., Amiel, A., Ravia, Y., Jaspers, N. G. J., and Shiloh,Y. (1989) Cellular and biochemical characteristics of an immortalized ataxia-telangiectasia (group AB) cell line. *Cancer Res.* **49,** 2495–2501.

19. Pear W. S., Nolan, G. P., Scott, M. L., and Baltimore, D. (1993) Production of high-titer helper-free retroviruses by transient transfection. *Proc. Natl. Acad. Sci. USA* **90,** 8392–8396.
20. Petersen, C. and Legerski, R. (1991) High-frequency transformation of human repair-deficient cell-lines by an epstein-barr virus-based cDNA expression vector. *Gene* **107,** 279–284.
21. Hopp, T. P., Prickett, T. S., Price, V. L., Libby, R. T., March, C. J., Ceretti, D. P., Urdal, D. L., Conion, P. J. (1988) A short polypeptide marker sequence useful for recombinant protein identification and purification. *Biotechnology* **6,** 1204–1210.

10

Detection and Purification of a Multiprotein Kinase Complex from Mammalian Cells

IKK Signalsome

Frank Mercurio, David B. Young, and Anthony M. Manning

1. Introduction

Steady progress has been made in our understanding of the mechanisms by which intracellular signals are relayed from the cell membrane to the nucleus. One theme that has emerged from studies of several different pathways involves the integration of key signaling proteins into multiprotein complexes. Such a mechanism organizes the proper repertoire of proteins into specific signaling pathways, preventing inappropriate activation of regulatory proteins such as transcription factors or cell cycle regulators. Varying combinations of related enzymes in a complex, often in a cell-specific manner, enables the propagation of distinct cellular responses. The precise mode of regulation can take many forms. For example, cell activation can initiate recruitment of key regulatory proteins into a higher-order complex, resulting in the sequential activation of a kinase cascade. Alternatively, enzymes may already exist as part of a multiprotein complex that functions to maintain their steady-state activity and, at the same time, properly position the enzymes so as to respond to incoming signals from the cellular environment. Such complexes could be regulated through control of their subcellular localization. Higher-order complex formation is mediated, in part, by small conserved protein–protein interaction motifs, including leucine zipper, helix–loop–helix, WW, WD-40, SH2, SH3, PH, PTB, and PDZ motifs *(1)*. Determination of the full complement of proteins comprising these signaling complexes would greatly enhance our understanding of how specificity is achieved in the regulation of cellular processes. Furthermore, the identification of one or more components comprising

From: *Methods in Molecular Biology, vol. 99: Stress Response: Methods and Protocols*
Edited by: S. M. Keyse © Humana Press Inc., Totowa, NJ

a multiprotein complex would enable the development of immunoaffinity purification protocols for the purification and characterization of additional unknown components within that complex.

The application of biochemical techniques specifically tailored for the purification of large multiprotein complexes has facilitated the elucidation of the signal transduction pathway responsible for activation of the nuclear transcription factor NF-κB (**Fig. 1**). Transcription factors of the NF-κB family are critical regulators of genes that function in inflammation, cell proliferation, and apoptosis *(2,3)* and are considered to be key regulators of the cellular stress response (*see also* Chapter 16, this volume). Activation of NF-κB is controlled by an inhibitory subunit, IκBα, which retains NF-κB in the cytoplasm. NF-κB activation requires sequential phosphorylation, ubiquitination, and degradation of IκBα, thereby freeing NF-κB to translocate to the nucleus. Ser32 and Ser36 of IkBα represent the residues that undergo stimulus-dependent phosphorylation leading to IκBα degradation (**Fig. 2**) *(4,5)*. We recently described the immunoaffinity purification of a large multiprotein complex (> 700 kDa), the IκB kinase (IKK) signalsome, containing a cytokine-inducible IκB kinase activity that phosphorylates IκBα *(6)*. The aim of this chapter is to describe the rationale and methods used to identify the IκB kinase and other components of the IKK signalsome.

Our initial objective was to identify the kinase subunit(s) responsible for direct phosphorylation of IκBα. To this end, whole-cell extracts (WCEs) from tumor necrosis factor-α (TNFα)-induced HeLa cells were fractionated by standard chromatographic methods. Each fraction was assayed for IκBα kinase activity by phosphorylation of a glutathione-*S*-transferase (GST)–IκBα (residues 1–54) fusion protein, which contains the sites of inducible phosphorylation. In parallel with a standard biochemical purification scheme to isolate the IκB kinase, we initiated a rational search to identify known proteins that might be components of the IKK signalsome. Antibodies directed against a component of the IKK signalsome would provide an extremely useful reagent for immunoaffinity purification of the IKK complex. Previous reports had indicated that NF-κB activation occurs under conditions that also stimulate mitogen-activated protein kinase (MAP kinase) pathways *(7–9)*. Therefore, we examined chromatographic fractions containing the IKK signalsome activity by immunoblotting for the presence of proteins associated with MAP kinase cascades, including protein phosphatases that are implicated in MAP kinase inactivation. Antibodies made against proteins that copurified with the IKK signalsome activity were then further examined for their ability to immunoprecipitate the IκB kinase activity.

Of a panel of antibodies tested, one of three anti-MKP1(CL100) polyclonal antibodies efficiently immunoprecipitated the high-molecular-weight IκBα

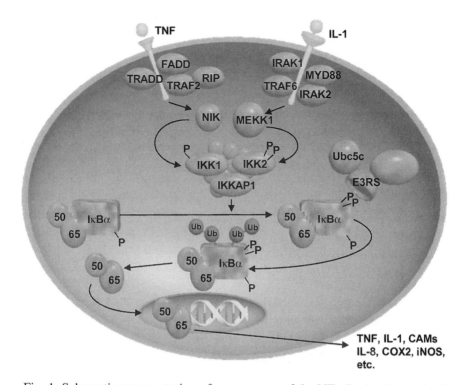

Fig. 1. Schematic representation of components of the NF-κB signal transduction pathway leading from the TNF-α and IL-1 receptors. A number of signal transduction proteins have been identified as associated with these receptors, including TNF-receptor associated factors 2 and 6 (TRAF2 and 6), death domain-containing proteins (TRADD and FADD), kinases associated with the IL-1 receptor (IRAK1 and 2, and MYD88). Other proteins such as ring finger interacting protein (RIP) have been identified based on their ability to interact with several of these proteins. Signals emanating from the TNF and IL-1 receptors activate members of the MEKK-related family, including NIK and MAKK1. These proteins are involved in activation of IKK1 and IKK2, the IκB kinase components of the IKK signalsome. These kinases phosphorylate members of the IκB family at specific serines within their N-termini, leading to site-specific ubiqitination and degradation by the 26S proteosome.

IκB Protein	Phosphorylation Sites													
IκBα	L	D	D	R	H	D	S	G	L	D	S	M	K	D	E	E
IκBβ	A	D	E	W	C	D	S	G	L	G	S	L	G	P	D	A
IκBε	E	E	S	Q	Y	D	S	G	I	E	S	L	R	S	L	R

Fig. 2. Amino acid sequence of the corresponding regions of IκBα, β, and ε. The conserved serine residues, shown in boxes, have been mapped as sites of stimulus-dependent phosphorylation that target the protein for degradation. For IkBα, these serines are amino acid residues 32 and 36.

kinase complex. In an effort to enrich for the IκB kinase component, we sought to selectively dissociate any nonkinase-associated component of the IKK signalsome from the immunocomplex. To this end, the immune precipitate was subjected to sequential elutions using increasing concentrations of the chaotropic agent, urea. The eluate and immunobeads were subsequently assayed for IκB kinase activity. Surprisingly, the IκB kinase activity remained bound to the immunocomplex even at concentrations of up to 6 M urea. Based on these findings we developed a two-step IKK signalsome purification method. Proteins from whole-cell lysates of TNF-α-stimulated HeLa cells were immunoprecipitated with anti-MKP1-antibodies; the immune complex was washed with 3 M urea, eluted with MKP1 peptide and further fractionated by anion-exchange chromatography. Fractions containing the IκB kinase activity were pooled and subjected to sodium dodecyl sulfate (SDS) gel electrophoresis. Two prominent bands of 85 and 87 kDa were excised, analyzed by high-mass accuracy matrix-assisted laser deposition and ionization (MALDI) peptide mass mapping, and identified as two closely related protein serine kinases termed IκB kinase-1 (IKK1) and IκB kinase-2 (IKK2), respectively *(6,10,11)*. They represent members of a new family of intracellular signal-transduction enzymes containing an *N*-terminal kinase domain and a C-terminal region with two protein interaction motifs, a leucine zipper, and a helix-loop-helix motif. IKK1 and IKK2 have been shown to form heterodimers as well as homodimers, in vitro and in vivo *(12)*. The relative contribution of IKK1 and IKK2 to NF-κB activation in response to a given stimuli has not been clearly defined. However, overexpression of IKK2, but not IKK1, dominant negative mutants completely block TNF-α-induced NF-κB activation *(6)*. Similar results were observed in studies using lipopolysaccharide (LPS)-stimulation of monocytes *(13)*. These finding suggest that IKK2 may play a more critical role in NF-κB activation in response to proinflammatory cytokines than does IKK1.

In an effort to better understand the nature of the IKK signalsome and the mechanism by which regulation is achieved, we developed an immunoaffinity purification protocol to identify additional components comprising this multiprotein complex. In contrast to the initial protocol used to identify the IKKs where 3 M urea wash conditions were employed, the complex was isolated under mild wash conditions so as to retain loosely bound regulatory components *(12)*. Here we purified and cloned a novel nonkinase component of the IKK signalsome, named IKK-associated protein-1 (IKKAP1). IKKAP1 associates with IKK2 in vitro and in vivo via sequences contained within the *N*-terminal coiled-coil repeat region of IKKAP1 *(12,14)*. Mutant versions of IKKAP1, which either lack the N-terminal IKK2-binding domain or contain only the IKK2-binding domain, disrupt the NF-κB signal-transduction cascade.

Additional putative IKK signalsome components were isolated using this method and are currently being evaluated.

We have also employed a similar approach to immunnoaffinity purify and clone the IκBα E3 ubiquitin ligase *(15)*. Activation of the IKK signalsome leads to IκBα phosphorylation at serine 32 and 36 and, consequently, ubiquitin degradation of phosphorylated IκBα (pIκBα). The phosphorylation of IκB serves as a recognition motif for the IκBα E3 ubiquitin ligase *(16)*. We previously observed that the NF-κB–pIκBα complex, which is generated in response to TNF-α stimulation, undergoes a dramatic increase in molecular mass and reasoned that this could be the result of recruitment of components of the ubiquitin machinery. Taking advantage of the high affinity that the IκBα E3 ubiquitin ligase displayed for pIκBα, a single-step immunoaffinity purification scheme was developed to isolate the ligase from HeLa cell whole-cell lysate *(15)*. These examples demonstrate the power of using immunoaffinity purification techniques to elucidate key regulatory components and thus the mechanism by which signal transduction pathways achieve specificity in regulation.

2. Materials

2.1. Equipment

1. Environmental Orbital Shaker (New Brunswick Scientific, Hatfield, UK).
2. Spectrophotometer, Genesys 5 (Spectronic Instruments).
3. Centrifuge RC 5C (Sorvall).
4. Rotor, GS3 (Sorvall).
5. Microcentrifuge, 5415 C (Eppendorf, Madison, WI).
6. Sonicator, Ultrasonic XL (Heat Systems, Farmingdale, NY).
7. Rotator, Hematology/Chemistry Mixer 346 (Fisher Scientific, Pittsburgh, PA).
8. Centrifuge, tabletop, RT 6000D (Sorvall).
9. FPLC: AKTA explorer FPLC system with Unicorn version 2.10 software (Pharmacia Biotech).

2.2. Purification Reagents and Resins

1. Glycerol stock of GST-IκBα 1–54 WT: Glycerol culture of *E. coli* transformed with pGEX-KG (Pharmacia Biotech) expressing a GST-IκBα 1–54 wild-type fusion protein.
2. Glycerol stock of GST-IκBα 1–54 S32/36>T: Glycerol culture of *E. coli* transformed with pGEX-KG (Pharmacia Biotech) expressing a GST-IκBα 1–54 S32/36>T (Ser 32/36 to Thr substitution) fusion protein.
3. LB (Luria Broth) medium: 85 mM NaCl, 10 mg/mL tryptone, 5 mg/mL yeast extract.
4. Ampicillin: Prepare as a 1-mg/mL stock in water and store at –20°C.
5. IPTG (isopropyl-1-thio-β-D-galactopyranoside) (Sigma, St. Louis, MO). Prepare as a 100-mM stock in water and store at –20°C.

6. SDS sample buffer (1X): 50 m*M* Tris-HCl, pH 6.8, 100 m*M* dithiothreitol (DTT), 5% glycerol, 1.7% SDS, 0.004% Bromphenol blue. Prepare as a 6X stock and store at −20°C.

7. Polyacrylamide gel electrophoresis (PAGE) gel running buffer: 25 m*M* Tris, 192 m*M* glycine, 0.1% SDS, pH 8.3 (available from Novex [San Diego, CA] as Tris–Glycine SDS running buffer, 10X stock).

8. Polyacrylamide gels (PAGE gels) (Novex): 10% gel—7.4% acrylamide, 2.6% bis-acryl, 25 m*M* Tris, 192 m*M* glycine, pH 8.6; 12% gel—9.4% acrylamide, 2.6% bis-acryl, 25 m*M* Tris, 192 m*M* glycine, pH 8.6.

9. GelCode Blue (Pierce, Rockford, IL).

10. NETN lysis buffer: 20 m*M* Tris-HCL, pH 8.0, 100 m*M* NaCl, 1 m*M* EDTA, 0.5% Nonidet P-40 (solution should be degassed and stored at 4°C). Just prior to use, the following inhibitors should be added: 5 m*M* benzamidine, 2 m*M* DTT, 4 μL/mL phenylmethylsulfonyl flouride (PMSF), 1 μL/mL aprotinin.

11. Lysozyme (Sigma). Prepare as a 20-mg/mL stock in water and store at −20°C.

12. Glutathione Sepharose 4B (Amersham Pharmacia Biotech).

13. GST elution buffer: 100 m*M* Tris-HCl, pH 8.0, 20 m*M* NaCl, 1 m*M* DTT, 0.1% Triton X-100, 20 m*M* glutathione (solution should be freshly prepared).

14. DMEM (Dulbecco's modified Eagle's medium) (1X), with 4.0 m*M* L-glutamine, with 4500 mg/mL glucose (Hyclone, Logan, UT).

15. Fetal bovine serum (FBS) (Hyclone).

16. L-Glutamine, 200 m*M* (Hyclone).

17. Penicillin-streptomycin solution, 10,000 U/mL each (Hyclone).

18. 1X Trypsin-EDTA solution (Mediatech): 0.05% trypsin, 0.53 m*M* EDTA-4Na in Hank's balanced salts solutions (HBSS) without Ca, Mg, or NaHCO$_3$.

19. TNF-α (PeproTech). Prepare as a 100 μg/mL stock in water and store at −80°C.

20. PBS (Dulbecco's phosphate-buffered saline) (1X) without calcium or magnesium (Hyclone).

21. 1X phosphatase inhibitors: 20 m*M* β-glycerophosphate, 10 m*M* NaF, 1 m*M* benzamidine, 0.5 m*M* EGTA, 0.3 m*M* Na$_3$VO$_4$. Prepare as a 20X stock in water and store at −20°C.

22. PNPP (*p*-nitrophenyl phosphate) (Sigma 104 phosphatase substrate). Prepare as a 0.5 *M* stock in water and store at −20°C.

23. Whole-cell extract (WCE) lysis buffer: 20 m*M* HEPES, pH 8.0, 0.5 *M* NaCl, 1 m*M* EDTA, 1 m*M* EGTA, 0.25% Triton X-100 (store at −20°C).

24. Protease inhibitor cocktail tablets (complete, EDTA-free, Boehringer Mannheim [Mannheim, Germany]). Prepare as a 100X stock in water and store at −20°C.

25. DTT (dithiothreitol) (Fisher Scientific, Pittsburgh, PA). Prepare as a 1 *M* stock in water and store at −20°C.

26. Anti-MKP-1 antibody, rabbit polyclonal (sc-1102) (Santa Cruz Biotechnology, Santa Cruz, CA).

27. Protein-A agarose (Calbiochem, La Jolla, CA).

28. Pull down (PD) buffer: 40 m*M* Tris-HCl, pH 8.0, 0.5 *M* NaCl, 6 m*M* EDTA, 6 m*M* EGTA, 0.1% Nonidet P-40 (store at 4°C).

29. RIPA buffer: 50 mM Tris-HCl pH 8.0, 150 mM NaCl, 1 mM EDTA, 0.4% deoxy-cholate, 1% Nonidet P-40, 10 mM β-glycerophosphate, 10 mM NaF, 10 mM PNPP, 0.3 mM Na$_3$VO$_4$, 1 mM benzamidine, 1 mM DTT, 2 μM PMSF, 10 μg/mL aprotinin, 1 μg/mL leupeptin, 1 μg/mL pepstatin.
30. PD-10 columns (Pharmacia Biotech).
31. FPLC running buffers: Q buffer: 20 mM Tris-HCl, pH 8.0, 0.2 mM EDTA, 0.2 mM EGTA, supplemented with 1X phosphatase inhibitors, 10 mM PNPP, 1 mM DTT. Buffer A (low salt): Q buffer + 50 mM NaCl, pH 8.0. Buffer B (high salt): Q buffer + 1 M NaCl, pH 8.0.
32. HiTrap Q anion exchange column, 1 mL (Pharmacia Biotech).
33. Kinase buffer: 20 mM HEPES, pH 7.7, 1 mM MgCl$_2$, 1 mM MnCl$_2$ (store at room temperature), supplemented with 1X phosphatase inhibitors, 20 mM PNPP, 1 mM DTT, and 0.5 mM benzamidine.
34. Reaction buffer: For each 20 μL of kinase buffer, supplement with 3.6 μCi of [γ^{32}P]ATP, 10 μM ATP, and 2 μg GST-IκBα (prepare just prior to use).
35. Centricon-30 concentrators (Amicon, Danvers, MA).
36. Colloidal Blue stain kit (Novex).

3. Methods
3.1. IκB Kinase Assay

The IκB kinase displays a high degree of specificity for serines 32 and 36 of IκBα. A mutant variant in which these residues were substituted with threonines shows markedly reduced phosphorylation and degradation in TNF-α-stimulated cells and interferes with endogenous NF-κB activation *(5)*. In an effort to enhance our ability to isolate the bona fide IκB kinase, we exploited this exquisite substrate specificity and developed a [^{32}P]-based in vitro kinase assay using glutatione-*S*-transferase (GST)-IκBα (1–54) WT and GST-IκBα (1–54) S32/36>T (serines 32 and 36 are substituted by threonines) as specific and control substrates, respectively. Phosphorylation of the GST-IκBα substrate is monitored by SDS-PAGE and subsequent autoradiography.

3.1.1. Production of GST IκBα Substrates
3.1.1.1. GROWTH AND INDUCTION OF BACTERIAL CELLS

1. Using an inoculating loop, transfer some of the GST-IκBα 1–54 glycerol culture to a 600-mL flask containing 100 mL of LB medium with ampicillin (50 μg/mL). Incubate the culture in an environmental shaker set at 250 rpm overnight at 37°C.
2. The following morning, add 30 mL of the overnight culture to a 2-L flask containing 500 mL of LB medium with ampicillin (50 μg/mL). Incubate the 500 mL culture on a shaker set at 250 rpm at 37°C until the optical density (OD)$_{550}$ is between 0.6 and 0.8 (*see* **Note 1**).
3. Remove a 1-mL aliquot from the culture and retain for subsequent SDS-PAGE analysis. Induce expression of the GST-IκBα 1–54 fusion protein by adding 2 mL

of 100 mM IPTG stock to the 500-mL culture (0.4 mM final). Incubate at 37°C for an additional 3 h (*see* **Note 2**).

4. Remove a 1-mL aliquot from the induced culture for SDS-PAGE analysis. Pour the culture into a large centrifuge bottle and centrifuge for 15 min at 5000g at 4°C. Remove the supernatant and freeze the cell pellet by placing on dry ice. Bacterial cell pellets can be stored at –80°C.

5. To test for protein expression, analyze the 1 mL aliquots from **steps 3** and **4** using SDS-PAGE. Gels are run using the method of Laemmli et al. *(17)*. In all experiments described in this chapter, the precast PAGE gel system from Novex was used, and the manufacturer's instructions were followed exactly (*see* **Note 3**). Centrifuge the 1-mL aliquots for 1 min at 3000g (microfuge), and remove the supernatant. Resuspend the cell pellet in 300 µL of 2X SDS sample buffer and heat at 95°C for 5 min. Pellet the cell debris by centrifuging the sample for 5 min at top speed (microfuge), and load approx 5–10 µL of the supernatant into separate lanes of a 12% SDS-PAGE gel, reserving one lane for prestained protein markers. GST IκBα 1–54 migrates as a 38-kDa protein and can be rapidly visualized by staining the gel with GelCode Blue (Pierce).

3.1.1.2. Affinity Purification of the Soluble GST-IκBα 1–54 Fusion Protein

Soluble GST-IκBα 1–54 is batch purified from cell lysate supernatant by affinity purification on glutathione–Sepharose 4B resin.

1. Quick thaw cell pellet at 37°C and place on ice. Resuspend cell pellet from 500-mL culture in 20 mL of ice-cold NETN lysis buffer, transfer to a 50-mL conical tube and place on ice.

2. Add 50 µL of a 20 mg/mL lysozyme stock per 20 mL of NETN lysis buffer (50 µg/mL final lysozyme concentration) and incubate on ice for 1 h.

3. With the sample tubes still on ice, sonicate the sample for 30-s intervals with the sonicator set at 3.5. Make sure that the sample does not become warm during the sonication. The sample will become lighter in color and slightly foamy when the lysis is complete (typically three to five rounds of sonication).

4. Transfer to a 30-mL plastic centrifuge tube and spin at 4°C for 10 min at approx 5000g.

5. Transfer supernatant to a 15-mL conical tube, and add 200 µL of packed GSH–Sepharose beads (*see* **Note 4**).

6. Rotate the sample at 4°C for at least 1 h, followed by centrifugation in a tabletop centrifuge at 1500g for approx 20 s at 4°C.

7. To wash the beads, pour off the supernatant, add 15 mL of cold NETN lysis buffer, invert the tube several times to resuspend the beads, and centrifuge in a tabletop centrifuge at 1500g for approx 20 s at 4°C. Perform this step three times.

8. To elute the GST IκBα 1–54 protein from the beads, add GST elution buffer to the sample at 2X the Sepharose bead volume and transfer the sample to microfuge tubes. Gently rotate the samples at 4°C for at least 20 min.

9. Spin the sample(s) briefly (10 s) at top speed in a microcentrifuge and transfer the eluate (supernatant) to a new microfuge tube(s). Be sure to save the supernatant, it contains the eluted GST IκBα 1–54 protein.

10. To quantitatively recover the GST IκBα 1–54 protein, subject the GSH–Sepharose beads to two additional elutions as described in **step 9**. Determine the protein concentration for each of the three eluates using Bradford protein analysis. Add glycerol to a final concentration of 10% to the protein samples and store at –80°C.

11. To establish the integrity and purity of the GST-IκBα 1–54 protein, analyze the samples by SDS-PAGE. Use 1–5 µg of protein for the analysis and add 6X SDS sample buffer to achieve a final concentration of 1X sample buffer. Heat the samples for 3–5 min at 95°C to denature and load the denatured samples into separate lanes of a 12% PAGE gel, reserving one lane for loading prestained markers (*see* **Note 5**). Run the gels for 15–20 min at 15 mA/gel, just enough to get the samples through the stacking gel, and then increase the current to 30 mA/gel and run until the dye front has reached the bottom of the gel (*see* **Note 6**). The protein bands are rapidly visualized by staining the gel with GelCode Blue.

3.2. Generation of HeLa WCE

3.2.1. Cell Culture for HeLa Cells

1. Maintain the HeLa cells in a growth medium of DMEM supplemented with 10% FBS, 2 mM L-glutamine, 100 U/mL penicillin, 100 µg/mL streptomycin at 5% CO_2 and 37°C in a humidified incubator. The HeLa cells are adherent and should be grown to confluency in a 150×25-mm dish (20 mL medium/dish).

2. When confluent, split the cells 1:3 for a 2-d growth period or 1:5 for a 3-d growth period.

3. To split the cells, remove the media and wash the dish once with 5 mL of warm PBS without calcium and magnesium.

4. Remove the PBS, add 4 mL of warm 1X trypsin–EDTA solution to the dish, and place the dish back in the incubator for 1–2 min.

5. Following incubation, the cells should detach from the plate surface (if necessary, try tapping gently on the side of the dish). Inactivate the trypsin with 4 mL of warm DMEM growth medium (containing the supplements listed above) and pipet the cells up and down several times to ensure that there are no cell aggregates.

6. Transfer the cells to a conical tube and spin at 200g for 5 min at room temperature. Remove the supernatant and resuspend the cells in DMEM growth medium by pipetting up and down four to six times to disrupt cell aggregates.

7. Aliquot the resuspended cells to new dishes for either a 1:3 or 1:5 split and bring the total volume to 20 mL per 150×25-mm dish with the supplemented DMEM growth medium.

3.2.2. Induction and Harvesting of HeLa Cells

1. Reduce the media volume to 10 mL/plate and stimulate the cells with 2 µL of a 100 µg/mL TNF-α stock (20 ng/mL final concentration) for 7 min at 37°C and 5% CO_2 in a humidified incubator (*see* **Note 7**).

2. Immediately following TNF-α induction, remove the media, put the plates on ice, and rinse the plates twice with 7 mL of ice-cold PBS containing 1X phosphatase inhibitors (*see* **Note 8**).
3. Scrape the cells from the dish with a plastic cell lifter (Costar, Cambridge, MA), transfer to pre-chilled conical tubes, and spin at 200 to 300*g* for 5 min at 4°C.
4. Decant and discard the supernatants, freeze the cell pellets on dry ice, and store these at –80°C.

3.2.3. Preparation of Whole-Cell Lysates

Throughout all remaining steps in the preparation of WCE as well as the subsequent purification steps, it is critical that the cells and cell lysate be kept cold at all times.

1. Quick-thaw the frozen pellet at 37°C and immediately place on ice (*see* **Note 9**). Resuspend the cell pellet with just enough ice-cold PBS supplemented with 1X phosphatase inhibitors to generate a thick slurry.
2. Add to the cells an amount of ice-cold WCE buffer supplemented with 1X phosphatase and protease inhibitors, and 1 m*M* DTT equal to three times the volume of the cell slurry. The WCE buffer should be added in 1 vol increments with immediate gentle mixing after each addition (*see* **Note 10**).
3. Rotate the extracts at 4°C for $^1/_2$ to 1 h, and spin the samples at 10,000*g* for 30 min at 4°C in the Sorvall centrifuge (*see* **Note 11**).
4. Transfer the supernatants to prechilled tubes and assay the protein content by the Bradford technique.

3.3. Immunoprecipitation of the IKK Signalsome

3.3.1. Immunoprecipitation of the IKK Complex with Anti-MKP1 Antibodies

1. Add anti-MKP1 antibody to the HeLa whole cell lysate at a ratio of 1 mg of antibody to 250 mg of lysate and rotate at 4°C for 2 h (*see* **Note 12**).
2. After 2 h, add Protein-A agarose to the sample at a ratio of 2.5 mL of packed bead volume per 1 mg of antibody and rotate the sample at 4°C for an additional 2 h (*see* **Note 13**).
3. Centrifuge the sample at 800*g* for 5 min at 4°C in a tabletop centrifuge to pack the agarose beads, and remove supernatant (*see* **Note 14**).

3.3.2. Washing of the IKK-Containing Immunocomplex

1. Perform the following series of washes, in order, with ice-cold buffers using 7–10 times the volume of the beads for each wash (*see* **Note 15**).
2. Wash the agarose beads four times with PD buffer, twice with RIPA buffer, twice with PD buffer, once with PD buffer + 3.0 *M* Urea, three times with PD buffer (*see* **Note 16**). For each wash, after the buffer is added, invert the tube several times to resuspend the beads; following this, spin the sample at 800*g* for 5 min at 4°C to pack the agarose beads, and pour off the wash buffer (*see* **Note 17**).

3.3.3. Specific Peptide Elution of IKK Complex from Agarose Beads

1. After washing, add an amount of PD buffer equal to approx $^1/_2$ the volume of the beads to make a thick slurry.
2. To elute the protein, add the MKP1 peptide to which the antibody was generated to the slurry, at a ratio of 5 mg of peptide to 1 mg of antibody. Rotate overnight at 4°C (*see* **Note 18**).
3. After rotating, spin the sample at 800g for 5 min at 4°C in a tabletop centrifuge, and transfer the supernatant to a new tube.

3.4. Anion Exchange Chromatography of Eluted IKK Signalsome

3.4.1. Sample Preparation for Mono Q Column

1. The eluted sample should be run over a PD-10 column to remove residual salt and establish the protein sample in an appropriate buffer for fast protein liquid chromatography (FPLC) purification.
2. Remove the top cap from a PD-10 column, pour off the excess liquid, and then remove the bottom cap.
3. Support the column over a receptacle large enough to hold at least 30 mL, and equilibrate the column with 25 mL of ice-cold buffer A.
4. After the equilibration buffer has completely run into the column, add the sample in a volume of 2.5 mL. If there are more than 2.5 mL of sample, use multiple columns. If there are less than 2.5 mL of sample, bring the volume up to 2.5 mL with buffer A. There is no need to collect the eluate from this step.
5. Add 3.5 mL of ice-cold buffer A to the column and collect the eluate in an appropriate sized container that is packed in ice. The eluate contains the IKK complex.

3.4.2. Mono Q Column Chromatography

The FPLC is set up in a cold box and all buffers have been prechilled to 4°C.

1. Set the detector wavelength at 280 nm and the flow rate at 1.0 mL/min.
2. Condition a new column by the following: Run through 5–10 mL of 100% buffer A, ramp up to 100% buffer B over 3–5 mL, maintain at buffer B for 5–10 mL, ramp back down to 100% buffer A over 3–5 mL, and maintain at 100% buffer A for 5–10 mL. This is done to remove the column storage buffer and any bound material that might elute off of the column with the running buffers.
3. Set up the gradient for the run as follows:
   - Equilibrate column with 1 mL of 100% buffer A before loading.
   - Load the sample onto the column followed by a column wash with 5 mL of 100% buffer A (final [NaCl] = 50 mM).
   - Ramp up to 80% buffer B over 20 mL (final [NaCl] = 0.81 M).
   - Ramp up to 100% buffer B over 3 mL (final [NaCl] = 1 M).
   - Hold at 100% buffer B for 5 mL (final [NaCl] = 1 M).
   - Ramp back down to 100% buffer A over 2 mL.
   - Hold at 100% buffer A for 3 mL (final [NaCl] = 50 mM).

4. During the sample loading and the 5-mL column wash, collect 3-mL fractions. During the gradients, collect 0.5-mL fractions (*see* **Note 19**).

3.4.3. Kinase Assay and SDS-Page Analysis of Mono Q Fractions (see **Note 20**)

1. For each sample, prepare 20 μL of reaction buffer.
2. Mix 10 μL of each of the odd numbered fractions from the FPLC run with 20 μL of the reaction buffer in separate wells of a 96-well plate.
3. Incubate the reactions at 30°C for 30–60 min and add 6 μL of 6X SDS sample buffer to a final concentration of 1X SDS buffer.
4. Denature the samples for 5 min in a heat block designed to hold a 96-well plate set at 95°C (*see* **Note 21**).
5. Load the reactions into separate lanes of a 12% PAGE gel(s), reserving one lane for a prestained protein marker. Run the gel(s) in 1X PAGE running buffer at 15 mA/gel until the samples migrate through the stacking gel. Increase the current to 30 mA/gel for approx 1 h or until the dye front from the sample buffer runs to the bottom of the gel (*see* **Note 22**).
6. Cut away the stacking section of the gel and the bottom strip containing the dye front (*see* **Note 23**). Rinse the gel briefly in deionized (DI) water, and fix the gel for 30 min in 10% acetic acid and 10% methanol (*see* **Note 24**). Again, rinse the gel briefly in DI water, place on 3-MM Whatman paper, cover with plastic wrap, and dry on a gel dryer for 1 h at 80°C. Expose the dried gels to film for autoradiographic analysis (**Fig. 3**).
7. Pool the Mono Q fractions that display IκBα kinase activity, and concentrate the pooled fractions for subsequent SDS-PAGE band isolation of the desired protein (*see* **Subheading 3.5.**).

3.5. Sample Preparation for Submission of Protein for Microsequencing

3.5.1. Concentration of IKK-Containing Samples

By centrifuging a sample through a centricon column, proteins below a specific cutoff size can pass through the column membrane, while larger proteins are retained. The membrane also allows the protein solvent to pass through but prevents the sample from being centrifuged to dryness. Centricon-30 columns have a cutoff of 30 kDa.

1. Load up to 2 mL of sample onto a Centricon-30 column and spin in a fixed angle rotor at 4500*g* for 30 min. at 4°C.
2. If the sample is greater than 2 mL, after the first spin, add more sample to the same column (not exceeding 2 mL) and spin again at 4500*g* for an additional 30 min at 4°C (*see* **Note 25**).
3. If necessary, continue to add sample to the column (not exceeding 2 mL/load) and centrifuge as before until the sample is concentrated into a final volume of 50–100 μL. This is necessary in order to load the entire sample into one or two wells of a PAGE gel.

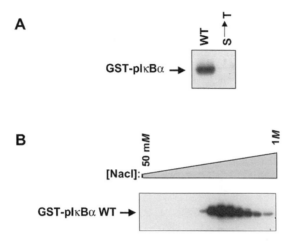

Fig. 3. IκBα kinase assay. (**A**) The specificity of the immunoprecipitated IκB kinase is shown by its ability to phosphorylate the wild ype GST-IκBα 1–54 substrate (WT), but not the mutant GST-IκBα 1-54 S32/36 >T substrate (S > T). IκB kinase activity was immunoprecipitated with anti-MKP1 antibodies from WCE of HeLa cells stimulated with TNFα and subsequently eluted with specific peptide. (**B**) The IκB kinase activity from the Mono Q fractions was monitored as described in **Subheading 3.4.3.** The major peak of kinase activity elutes from the Mono Q column at approx 0.4 *M* NaCl.

3.5.2. Preparative SDS-PAGE Analysis

1. In a microfuge tube, mix the concentrated samples with 6X SDS sample buffer to obtain a final concentration of 1X SDS sample buffer.
2. Denature the samples in a 95°C heat block for 3–5 min and load into separate lanes of a 10% PAGE gel(s), reserving one lane for a prestained protein marker.
3. Run the gel in 1X PAGE gel running buffer at 15 mA until the samples migrate through the stacking gel. Increase the current to 30 mA/gel for approx 1 h or until the dye front from the sample buffer runs to the bottom of the gel.

3.5.3. Colloidal Stain and Band Isolation

1. After running the samples, remove the stacking portion of the gel along with bottom strip of the separating gel containing any of the dye front.
2. Rinse the gel briefly with DI water, and stain following the protocol for the Colloidal Blue stain kit from Novex.
3. After staining for the minimum amount of time to visualize the protein bands, destain the gel in water for the minimum amount of time required to reduce background staining (**Fig. 4**).
4. Cut the desired band from the gel with a clean razor blade or scalpel, cutting as close to the band as possible to avoid contamination with proteins migrating nearby (*see* **Note 26**).

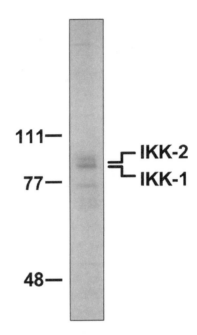

Fig. 4. Colloidal Blue stain of the purified IKK signalsome. The purified IKK complex was separated on a 10% PAGE gel. The bands were visualized by Colloidal Blue staining of the gel. The bands were excised from the gel and analyzed by nanospray mass spectroscopy. The bands corresponding to the IKK1 and IKK2 kinases are indicated.

5. Place the excised band in a microfuge tube, add 1 mL of 50% acetonitrile, 50% purified water and rotate at room temperature for approx 15 min.
6. Remove the liquid with a pipet, add another 1 mL of 50% acetonitrile and 50% purified water, and rotate 15 min at room temperature.
7. Remove the liquid and freeze the gel slice at –80°C

4. Notes

1. At 37°C, the cultures will typically reach the 0.6–0.8 OD_{550} range in 3–4 h. Alternatively, the cultures can be grown overnight at 30°C, which should prevent them from growing past the log phase density.
2. 3 h was determined to be the optimal induction period for preparation of GST IκBα 1–54 proteins. Shorter timepoints will decrease the yield. Longer time points (up to 5 h) have been used without a significant decrease in yield.
3. Care should be taken with SDS-PAGE gels in that they may still contain acrylamide and bisacrylamide that has not polymerized, both of which are neurotoxins.
4. The glutathione–Sepharose 4B beads are purchased as a slurry, but the beads must be pretreated by removing the suspending liquid and washing with PBS

before use. Gently shake the bottle of glutathione–Sepharose 4B to resuspend the matrix and transfer an appropriate amount of the matrix to a 15-mL conical tube (1.33 mL of the slurry contains approx 1 mL of packed beads). Centrifuge at lowest setting on a tabletop centrifuge (approx 100g) for 5 min. Pour off the supernatant and add 10 mL of cold PBS per 1 mL of packed beads, invert the tube several times to mix, and centrifuge as before for 3 min (if more than 1 mL of beads are being washed, it may be necessary to do more than one wash, for example, 2 mL of beads, two washes with 10 mL PBS each). Add 1 mL of cold PBS per 1 mL of beads to make a 50% slurry. The washed beads can be stored this way for 1 mo at 4°C.

5. A heat block or heated water bath may be used.

6. The run time at 30 mA should be determined by the size of the proteins to be separated, for GST-IκBα it was necessary to keep proteins of >30 kDa on the gel.

7. Reducing the volume to 10 mL helps conserve on the amount of cytokine used. This is particularly important for large-scale experiments.

8. TNF-α-stimulated activation of the IκBα kinase is very transient, peak activity is achieved within 7 min and returns to near-basal levels by 20 min. Therefore, it is essential that the cells be brought to 4°C immediately upon 7 min postinduction, so as to capture the IκBα kinase in its highest activity state. The inclusion of phosphatase inhibitors in the ice-cold PBS buffer helps prevent inactivation of the kinase activity.

9. Be extremely careful not to allow the cell pellet to become warm during the thawing step, this will result in the downregulation of the IκBα kinase activity.

10. Addition of all of the WCE buffer at one time can lead to a local high concentration of salt/detergent resulting in nuclear lysis, thereby increasing the viscocity of the extract and generating conditions for extremely low recovery of the protein.

11. The presence of PNPP causes the extract to turn a yellow-orange color.

12. The anti-MKP1 antibody is available conjugated to agarose beads; however, we found that allowing the nonconjugated antibody to interact with the protein first, followed by binding to protein-A agarose, enables dramatically higher recovery of the IκBα kinase activity. Most likely, this is the result of steric hindrance imposed by such a large complex (> 700 kDa) interacting with an antibody crosslinked to agarose beads.

13. The protein-A agarose beads are purchased as a slurry, but the liquid must be removed and the beads washed with PD buffer before being used. Swirl the bottle containing the protein A agarose several times to resuspend the beads and transfer an aliquot to a 15-mL conical tube (1 mL of the slurry contains approx 0.4 mL of packed beads). In a tabletop centrifuge, spin the beads at 1000g for 1 min. Remove the liquid and add PD buffer with 1X phosphatase inhibitors at a ratio of 2 mL of buffer to 1 mL of packed beads. Invert the tube several times to resuspend the beads, centrifuge as before, and remove and discard the supernatant. Repeat the PD buffer wash two additional times. Resuspend the beads using an aliquot of the cell lysate in order to add the protein-A agarose to the sample.

14. The supernatant should be made 10% glycerol and stored at –80°C. The WCE's can be used to isolate other protein complexes, and so forth.

15. It is important to use a large excess of buffer for the washings. If the volume of beads prevents this, the sample should be split.
16. It is critical that the IKK-containing immune complex not be left in the 3 *M* urea buffer for too long, as this will inactivate the kinase activity. Typically, the immunobead–3 *M* urea slurry is immediately placed in the tabletop centrifuge, pelleted as described, and 7–10 vol of ice-cold PD buffer added quickly to the pellet.
17. During the final wash, the beads can be transferred to a new tube to prevent eluting any contaminating proteins that may have adhered to the plastic.
18. We observe near quantitative recovery of the IκBα kinase activity with overnight elution, however, we have not tested whether shorter elution times would achieve the same results.
19. The major peak of IKK activity elutes at around 0.4 *M* NaCl. We sometimes observe a minor amount of IKK activity in the flow-through fractions.
20. One must follow standard laboratory radiation safety procedures when working with [^{32}P].
21. Alternatively, the samples can be denatured using a shallow water bath at 80°C for 5–10 min.
22. Caution should be exercised, as the running buffer and the gels will be radioactive. Appropriate disposal procedures should be followed for radioactive liquids and solids.
23. The bottom portion of the gel is extremely radioactive as it contains the major band of unincorporated [^{32}P]ATP that migrates just below the dye front. It is important to remove this portion of the gel so that the signal from the unincorporated [^{32}P] does not interfere with the signal from the protein bands.
24. This step helps to reduce the background signal by further removal of unincorporated [^{32}P].
25. The protein concentration of the IKK-containing sample is extremely low and is therefore sensitive to protein loss via nonspecific binding to the plastic and membrane components of the column. To minimize the loss of protein, we pass the entire sample (up to 12 mL) through the same Centricon column. This takes a considerable amount of time; however, when the goal is to isolate enough protein for microsequencing, we find that it is worth the effort.
26. Try to cut as close as possible to the protein band, as excess gel reduces recovery of the protein for microsequencing.

References

1. Sudol, M. (1998) From Src homology domains to other signaling modules: proposal of the "protein recognition code." *Oncogene* **17**, 1469–1474.
2. Baldwin, A. S. (1996) The NF-κB and IκB proteins: New discoveries and insights. *Annu. Rev. Immunol.* **14**, 649–681.
3. May, M. J. and Ghosh, S. (1998) Signal transduction through NF-κB. *Immunology Today* **19**, 80–88.

4. Brown, K., Gerstberger, S., Carlson, L., Franzoso, G., and Siebenlist, U. (1995) Control of I kappa B-alpha proteolysis by site-specific, signal-induced phosphorylation. *Science* **267**, 1485–1488.

5. DiDonato, J., Mercurio, F., Rosette C., Wu-Li, J., Suyang H., Ghosh, S., and Karin, M. (1996) Mapping of the inducible IkappaB phosphorylation sites that signal its ubiquitination and degradation. *Mol. Cell. Biol.* **16**, 1295–1304.

6. Mercurio F., Zhu H., Murray, B. W., Shevchenko, A., Bennett, B. L., Li, J., Young, D. B., Barbosa, M., Mann, M., Manning, A. M., and Rao, A. (1997) IKK-1 and IKK-2: Cytokine-activated IκB kinases essential for NF-κB activation. *Science* **278**, 860–866.

7. Lee, F. S., Hagler, J., Chen, Z. J., and Maniatis, T. (1997) Activation of the IκBα complex by MEKK1, a kinase of the JNK pathway. *Cell* **88**, 213–222.

8. Nakano, H., Shindo, M., Sakon, S., Nishinaka, S., Mihara, M., Yagita, H., and Okumura, K. (1998) Differential regulation of IκB kinase α and β by two upstream kinases, NF-κB-inducing kinase and mitogen-activated protein kinase/ERK kinase kinase-1. *Proc. Natl. Acad. Sci. USA* **95**, 3537–3542.

9. Malinin, N. L., Boldin, M. P., Kovalenko, A. V., and Wallach, D. (1997) MAP3K-related kinase involved in NF-kappaB induction by TNF, CD95 and IL-1. *Nature* **385**, 540–544.

10. DiDonato, J. A., Hayakawa, M., Rothwarf, D. M., Zandi, E., and Karin, M. (1997) A cytokine-responsive IκB kinase that activates the transcription factor NF-κB *Nature* **388**, 853–862.

11. Zandi, E., Rothwarf, D. M., Delhasse, M., Hayakawa, M., and Karin, M. (1997) The IkB kinase complex (IKK) contains two kinase subunits, IKKα and IKKβ, necessary for IκB phosphorylation and NF-κB activation. *Cell* **91**, 243–252.

12. Mercurio, F., Murray, B., Bennett, B. L., Pascual, G., Shevchenko, A., Zhu, H., Young, D. B., Li, J., Mann, M., and Manning, M. (1999) IKKAP-1, a novel regulator of NF-κB activation, reveals heterogeneity in IκB complexes. *Mol. Cell. Biol.* **19**, 1526–1538.

13. O'Connell, M. A., Bennett, B. L., Mercurio, F., Manning, A., and Mackman, N. (1998) Role of IKK1 and IKK2 in lipopolysaccharide signaling in human monocytic cells. *J. Biol. Chem.* **273**, 30,410–30,414.

14. Rothwarf, D. M., Zandi, E., Natoli, G., and Karin, M. (1998) IKK-gamma is an essential regulatory subunit of the IkappaB kinase complex. *Nature* **395**, 297–300.

15. Yaron, A., Hatzubai, A., Davis, M., Lavon, I., Amit, S., Manning, A., Andersen, J., Mann, M., Mercurio, F., and Ben-Neriah, Y. (1998) Identification of the receptor component of the IκBα-ubiquitin ligase. *Nature* **396**, 590–594.

16. Yaron, A., Alkalay, I., Gonen, H., Hatzubai, A., Jung, S., and Beyth, S. (1997) Inhibition of NF-κB cellular function via specific targeting of the IκB ubiquitin ligase. *EMBO J.* **16**, 101–107.

17. Laemmli, U.K. (1970) Cleavage of structural proteins during the assembly of the head of bacteriophage T4. *Nature* **227**, 680–695.

11

Methods to Assay Stress-Activated Protein Kinases

Ana Cuenda

1. Introduction

Mitogen-activated protein (MAP) kinases play an essential role in regulating diverse cellular processes in mammalian cells in response to many extracellular signals (*see also* Chapter 12, this volume).

Five MAP kinase family members have been identified so far in mammalian cells that are activated by cellular stresses (e.g., chemicals, heat and osmotic shock, ultraviolet radiation and inhibitors of protein synthesis), by bacterial lipopolysaccharide and by the cytokines interleukin-1 (IL-1) and tumor necrosis factor (TNF), and have therefore been termed stress-activated protein kinases (SAPKs) (*1*). Isoforms of SAPK1/JNK (*2*), SAPK2a/p38 (*3*), SAPK2b/p38β (*4*); SAPK3 (also termed ERK6 and p38γ) (*5–8*), SAPK4 (also called p38δ) (*1,9*), and ERK5/BMK1 (*10,11*), phosphorylate various cellular proteins, including other protein kinases and a number of transcription factors. MAP kinase-activated protein kinase-2 (MAPKAP kinase-2), MAPKAP kinase-3 and PRAK/MAPKAP kinase-5 are three of the kinases that are phosphorylated and activated activated physiologically by SAPK2a/p38 and SAPK2b/p38β (SAPK2/p38). Physiological substrates of MAPKAP kinase-2 and -3 include the heat shock protein 27/25 (HSP27/25) (*1; see also* Chapter 26, this volume). Recently, two new SAPK2/p38 substrates have been described, the MAPK-interacting kinases (Mnk1, Mnk2) and the Mitogen- and stress-activated protein kinase-1 (Msk1), these can also be activated in vivo by MAPK (*1,12*) (**Fig. 1**). SAPKs are activated by dual specificity kinases bearing homology to MAP kinase kinase that are termed SAP kinase kinases (SKKs). These enzymes phosphorylate both threonine and tyrosine residues in TPY (SAPK1/JNK), TGY (SAPK2a/p38, SAPK2b/p38β, SAPK3, and SAPK4), and TEY (ERK5/BMK1) sequence motifs, in the activation domain (*2,6,9–11*).

From: *Methods in Molecular Biology, vol. 99: Stress Response: Methods and Protocols*
Edited by: S. M. Keyse © Humana Press Inc., Totowa, NJ

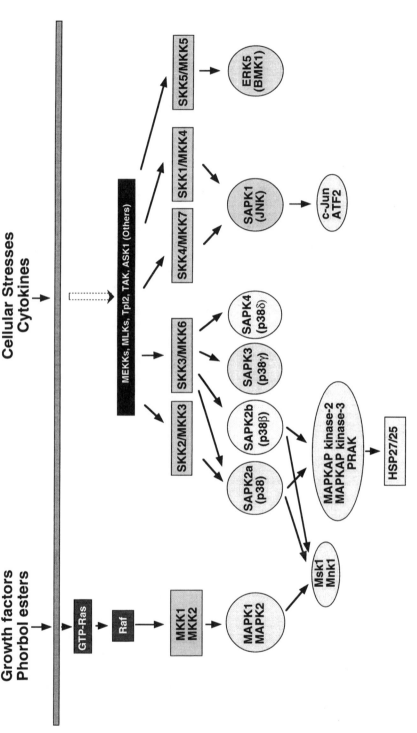

Fig. 1. Schematic representation of mammalian SAP kinase signal transduction pathways. Following stimulation by cellular stress or cytokines, the SKKs activate the SAPK group of kinases. These SAPKs phosphorylate a number of substrates including other kinases and transcription factors.

Several SKKs have been cloned and identified in mammalian cells *(1)*. In vitro, SKK1/MKK4 *(13–15)* and SKK4/MKK7 *(16–18)* phosphorylate, and activate SAPK1/JNK. SKK2/MKK3 *(13)* and SKK3/MKK6 *(19–22)* phosphorylate and activate SAPK2a/p38, SAPK2b/p38β, SAPK3, and SAPK4, but not SAPK1 (**Fig. 1**). The identity of the protein kinases, which activate these different SKKs in vivo, is unclear. More than ten enzymes capable of activating these SKKs in vitro and/or in cotransfection experiments have been identified so far. This includes members of the MEKK and MLK family as well as the TAK, Tpl2, and ASK-1 protein kinases *(23)*.

The ability to measure activation of specific protein kinases in response to a variety of extracellular stimuli is vital to gaining an understanding of the physiological role of signaling through this complex network of pathways in mammalian cells. In this chapter, are presented a series of methods for the assay of SAPKs, their upstream activators (SKKs) and certain of the protein kinases that are phosphorylated and activated by SAPKs in mammalian cells. Wherever possible, these assays are designed in such a way as to be specific, sensitive and quantitative.

2. Materials

1. Buffer A: 50 mM Tris-HCl, pH 7.5, 0.27 M sucrose, 1 mM EDTA, 1 mM EGTA, 1 mM orthovanadate, 10 mM sodium β-glycerophosphate, 50 mM sodium fluoride, 5 mM sodium pyrophosphate, 1% (v/v) Triton X-100, 0.1% (v/v) 2-mercaptoethanol, 1 mM benzamidine, 0.2 mM phenylmethylsulfonyl fluoride (PMSF) and leupeptin (5 μg/mL).
2. Buffer B: 20 mM Tris-HCl, pH 7.5, 1 mM EDTA, 0.01% (v/v) Brij-35, 5% (v/v) glycerol, 0.1% (v/v) 2-mercaptoethanol, and 0.2 mg/mL bovine serum albumin (BSA).
3. PKI (a 20-residue peptide [TTYADFIASGRTGRRNAIHD] that is a specific inhibitor of cyclic AMP-dependent protein kinase) can be purchased from Sigma (St. Louis, MO).
4. 50 mM magnesium acetate; 0.5 mM [γ-^{32}P]ATP (2×10^5 cpm/nmol diluted from a 10 μCi/mL stock from Amersham [Arlington Heights, IL]).

2.1. Coupling Antibodies to Protein-G–Sepharose for Immunoprecipitation of Kinases

1. Protein-G–Sepharose from Pharmacia Biotech.
2. Shaking platform (IKA) from Merck (Rahway, NJ).

2.2. Cell Stimulation and Cell Lysis

TNF and anisomycin can be purchased from Sigma and IL-1 from Boehringer Mannheim (Mannheim, Germany).

2.3. MAPKAP Kinase-2 and MAPKAP Kinase-3 Assays

1. MAPKAP kinase reaction buffer: 100 mM sodium β-glycerophosphate, pH 7.5, 0.2 mM EDTA, 5 μM PKI and 60 μM peptide KKLNRTLSVA (which is the standard substrate for MAPKAP kinases, from Upstate Biotechnology; *see* **Notes 1** and **2**).
2. Anti-MAPKAP kinase-2 and anti-MAPKAP kinase-3 polyclonal antibodies from Upstate Biotechnology. These antibodies have been previously characterized and shown to be specific for MAPKAP kinase-2 and MAPKAP kinase-3 in immuno-precipitations *(24)*.
3. Active MAPKAP kinase-2 that can be used as a control in these experiments can be purchased from Upstate Biotechnology.
4. Incubators for assaying immunoprecipitated kinases from cell extracts (Stuart Scientific) from Merck.
5. Phosphocellulose P81 paper (Whatman, Cligton, NJ).
6. Phosphoric acid, 75 mM.
7. Acetone.

2.4. Msk1 Assays

1. Msk1 reaction buffer: 50 mM Tris-HCl pH 7.5, 0.2 mM EGTA, 0.2% (v/v) 2 mercaptoethanol, 5 μM PKI, and 60 μM Crosstide (GRPRTSSFAEG) from Upstate Biotechnology.
2. Anti-Msk1 antibodies, active and inactive recombinant Msk1 from Upstate Biotechnology.

2.5. Assaying SAPK2a/p38, SAPK2b/p38β, SAPK3, SAPK4, and ERK5/BMK1

1. SAPK reaction buffer: 50 mM Tris-HCl, pH 7.5, 0.2 mM EGTA, 0.2 mM sodium orthovanadate, 2 μM PKI, and 0.66 mg/mL MBP (from Gibco-BRL, Gaithersburg, MD).
2. Anti-SAPK2a/p38, anti-SAPK3, and anti-SAPK4 polyclonal antibodies from Upstate Biotechnology. Anti-ERK5/BMK1 antibody from Calbiochem (La Jolla, CA) or Santa Cruz Biotechnology (Santa Cruz, CA). Some of these antibodies, including anti-SAPK1/JNK and SAPK2b/p38β, are also available from New England Biolabs (Beverly, MA).
3. Inactive GST-MAPKAP kinase-2 from Upstate Biotechnology.
4. Magnesium acetate 60 mM; cold ATP 0.6 mM.

2.6. SAPK1/JNK Activity Assays

1. SAPK1/JNK reaction buffer: 50 mM Tris-HCl, pH 7.5, 0.2 mM EGTA, 2 mM sodium orthovanadate, 5 μM PKI, 0.2% (v/v) 2-mercaptoethanol, and 0.4 mg/mL GST-ATF2 or 0.4 mg/mL GST c-Jun (which are standard substrates for SAPK1/JNK, from Upstate Biotechnology, Santa Cruz Biotechnology, or New England Biolabs) *(25,26)*.

2. SDS-PAGE sample buffer: 6% (w/v) SDS, 400 mM Tris-HCl, pH 6.8, 50% (v/v) glycerol, 5% (v/v) 2-mercaptoethanol, and 0.2% (v/v) Bromophenol Blue.
3. SDS-PAGE 10% running gel mix: 0.38 M Tris-HCl, pH 8.8, 0.1% SDS, 10% acrylamide, 0.3% bisacrylamide, 0.1% (w/v) ammonium persulfate (APS). Polymerize by adding 0.1% (v/v) N',N',N',N'-tetramethylendiamine (TEMED).
4. SDS-PAGE stacking gel mix: 0.125 M Tris-HCl, pH 6.8, 0.1% SDS , 3.9% acrylamide, 0.1% bisacrylamide, 0.05% APS. Polymerize by adding 0.1% TEMED.
5. Stain: Coomasie Blue 0.2% in 50% methanol and 10% acetic acid.
6. Destain: 50% methanol and 10% acetic acid.
7. Glutathione-S-Sepharose (GSH-Sepharose) from Pharmacia Biotech.
8. Washing buffer: 20 mM Tris-HCl, pH 7.5, 50 mM NaCl, 2.5 mM magnesium acetate, 0.1 mM EGTA, 0.05% (v/v) Triton X-100, and 0.1% (v/v) 2-mercaptoethanol.
9. Buffer C: 25 mM Tris-HCl, 20 mM β-glycerophosphate, pH 7.4, 10 mM magnesium acetate, 0.1 mM EGTA, and 1 mM sodium orthovanadate.

2.7. Other SAPK Assays

1. Anti-phosphotyrosine, anti-phospho-p38 and anti-phospho-JNK antibodies can be purchased from New England Biolabs.
2. 2-propanol 20% (v/v), 50 mM HEPES, pH 7.6.
3. 50 mM HEPES, pH 7.6, 5 mM 2-mercaptoethanol.
4. 50 mM HEPES, pH 7.6, 5 mM 2-mercaptoethanol, 6 M urea.
5. Buffer D: 50 mM HEPES, pH 7.6, 5 mM 2-mercaptoethanol, 0.05% (v/v) Tween-20.
6. Buffer E: 25 mM Tris-HCl, pH 7.5, 0.1 mM EGTA, 1 mM sodium orthovanadate, 1 μM PKI, 0.1% (v/v) 2-mercaptoethanol; and 10 mM magnesium acetate; 0.1 mM [γ-^{32}P]ATP (2×10^5 cpm/nmol).
7. Trichloroacetic acid (TCA) 5% containing 1% sodium pyrophosphate.
8. PAGE 10% running gel mix (minus SDS): 0.38 M Tris-HCl, pH 8.8, 10% (w/v) acrylamide, 0.166% (w/v) bis-acrylamide and 0.05% (w/v) APS. Initiate polymerization by adding 0.05% (v/v) TEMED.
9. PAGE stacking gel mix (minus SDS): 0.25 M Tris-HCl, pH 6.7, 4.8% (w/v) acrylamide, 0.2% (w/v) bis-acrylamide, and 0.1% (w/v) APS. Initiate polymerization by adding 0.1% (v/v) TEMED.
10. Active SAPKs, which can be used as a control in these experiments, can be purchased from Upstate Biotechnology.

2.8. SKK/MKK Assays

1. Inactive GST-SAPK2/p38 and inactive SAPK1/JNK from Upstate Biotechnology.
2. 40 mM magnesium acetate; 0.4 mM cold ATP.
3. Buffer F: 50 mM Tris-HCl, pH 7.5, 0.1 mM EGTA, 0.03% (v/v) Brij-35, 0.1% (v/v) 2-mercaptoethanol, and 5% (v/v) glycerol.
4. All anti-SKK/MKKs polyclonal antibodies from Upstate Biotechnology. These antibodies have been previously characterized and shown to be specific in immunoprecipitation and immunoblots *(16,19)*.
5. Active SKKs, which can be used as controls in these experiments, can be purchased from Upstate Biotechnology.

3. Methods

3.1. Coupling Antibodies to Protein-G–Sepharose for Immunoprecipitation of Kinases

1. Equilibrate the Protein-G–Sepharose with buffer A. Prepare a slurry that is 50% Protein-G–Sepharose beads and 50% buffer.
2. Incubate 10 μL of the Protein-G Sepharose slurry with the required amount of antibody (*see below*), for 30 min at 4°C on a shaking platform.
3. Wash four times with 10 vol excess of buffer A, to remove uncoupled antibodies that could interfere with the assay.
4. A stock of antibody coupled to Protein-G–Sepharose beads, which can be used for a large number of assays, may be prepared and stored for up to 1 mo at 4°C.

3.2. Cell Stimulation and Cell Lysis

1. Cells are grown as usual on a 6- or 10-cm dish to near confluency. It is recommended that cells are not serum starved overnight before stimulation because this may cause some activation of the stress pathways in certain cell lines.
2. In order to activate stress pathways, cells can be exposed for up to 1 h to some of the following stimuli: 0.5 mM sodium arsenate; 0.5 M sorbitol (for osmotic shock 0.5–0.7 M NaCl can also be used); 50 ng/mL to 10 μg/mL anisomycin; for ultraviolet (UV) irradiation use 60–200 J/m^2 and then incubate cells at 37°C (do not forget to remove the medium before irradiation, as this will absorb the UV radiation, and add the medium back after the stimulation); 1 mM H$_2$O$_2$; 100 ng/mL TNF, and 20 ng/mL IL-1.
3. Following the required stimulation, remove the medium from the dishes and place cells on ice.
4. Prepare cell extracts by adding 1.0 mL (for a 10-cm dish) or 0.5 mL (for a 6-cm dish) of ice-cold buffer A to the cells and resuspend the solubilized cells using a plastic scraper. Transfer the lysate to an Eppendorf tube on ice.
5. Centrifuge the lysates for 5 min at 14,000g at 4°C, take the supernatant and determine the protein concentration.
6. At this stage, the lysate can be snap-frozen in liquid nitrogen and stored at –80°C without significant loss of activity.

3.3. MAPKAP Kinase-2 and MAPKAP Kinase-3 Assays

3.3.1. Assaying Recombinant or Purified MAPKAP Kinase-2 and MAPKAP Kinase-3

1. Dilute active MAPKAP kinase-2 or MAPKAP kinase-3 in cold-ice buffer B, to a concentration at which the enzyme activity will be linear (*see* **Subheading 3.9.1.**).
2. Incubate 5 μL diluted MAPKAP kinase-2 or MAPKAP kinase-3 with a mixture containing 10 μL distilled water and 25 μL of MAPKAP kinase reaction buffer (containing the substrate) for 3 min at 30°C.
3. Start the reaction by the addition of 10 μL of 50 mM magnesium acetate; 0.5 mM [γ-^{32}P]ATP (approx 2 × 10^5 cpm/nmol).

4. After 10 min at 30°C, stop the reaction by pipetting 40 µL of the assay mixture on to a 2 cm × 2 cm square of phosphocellulose P81 paper (P81; Whatman), this binds the MAPKAP kinase peptide substrate, but not [γ-^{32}P]ATP.
5. Immerse the P81 papers in a beaker containing 75 mM phosphoric acid (5–10 mL/paper), and stir for approx 2 min.
6. Wash the papers five times in 75 mM phosphoric acid to remove the ATP (approx 2 min for each wash).
7. Wash once in acetone to remove the phosphoric acid.
8. Dry the P81 papers (using a hair drier) and place each one into a 1.5-mL plastic microcentrifuge tube.
9. Add 1.0 mL of scintillation fluid per tube and analyze the amount of [^{32}P] incorporated into the MAPKAP kinase peptide substrate by scintillation counting.

3.3.2. Immunoprecipitation and Assay of MAPKAP Kinase-2 and MAPKAP Kinase-3 from Cell Extracts

1. Prepare cell extracts as described in **Subheading 3.2.**
2. Mix 50 µg cell extract protein with 5 µL protein G-Sepharose conjugated to 3 µg of anti-MAPKAP kinase-2 or 10 µg of anti-MAPKAP kinase-3 antibody (as described in **Subheading 3.1.**) and incubate for 90 min at 4°C on a shaking platform.
3. Centrifuge the suspension for 1 min at 14,000g, discard the supernatant and wash the pellet twice with 1.0 ml of buffer A containing 0.5 M NaCl, and twice with buffer A.
4. Assay MAPKAP kinase activity as in **Subheading 3.3.1.**, except that the reaction is carried out on a shaking platform at 30°C to keep the immunoprecipitated MAPKAP kinase/protein-G–Sepharose complex in suspension.

3.4. Msk1 Assay

3.4.1. Assaying Recombinant or Purified Msk1

1. Dilute active Msk1 in cold-ice buffer B, to a concentration at which the enzyme activity will be linear (*see* **Subheading 3.9.1.**).
2. Incubate 5 µL diluted Msk1 with a mixture containing 10 µL distilled water and 25 µL of Msk1 reaction buffer (containing the substrate) for 3 min at 30°C.
3. Start the reaction by the addition of 10 µl of 50 mM magnesium acetate; 0.5 mM [γ-^{32}P]ATP (approx 2 × 10^5 cpm/nmol).
4. After 15 min at 30°C stop the reaction by removing 40 µL of the assay mixture and spotting it on to P81 paper, then follow the procedure exactly as described in **Subheading 3.3.1.**

3.4.2. Immunoprecipitation and Assay of Msk1 From Cell Extracts

1. Prepare cell extracts exactly as described in **Subheading 3.2.**
2. Mix a volume of cell extract containing 500 µg protein with 5 µL protein-G-Sepharose conjugated to 5 µg of anti-Msk1 antibody (as described in **Subheading 3.1.**) and incubate for 90 min at 4°C on a shaking platform.
3. Centrifuge the suspension for 1 min at 14,000g, discard the supernatant and wash the pellet twice with 1.0 mL of buffer A containing 0.5 M NaCl, and twice with buffer A.

4. Assay Msk1 activity as in **Subheading 3.4.1.**, except that the reaction is carried out on a shaking platform at 30°C to keep the immunoprecipitated Msk1/protein-G–Sepharose complex in suspension.

3.5. Assaying SAPK2a/p38, SAPK2b/p38β, SAPK3, SAPK4, and ERK5/BMK1

All of these SAPKs are able to phosphorylate myelin basic protein (MBP), which can be used as a non-specific substrate in vitro. However, to assay SAPK2a/p38 and SAPK2b/p38β (SAPK2/p38) activities a more specific and sensitive assay can be used, namely the activation of MAPKAP kinase-2 (*see* **Subheading 3.5.3.**) as SAPK3 and SAPK4 cannot use MAPKAP kinase-2 as an effective substrate *(9)*.

3.5.1. Assaying Purified and Recombinant SAPKs

1. Dilute active SAPKs or ERK5/BMK1 in cold-ice buffer B, to a concentration at which the enzyme activity will be linear (*see* **Subheading 3.9.1.**).
2. Incubate 5 μL diluted SAPK with a mixture containing 10 μL distilled water and 25 μL of SAPK reaction buffer (containing the substrate) for 3 min at 30°C.
3. Start the reaction by the addition of 10 μL of 50 mM magnesium acetate; 0.5 mM [γ-^{32}P]ATP (approx 2×10^5 cpm/nmol).
4. After 20 min at 30°C stop the reaction by pippeting 40 μL of the assay mixture and spotting it onto phosphocellulose P81 paper. Then follow the same procedure described in **Subheading 3.3.1.**

3.5.2. Immunoprecipitation and Assay of SAPKs or ERK5/BMK1 from Cell Extracts

1. Prepare cell extracts exactly as described in **Subheading 3.2.**
2. Mix 100–200 μg cell extract protein with 5 μL protein-G–Sepharose conjugated (as described in **Subheading 3.1.**) to either 5 μL of anti-SAPK2/p38 serum, 5 μg anti-ERK5 serum, 5 μg of anti-SAPK3 or 5 μg of anti-SAPK4 affinity purified antibody and incubate for 90 min at 4°C on a shaking platform.
3. Centrifuge the suspension for 1 min at 14,000g, discard the supernatant and wash the pellet twice with 1.0 mL of buffer A containing 0.5 M NaCl, and then twice with buffer A.
4. Assay kinase activity as in **Subheading 3.5.1.**, except that the reaction is carried out on a shaking platform at 30°C to keep the immunoprecipitated kinase/protein-G–Sepharose complex in suspension.

3.5.3. SAPK2a/p38 and SAPK2b/p38β (SAPK2/p38) Assays using MAPKAP Kinase-2 as Substrate

1. Dilute active SAPK2/p38 (recombinant or from cell lysate) in cold-ice buffer B, to a concentration at which the enzyme activity will be linear (*see* **Subheading 3.9.1.**).

2. Incubate 4 μL of diluted SAPK2/p38 with 6 μL of 0.4 μM inactive GST-MAPKAP kinase-2 for 3 min at 30°C.
3. Initiate MAPKAP kinase-2 activation by adding 2 μL of 60 mM magnesium acetate and 0.6 mM unlabeled ATP.
4. After 30 min at 30°C, remove a 5-μL aliquot, dilute it 10-fold in buffer B, and then take 5 μL of the diluted enzyme to assay for MAPKAP kinase-2 activity exactly as described in **Subheading 3.3.1.** When SAPK2/p38 that has been immunoprecipitated from cell lysates is assayed using this procedure, the first part of this assay should be done on a shaking platform to allow the immunoprecipitated kinase to mix with the substrate during the assay.

3.6. SAPK1/JNK Activity Assay

The activity of purified (from tissues/cells) or recombinant SAPK1/JNK can be assayed by the phosphorylation of the transcription factors ATF2 (*see* **Note 3**) or c-Jun.

3.6.1. Assaying Activated SAPK1/JNK

1. Dilute active SAPK1/JNK in cold-ice buffer B, to a concentration at which the enzyme activity will be linear (*see* **Subheading 3.9.1.**).
2. Incubate 5 μl diluted SAPK1/JNK with a mixture containing 10μL distilled water and 25 μL of SAPK1/JNK reaction buffer (containing the appropriate substrate) for 3 min at 30°C.
3. Start the reaction by the addition of 10 μl of 50 mM magnesium acetate; 0.5 mM [γ-^{32}P]ATP (2×10^5 cpm/nmol).
4. After 30 min at 30°C stop the reaction by adding 5 μL SDS-PAGE sample buffer.
5. Incubate at approx 95°C for 2–3 min and then load the samples onto a 10% SDS-PAGE gel (prepared as detailed in **Subheading 2.6.**) to resolve the proteins.
6. Stain the gel with 0.2% Coomassie Blue in 50% methanol/10% acetic acid, and destain in 50% methanol/10% acetic acid.
7. Dry the gel and then detect the labeled proteins by autoradiography. Excise the bands corresponding to [^{32}P]-labeled-ATF2 or c-Jun protein and measure the [^{32}P] incorporated into them by scintillation counting.

3.6.2. Assay of SAPK1/JNK From Cell Extracts

1. Prepare cell extracts exactly as described in **Subheading 3.2.**
2. Mix 100–200 μg of cell extract protein with 10 μg GST-c-Jun bound to 5 μL packed GSH-Sepharose beads and incubate for 3 h at 4°C on a shaking platform, to allow SAPK1/JNK to bind to its substrate c-Jun (*see* **Note 4**).
3. Centrifuge the suspension for 1 min at 14,000g, discard the supernatant and wash the pellet four times with 1.0 mL of washing buffer and once with buffer C.
4. Resuspend the pellet in 27 μL buffer C plus 2.5 μM PKI. After 3 min at 30°C initiate the reaction by adding 3 μL 1mM [γ-^{32}P]ATP (2×10^5 cpm/nmol).
5. Stop the reaction after a further 20 min at 30°C by adding 5 μL SDS-PAGE sample buffer.

6. Incubate at approx 95°C for 2–3 min and then load the samples onto a 10% SDS-PAGE gel (prepared exactly as described in **Subheading 2.6.**) to resolve the proteins.
7. Stain and destain the gel as described in **Subheading 3.6.1.** This staining will visualize the precipitated c-Jun and thus confirm equal precipitation/loading.
8. Dry the gel and then autoradiograph. Excise the [^{32}P]-labeled c-Jun band and measure the [^{32}P] incorporation by scintillation counting.

3.7. Other SAPK Assays

The activation of SAPKs by cytokines and cellular stress has also been assessed in cell extracts by several other procedures. These are useful but have some limitations and potential drawbacks (*see* **Notes 5–8**).

3.7.1. Immunoblotting With Anti-Phosphotyrosine Antibodies

The activation of SAPK is accompanied by the phosphorylation of a tyrosine residue (*see* **Subheading 1.**), which can be detected by immunoblotting with a suitable anti-phosphotyrosine antibody or an antiphospho-SAPK antibody that only recognizes the phosphorylated and active form of the SAPK (*see* **Note 5**).

1. Prepare cell extracts exactly as described in **Subheading 3.2.**
2. Incubate 100–200 µg cell extract protein with 5 µL protein-G–Sepharose conjugated to the anti-SAPK antibody (as indicated in **Subheading 3.1.**) for 90 min at 4°C on a shaking platform.
3. Centrifuge the suspension for 1 min at 14,000*g*, discard the supernatant and wash the pellet twice with 1.0 mL of buffer A containing 0.5 *M* NaCl, and twice with buffer A.
4. Add 5 µL of SDS-PAGE sample buffer to the pellets, heat them at approx 95°C for 2–3 min and then load onto a 10% SDS-PAGE gel (prepared exactly as described in **Subheading 2.6.**) to resolve the proteins.
5. Perform a Western blot according to standard techniques *(27)*, using either anti-phosphotyrosine antibodies or antiphospho-SAPK antibodies (these are commercially available), together with a control using antibodies (that recognize both the phosphorylated and the dephosphophosphorylated form of the kinase) specific against the SAPKs immunoprecipitated to confirm equal precipitation/loading.

3.7.2. In-Gel Kinase Assays

In these assays, the kinase is first denatured by dissolving cell extracts in SDS and then subjected to electrophoresis on an acrylamide gel polymerized in the presence of MBP or c-Jun substrate. Following electrophoresis, the kinase is then renatured and phosphorylation of the substrate is initiated by incubating the gel with Mg-[γ-^{32}P]ATP. After washing to remove the unincorporated ATP, the position of the [^{32}P]-labeled substrate is located by autoradiography (*see* **Note 6**).

1. When assaying SAPK2/p38, SAPK3 and SAPK4 it is recommended that these proteins are first immunoprecipitated from the cell extract as described in **Subheading 3.5.** For SAPK1/JNK, total cell extracts can be used provided that c-Jun is employed as the substrate (*see* **Note 4**).
2. Prepare a gel mix for a standard 10% SDS-PAGE gel (*see* **Subheading 2.6.**) but, before polymerisation of the running gel, add 0.5 mg/mL SAPK substrate: MBP (for SAPK2/p38, SAPK3, and SAPK4) or c-Jun (for SAPK1/JNK).
3. Add SDS-PAGE sample buffer to the immunoprecipitated SAPK or to the cell extract, load the samples (without boiling) onto the gel, and run it in the cold room at 100 V. It is important not to boil the samples as this can cause problems in renaturing the protein in the following steps.
4. Wash the gel twice for 30 min at room temperature in 100 mL of 20% (v/v) 2-propanol, 50 mM HEPES, pH 7.6, to remove the SDS.
5. Wash the gel twice for 30 min at room temperature in 100 mL of 50 mM HEPES, pH 7.6 and 5 mM 2-mercaptoethanol.
6. Incubate at 25°C for 1 h in 200 mL of 50 mM HEPES, pH 7.6, 5 mM 2-mercaptoethanol, and 6 M urea.
7. Incubate at 25°C for 1 h in 200 mL of buffer D containing 3 M urea. Change the buffer three times during the incubation time.
8. Incubate at 25°C for 1 h in 200 ml of buffer D containing 1.5 M urea. Change the buffer three times during the incubation time.
9. Incubate at 25°C for 1 h in 200 mL of buffer D containing 0.75 M urea. Change the buffer three times during the incubation time.
10. Incubate at 4°C overnight in 100 mL of buffer D.
11. Perform the kinase assay by incubating the gel in 10–20 mL of buffer E for 1 h.
12. Wash the gel with 100 mL of 5% trichloroacetic acid (TCA) containing 1% sodium pyrophosphate at 25°C for 2 h with at least five changes of wash buffer, or until no radioactivity could be detected in the wash.
13. Stain and destain as in **Subheading 3.5.1.** Dry the gel and then autoradiograph.

3.7.3. Decrease of Electrophoretic Mobility

The phosphorylation of SAPK by SKK is accompanied by a decrease in its electrophoretic mobility on PAGE gels and this can be detected in cell extracts by immunoblotting with specific anti-SAPK antibodies (*see* **Note 7**).

1. Prepare cell extracts exactly as described in **Subheading 3.2.**
2. Prepare a 10% PAGE separating gel mixture exactly as described in **Subheading 2.7.** This does not contain SDS in order to accentuate "band shifts."
3. Prepare a stacking gel mixture as described in **Subheading 2.7.**
4. Load 20–50 μg protein of cell extract in SDS-PAGE buffer after boiling at approx 95°C for 2 min and run gel normally.
5. Perform a Western blot according to standard techniques *(27)*, using antibodies specific for the SAPK isoform under study.

3.8. SKK/MKK Assays

SKK/MKK activities can be assayed by their ability to specifically phosphorylate and activate the appropriate downstream substrate, SAPK2/p38 or SAPK1/JNK (*see* **Note 8**).

3.8.1. Measuring SKK2/MKK3 and SKK3/MKK6 Activities Using Activation of SAPK2/p38

1. Dilute purified or recombinant SKK2/MKK3 or SKK3/MKK6 in cold-ice buffer B, to a concentration at which the activity of the enzyme will be linear (*see* **Subheading 3.9.1.**).
2. Incubate 3 μL of diluted SKK2/MKK3 or SKK3/MKK6 with 6 μL of 0.4 μM inactive GST-SAPK2/p38 for 3 min at 30°C.
3. Initiate SAPK2/p38 activation by adding 3 μL of 40 mM magnesium acetate and 0.4 mM unlabeled ATP.
4. After 30 min at 30°C stop the reaction by diluting with 24 μL of ice-cold buffer B.
5. Withdraw 5-μL aliquots and assay for SAPK2/p38 activity as described in **Subheadings 3.5.1.** or **3.5.3.**

3.8.2. Measuring SKK1/MKK4 and SKK4/MKK7 Activities Using Activation of SAPK1/JNK

1. Dilute purified or recombinant SKK1/MKK4 or SKK4/MKK7 in ice-cold buffer B, to a concentration at which the activity of the enzyme will be linear (*see* **Subheading 3.9.1.**).
2. Incubate 5 μL diluted SKK1/MKK4 or SKK4/MKK7 with 2.5 μL of 10 μM inactive SAPK1/JNK diluted in buffer F for 3 min at 30°C.
3. Start the SAPK1/JNK activation by adding 2.5 μL of 40 mM magnesium acetate; 0.4 mM unlabeled ATP.
4. After 30 min at 30°C, stop the reaction by adding a mixture containing 5 μL distilled water and 25 μL SAPK1/JNK reaction buffer (**Subheading 2.6.**) and add 10 μL 50 mM magnesium acetate; 0.5 mM [γ-^{32}P]ATP to initiate the SAPK1/JNK assay which is then performed exactly as described in **Subheading 3.6.1.**

3.8.3. Assaying SKK/MKKs From Cell Extracts

1. Prepare cell extracts exactly as described in **Subheading 3.2.**
2. Mix 100–200 μg cell extract protein with 5 μL protein-G–Sepharose conjugated (as described in **Subheading 3.1.**) to either 10 μg of anti-SKK1/MKK4, 10 μg of anti-SKK4/MKK7, 2 μg of anti-SKK3/MKK6, or 5 μg of anti-SKK2/MKK3 affinity purified antibody, and incubate for 90 min at 4°C on a shaking platform.
3. Centrifuge the suspension for 1 min at 14,000g, discard the supernatant and wash the pellet twice with 1.0 mL of buffer A containing 0.5 M NaCl, and twice with buffer A.
4. Assay SKK2/MKK3 and SKK3/MKK6 kinase activity as described in **Subheading 3.8.1.** or SKK1/MKK4 and SKK4/MKK7 as described in **Subheading 3.8.2.**

3.8.4. Assaying SKK/MKKs by Phosphorylation of SAPKs

1. Dilute active recombinant SKK/MKK or SKK/MKK immunoprecipitated from cell extracts in ice-cold buffer F to a concentration at which the enzyme activity is linear (*see* **Subheading 3.9.1.**).
2. Incubate 5 µL of the diluted SKK/MKK, for 3 min at 30°C, with a reaction mixture containing 10 µL distilled water and 25 µL of 2X buffer F containing 1 µ*M* SAPK.
3. Start the reaction by the addition of 10 µL of 50 m*M* magnesium acetate; 0.5 m*M* [γ-^{32}P]ATP (2×10^5 cpm/nmol).
4. After 30 min at 30°C stop the reaction by adding 5 µL SDS-PAGE sample buffer.
5. Incubate at approx 95°C for 2–3 min and then load onto a 10% SDS-PAGE gel (prepared as described in **Subheading 2.6.**).
6. Stain and destain the gel as indicated in **Subheading 3.6.1.** Dry the gel and then autoradiograph. Excise the [^{32}P]-labeled-SAPK band and measure the [^{32}P] incorporation by scintillation counting.

3.9. Controls and the Calculation of Unit Activities

3.9.1. Controls

1. In order to ensure that the activity of MAPKAP kinase, Msk1, SAPK, or SKK is linear over the period of assay, it is important to establish that the addition into the assay of half the quantity of enzyme (by diluting the kinase in the appropriate buffer, as indicated in Subheading 2.) reduces the measured activity by 50%.
2. In all assays, control reactions must be done in which the MAPKAP kinase, Msk1, SAPK, or SKK is omitted. This gives the blank value that must be subtracted from the activity of each of the samples. In order for any measured activity to be reliable, it needs to be at least fivefold higher than this value.

3.9.2. Calculation of Unit

1. One unit of MAPKAP kinase, Msk1, SAPK, or SKK activity is the amount of enzyme that incorporates 1 nmol of phosphate into the substrate (MAPKAP kinase peptide, MBP, ATF2, c-Jun, or SAPK) in one min. The MAPKAP kinase, SAPK or SKK activity in units per milliliter (U/mL) is calculated using the formula 25 CD/ST, where C is the [^{32}P] radioactivity incorporated into the substrate (in cpm) after the blank has been subtracted, D is the fold dilution of the MAPKAP kinase, SAPK or SKK solution before assay, *S* is the specific activity of the [γ-^{32}P]ATP (cpm/nmol) and T is the time of the reaction.
2. When SAPK2/p38 activity is measured using the MAPKAP kinase-2 activation assay, one unit of SAPK2/p38 is that amount of enzyme, which increases the activity of MAPKAP kinase-2 by 1 U/min.
3. When SKK activity is measured using the SAPK activation assay, one unit of SKK activity is that amount of enzyme, which increases the activity of SAPK by 1 U/min.

4. Notes

1. MAPKAP kinase-2 and MAPKAP kinase-3 can also be assayed by the phosphorylation of their physiological substrate the small heat shock protein HSP25/HSP27 (obtained from StressGen Biotechnologies Corp). To detect and quantify the phosphorylation of HSP27/25 the protein is analysed by SDS-PAGE, followed by Coomassie staining and autoradiography (as described in **Subheading 3.6.1.**) *(28)*.

2. PRAK/MAPKAP kinase-5 activity can also be assayed as described for MAPKAP kinase-2 and MAPKAP kinase-3 in this chapter *(29)*.

3. We have shown that SAPK2/p38, SAPK3, and SAPK4 can also phosphorylate the transcription factor ATF2 *(6,9)*.

4. To "pull down" SAPK1/JNK specifically from cell extracts we use GST-c-Jun bound through its GST moiety to glutathione (GSH)-Sepharose beads to generate an affinity matrix for SAPK1/JNK. Like c-Jun, the *N*-terminus of ATF2 also binds to SAPK1/JNK and this property could also be used to pull down the kinase from cell extracts. However, ATF2 (unlike c-Jun) also interacts with other SAPKs that are activated by the same stimuli that activate SAPK1/JNK and can phosphorylate ATF2 (*see* **Note 3**). Therefore, it is recommended that only c-Jun is used to assay SAPK1/JNK activity in cell extracts.

5. One potential pitfall of assaying the SAPK activity directly from cell extracts using anti-phosphotyrosine antibodies or phospho-specific (anti-phospho-SAPK) antibodies raised specifically against the phosphotyrosine in the phosphorylation site, is that SAPK2a/p38, SAPK2b/p38β, SAPK3, and SAPK4, all possess nearly identical residues in the sequence surrounding their activating phosphorylation sites, and all the SAPKs are similar in size. We have shown that the commercially available phospho-specific p38 antibodies also recognize the active form of SAPK2b/p38β, SAPK3, and SAPK4. If this method is used to assay SAPKs it is essential that these enzymes are first immunoprecipitated from the cell extract using specific SAPK antibodies, followed by blotting the immunoprecipitated SAPK using phosphotyrosine or the phosphospecific antibodies. There are also anti-phospho-SKK antibodies from New England Biolabs that can be used to detect active SKKs. However, it is best to immunoprecipitate the SKK of interest prior immunoblotting with these antibodies. Another serious disadvantage of using this method is that it is not quantitative.

6. This assay assumes that no other kinases (that also phosphorylate the same substrate) comigrate with the kinase that is being assayed and that the renaturation of the kinase is uniform throughout the gel. However, not all kinases are efficiently renatured and different kinases may renature to differing extents. It should also be noted that the in gel kinase assay is time consuming and also more expensive than the normal assay because of the amount of radioactive ATP which is required.

7. The phosphorylation of SAPKs by SKKs is accompanied by a decrease in their electrophoretic mobility on SDS-PAGE gels and this can be detected in cell extracts by immunoblotting with specific anti-SAPK antibody. This assay can be

used to detect activation, but is only semiquantitative and unsuitable for detecting low levels of activation of SAPKs. There is also the potential danger that the electrophoretic mobility may be decreased by phosphorylation at sites other than those labeled by SKKs, and which do not lead to the activation of the SAPK, thus invalidating the assay.

8. SKK (and SAPK) activities can also be measured by autophosphorylation after immunoprecipitation from cell extracts. This method of assaying kinases is not very reliable as the immunoprecipitated protein can become contaminated by other kinase(s) that may also phosphorylate the SAPK or SKK. Thus, the incorporation of phosphate measured in this type of assay may not be a reflection of the true SAPK or SKK activity.

References

1. Cohen, P. (1997) The search for physiological substrates of MAP and SAP kinases in mammalian cells. *Trends Cell Biol.* **7**, 353–361.
2. Kyriakis, J. M., Banerjee, P., Nikolakaki, E., Dai, T., Rubie, E. A., Ahmad, M. F., Avruch, J., and Woodgett, J. R. (1994) The stress-activated protein kinase subfamily of c-Jun kinases. *Nature* **369**, 156–160.
3. Han, J., Lee, J. D., Bibbs, L., and Ulevitch, R. J. (1994) A MAP kinase targeted by endotoxin and hyperosmolarity in mammalian cells. *Science* **265**, 808–811.
4. Jiang, Y., Chen, C., Li, Z., Guo, W., Gegner, J. A., Lin, S., and Han, J. (1996) Characterization of the structure and function of a new mitogen activated protein kinase (p38β). *J. Biol. Chem.* **271**, 17,920–17,926.
5. Mertens, S., Craxton, M., and Goedert, M. (1996) SAP kinase-3, a new member of the family of mammalian stress-activated protein kinases. *FEBS Lett.* **383**, 273–276.
6. Cuenda, A., Cohen, P., Buee-Scherrer, V., and Goedert, M. (1996) Activation of stress-activated protein kinase-3 (SAPK3) by cytokines and cellular stresses is mediated via SAPKK3 (MKK6); comparison of the specificities of SAPK3 and SAPK2 (RK/p38). *EMBO J.* **16**, 295–305.
7. Lechner, C., Zahalka, M. A., Giot, J. F., Moller, M. P., and Ullrich, A. (1996) ERK6, a mitogen-activated protein kinase involved in C2C12 myoblast differentiation. *Proc. Natl. Acad. Sci. USA* **93**, 4355–4359.
8. Li, Z., Jiang, Y., Ulevitch, R. J., and Han, J. (1996) The primary structure of p38γ: a new member of the p38 group of MAP kinases. *Biochem. Biophys. Res. Commun.* **228**, 334–340.
9. Goedert, M., Cuenda, A., Craxton, M., Jakes, R., and Cohen, P. (1997) Activation of novel stress-activated protein kinase (SAPK4) by cytokines and cellular stresses is mediated by SAPKK3 (MKK6); comparison of the specificity with that of other SAP kinases. *EMBO J.* **16**, 3563–3571.
10. Abe, J. I., Kusuhara, M., Ulevitch, R. J., Berk, B. C., and Lee, J. D. (1996) Big mitogen-activated protein kinase 1 (BMK1) is a redox-sensitive kinase. *J. Biol. Chem.* **271**, 16,586–16,590.
11. Zhou, G., Bao, Z. Q., and Dixon, J. E. (1995) Components of a new human protein kinase signal transduction pathway. *J. Biol. Chem.* **270**, 12,665–12,669.

12. Deak, M., Clifton, A. D., Lucocq, J. M., and Alessi, D. R. (1998) Mitogen- and stress-activated protein kinase-1 (MSK1) is directly activated by MAPK and SAPK2/p38, and may itself mediate activation of CREB. *EMBO J.* **17,** 4426–4441.
13. Derijard, B., Raingeaud, J., Barrett, T., Wu, I-H., Han, J., Ulevitch, R. J., and Davis, R. J. (1995) Independent human MAP kinase signal transduction pathways defined by MEK and MKK isoforms. *Science* **267,** 682–684.
14. Sanchez, I., Hughes, R. T., Mayer, B. J., Yee, K., Woodgett, J. R., Avruch, J., Kyriakis, J. M., and Zon, L. I. (1994) Role of SAPK/ERK kinase-1 in the stress-activated pathway regulating transcription factor c-Jun. *Nature* **372,** 794–798.
15. Lin, A., Minden, A., Martinetto, H., Claret, F-X., Lange-Carter, C., Mercurio, F., Johnson, G. L., and Karin, M. (1995) Identification of a dual specificity kinase that activates the Jun kinase and p38-Mpk2. *Science* **268,** 286–290.
16. Lawler, S., Cuenda, A., Goedert, M., and Cohen, P. (1997) SKK4, a novel activator of stress-activated protein kinase-1 (SAPK1/JNK). *FEBS Lett.* **414,** 153–158.
17. Holland, P. M., Suzanne, M., Campbell, J. S., Noselli, S., and Cooper, J. A. (1997) MKK7 is a stress-activated mitogen-activated protein kinase kinase functionally related to hemipterous, *J. Biol. Chem.* **272,** 24,994–24,998.
18. Tournier, C., Whitmarsh, A. J., Cavanagh, J., Barrett, T., and Davis, R. J. (1997) Mitogen-activated protein kinase kinase 7 is an activator of the c-Jun NH2-terminal kinase. *Proc. Natl. Acad. Sci. USA* **94,** 7337–7342.
19. Cuenda, A., Alonso, G., Morrice, N., Jones, M., Meier, R., Cohen, P., and Nebreda, A. R. (1996) Purification and cDNA cloning of SAPKK3, the major activator of RK/p38 in stress- and cytokines- stimulated monocytes and epithelial cells. *EMBO J.* **16,** 4156–4164.
20. Raingeaud, J., Whitmarsh, A. J., Barret, T., Derijard, B., and Davis, R. J. (1996) MKK3- and MKK6-regulated gene expression is mediated by the p38 mitogen-activated protein kinase signal transduction pathway. *Mol. Cell. Biol.* **16,** 1247–1255.
21. Moriguchi, T., Kuroyanagi, N., Yamaguchi, K., Gotoh, Y., Irie, K., Kano,T., Shirakabe, K., Muro,Y., Shibuya, H., Matsumoto, K., Nishida, E., and Hagiwara, M. (1996) A novel kinase cascade mediated by mitogen-activated protein kinase kinase 6 and MKK3. *J. Biol. Chem.* **271,** 13,675–13,679.
22. Han, J., Lee, J. D., Jiang, Y., Li, Z., Feng, L., and Ulevitch, R. J. (1996) Characterization of the structure and function of a novel MAP kinase kinase (MKK6). *J. Biol. Chem.* **271,** 2886–2891.
23. Fanger, G. R., Gerwins, P., Widmann, C., Jarpe, M. B., and Johnson, G. L. (1997) MEKKs, GCKs, MLKs, PAKs, TAKs and Tpls: upstream regulators of the c-Jun amino-terminal kinases? *Curr. Opin. Gen. Dev.* **7,** 67–74.
24. Clifton, A. D., Young, P. R., and Cohen, P. (1996) A comparison of the substrate specifity of MAPKAP kinase-2 and MAPKAP kinase-3 and their activation by cytokines and cellular stress. *FEBS Lett.* **392,** 209–214.
25. Pulverer, B. J., Kyriakis, J. M., Avruch, J., Nikolakaki, E., and Woodgett, J. R. (1991) Phosphorylation of c-Jun is mediated by MAP kinases. *Nature* **353,** 670–674.

26. Livingston, C., Patel, G., and Jones, N. (1995) ATF-2 contains a phosphorylation-dependent transcriptional activation domain. *EMBO J.* **14,** 1785–1797.
27. Harlow, E., and Lane, D. (1988) *Antibodies: A Laboratory Manual* Cold Spring Harbor Laboratory, Cold Spring Harbor, N Y.
28. Cuenda, A., Rouse, J., Doza, Y. N., Meier, R., Cohen, P., Gallagher, T. F., Young, P. R., and Lee, J. C. (1995) SB 203580 is a specific inhibitor of a MAP kinase homologue which is stimulated by cellular stresses and interleukin-1. *FEBS Lett.* **364,** 229–233.
29. New, L., Jiang, Y. Zhao, M., Liu, K., Zhu, W., Flood, L. J., Kato, Y., Parry, G. C. N., and Han, J. (1998) PRAK, a novel protein kinase regulated by the p38 MAP kinase. *EMBO J.* **17,** 3372–3384.

12

Monitoring the Activation of Stress-Activated Protein Kinases Using GAL4 Fusion Transactivators

Chao-Feng Zheng and Li Xu

1. Introduction
1.1. Stress-Activated Protein Kinase Cascades

A multicellular organism is composed of many types of cells performing specialized functions. Cells have to communicate with each other for the organism to function as a whole. They do so at many levels and by various mechanisms. A cell's identity is determined by the proteins synthesized within it. Therefore, the regulation of gene expression, especially at the transcriptional level, is a key mechanism of controlling cell growth and differentiation. To control transcription in response to extracellular stimuli originating either from other cells or the surrounding environment, signals from outside the cell are transmitted to the transcription machinery inside the nucleus via a variety of signaling molecules. These include receptors, adaptor proteins, G-proteins, protein kinases, and protein phosphatases, which form intricate networks known as signal transduction pathways (1–4).

The mitogen-activated protein kinase (MAPK) pathways that mediate signals from growth factors (e.g., epidermal growth factor [EGF] and nerve growth factor [NGF]) and cellular stress such as heat, ultraviolet (UV), oxidative stress, and protein synthesis inhibitors are among the best characterized and most highly conserved signal transduction pathways so far discovered (**Fig. 1**; see also Chapter 11, this volume). Since the identification of the first member of the MAPK family 11 yr ago (5), more than 100 MAPK family members have been cloned from a wide variety of organisms (6). These signaling pathways all use a three-component protein kinase cascade consisting of MAPK/MAPK kinase/MAPK kinase kinase, but different MAP kinase modules receive diverse upstream signals and cause distinct downstream changes (**Fig. 1**). In the budding yeast

From: *Methods in Molecular Biology, vol. 99: Stress Response: Methods and Protocols*
Edited by: S. M. Keyse © Humana Press Inc., Totowa, NJ

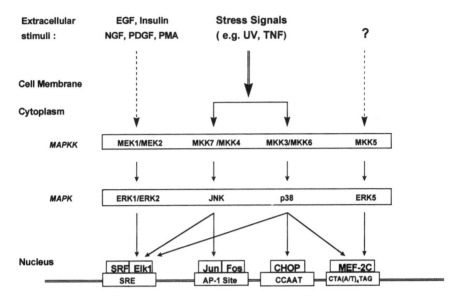

Fig. 1. MAPK and SAPK signal-transduction pathways in mammalian cells. Only selected pathways are shown. As shown in the figure by solid arrows, MAPKs are phosphorylated and activated by MAPKKs directly, and they can then directly phosphorylate and activate downstream transcriptional activators such as ELK1, CHOP, or c-Jun. On the other hand, there may be many steps from the cell surface or other part of the cell to the activation of MAPKKs.

S. cerevisiae, four MAP kinase modules have been identified to function in mating/pseudohyphal development/invasive growth (Fus3p/Kss1p), cell wall biosynthesis (Mpk1p), osmosensing (Hog1p), and sporulation (SmK1p) *(7,8).*

In mammalian cells, at least 10 MAPK isoforms have been cloned and characterized *(6,8).* These include the "classical" MAPKs ERK1 and ERK2, abbreviated from extracellular signal-regulated kinases *(5,9,10).* They were first identified in insulin-treated adipocyte cells and relay signals from growth factors and the cancer-promoting agent PMA. Another group of MAPKs in mammalian cells comprises the stress-activated protein kinases (SAPKs), including c-Jun *N*-terminal kinase (JNK) and p38 MAPK, each of which has multiple isoforms and splice variants *(6,8,11–13).* SAPKs, as the name suggests, relay signals in response to diverse cellular stresses such as UV radiation, heat shock, osmotic and oxidative stress, cytokines, and inhibitors of protein synthesis (**Fig. 1**) *(11–13).*

Once activated by phosphorylation on conserved threonine and tyrosine residues by MAPK kinases, MAPK/SAPKs are able to phosphorylate and modulate the activity of many cellular proteins. These include protein kinases, regulators, metabolic enzymes, receptors, and transcription factors. The latter targets mediate the ability of MAPKs to transmit signals from outside the cell to the transcription machinery in the cell nucleus (**Fig. 1**) *(1–4,14,15)*. The phosphorylation of the activation domains of certain transcription factors by SAPK/MAPKs reflects the activation status of the respective kinases and upstream signaling molecules within that MAPK pathway.

1.2. The Use of GAL4 Fusions as Sensors for Kinase Activation

Initially demonstrated with the yeast GAL4 protein, the DNA-binding domain (dbd) and the activation domain (ad) of eukaryotic transactivators can be structurally and functionally separated *(16)*. The DNA binding domain will retain its ability to bind the specific DNA sequence when separated from the rest of the transactivator; the activation domain, on the other hand, will confer trans-activating activity to a fusion protein containing a different DNA binding domain. The amino terminal DNA binding domain (amino acid residues 1–92 or 1–147) of yeast GAL4 protein has been used to build hundreds of such fusion proteins *(17)*. These chimeric proteins have been used to clone interacting proteins using various two-hybrid systems and to determine if a protein has trans-activating or trans-suppressing activity on transcription, and if so, which part of the protein is responsible for this activity *(18–20)*. In mammalian cells, GAL4 fusions have been applied to test protein–protein interaction, to study chromatin structure and function, and to serve as inducible transcription factors for protein expression and the measurement of the biological activities of steroid hormones *(21–24)*.

When activated by upstream signals, the stress-activated protein kinases JNK or p38 and other MAPKs are able to translocate into the nucleus and phosphorylate critical residues that activate transcription factor(s) such as c-Jun, Elk1, and CHOP/GADD153 (**Fig. 1**). Fusion transactivators consisting of the DNA binding domain of yeast GAL4 protein or *Escherichia coli* LexA and the activation domains of mammalian transcription activators including CREB *(25–27)*, Elk1 *(28–30)*, c-Jun *(31–33)*, and CHOP/GADD153 *(34)* have all been used successfully as sensors for cAMP-dependent protein kinase (PKA) and MAP kinase pathways in the literature and are now available commercially (PathDetect™ Trans-Reporting Systems from Stratagene [La Jolla, CA]) (**Fig. 2**; *see* **Notes 1–4**) *(27,35–37)*.

Each trans-reporting system includes a unique fusion trans-activator plasmid that expresses a fusion protein consisting of the activation domain of either the c-Jun, CHOP, Elk1, or CREB protein and the DNA binding domain of yeast GAL4 (residues 1–147, *see* **Note 2**). The activation moiety of these fusion transactivators are phosphorylated and activated by stress-activated protein kinases JNK or p38 MAPK, ERK1/2 or cyclic AMP-dependent kinase (PKA), respectively. The activity of the fusion activators, therefore, reflects the in vivo

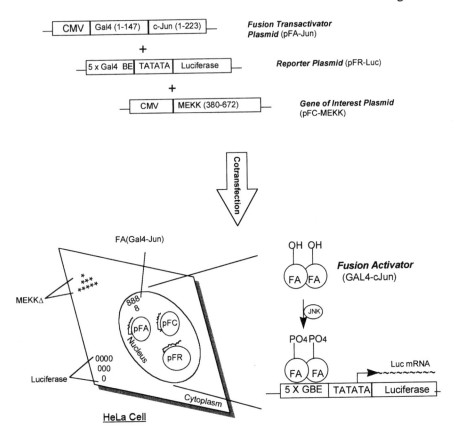

Fig. 2. Monitoring JNK activation using GAL4-c-Jun as the sensor. When trans-fected into mammalian cells such as HeLa cells, pFC-MEKK expresses active MEKK which activates JNK. Activated JNK phosphorylates and activates the fusion activa-tor (FA) expressed from pFA-Jun. The activated FA (GAL4-Jun) binds to GAL4 bind-ing elements (GBE) in pFR-Luc and stimulates luciferase expression. Other reporting systems work the same way except that different activators were used to monitor the activation of different protein kinases (**Fig. 1**). These systems can also be used to monitor the activation of these protein kinases and their respective signaling pathways caused by extracellular stimuli such as growth factors and cellular stress signals. In the latter case, only the fusion transactivator plasmid and the reporter plasmid are cotransfected into the cell.

activation of these kinases and their corresponding signal-transduction path-ways. The GAL4 dbd moiety enables the fusion activators to bind the 5 tandem repeats of the 17 mer GAL4 binding element upstream of the firefly luciferase gene in a separate reporter vector pFR-Luc (*see* **Note 1**) (**Fig. 2**). When a fusion trans-activator plasmid, reporter plasmid, and an uncharacterized gene are cotransfected into mammalian cells, direct or indirect phosphorylation of the transcription activation domain of the fusion trans-activator protein by the uncharacterized gene product will activate transcription of the luciferase gene from the reporter plasmid (**Fig. 2**

2. Materials

2.1. Plasmids

All plasmids used for transfection were supercoiled DNAs purified by Qiagen's (Chatsworth, CA) Maxiprep kit or Stratagene's Strataprep kit from *E. coli* XL1-Blue (Stratagene). The pFR-Luc reporter plasmid contains a synthetic promoter with five tandem repeats of the yeast GAL4 binding sites that control expression of the *Photinus pyralis* (American firefly) luciferase gene (*see* **Note 1**) *(27)*. The fusion transactivator plasmid (*see* **Note 2**) expresses a fusion protein of the DNA binding domain of yeast GAL4 (1–147) and an activation domain from a transcription factor that may be activated by a protein kinase. All the fusion trans-activator proteins bind to the reporter plasmid at the GAL4 binding sites, but are activated by, and hence serve as sensors for, different kinases and signaling pathways (**Fig. 1**; *see* **Notes 1** and **2**) *(27)*. Various control plasmids are also used (*see* **Note 3**).

In a situation where a transcription factor is suspected of receiving signals from upstream signal transduction pathways, but no immediate upstream kinase(s) has yet been identified, there may be no ready-made fusion transactivator expression vector (such as pFA2-cJun) available. In order to study these pathways, the pFA-CMV plasmid (*see* **Note 4**) can be used to construct suitable fusion transactivator plasmids *(36)*. The pFA-CMV plasmid is designed for convenient insertion of the activation domain sequence of any transcription activator, c-terminal to the DNA binding domain of the yeast GAL4 protein (amino acid residues 1–147).

2.2. Mammalian Cells

Many of the commonly used cell lines such as HeLa, CHO, CV-1, and NIH3T3 can be purchased from American Type Culture Center (ATCC, Manasses, VA). Certain primary cells or cell lines developed in individual labs may also be used.

2.3. Reagents

1. Cell culture medium (e.g., Dulbecco's minimum essential medium [DMEM]) and calcium- and magnesium-free phosphate-buffered saline (PBS) were from

Gibco-BRL (Gaithersburg, MD). Complete medium refers to DMEM containing 10% fetal bovine serum (FBS), 1% L-glutamine, 1% penicillin, and streptomycin, and 50 μM β-mercaptoethanol (*see* **Note 5**).

2. Luciferase assay kits were obtained from Stratagene.
3. LipofectAMINE was from GIBCO BRL and LipoTaxi was from Stratagene.
4. Cell lysis buffer (5 X stock): 40 mM Tricine, pH 7.8; 50 mM NaCl; 2 mM EDTA; 1 mM MgSO$_4$; 5 mM dithiothreitol (DTT); 1% Triton X-100.
5. Luciferase assay reagent: 40 mM Tricine, pH 7.8; 0.5 mM ATP; 10 mM MgSO$_4$; 0.5 mM EDTA; 10 mM DTT; 0.5 mM coenzyme A; 0.5 mM luciferin.

2.4. Equipment and Labwares

1. Polypropylene tubes (2054) were from Falcon (Los Angeles, CA), bottles and tissue culture dishes (6-, 12-, or 24-well plates) were from various suppliers of tissue-culture-grade plasticware.
2. A luminometer is needed for luciferase activity assays. Although an old model of a single-tube luminometer from Tropix (Bedford, MA) was used for most of the experiments described here, newer models of luminometer are available from several vendors such as EG&G, Wallac, Lab Systems, BMG, or Packard.

3. Methods
3.1. Preprotocol Considerations
3.1.1. Choosing a Cell Line

The PathDetect signal transduction pathway trans-reporting systems may be used with various mammalian cell lines, provided the cells contain the protein kinases that activate the fusion trans-activator protein. The endogenous protein kinases and transcription activator activities in the cell line used will determine the background and, hence, the sensitivity of the assay. Although these systems have been found to work in various cell lines, including HeLa, CHO, CV-1, and NIH3T3, a given reporting system might work better in one cell line under defined conditions. Whether these systems will give satisfactory results for a cell line of particular interest will need to be determined experimentally.

3.1.2. Choosing a Transfection Method

As with all transfection assays, the sensitivity of an assay using a PathDetect signal-transduction pathway reporting system is greatly influenced by the transfection efficiency. A high transfection efficiency generally provides a more sensitive assay that requires a smaller volume of sample. Transfection conditions should be optimized before performing the assays with a reporter plasmid.

Because the luciferase assay is very sensitive, various transfection methods, such as calcium phosphate precipitation and lipid-mediated transfection, may be used. Lipid-mediated transfection generally results in higher and more consistent transfection efficiency than other chemical methods in many cell lines.

3.1.3. Tissue Cultureware

The protocols given are based on six-well tissue culture dishes with a well diameter of approx 35 mm and a surface area of approx 9.4 cm². When dishes with smaller wells are used, decrease the number of cells per well and the volume of reagents according to the surface area of the wells. Both 12- and 24-well plates have been used and found to give consistent results.

3.1.4. Designing the Experiments

3.1.4.1. STUDYING THE EFFECTS OF A GENE PRODUCT

To study the effect of a given gene product (e.g., MEKK) on a particular signaling pathway (e.g., JNK pathway) using GAL4 fusion transactivators, the gene of interest should be cloned into a mammalian expression vector such as pCMV-Script (Stratagene) or pcDNA3 (Invitrogen, San Diego, CA) The experimental plasmid without the gene of interest that may lead to activation of the fusion trans-activator protein should be used as a negative control to ensure that the effect observed is not caused by the introduction of viral promoters (e.g., CMV, RSV, or SV40) or other proteins expressed from the plasmid. Depending on the purpose of the experiment, other controls such as a nonactivatable mutant of the fusion trans-activator protein might be required.

Typical initial experimental conditions for the PathDetect trans-reporting system are outlined in **Tables 1** and **2**. As all assays are to be run in triplicate, eight samples will utilize four six-well tissue culture dishes. Sample numbers are indicated in Column A. Column B indicates the amount (volume) of reporter plasmid to use. Column C indicates the amount of fusion trans-activator plasmid (pFA2–c-Jun, pFA2–Elk1, pFA2–CREB, pFA–CHOP, or pFC2–dbd plasmids) to be used in each sample. Column D indicates the amount of pFC2–dbd (negative control for the pFA plasmid to ensure the effects observed are not due to the GAL4 DNA binding domain) to be used in each sample. Column E indicates the appropriate amount (volume) of positive control to be used (pFC–MEKK for the c-Jun, pFC–MEK3 for CHOP, pFC–MEK1 for Elk1, or pFC–PKA for CREB). Column F indicates the amount (volume) of the experimental mammalian expression plasmid containing the gene of interest. Column G indicates amounts of the negative control for the promoter used to express the gene of interest. Finally, column H indicates the amount of unrelated plasmid DNA containing no mammalian promoters or other elements to be used to keep the amount of DNA in each sample constant.

3.1.4.2. STUDYING THE EFFECTS OF AN EXTRACELLULAR STIMULUS

The PathDetect reporting systems may also be used to study the effects of extracellular stimuli, such as growth factors, cellular stresses or drugs, on cor-

Table 1
Sample Experiment to Study the Effects of a Gene Product

A Sample no.	B pFR-Luc plasmid (reporter plasmid)	C Fusion trans-activator plasmid[a]	D pFC2-dbd (negative control for pFA plasmid)	E Positive control	F Experimental plasmid with gene of interest	G Experimental plasmid without insert	H Plasmid DNA
1[b]	1.0 µg (1 µL)	50 ng (2 µL)	—	—	—	50 ng	950 ng
2[c]	1.0 µg (1 µL)	50 ng (2 µL)	—	—	—	100 ng	900 ng
3[d]	1.0 µg (1 µL)	50 ng (2 µL)	—	—	—	1000 ng	—
4[e]	1.0 µg (1 µL)	50 ng (2 µL)	—	—	50 ng	—	950 ng
5[f]	1.0 µg (1 µL)	50 ng (2 µL)	—	—	100 ng	—	900 ng
6[g]	1.0 µg (1 µL)	50 ng (2 µL)	—	—	1000 ng	—	—
7[h]	1.0 µg (1 µL)	50 ng (2 µL)	—	50 ng (2 µL)	—	—	950 ng
8[i]	1.0 µg (1 µL)	—	50 ng (2 µL)	—	100 ng	—	900 ng

[a]This quantity may need to be optimized, usually within the range of 1–100 ng.
[b]Sample 1 lacks the gene of interest and, therefore, controls for sample 4.
[c]Sample 2 lacks the gene of interest and, therefore, controls for sample 5.
[d]Sample 3 lacks the gene of interest and, therefore, controls for sample 6.
[e]Sample 4 measures the effect of the gene product on the signal transduction pathway involved.
[f]Sample 5 measures the effect of the gene product on the signal transduction pathway involved.
[g]Sample 6 measures the effect of the gene product on the signal transduction pathway involved.
[h]Sample 7 measures the efficacy of the assay for the cell line chosen.
[i]Sample 8 does not contain an activation domain and should show results similar to samples 1–3.

Table 2
Sample Experiment to Study the Effects of Extracellular Stimuli

Sample no.	pFR-Luc Plasmid (reporter plasmid)	Fusion trans-activator plasmid[a]	pFC2-dbd (negative control)	Positive control	Extracellular stimuli
[b]	1.0 µg (1 µL)	—	50 ng (2 µL)	—	Serum (10%)
[c]	1.0 µg (1 µL)	50 ng (2 µL)	—	—	Serum (10%)
[d]	1.0 µg (1 µL)	—	50 ng (2 µL)	—	EGF (100 ng/mL)
[e]	1.0 µg (1 µL)	50 ng (2 µL)	—	—	EGF (100 ng/mL)
[f]	1.0 µg (1 µL)	—	50 ng (2 µL)	—	Medium
[g]	1.0 µg (1 µL)	50 ng (2 µL)	—	—	Medium
[h]	1.0 µg (1 µL)	50 ng (2 µL)	—	50 ng (2 µL)	—
[i]	1.0 µg (1 µL)	—	50 ng (2 µL)	50 ng (2 µL)	—

[a]This quantity may need to be optimized, usually within the range of 1–100 ng.
[b]Sample 1 lacks the fusion trans-activator protein and, therefore, controls for sample 2.
[c]Sample 2 measures the effect of FBS on kinase activation.
[d]Sample 3 lacks the fusion trans-activator protein and, therefore, controls for sample 4.
[e]Sample 4 measures the effect of EGF on kinase activation.
[f]Sample 5 controls for the extracellular stimulus as well as the fusion *trans*-activator protein.
[g]Sample 6 controls for the extracellular stimulus.
[h]Sample 7 measures the efficacy of the assay for the cell line chosen.
[i]Sample 8 does not contain an activation domain and should show results similar to samples 1,3 and 5.

responding signal transduction pathways (*see* **Table 2**). Cells are transfected with the fusion trans-activator plasmid and then treated with the stimulus of interest. Luciferase expression from the reporter plasmid indicates the activation of the fusion trans-activator protein and, therefore, the presence of the endogenous protein kinase (e.g., MAPK, JNK, PKA, p38 kinase, or an uncharacterized upstream activator).

3.2. Cell Culture and Transfection

3.2.1. Growing the Cells (see **Note 5**)

1. Thaw and seed frozen cell stocks in complete medium in 50-mL or 250-mL tissue culture flasks.
2. Split the cells when they just become confluent.
3. Subculture the cells at an initial density of approx 1×10^5–2×10^5 cells/mL every 3–4 d.

3.2.2. Preparing the Cells

1. Seed 3×10^5 cells in 2 mL of complete medium in each well of a six-well tissue culture dish.
2. Incubate the cells at 37°C in a CO_2 incubator for 24 h.

3.3. Preparing the DNA Mixtures for Transfection

3.3.1. Studying the Effects of a Gene Product

Combine the plasmids to be cotransfected in a sterile Falcon 2054 polypropylene tube as indicated for samples 1–8 in **Table 1**. As each assay is run in triplicate, the amount of plasmid DNA in each tube should be sufficient for three transfections (*see* **Table 1** for the appropriate amounts). For example, to prepare sample 1 in triplicate as indicated in **Table 1**, combine the following components in an Eppendorf microcentrifuge tube and then proceed to **Subheading 3.4.**:

> 3 μL (3 μg) of pFR-Luc
> 6 μL (150 ng) of fusion trans-activator plasmid
> 150 ng of experimental plasmid without an insert
> 2.85 μg of unrelated plasmid DNA

3.3.2. Studying the Effects of Extracellular Stimuli

Combine the plasmids to be cotransfected in a sterile Falcon 2054 polypropylene tube as indicated by samples 1–6 in Luciferase **Table 2**. As each assay is run in triplicate, the amount of plasmid DNA in each tube should be sufficient for three transfections (*see* **Table 2** for appropriate amounts). For example, to prepare sample 1 in triplicate as indicated in **Table 2**, combine the following components in a microcentrifuge tube and then proceed to **Subheading 3.4.**:

> 3 mL (3 μg) of pFR–Luc
> 6 mL (150 ng) of pFC2–dbd

3.4. Transfecting the Cells

A number of transfection methods, including calcium phosphate precipitation and lipid-mediated transfection, may be used. Transfection efficiencies vary between cell lines and according to experimental conditions. The following protocol utilizes HeLa cells and a lipid-mediated transfection method using either LipoTaxi (Stratagene) or LipofectAMINE (Life Technology, Bethesda, MD). Transfection procedures should be optimized for the cell line chosen.

1. Using a lipid transfection kit, prepare a DNA–lipid complex according to the manufacturer's instructions.
2. Dilute the DNA–lipid complex to a final volume of 3 mL (for a triplicate sample) with serum-free medium.
3. Rinse the cells (from **step 2** of **Subheading 3.2.2.**) with 2 mL of serum-free medium.
4. Mix gently and overlay 1 mL of the diluted DNA–lipid complex onto the cells in each of the three wells.

5. Incubate the cells with the DNA–lipid complex for 5–7 h or as recommended by the manufacturer.
6. After incubation, add 1.2 mL of complete medium containing 1% FBS to each well (*see* **Note 6**).
7. **If studying the effects of a gene product,** perform **step 7a. If studying the effects of extracellular stimuli,** perform **step 7b.**
   a. Replace the medium with fresh complete medium containing 0.5% FBS (*see* **Note 6**) 18–24 h after the beginning of transfection. After incubating an additional 18–24 h, proceed to **Subheading 3.5.**
   b. Replace the medium with fresh medium containing the appropriate extracellular stimuli (e.g., EGF) 18–24 h after the beginning of transfection. After incubating an additional 5–7 h, proceed to **Subheading 3.5.**

3.5. Extracting the Luciferase

1. Remove the medium from the cells and carefully wash the cells twice with 2 mL of PBS buffer.
2. Remove as much PBS as possible from the wells with a Pasteur pipet. Add 400 µL of cell lysis buffer to the wells and swirl the dishes gently to ensure uniform coverage of the cells.
3. Incubate the dishes for 15 min at room temperature. Swirl the dishes gently midway through the incubation. Assay for luciferase activity directly from the wells within 2 h.
   *To store for later analysis, transfer the solutions from each well into a separate microcentrifuge tube. Spin the samples in a microcentrifuge at full speed. Store at –80°C (see **Note 7**).*

3.6. Performing the Luciferase Activity Assay

1. Mix 5–20 µL of cell extract with 100 µL of luciferase assay reagent, both equilibrated to room temperature, in a Falcon 2054 polypropylene tube.
2. Measure the light emitted from the reaction with a luminometer using an integration time of 10–30 s (*see* **Note 8** and **refs.** *38* and *39*). Luciferase activity may be expressed in relative light units (RLU) as detected by the luminometer from the 20-µL sample. The activity may also be expressed as RLU/well, RLU/number of cells, or RLU/mg of total cellular protein.

4. Notes

1. pFR-Luc plasmid (**Fig. 3**).
2. Fusion transactivator plasmids (**Fig. 4**).
3. Control Plasmids (**Fig. 5**).
4. pFA-CMV Plasmid (**Fig. 6**).
5. The protocols given are designed for adherent cell lines such as HeLa and NIH3T3. Optimization of media and culture conditions may be required for other cell lines.

Sequence of GAL4 Binding Element in the pFR-Luc Plasmid

```
GT CGGAGTACTGTCCTCCG AG CGGAGTACTGTCCTCCG
AG CGGAGTACTGTCCTCCG AG CGGAGTACTGTCCTCCG
AG CGGAGTACTGTCCTCCG AG CGGAGACTCTAGAGGG
TATATA ATGGATCCCCGGGT AC CGAGCTCGAATTC--
--CAGCTTGGCATTCCGGTACTGTTGGTAAA ATG--Luciferase
```

Fig. 3. pFR-Luc Plasmid.

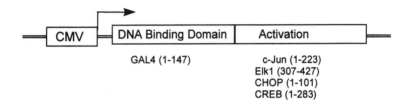

GAL4 (1-147)   c-Jun (1-223)
                                      Elk1 (307-427)
                                      CHOP (1-101)
                                      CREB (1-283)

Fig. 4. Fusion transactivator plasmids.

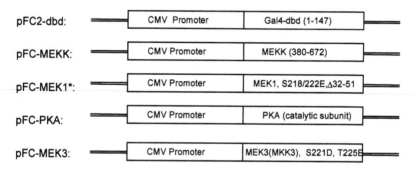

Fig. 5. Control plasmids.

6. Due to the possibility that uncharacterised factors in the serum may activate signal transduction pathways, low serum concentrations are used; however, the use of 10% serum has also yielded satisfactory results in some cases.

7. If this passive lysis method does not yield satisfactory results, perform the following active lysis. Scrape all surfaces of the tissue culture dish, pipet the cell lysate to microcentrifuge tube and place on ice. Lyse the cells by brief sonication with the microtip set at the lowest setting **or** freeze the cells at –80°C for 20 min and then thaw in a 37°C water bath and vortex 10–15 s. Spin the tubes in a microcentrifuge at maximum speed for 2 min. Use the supernatant for the luciferase activity assay.

pFA-CMV Plasmid

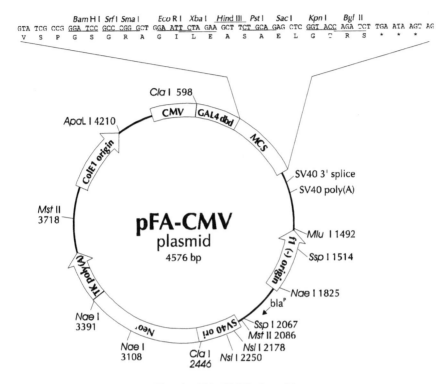

Fig. 6. pFA-CMV plasmid.

8. Each freeze-thaw cycle of cell lysates results in a significant loss of luciferase activity (as much as 50%). Make sure the assay reagents are equilibrated to room temperature before performing the assays as the luminescent reaction catalysed by luciferase is greatly affected by temperature. If very low luciferase activity is present in the samples, an integration period as long as a few minutes can be used, although doing so will greatly increase the time taken to do the assays.

Acknowledgments

The authors wish to thank John Bauer, Mary Buchanan, and Joe Sorge for their support of this project, Cathy Chang, Peter Vaillancourt, and Alan Greener for valuable suggestions and stimulating discussion, Maurina Sherman and David Boe for critical reading of the protocols, and many others in Stratagene for various favors. We would also like to thank Dr. C. Hauser at La Jolla Cancer Foundation, Drs. M. Karin and F. X. Claret at UCSD, Dr. J. S. Gutkind at the NIH, Dr. A. Lin at the University of Alabama, Dr. T. Deng at the University of Florida, and Dr. K.-L. Guan at the University of Michigan.

References

1. Boulikas, T. (1995) Phosphorylation of transcription factors and control of the cell cycle. *Crit. Rev. Eukaryotic Gene Express.* **5**, 1–77.
2. Hunter, T. and Karin, M. (1992) The regulation of transcription by phosphorylation. *Cell* **70**, 357–387.
3. Karin, M. and Hunter, T. (1995) Transcriptional control by protein phosphorylation: signal transmission from the cell surface to the nucleus. *Curr. Biol.* **5**, 747–757.
4. Treisman, R. (1996) Regulation of transcription by MAP kinase cascades. *Curr. Opin. Cell Biol.* **8**, 205–215.
5. Ray, L. B. and Sturgill, T. W. (1987) Rapid stimulation by insulin of a serine/threonine kinase in 3T3-L1 adipocytes that phosphorylates microtubule-associated protein 2 in vitro. *Proc. Natl. Acad. Sci. USA* **84**, 1502–1506.
6. Kultz, D. (1998) Phylogenetic and functional classification of mitogen- and stress-activated protein kinases. *J. Mol. Evol.* **46**, 571–588.
7. Levin, D. E. and Errede, B. (1995) The proliferation of MAP kinase signaling pathways in yeast. *Curr. Opin. Cell Biol.* **7**, 197–202.
8. Waskiewicz, A. J. and Cooper, J. A. (1995) Mitogen and stress response pathways: MAP kinase cascades and phosphatase regulation in mammals and yeast. *Curr. Opin. Cell Biol.* **7**, 798–805.
9. Cobb, M. and Coldsmith, E. J. (1995) How MAP kinases are regulated. *J. Biol. Chem.* **270**, 14,843–14,846.
10. Boulton, T. G., Nye, S. H., Robbins, D. J., Ip, N. Y., Radziejewska, E., Morgenbesser, S. D., DePinho, R. A., Panayotatos, N., Cobb, M. H., and Yancopoulos, G. D. (1991) ERK's: a family of protein serine/threonine kinases that are activated and tyrosine phosphorylated in response to insulin and NGF. *Cell* **65**, 663–675.
11. Kyriakis J. M. et al. (1994) The stress-activated protein kinase subfamily of c-Jun kinases, *Nature* **369**, 156–160.
12. Derijard B., Hibi, M., Wu, I. -H., Barrett, T., Su, B., Deng, T., Karin, M., and Davis, R. (1994) JNK1: a protein kinase stimulated by UV light and Ha-Ras that binds and phosphorylates the c-Jun activation domain. *Cell* **76**, 1025–1037.
13. Han, J., Bibbs, L., and Ulevitch, R. J. (1994) A MAP kinase targeted by endotoxin and hyperosmolarity in mammalian cells. *Science* **265**, 808–811.
14. Chen, R. H., Sarnecki, C., and Blenis, J. (1992) Nuclear localization and regulation of erk- and rsk-encoded protein kinases. *Mol. Cell. Biol.* **12**, 915–927.
15. Traverse, S., Gomez, N., Paterson, H., Marshall, C., and Cohen, P. (1992) Sustained activation of the mitogen-activated protein (MAP) kinase cascade may be required for differentiation of PC12 cells. Comparison of the effects of nerve growth factor and epidermal growth factor. *Biochem. J.* **288**, 351–355 .
16. Ma, J. and Ptashne, M. (1987) Deletion analysis of GAL4 defines two transcriptional activating segments. *Cell* **48**, 847–853.
17. Sadowski, I. and Ptashne, M. (1989) A vector for expressing GAL4(1-147) fusions in mammalian cells. *Nucleic Acids Res.* **17**, 7539.

18. Fields, S. and Song, O. (1989) A novel genetic system to detect protein-protein interactions. *Nature* **340**, 245,246.
19. Mendelsohn, A. R. and Brent. R., (1994) Applications of interaction traps/two-hybrid systems to biotechnology research. *Curr. Opin. Biotech.* **5**, 482–486.
20. Allen, J. B., Walberg, M. W., Edwards, M. C., and Elledge, S. J. (1995) Finding prospective partners in the library: the two-hybrid system and phage display find a match. *Trends Biochem. Sci.* **20**, 511–517.
21. Jausons-Loffreda, N., Balaguer, P., Roux, S., Fuentes, M., Pons, M., Nicolas, J.-C., Gelmini, S., and Pazzagli, M. (1994) Chimeric receptors as a tool for lumines-cent measurement of biological activities of steroid hormones. *J. Biolumin. Chemilumin.* **9**, 217–221.
22. Braselmann, S., Graninger, P., and Busslinger, M. (1993) A selective transcrip-tional induction system for mammalian cells based on Gal4-estrogen receptor fusion proteins. *Proc. Natl. Acad. Sci. USA* **90**, 1657–1661.
23. Louvion, J. -F., Havaux-Corpf, B., and Picard, D. (1993) Fusion of GAL4-VP16 to a steroid-binding domain provides a tool for gratuitous induction of galactose-responsive genes in yeast. *Gene* **131**, 129–134.
24. Dang, C. V., Barrett, J., Billa-Garcia, M., Resar, L. M. S., Kato, G., and Fearon, E. R. (1991) Intracellular leucine zipper interactions suggest c-Myc hetero-oligo-merization. *Mol. Cell. Biol.* **11**, 945–962.
25. Flint, K. J. and Jones, N. C. (1991) Differential regulation of three members of the ATF/CREB family of DNA- binding proteins. *Oncogene* **6**, 2019–1026.
26. Xing, J., Ginty, D. D., and Greenberg, M. E. (1996) Coupling of the RAS-MAPK pathway to gene activation by RSK2, a growth factor-regulated CREB kinase. *Science,* **273**, 959–963.
27. Xu, L., Sanchez, T., and Zheng, C. -F. (1997) In vivo signal transduction pathway reporting systems. *Strategies* **10**, 1–3.
28. Enslen, H., Tokumitsu, H., Stork, P. J. S., Davis, R. J., and Soderling, T. R. (1996) Regulation of mitogen-activated protein kinases by a calcium/calmodulin-dependent protein kinase cascade. *Proc. Natl. Acad. Sci. USA* **93**, 10,803–10,808.
29. Marais R., Wynne, J., and Treisman, R. (1993) The SRF accessory protein Elk-1 contains a growth factor-regulated transcriptional activation domain. *Cell* **73**, 381–393.
30. Minden, A., Lin, A., Claret, F. X., Abo, A., and Karin, M. (1995) Selective activa-tion of the JNK signaling cascade and c-Jun transcriptional activity by the small GTPase Rac and cdc42Hs. *Cell* **81**, 1147–1157.
31. Lin A., Minden, A., Martinetto, H., Claret, F. X., Lange-Carter, C., Mercurio, F., Johnson, G. L., and Karin, M. (1995) Identification of a dual specificity kinase that activates the Jun kinases and p38-Mpk2. *Science* **268**, 286–289.
32. Smeal, T., Hibi, M., and Karin, M. (1994) Altering the specificity of signal trans-duction cascades: positive regulation of c-Jun transcriptional activity by protein kinase A. *EMBO J.* **13**, 6006–6010 .

33. Lee, J.-S., See, R. H., Deng, T., and Shi, Y. (1996) Adenovirus E1A downregulates cJun- and JunB-mediated transcription by targeting their coactivator p300. *Mol. Cell. Biol.* **16,** 4312–4326.

34. Wang, X. Z. and Ron, D. (1996) Stress-induced phosphorylation and activation of the transcription factor CHOP (GADD 153) by p38 MAP kinase. *Science* **272,** 1347–1349.

35. Xu, L., Sanchez, T., and Zheng, C. -F. (1997) Signal Transduction Pathway reporting systems using cis-acting enhancer elements. *Strategies* **10,** 79–80.

36. Xu, L. and Zheng, C. -F. (1997) New fusion trans-activator plasmids for studying signal transduction pathways. *Strategies,* **10,** 81–83.

37. Sanchez, T., Xu, L., Buchanan, M., and Zheng, C. -F. (1998) Optimizing transfection conditions for studying signal transduction pathways. *Strategies* **11,** 52,53.

38. de Wet, J. R., Wood, K. V., DeLuca, M., Helinski, D. R., and Subramani, S. (1987) Firefly luciferase gene: Structure and expression in mammalian cells. *Mol. Cell. Biol.* **7,** 725–737.

39. Thompson, J. F., Hayes, L. S., and Lloyd, D. B. (1993) Modulation of firefly luciferase stability and impact on studies of gene expression. *Gene* **103,** 171–177.

13

Use of Kinase Inhibitors
to Dissect Signaling Pathways

Ana Cuenda and Dario R. Alessi

1. Introduction

Protein kinases form one of the largest families of proteins encoded in the human genome, and these enzymes have critical roles in controlling all cellular processes. The abnormal phosphorylation state of proteins is the cause or consequence of many diseases and, for this reason, protein kinases have become attractive drug targets for the treatment of cancer, diabetes, hypertension, inflammation, and other disorders *(1–3)*.

There are estimated to be approx 2,000 protein kinases encoded by the human genome, phosphorylating a total of approx 30,000 proteins. One of the major challenges of current research is to be able to establish the downstream physiological processes that each kinase is regulating in a cell, as well as determining which upstream signal-transduction components regulate its activity. One of the most successful strategies to achieve this goal has been the use of small-cell permeant compounds that are specific inhibitors of particular protein kinases. Despite the difficulties in obtaining such compounds, there are now several relatively specific protein kinase inhibitors that are readily available to inhibit the classical mitogen-activated protein kinase (MAPK), the stress-activated protein kinase-2 (SAPK2/p38), the phosphoinositide 3-kinase (PI 3-kinase) and the mTOR/p70 S6 kinase pathways (**Fig. 1**). Over the past few years, these inhibitors have revolutionized our understanding of the physiological processes that are regulated by these signaling pathways. In this chapter, we discuss the use of these compounds in mammalian cells and the known pitfalls of each inhibitor. We also provide assay procedures that should be used to demonstrate that the inhibitors are preventing the activation of the desired kinase in cells.

From: *Methods in Molecular Biology, vol. 99: Stress Response: Methods and Protocols*
Edited by: S. M. Keyse © Humana Press Inc., Totowa, NJ

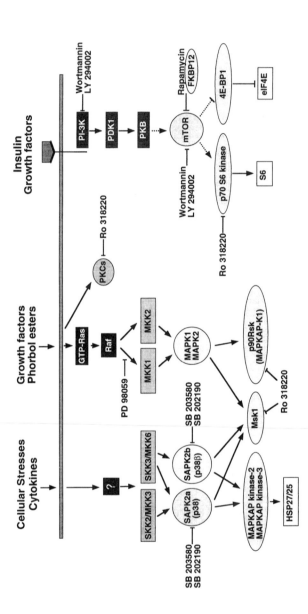

Fig. 1. Schematic representation of the blockade by different inhibitors of the stress and mitogen-induced signaling pathways in mammalian cells. When cells are subjected to different stresses or cytokines, SKK2/MKK3 and SKK3/MKK6 mediate the activation of SAPK2a/p38 and SAPK2b/p38β, whose activity is blocked by SB 203580/SB202190. Members of the Raf family of protein kinase mediate the activation of MKK1 and MKK2 by growth factors. PD 98059 prevents the activation of MKK1 by Raf, but the activation of MKK2 is inhibited much more weakly. Ro-318220 not only inhibits PKC activity, but also blocks p70 S6 kinase p90 Rsk1/2 and Msk1 activities. PI 3-kinase (PI-3K) is activated by a ligand (growth factor)-bound tyrosine kinase receptor. PI 3-kinase activity is inhibited by wortmannin and LY 294002. Activated PI-3K phosphorylates phosphatidylinositol 4,5-P$_2$ (PIP2) to produce phosphatidylinositol 3,4,5-P$_3$ (PIP3), which binds to PKB to allow it to be phosphorylated and activated by PDK1. Active PKB may activate mTOR, whose activity is blocked by the inhibitory complex rapamycin-FKBP12. Wortmannin (at high concentration) and LY 294002 also inhibit mTOR activity. An arrow indicates activation and a bar indicates repression. A broken arrow or bar indicates a link that is yet to be resolved.

162

2. Materials

1. All the inhibitors should be dissolved in hygroscopic and sterile dimethylsulphoxide (DMSO) from Sigma (St. Louis, MO). The stock concentration should be 1000-fold the final concentration used to treat cells and solutions should be stored in aliquots at $-20°C$. These compounds should be thawed immediately before use and repeated freezing and thawing cycles should be avoided. The inhibitors should be added directly to the tissue culture medium of cells without further dilution. For control experiments (without inhibitor), the equivalent volume of sterile DMSO should be added to the cells. The volume of DMSO added to the culture medium should always equal or 0.1% of the total volume of medium in which the cells are incubated as this amount is not toxic.

2. Buffer A: 50 mM Tris-HCl, pH 7.5, 0.27 M sucrose, 1 mM EDTA, 1 mM EGTA, 1 mM sodium orthovanadate, 10 mM sodium β-glycerophosphate, 50 mM sodium fluoride, 5 mM sodium pyrophosphate, 1 % (w/v) Triton X-100, 0.1% (v/v) 2-mercaptoethanol, 1 mM benzamidine, 0.2 mM phenylmethylsulfonyl fluoride (PMSF) and leupeptin (5 µg/mL).

3. Sodium dodecyl sulfate-polyacrylamide gel electrophoresis (SDS-PAGE) sample buffer: 6% (w/v) SDS, 400 mM Tris (pH 6.8), 50% (v/v) glycerol, 5% (v/v) 2-mercaptoethanol, and 0.2% (v/v) bromophenol blue.

4. PKI, a 20-residue peptide (TTYADFIASGRTGRRNAIHD) which is a specific inhibitor of cyclic AMP-dependent protein kinase, can be purchased from Sigma.

5. Crosstide (GRPRTSSFAEG *[4]*) from Upstate Biotechnology.

6. 50 mM magnesium acetate; 0.5 mM [γ-^{32}P]ATP (2×10^5 cpm/nmol diluted from a 10 µCi/µL stock from Amersham).

7. 12-O-Tetradecanoylphorbol 13-acetate (TPA) and anisomycin can be purchased from Sigma, epidermal growth factor (EGF) and platelet-derived growth factor (PDGF-B/B) form are purchased from Gibco/BRL or Boehringer Mannheim (Mannheim, Germany). Insulin can be purchased from Norvo Nordisk.

2.1. Inhibitors of SAPK2/p38: SB 203580 and SB 202190

1. SB 203580 and SB 202190 can be purchased from Calbiochem. A stock solution of 20 mM should be prepared.

2. Phosphate-free Dulbecco's modified Eagle medium (DMEM) from ICN.

3. Dialyzed fetal calf serum (FCS) is prepared by dialysing it against phosphate-buffered saline (PBS) (prepared from Oxoid tablets, cat. no. BR 14a).

4. [^{32}P]-orthophosphate from Amersham (10×10^6 cpm/µL).

5. Anti-HSP27/25 antibody can be purchased from Stressgen (York, UK).

2.2. An Inhibitor of MKK1/MEK1 Activation: PD 98059

1. PD 98059 can be purchased from Calbiochem or from New England Biolabs (Beverly, MA). A stock solution of 50 mM should be prepared. PD 98059 is soluble in aqueous solutions but only up to a concentration of 50 µM.

2. MAPK reaction buffer: 50 mM Tris-HCl, pH 7.5, 0.2 mM EGTA, 0.2 mM sodium orthovanadate, 2 µM PKI and 0.66 mg/mL myelin basic protein (MBP) from Gibco-BRL.

3. p90Rsk (MAPKAP kinase-1) reaction buffer: 50 m*M* Tris-HCl, pH 7.5, 0.2 m*M* EGTA, 0.2% (v/v) 2-mercaptoethanol, 5 μ*M* PKI, and 60 μ*M* Crosstide.
4. Anti-p42MAPK and anti-p90Rsk (MAPKAP kinase-1) polyclonal antibodies can be purchased from Upstate Biotechnology.

2.3. Rapamycin

1. Rapamycin can be purchased from Calbiochem. A stock solution of 1–10 m*M* should be prepared.
2. p70 S6 kinase reaction buffer: 50 m*M* Tris-HCl, pH 7.5, 0.2 m*M* EGTA, 0.2% (v/v) 2-mercaptoethanol, 5 μ*M* PKI and 200 μ*M* of the peptide KKRNRTLTV or 60 μ*M* Crosstide.
3. Anti-p70 S6 kinase polyclonal antibody can be purchased from Upstate Biotechnology.

2.4. Wortmannin and LY 294002

1. Wortmannin can be purchased from Sigma and LY 294002 from Calbiochem. Stock solutions of 0.1 m*M* (wortmannin) or 100 m*M* (LY 294002) should be prepared. Wortmannin is extremely toxic and it should be handled and dissolved in a fume hood. Wortmannin, unlike LY 294002, is also extremely unstable and has a half-life of only a few minutes in aqueous solution. Once thawed, wortmannin should be used immediately and never refrozen. If incubations of longer than 20 min are required, it is recommended that LY294002 is used rather than wortmannin, because of its much higher stability in aqueous solution. Stock wortmannin solutions in DMSO should be completely transparent. If they are a pale yellow color, this is a sign that they have degraded and should not be used.
2. PKB reaction buffer: 50 m*M* Tris-HCl, pH 7.5, 0.2 m*M* EGTA, 0.2% (v/v) 2-mercaptoethanol, 5 μ*M* PKI, and 60 μ*M* Crosstide.
3. Polyclonal antibodies which specifically recognise the three distinct isoforms of PKB (α, β, and γ) can be obtained from Upstate Biotechnology.

2.5. Ro 318220 and GF 109203X

Ro 318220 and GF 109203X can be purchased from Calbiochem (Nottingham, UK). Stock solutions of 5 m*M* should be prepared.

3. Methods
3.1. Inhibitors of SAPK2/p38: SB 203580 and SB 202190

These two compounds belong to a class of pyridinyl imidazoles (**Fig. 2**) developed at SmithKline Beecham as inhibitors of the LPS-induced synthesis of proinflammatory cytokines (interleukin-1 [IL-1] and tumor necrosis factor [TNF]) in monocytes *(5)*. SB 203580 and SB 202190 are relatively specific inhibitors of SAP2a/p38 and SAPK2b/p38β (SAPK2/p38) *(6,7)*. They inhibit SAPK2/p38 with an IC_{50} in the submicromolar range and do not significantly

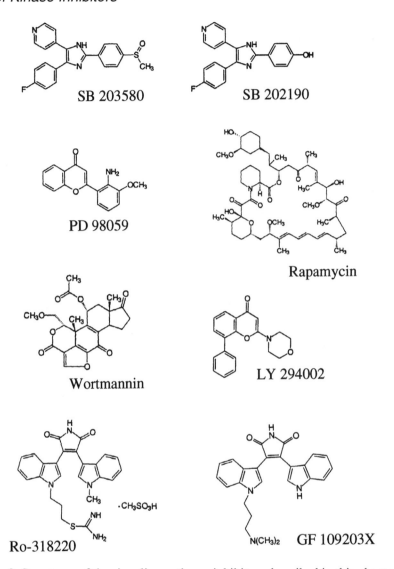

Fig. 2. Structures of the signaling pathway inhibitors described in this chapter.

inhibit many other protein kinases tested in vitro (including other closely related SAPKs like SAPK3/p38γ, SAPK4/p38δ) at a concentration of 10 μM *(6–8;see* **Note 1**). SB 203580 has been used to elucidate some of the physiological processes in which SAPK2/p38 is involved *(9)* and to identify physiological substrates of SAPK2/p38, such as the protein kinases MAPKAP kinase-2, MAPKAP kinase-3, PRAK/MAPKAP kinase-5, Mnk1/2, and Msk1, and several transcription factors *(9–11)*.The specificity of SB 203580 in cell-based assays appears high, where it suppresses the phosphorylation of substrates of SAPK2/p38 but not the phosphorylation of other proteins mediated by the MAPK, SAPK1/JNK, p70 S6 kinase, or protein kinase B (PKB) signaling pathways *(9)*.

It is recommended that for initial experiments SB 203580/SB 202190 is added to cells at a final concentration of 10 μM for 30–60 min prior to subjecting the cells to a cellular stress (e.g., UV irradiation, arsenite or osmotic shock) or cytokine treatment (e.g., IL-1 or TNF). In order to verify that the SB203580/SB202190 inhibitors have worked it is essential to assay MAPKAP kinase-2 or -3 activity or HSP27/25 (which is a physiological substrate for MAPKAP kinase-2) phosphorylation in cells that have been stimulated in the presence or absence of these compounds.

3.1.1. Inhibition of MAPKAP Kinase-2 (or MAPKAP Kinase-3) Activation by SB 203580 /SB 202190 In Vivo

Incubation of cells with SB203580/SB202190 should inhibit by >80% the activation of MAPKAP kinase-2 or MAPKAP kinase-3 induced by cytokines and stressful stimuli.

1. Incubate cells for 30–60 min (in the culture medium) in the presence of 10 μM SB 203580/SB 202190 or with the equivalent volume of DMSO (as a control) before stimulation.
2. Stimulate cells for the required length of time with cellular stress or cytokines (as described in Subheading 3.2. of Chapter 11, this volume) in the continuous presence of SB 203580/SB 202190 or DMSO.
3. After stimulation, remove the medium, place the cells on ice and add 1.0 mL (for a 10-cm dish) or 0.5 mL (for a 6-cm dish) of ice-cold buffer A, scrape the cells with a cell scraper and transfer the cell lysate to an Eppendorf tube.
4. Centrifuge the lysate 5 min at 14,000g at 4°C, remove the supernatant and determine its protein concentration. At this stage the samples can be snap frozen in liquid nitrogen and stored at –80°C for up to 4 mo without any loss of activity.
5. Immunoprecipitate and determine the MAPKAP kinase-2 or -3 activity as described in Subheading 3.3.2. of Chapter 11, this volume.

3.1.2. Inhibition of HSP27/25 Phosphorylation by SB203580 In Vivo

Incubation of cells with SB 203580 should inhibit the stress/cytokine induced phosphorylation of HSP27/25 by >80%.

1. Wash cells (cultured on a 10-cm or 6-cm dish) four times with phosphate-free DMEM containing 10% dialysed FCS, and then incubate for 3 h in the same medium supplemented with 0.5 mCi of [^{32}P]-orthophosphate.
2. Incubate cells for 30 min (without changing the medium from **step 1**) in the presence of 10 μM SB 203580 or the equivalent volume of DMSO prior to stimula-

tion with cellular stresses or cytokines (as described in Subheading 3.2. of Chapter 11, this volume) in the continuous presence of SB 203580 or DMSO.

3. After stimulation, place the cells on ice and wash three times with ice cold PBS. Lyse the cells as in **Subheading 3.1.1.**, centrifuge for 5 min at 14,000g (4°C) and remove the supernatant for analysis.

4. Incubate a suitable volume of cell lysate (i.e., 100 μg protein) with 10 μL protein G-Sepharose conjugated to 10 μg of anti-HSP27/25 antibody (to conjugate Protein G-Sepharose to any antibody, *see* Subheading 3.1. of Chapter 11, this volume).

5. Incubate for 60 min at 4°C on a shaking platform and then centrifuge the suspension at 14,000g for 2 min.

6. Discard the supernatant and wash the antibody-protein-G–Sepharose pellet four times with buffer A containing 0.5 M NaCl, and twice with buffer A.

7. Add 5 μL of SDS-PAGE sample buffer to the pellets (40 μL final volume) and incubate for 5 min at 100°C.

8. Load the samples onto a standard 15% SDS-PAGE gel to resolve the proteins *(12)*.

9. Following electrophoresis, dry the gel.

10. Autoradiograph the gel to localise HSP27/25. HSP27/25 phosphorylation can then be quantified by excision of the band corresponding to HSP27/25 and measurement of ^{32}P incorporation by scintillation counting.

3.2. Inhibition of MKK1/MEK1 Activation: PD 98059

PD 98059 is a flavone compound (**Fig. 2**) developed at Park-Davis that binds to the inactive form of MAPK kinase-1 (MKK1), preventing its activation by c-Raf and other upstream activators *(13)*. It does not compete with ATP and does not inhibit the phosphorylated (fully activated) form of MKK1. This unique mechanism of action probably explains the exquisite specificity of PD 98059 in vitro and in vivo *(13)*. Although the amino acid sequence of MKK2 is 90% identical to that of MKK1, PD 98059 is at least 10-fold less effective in preventing MKK2 activation in vitro. The PD 98059 inhibitor has been shown not to inhibit any other kinase tested in vitro, or affect the activation of the SAPK1/JNK, SAPK2/p38, SAPK3/p38γ, SAPK4/ p38δ, PI 3-kinase, PKB, and p70 S6 kinase signaling pathways (*see* **Note 2**). It is recommended that PD 98059 is added to cells at a concentration of 50 μM, 30–60 min prior to subjecting the cells to stimulation with phorbol ester or growth factors (e.g., TPA, EGF, PDGF). In order to verify that the PD 98059 has worked it is essential to assay for the activation of either p42/p44MAPK or p90Rsk1/2 in cells stimulated in the presence or absence of the inhibitor.

3.2.1. Inhibition of MAPK Activation by PD 98059 In Vivo

Incubation of cells with PD 98059 typically results in over 90% inhibition of the activation of MAPK (*see* **Note 3**).

1. Incubate cells for 30–60 min (in the culture medium) in the presence of 50 μ*M* PD98059 or with the equivalent volume of DMSO (as a control) before stimulation.
2. In order to activate the MAPK pathway, cells should be stimulated for 5 to 15 min (in the continuous presence of PD 98059 or DMSO) with either of the following stimuli: 200 ng/mL TPA, 100 ng/mL EGF or 50 ng/mL PDGF.
3. Lyse the cells in buffer A and centrifuge as described in **Subheading 3.1.1.**
4. Incubate 50–100 μg protein cell extract with 5 μL of protein G-Sepharose conjugated to 5 μg of anti-p42MAPK antibody for 60 min at 4°C on a shaking platform (to conjugate protein-G–Sepharose to any antibody, *see* Subheading 3.1. of Chapter 11, this volume).
5. Centrifuge the suspension for 1 min at 14,000*g*, discard the supernatant and wash the pellet twice with 1.0 mL of buffer A containing 0.5 *M* NaCl and twice with buffer A. Assay the immunoprecipitated MAP kinase activity by incubating the pellets with a mixture containing 10 μL distilled water and 25 μL of MAPK reaction buffer for 3 min at 30°C, prior starting the reaction by the addition of 10 μL of 50 m*M* magnesium acetate; 0.5 m*M* [γ-^{32}P]ATP (approx 2×10^5 cpm/nmol). This reaction must be carried out on a shaking platform at 30°C to keep the immunoprecipitated MAPK/protein-G–Sepharose complex in suspension.
6. After 15 min at 30°C, stop the reaction by pipetting 40 μL of the assay mixture on to phosphocellulose P81 paper, wash these papers and quantify the radioactivity exactly as described in Subheading 3.3.1. of Chapter 11, this volume.

3.2.2. Inhibition of p90Rsk (MAPKAP kinase-1) Activation by PD 98059 In Vivo

Incubation of cells with PD 98059 typically results in over 90% inhibition of the activation of p90Rsk (*see* **Note 3**)

1. Incubate cells for 30–60 min (in the culture medium) in the presence of 50 μ*M* PD 98059 or with the equivalent volume of DMSO (as a control) before stimulation.
2. Stimulate, lyse the cells in buffer A and centrifuge as described in **Subheading 3.1.1.**
3. Incubate 50 μg protein cell extract with 5 μL of protein-G–Sepharose conjugated to 5 μg of p90Rsk1/2 antibody for 60 min at 4°C on a shaking platform (to conjugate protein-G–Sepharose to any antibody, *see* **Subheading 3.1.** of Chapter 11, this volume).
4. Centrifuge the suspension for 1 min at 14,000*g*, discard the supernatant and wash the pellet twice with 1.0 mL of buffer A containing 0.5 *M* NaCl, and twice with buffer A.
5. Assay the immunoprecipitated p90Rsk activity by incubating the pellets with a mixture containing 10 μL distilled water and 25 μL of p90Rsk 1/2 reaction buffer for 3 min at 30°C, prior to starting the reaction by the addition of 10 μL of 50 m*M* magnesium acetate; 0.5 m*M* [γ-^{32}P]ATP (approx 2×10^5 cpm/nmol). This reaction must be carried out on a shaking platform at 30°C to keep the immunoprecipitated kinase/protein-G–Sepharose complex in suspension.

6. After 15 min at 30°C stop the reaction by pipetting 40 μL of the assay mixture on to phosphocellulose P81 paper, wash the papers and quantify the radioactivity as described in Subheading 3.3.1. of Chapter 11, this volume.

3.3. Rapamycin

Rapamycin (**Fig. 2**) is a potent immunosuppressant and at concentrations of 100 nM will lead to a rapid inactivation of p70 S6 kinase in all cells studied. Rapamycin will also prevent the activation of p70 S6 kinase by all known agonists *(14–16)*. Rapamycin does not inactivate the p70 S6 kinase directly. Instead, it interacts with the immunophilin FK506-binding protein (FKBP) which is a peptidyl prolyl isomerase involved in regulating protein folding. The rapamycin-FKBP complex then interacts with the protein kinase mTOR/FRAP, thereby inactivating it (**Fig. 1**). This leads to the inactivation of the p70 S6 kinase by an unknown mechanism (*see* **Note 4**). Using cells which overexpress a mutant mTOR protein that cannot interact with the FKBP-rapamycin complex it has been shown that p70 S6 kinase activity or its activation is no longer sensitive to rapamycin. As a positive control to demonstrate that rapamycin is effective in cells, we recommend that TPA, growth factor, or insulin induced activation of the p70 S6 kinase should be determined in the presence and absence of rapamycin.

3.3.1. Inhibition of p70 S6 Kinase Activation by Rapamycin In Vivo

Incubation of cells with rapamycin typically results in over 95% inhibition of the activation of p70 S6 kinase, as well as decreasing the p70 S6 kinase activity in unstimulated cells to undetectable levels.

1. Incubate cells for 10–30 min (in the culture medium) in the presence of 100 nM rapamycin or with the equivalent volume of DMSO (as a control) before stimulation.
2. Stimulate cells for between 10 and 30 min with 200 ng/mL TPA, 100 ng/mL EGF, or 100 nM insulin (in the continued presence of rapamycin or DMSO). Lyse the cells in buffer A and centrifuge as described in **Subheading 3.1.1.**
3. Incubate a volume of cell extract containing 50 μg protein with 5 μL of protein G-Sepharose conjugated to 5 μg of anti-p70 S6 kinase antibody for 60 min at 4°C on a shaking platform (to conjugate protein-G-Sepharose to any antibody, *see* Subheading 3.1. of Chapter 11 this volume).
4. Centrifuge the suspension for 1 min at 14,000g, discard the supernatant and wash the pellet twice with 1.0 mL of buffer A containing 0.5 M NaCl and twice with buffer A.
5. Assay the immunoprecipitated p70 S6 kinase activity by incubating the pellets with a mixture containing 10 μL distilled water and 25 μL of p70 S6 kinase reaction buffer (containing the substrate) for 3 min at 30°C, prior starting the reaction by the addition of 10 μL of 50 mM magnesium acetate; 0.5 mM [γ-^{32}P]ATP (approx 2×10^5 cpm/nmol). This reaction must be carried out on a shaking plat-

form at 30°C to keep the immunoprecipitated kinase/protein-G–Sepharose complex in suspension.

6. After 15 min at 30°C stop the reaction by pipetting 40 µL of the assay mixture on to phosphocellulose P81 paper, wash the papers and quantify the radioactivity as described in Subheading 3.3.1. of Chapter 11, this volume.

3.4. Wortmannin and LY 294002

Wortmannin *(17)* and LY 294002 *(18)* (**Fig. 2**) are cell-permeant inhibitors of PI 3-kinase. Wortmannin has an IC_{50} in the low nanomolar range for mammalian class 1a, class 1b and class 3 PI 3-kinases, although the IC_{50} for class 2 PI 3-kinases is greater *(19)*. This inhibitor is only specific for the class 1 PI 3-kinases when used at a concentration of <100 n*M* (*see* **Note 5**). At higher concentrations, wortmannin inhibits a number of other kinases, including the class 2 PI 3-kinases and mTOR/DNA-dependent protein kinases, as well as some other lipid kinases *(19)*. Wortmannin interacts both noncovalently (reversibly) as well as covalently (irreversibly) with PI 3-kinases, both types of interaction lead to the inhibition of PI 3-kinases *(20)*. LY 294002 is a structurally unrelated inhibitor of class 1 PI 3-kinases, which is completely stable in aqueous solution, although class 2 PI 3-kinase as well as mTOR/DNA-dependent protein kinases are also inhibited by this compound *(21)* at similar concentrations to those that inhibit class 1 PI 3-kinases. To check that wortmannin and LY 294002 are working as specific PI 3-kinase inhibitors, cells should be stimulated with insulin or a growth factor (e.g., EGF) and the activation of PKB or p70 S6 kinase assayed in the presence or absence of the inhibitor.

3.4.1. Inhibition of PKB or p70 S6 Kinase Activation by Wortmannin or LY 294002 In Vivo

1. Incubate cells for 1 to 16 h with serum-free medium.
2. After starvation, incubate cells for 10 min in the presence of 100 n*M* wortmannin or 10–30 min in the presence of 10–100 µ*M* LY 294002 (or the equivalent volume of DMSO as a control) before stimulation.
3. Stimulate cells for between 5 and 10 min (for PKB) or 10 and 30 min (for p70 S6 kinase) with 100 n*M* insulin or 100 ng/mL EGF (in the continued presence of the inhibitors or DMSO). Lyse the cells as described in Section 3.1.1.
4. To assay p70 S6 kinase activity, use the procedure described in **Subheading 3.3.1.**
5. To assay PKB activity incubate 50 µg protein cell extract with 5-µL aliquot of protein-G–Sepharose conjugated to 5 µg of anti-PKB antibody for 60 min at 4°C on a shaking platform (to conjugate Protein G-Sepharose to any antibody, *see* **Subheading 3.1.** of Chapter 11, this volume).
6. Centrifuge the suspension for 1 min at 14,000*g*, discard the supernatant and wash the pellet twice with 1.0 mL of buffer A containing 0.5 *M* NaCl, and twice with buffer A.

7. Assay the immunoprecipitated PKB activity by incubating the pellets with a mixture containing 10 μL distilled water and 25 μL of PKB reaction buffer for 3 min at 30°C and then start the reaction by adding 10 μL of 50 m*M* magnesium acetate; 0.5 m*M* [γ-^{32}P]ATP (approx 2×10^5 cpm/nmol). The reaction must be carried out on a shaking platform at 30°C to keep the immunoprecipitated kinase/protein-G–Sepharose complex in suspension.

8. After 15 min at 30°C the reaction is terminated by pipetting 40 μL of the assay mixture on to phosphocellulose P81 paper. Wash the papers and quantify the radioactivity exactly as described in Subheading 3.3.1. of Chapter 11 this volume.

3.5. Ro 318220 and GF 109203X

Protein kinase C (PKC) isoforms are thought to mediate numerous signal transduction processes including the control of cell growth, differentiation and homeostasis. Much of the evidence implicating PKC in these events has been based on the use of either tumour promoting phorbol esters, which are thought to activate PKC by mimicking the second messenger diacylglycerol, and the use of two small cell permeable inhibitors of PKC namely Ro 318220 *(22)* and GF 109203X *(23)*. These two compounds are bisindoylmaleimides which differ from each other in two functional groups and are analogues of staurosporine (**Fig. 2**). They are both potent inhibitors of the α,β and γ isoforms of PKC, with IC$_{50}$ values in the nanomolar range in standard assays in vitro, and compete for the ATP binding site on PKC *(23–25)*. These inhibitors were reported to be specific for PKC isoforms as they only inhibited protein kinase A (PKA), calcium/calmodulin dependent protein kinase(s) and several receptor protein tyrosine kinases at 1000-fold higher concentrations than was required to inhibit PKC *(22,23)*. It has recently been observed that both Ro 318220 and GF 109203X potently inhibit the activity of p90Rsk, Msk1 and the p70 S6 kinase at concentration similar to those which inhibit PKC isoforms (*see* **Note 6**; *10,26*). Although these compounds are not as specific as originally believed, neither Ro 318220 or GF 109203X (5 μ*M*) affect the growth factor induced activation of the MAP kinase pathway, or the stress/cytokine induced activation of the SAPK1/JNK or SAPK2/p38-MAPKAP kinase-2/3 pathways in cells. Knowing the sensitivity of a physiological process in a cell to inhibition by Ro 318220 of GF 109203X is therefore useful as it will provide evidence as to whether this process is likely to be downstream of PKC, p70S6K, Msk1 or p90Rsk. These inhibitors can thus yield important information, particularly when used in conjunction with the other kinase inhibitors described in this chapter.

As a control, to demonstrate that the Ro 318220 or GF 109203X inhibitors are functioning, cells should be incubated, stimulated with TPA, and the activation of MAPK or p90Rsk established in the presence and absence of these inhibitors.

3.5.1. Inhibition MAPK or p90Rsk Activation by Ro 318220 or GF 109203X In Vivo

Under these conditions Ro 318220 or GF 109203X will inhibit MAPK and p90Rsk activation by >90%.

1. Incubate cells for 30–60 min in the presence of 1–5 μM Ro 318220 or GF 109203X (or the equivalent volume of DMSO as a control) in the culture medium before stimulation.
2. Stimulate cells for 10–15 min with 200 ng/mL TPA (in the continued presence of the inhibitors or DMSO). Lyse the cells in buffer A as described in **Subheading 3.1.1.**
3. Centrifuge as described in **Subheading 3.1.1.** and remove the supernatant for analysis.
4. Assay MAP kinase or p90Rsk activity as described in **Subheading 3.2.**

4. Notes

1. SB 203580 inhibits SAPK2/p38 (SAPK2a/p38 and SAPK2b/p38β) by binding to the same site as ATP in the ATP-binding pocket of SAPK2/p38 *(8)*. The reason that SAPK2/p38 activity is inhibited by SB203580 has recently been elucidated. It has been found that inhibition of a protein kinase by this drug depends on the size of the amino-acid residue at the position equivalent to residue 106 in the ATP-binding region of SAPK2/p38 *(8)*. SAPK2/p38 has a small residue, a threonine, in this position. If this residue is mutated to a larger residue, then SB 203580 is ineffective in inhibiting the mutant protein. Most other kinases possess a large hydrophobic residue at this position, explaining why they are not inhibited. Raf is one of the few protein kinases that possesses threonine in this position, and it has been shown that SB 203580 inhibits c-Raf with an IC_{50} of 2 μM in vitro *(27)*. However, SB 203580 does not suppress either growth factor or phorbol ester-induced activation of the classical MAPK cascade in mammalian cells *(27)*. Two other protein kinases (the type II TGFβ receptor and the tyrosine protein kinase Lck) possess small residues at this position in their kinase domains and have been found to be sensitive to inhibition by SB 203580, although the IC_{50} values are 400–800 times higher than the IC_{50} values for SAPK2/p38 *(8)* . It should also be noted that SB 203580 is reported to inhibit two alternatively spliced forms of SAPK1c/JNK2 (JNK2β1 and JNK2β2) with only an approx 10-fold lower potency than SAPK2/p38 *(9)*. Therefore, it is important that this inhibitor is not used at a concentration of over 10 μM in cells. In order to ascribe any cellular action of the drug to the inhibition of SAPK2/p38, it is essential to demonstrate that the observed effect occurs at the same concentration of SB 203580 that prevents the phosphorylation of a physiological substrate of SAPK2/p38, such as MAPKAP kinase-2.
2. Although the PD98059 inhibitor is believed to possess very high specificity for preventing the activation of MKK1, it has recently been shown that PD 98059 is a ligand for the aryl hydrocarbon receptor (AHR) and functions as an AHR antagonist at concentrations commonly used to inhibit MKK activation *(28)*.

3. Another pitfall in using PD 98059 is that this compound is not always completely effective at inhibiting the activation of MKK1 by some agonists that induce a very high level of activation of this pathway such as EGF stimulation of fibroblast cells *(13)*. In order to overcome this problem it may be necessary to reduce the concentration of agonist to a level where the PD 98059 can inhibit MAP kinase/p90Rsk activation effectively.

4. The pitfall in using rapamycin is that the mechanism by which it is inhibiting signal transduction is not yet established. It is important to bear in mind that an effect of rapamycin in cells does not imply that a signaling effect is downstream of p70 S6 kinase, but only that it may be regulated in some way by a signaling pathway downstream of mTOR. It must also be considered possible that rapamycin is functioning by activating a protein phosphatase which may dephosphorylate a number of cellular substrates *(29)*.

5. At concentrations of 100 n*M*, wortmannin will also inhibit some isoforms of PI4-kinase *(30)*, phospholipase A2 *(31)* and mTOR *(21)*. Furthermore, mTOR is also sensitive to LY 294002 at a concentration nearly identical (approx 30 μ*M*) to that required for the inhibition of PI 3-kinases *(21)*. Therefore some caution must be taken in interpreting results based solely on the use of these reagents. It is important to perform a titration of wortmannin or LY 294002 and determine the IC_{50} value. This should be approx 10 n*M* for wortmannin and 5–30 μ*M* for LY 294002 for any effect that is mediated by the class 1 PI 3-kinases.

6. The PKC inhibitors Ro 318220 and GF 109203X have been used in over 350 published studies to investigate the physiological roles of PKC. However, recent studies demonstrating that the p70 S6 kinase, p90Rsk1/2 and Msk1 kinases (which are also potently activated by phorbol ester stimulation of cells) are inhibited by these compounds with similar potency to PKC isoforms, means that some of this data needs to be re-evaluated (*see* **Subheading 3.5.**).

References

1. Ishii, H., Jirousek, M. R., Koya, D., Takagi, C., Xia, P., Clermont, A., Lermont, A., Bursell, S. E., Kern, T. S., Ballas, L. M., Heath, W. F., Stramm, L. E., Feener, E. P., and King, G. L. (1996) Amelioration of vascular dysfunctions in diabetic rats by an oral PKCβ inhibitor. *Science* **272,** 728–731.

2. Uehata, M., Ishizaki, T., Satoh, H., Ono, T., Kawahara, T., Morishita, T., Tamakawa, H., Yamagami, K., Inui, J., Maekawa, M., and Narumiya, S. (1997) Calcium sensitization of smooth muscle mediated by a Rho-associated protein kinase in hypertension. *Nature* **389,** 990–994.

3. Badger, A. M., Bradbeer, J. N., Votta, B., Lee, J. C., Adams, J. L., and Griswold, D. E. (1996) Phamacological profile of SB 203580, a selective inhibitor of cytokine suppessive binding protein/p38 kinase, in animal models of arthritis, bone resorption, endotoxin shock and immune function. *J. Pharmacol. Exp. Ther.* **279,** 1453–1461.

4. Cross, D. A. E., Alessi, D. R., Cohen, P., Andjelkovic, M., and Hemmings, B. A. (1995) Inhibition of glycogen-synthase kinase-3 by insulin is mediated by protein kinase B. *Nature* **378,** 785–789.

5. Lee, J. C. and Young, P. R. (1994) A protein kinase involved in the regulation of inflammatory cytokine biosynthesis. *Nature* **372,** 739–746.
6. Cuenda, A., Rouse, J., Doza, Y. N., Meier, R., Young, P. R., Cohen, P., and Lee, J. C. (1995) SB 203580 is a specific inhibitor of a MAP kinase homologue which is stimulated by cellular stresses and interleukin-1. *FEBS Lett.* **364,** 229–233.
7. Goedert, M., Cuenda, A., Craxton, M., Jakes, R., and Cohen, P. (1997) Activation of the novel stress-activated protein kinase SAPK4 by cytokines and cellular stresses is mediated by SKK3/MKK6; comparison of its substrate specificity with that of other SAP kinases. *EMBO J.* **16,** 3563–3571.
8. Eyers, P. A., Craxton, M., Morrice, N., Cohen, P., and Goedert, M. (1998) Conversion of SB 203580-insensitive MAP kinase family members to drug-sensitive forms by a single amino-acid substitution. *Chem. Biol.* **5,** 321–328.
9. Cohen, P. (1997) The search for physiological substrates of MAP and SAP kinases in mammalian cells. *Trends Cell Biol.* **7,** 353–361.
10. Deak, M., Clifton, A. D., Lucocq, J. M., and Alessi, D. R. (1998) Mitogen- and stress-activated protein kinase-1 (MSK1) is directly activated by MAPK and SAPK2/p38, and may itself mediate activation of CREB. *EMBO J.* **17,** 4426–4441.
11. New, L., Jiang, Y., Zhao, M., Liu, K., Zhu, W., Flood, L. J., Kato, Y., Parry, G. C. N., and Han, J. H. (1998) PRAK, a novel protein kinase regulated by the p38 MAP kinase. *EMBO J.* **17,** 3372–3384.
12. Laemmli, U. (1970) Cleavage of structural proteins during the assembly of the head of bacteriophage T4. *Nature* **277,** 680–685.
13. Alessi, D. R., Cuenda, A., Cohen, P., Dudley, D. T., and Saltiel, A. R. (1995) PD 098059 is a specific inhibitor of the activation of mitogen-activated protein kinase kinase *in vitro* and *in vivo*. *J. Biol. Chem.* **270,** 27,489–27,494.
14. Ballou, L. M., Luther, H., and Thomas, G. (1991) MAP2 kinase and 70K-S6 kinase lie on distinct signalling pathways. *Nature* **349,** 348–350.
15. Price, D. J., Grove, J. R., Clavo, V., Avruch, J., and Bierer, B. E. (1992) Rapamycin-induced inhibition of the 70-kilodalton S6 protein-kinase. *Science* **257,** 973–977.
16. Kuo, C. J., Chung, J., Fiorentino, D. F., Flanagan, W. M., Blenis, J., and Crabtree, G. R. (1992) Rapamycin selectively inhibits interleukin-2 activation of p70 S6 kinase. *Nature* **358,** 70–73.
17. Ui, M., Okada, T., Hazeki, K., and Hazeki, O. (1995) Wortmannin as a unique probe for an intracellular signalling protein, phosphoinositide 3-kinase. *Trends Biochem. Sci.* **20,** 303–307.
18. Vlahos, C. J., Matter, W. F., Hui, K. Y., and Brown, R. F. (1994) A specific inhibitor of phosphatidylinositol 3-kinase, 2-(4-morpholinyl)-8-phenyl-4H-1-benzopyran-4-one (LY294002). *J. Biol. Chem.* **269,** 5241–5248.
19. Shepherd, P. R., Withers, D. J., and Siddle, K. (1998) Phosphoinositide 3-kinase: the key switch mechanism in insulin signalling. *Biochem. J.* **333,** 471–490.
20. Wymann, M. P., BulgarelliLeva, G., Zvelebil, M. J., Pirola, L., Vanhaesebroeck, B., Waterfield, M. D., and Panayotou, G. (1996) Wortmannin inactivates phosphoinositide 3-kinase by covalent modification of

Lys-802, a residue involved in the phosphate transfer reaction. *Mol. Cell. Biol.* **16,** 1722–1733.

21. Brunn, G. J., Williams, J., Sabers, C., Wiederrecht, G., Lawrence, J. C. Jr., and Abraham, R. T. (1996) Direct inhibition of the signaling functions of the mammalian target of rapamycin by the phosphoinositide 3-kinase inhibitors, wortmannin and LY 294002. *EMBO J.* **15,** 5256–5267.

22. Davis, P. D., Hill, C. H., Keech, E., Lawton, G., Nixon, J. S., Sedgwick, A. D., Wadsworth, J., Westmacott, D., and Wilkinson, S. E. (1989) Potent selective inhibitors of protein kinase-C. *FEBS Lett.* **259,** 61–63.

23. Toullec, D., Pianetti, P., Coste, H., Bellevergue, P., Grandperret, T., Ajakane, M., Baudet, V., Boissin, P., Boursier, E., Loriolle, F., Duhamel, L., Charon, D., and Kirilovsky, J. (1991) The bisindolylmaleimide GF-109203X is a potent and selective inhibitor of protein kinase C. *J. Biol. Chem.* **266,** 15,771–15,781.

24. Bradshaw, D., Hill, C. H., Nixon, J. S., and Wilkinson, S. E. (1993) Therapeutic potential of Protein kinase C inhibitors. *Agents Actions* **38,** 135–147.

25. Nixon, J. S., Bishop, J., Bradshaw, D., Davis, P. D., Hill, C. H., Elliott, L. H., Kumar, H, Lawton, G., Lewis, E. J., Mulqueen, M., Westmacott, D., Wadworth, J., and Wilkinson, S. E. (1992) The design and biological properties of potent and selective inhibitors of Protein kinase C. *Biochem. Soc. Trans* **20,** 419–425.

26. Alessi, D. R. (1997) The protein kinase C inhibitors Ro 318220 and GF 109203X are equally potent inhibitors of MAPKAP kinase-1β (Rsk-2) and p70 S6 kinase *FEBS Lett.* **402,** 121–123.

27. Hall-Jackson, C. A., Goedert, M., Hedge, P., and Cohen, P. (1999) Effect of SB 203580 on the activity of c-Raf in vitro and in vivo. *Oncogene* **18,** 2047–2054.

28. Reiners, J. J., Lee, J. Y. Jr., Clift, R. E., Dudley, D. T., and Myrand, S. P. (1998) PD98059 is an equipotent antagonist of the aryl hydrocarbon receptor and inhibitor of mitogen-activated protein kinase kinase. *Mol. Pharmacol.* **53,** 438–445.

29. DiComo, C. J. and Arndt, K. T. (1996) Nutrients, via the Tor proteins, stimulate the association of Tap42 with type 2A phosphatases. *Genes Dev.* **10,** 1904–1916.

30. Nakanishi, S., Catt, K. J., and Balla, T. (1995) A wortmannin-sensitive phosphatidylinositol 4-kinase that regulates hormone-sensitive pools of inositolphospholipids. *Proc. Natl. Acad. Sci. USA* **92,** 5317–5321.

31. Cross, M. J., Stewart, A., Hodgkin, M. N., Kerr, D. J., and Wakelam, M. J. (1995) Wormannin and its structural analog demethoxyviridin inhibit stimulated phospholipase A2 activity in swiss 3T3 cells- Wortmannin is not a specific inhibitor of phosphatidylinositol 3-kinase. *J. Biol. Chem.* **270,** 25,352–25,355.

14

The Development and Use of Phospho-Specific Antibodies to Study Protein Phosphorylation

Jeremy P. Blaydes, Borek Vojtesek, Graham B. Bloomberg, and Ted R. Hupp

1. Introduction

The reversible phosphorylation of proteins is a key mechanism whereby signalling cascades involved in the response to extracellular stimuli bring about changes in cellular function. These proteins include the kinases/phosphatases that form such signaling pathways as well as the transcription factors involved in inducible changes in gene expression *(1)*. Phosphorylation induces changes in the function of these proteins either by induction of allosteric conformational changes in the protein itself or in the regulation of its interaction with other cellular factors.

The study of phosphorylation in cells can involve a range of methods, but the primary technique for determining the extent of phosphate incorporation into specific sites in vivo involves labeling cells with [^{32}P] phosphate followed by phospho-amino acid analysis or peptide sequencing of the protein of interest *(2)*. Apart from the practical problems associated with this technique, the incubation of cells with radioactive precursors (^{32}P, ^{35}S, and ^{3}H) can in itself activate growth arrest and stress-responsive signaling pathways *(3–5)*, which obviously perturb protein phosphorylation. Thus, unless one is lucky enough to be studying a protein for which phosphorylation at specific sites can be studied indirectly, such as for example a mobility shift on an sodium dodecyl sulfate-polyacrylamide gel electrophoresis (SDS-PAGE) gel, there is a clear need for noninvasive methodologies that can be used to complement the well-established radiolabeling techniques.

From: *Methods in Molecular Biology, vol. 99: Stress Response: Methods and Protocols*
Edited by: S. M. Keyse © Humana Press Inc., Totowa, NJ

Given the sensitivity and specificity of immunochemical techniques, the ability to generate antibodies that are specific to either phosphorylated or unphosphorylated epitopes within the target protein provides a powerful alternative to the above techniques. Such antibodies are most effectively raised against immunogen-coupled phosphorylated peptides corresponding to amino acids surrounding the phosphorylation site *(6)*. Recent advances in peptide synthesis mean that such peptides can be synthesized chemically, rather than having to be phosphorylated after synthesis, and, furthermore, they can be protected against dephosphorylation in the host animal *(7)*. Both monoclonal and polyclonal antibodies can be generated to these peptides, although because polyclonal antibodies must normally be affinity purified to remove non-phospho-specific IgG, the use of monoclonals is preferable and obviously more reproducible.

Phospho-specific antibodies have been used in the analysis of several key proteins involved in intracellular signaling and growth regulation and examples include: the site of phosphorylation of CREB by cAMP-regulated protein kinase *(8)*, multisite phosphorylation of p70 S6 kinase *(9)*, and serine 780 on pRB, the site of phosphorylation by cyclin D-cdk4 *(10)*. More recently, stress-induced signaling pathways have been identified that target the tumor-suppressor protein p53 leading to the modulation of both the allosteric activation of this protein by C-terminal phosphorylation *(11–14)* and its interaction with its negative regulatory partner, mdm-2, by phosphorylation at the N-terminal DNA-PK site *(15,16)*. In this chapter, we describe how phospho-specific antibodies can be generated and, using antibodies to the C-terminal regulatory domain of p53 as a specific example, some of the biochemical assays that can be used to study stress-regulated phosphorylation events.

2. Materials

2.1. Reagents

All chemicals are obtained from Sigma (St. Louis, MO) unless indicated otherwise.

1. Keyhole Limpet Hemocyanin-KLH (Sigma, H2133), prepared at 10 mg/mL in H_2O.
2. Synthetic phospho-peptides are prepared at 4 mg/mL in H_2O.
3. Glutaraldehyde is stored as a 70% solution at –20°C.
4. Reacti-gel (6X) is obtained from Pierce (Rockford, IL) (product no. 20259).
5. Okadaic acid (O-7760, Sigma, sodium salt) is prepared as a 120 mM stock solution in ethanol and stored at –20°C.
6. Bradford assay reagent is from Bio-Rad (Hercules, CA).
7. Nonfat dry milk (Marvel™).
8. HRP-conjugated secondary antibodies to rabbit or mouse IgG are obtained from Dako (Santa Barbara, CA).
9. ECL–chemiluminescence solution and Hyperfilm is obtained from Amersham.

10. Laemmli buffer: 2% SDS, 25 mM Tris-HCl, pH 6.8, 10% glycerol, and 0.02% Bromophenol blue.
11. SDS-Polyacrylamide gels (10%) are prepared as described by Harlow and Lane *(17)*.
12. Blotting buffer: 150 mM glycine, 20 mM Tris, 0.1% SDS (w/v), and 20% methanol. The pH of the solution is 8.3.
13. Enzyme-linked immunosorbent assay (ELISA) coating buffer: 0.1 M sodium carbonate, pH 9.0.
14. Antibody elution buffer: 0.1 M glycine, pH 2.5.
15. TMB Liquid Substrate system for developing the ELISA is from Sigma (T-8540).
16. The ELISA plate reader used is a Dynatech Laboratories (Chantilly, VA) Model 4000.
17. Nitrocellulose membranes (HyBond C) are from Amersham.
18. Protein G beads for IgG purification are from Pharmacia (suspension in 50% ethanol).
19. ELISA plates (Falcon, product no. 3912) are from Becton-Dickinson (Rutherford, NJ).
20. Tissue culture plates are from Nunc (Weisbaden-Biebrich, Germany).
21. HiTrap MonoQ (5/5) columns are obtained from Pharmacia (Uppsala, Sweden).
22. Phospho-peptides are synthesized by standard methods *(18)* and contain a C-terminal amino acid with a free primary amine to facilitate glutaraldehyde crosslinking to carrier protein prior to immunization.
23. Urea lysis buffer: 7.0 M urea; 0.1 M dithiothreitol; 0.05% Triton X-100; 25 mM NaCl; 50 mM NaF; and 0.02 M HEPES, pH 7.6.
24. Phosphate-buffered saline (PBS): 140 mM NaCl, 2.6 mM KCl, 10 mM Na$_2$HPO$_4$, and 1.7 mM KH$_2$PO$_4$.
25. PBS-MTF buffer is PBS containing 5 % nonfat milk, 0.1 % Tween-20, and 50 mM NaF.
26. 0.1 M Carbonate buffer is at pH 9.0.
27. Kinase buffer: 10% glycerol; 25 mM HEPES, pH 7.6, 1 mM DTT, 0.02% Triton X-100, 0.025 M KCl, 1 mM benzamidine, 10 mM NaF, 10 mM MgCl$_2$, and 1 mM ATP.
28. Buffer A: 15% glycerol; 25mM HEPES, pH 8.0, 0.02% Triton X-100, 5 mM dithiothreitol (DTT), 1 mM benzamidine, and 50 mM NaF.
29. 0.22-μM Filters are from Amicon (Danvers, MA).

3. Methods

3.1. Immunization with Phospho-Peptides and Antibody Generation

1. Incubate approx 0.5 mL of peptide and 0.5 mL of KLH with glutaraldehyde at a final concentration of 0.1%. Crosslinking is performed at 37°C for 1 h.
2. Neutralize the glutaraldehyde by adding 0.1 mL of 1.5 M Tris-HCl, pH 8.0.
3. Inject the antigen into mice (50 μg of the peptide-KLH conjugate) for monoclonal antibody generation or into rabbits (500 μg of the peptide-KLH conjugate) for polyclonal antibody generation as described *(17)*.
4. Purify the monoclonal antibodies from hybridoma serum supernatant or from ascites tumors using a protein-G column as described *(17)*.

5. Dialyze the purified IgG into PBS to stabilize the monoclonal antibody for storage.
6. In the case of polyclonal sera, purify the IgG in the rabbit serum (25 mL) by applying the serum (containing NaF to a final concentration of 50 mM [*see* Note 1]) to a phospho-peptide column equilibrated in PBS (0.1 mg of peptide coupled to 0.1 mL of Pierce Reacti-Gel (6X) resin according to the manufacturer's instructions).
7. Wash the column with PBS and elute the IgG bound to the phospho-peptide using 0.5 mL of antibody elution buffer.
8. Dialyze the purified IgG into PBS to stabilize the affinity-purified polyclonal antibody for storage.

3.2. Analysis of Steady-State Levels of Protein Phosphorylation In Vivo

We have been able to generate phospho-specific polyclonal and monoclonal antibodies to several known and novel phosphorylation sites on the p53 molecule (**Fig. 1**) and have used these to study stress responses in two main types of assay: the quantitation of steady-state levels of p53 phosphorylation in vivo in response to different types of cellular stress (*see* **Note 2**) and the assay in vitro of the enzymes that target these sites.

The steady-state levels of phosphorylation at specific sites can be analyzed by either semiquantitative denaturing immunoblots, or by native ELISA. For example we have used antibodies raised against the KLH-linked phosphopeptide SRHKKLMFKTEGPDS[PO$_4$]D to study phosphorylation at serine 392 on human p53. Conventional mapping experiments have shown this site to be phosphorylated in vivo *(27)*. This residue can be phosphorylated in vitro by casein kinase-2 (CK2), and is a major regulatory site for modulating the conformation and biochemical activity of p53 *(28,29)*. The data shown in this chapter were obtained using a polyclonal antiserum, αp53-Pser392, but a monoclonal antibody, FP3, has also been generated against this site (**Fig. 1**).

3.2.1. Immunoblotting

1. Prepare extracts from tissue culture cells seeded in 75-cm^2 dishes.
2. Scrape cells off the dishes in ice-cold PBS, pellet by centrifugation, and snap-freeze in liquid nitrogen.
3. Lyse the pellets by incubating with 2–3 vol of urea lysis buffer for 15 min on ice.
4. Clarify the lysates by centrifugation at 11,000g for 10 min.
5. Determine the protein concentration by Bradford assay and boil equal amounts of protein in Laemmli buffer before loading onto a 10% SDS polyacrylamide gel.
6. Transfer the protein to nitrocellulose following electrophoresis and block the membranes in PBS-MTF buffer for 1 h.
7. Probe the blots with the affinity-purified primary antibody diluted 1:1000 in PBS-MTF buffer.

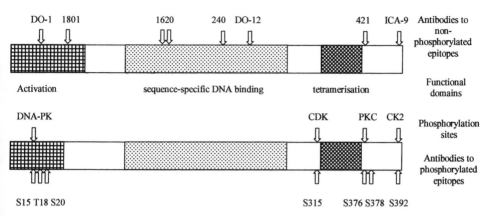

Fig. 1. Antibody reagents used in the study of human p53 and its regulation by phosphorylation. The top panel summarizes the amino acid positions of unphosphorylated epitopes and the bottom panel the positions of the monoclonals specific for phospho-epitopes on human p53 protein. Unmodified epitopes include: Mab-DO1 (amino acids 20–26 *[19]*), Mab-1801(amino acids 46–55], Mab-1620 [amino acids 106–113 and 146–156 *[20]*), Mab-240 (amino acids 213–217 *[21]*), Mab-DO-12 (amino acids 256–270 *[22]*), Mab-421 (amino acids 372–381 *[23]*), and Mab-ICA-9 (amino acids 383-393 *[24]*). Phosphorylated epitopes include: Polyclonal antibody to the DNA-PK site (phospho-serine 15; *[15]*), Mab-FPT18 (phospho-threonine 18 *[25]*), Mab-FPS20 (phospho-serine 20 *[26]*), Mab-FP1 (for the CDC2 site at phospho-315 [Blaydes et al., in preparation]), polyclonal antibody for the PKC site (for phospho-serines 376 and 378 *[14]*), and Mab-FP3 (for the CK2 site at phospho-serine 392 *[11]*).

8. Detect the antigen using the HRP-conjugated secondary antibody and visualize the HRP using ECL (Amersham).

The Western blot shown in **Fig. 2a** demonstrates that the αp53-Pser392 antiserum (1:1000 dilution) is specific for recombinant human p53 protein only when it has been phosphorylated in vitro by CK2. In **Fig. 2b** the antibody has been used on blots of lysates from a human melanoma cell line to demonstrate that exposure to ultraviolet (UV)-C irradiation induces a marked increase in the levels of p53 protein phosphorylated at serine 392, without increases in p53 protein levels *(11)*.

3.3. Development of Nonradioactive Kinase Assays In Vitro

Kinase assays are most conveniently and accurately performed using [γ–^{32}P]ATP as a substrate as it provides a very sensitive method for detecting protein phosphorylation. However, normal radioactive-based assays using crude lysates are not reliable because of the high background from contaminating

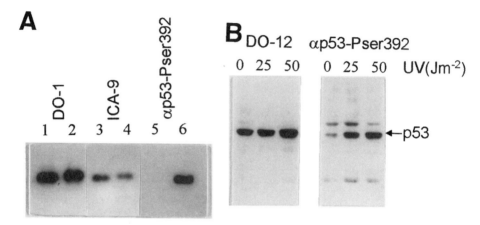

Fig. 2. Denaturing immunoblots to study p53 phosphorylation. (**A**) Full length human p53 was expressed in *Escherichia coli* and partially purified by heparin-sepharose chromatography *(23)*. p53 (5 ng/lane) was incubated in kinase buffer for 60 minutes at 30°C in the absence (lanes 1, 3, 5) or presence (lanes 2, 4, 6) of CK2, and then analyzed by Western blotting with the indicated antibodies. DO-1 recognizes an *N*-terminal epitope (**Fig. 1**) that is not phosphorylated by CK2 and the epitope for ICA-9 (a.a.388-393; Figure 1) is destabilized by phosphorylation at serine 392. αp53-Pser392 is specific for CK2-phosphorylated p53. (**B**) Human A375 melanoma cells were irradiated with the indicated doses of UV-C and lyzed 24 h later. Protein (25 µg of nuclear extract per lane) was then analyzed by blotting with either DO-12, which recognized a phosphorylation-insensitive epitope on denatured p53 (**Fig. 1**) or αp53-Pser392 antibody.

kinases, and as a result, most kinase assays using crude lysates (or partially pure samples) are done with small synthetic peptides to ensure specificity. Nonradioactive assays are useful for complementing such studies for a variety of reasons. The primary advantages of the immunochemical method over the radioactive method to study phosphorylation is that the enzyme assay uses full-length protein as a substrate permitting a more detailed analysis of kinase regulation. Second, the assay can be performed in crude lysates that are normally a source of contaminating kinases. In addition, the ELISA is as quantitative as radiolabeling methods and high-quality data can be obtained in a similar time frame to standard radiolabeling assays.

To perform a two-site ELISA analysis of p53 phosphorylation (**Fig. 3**), following the kinase assay, the p53 protein is first captured using an antibody to a site distinct to the phosphorylation site of interest and then detected with the phospho-specific antibody. In the case of the αp53-Pser392 polyclonal, the capturing antibody is a monoclonal antibody (DO-1) directed toward the *N*-terminus of p53 (*see* **Note 3**). The monoclonal antibody FP3 (**Fig. 1**) can also

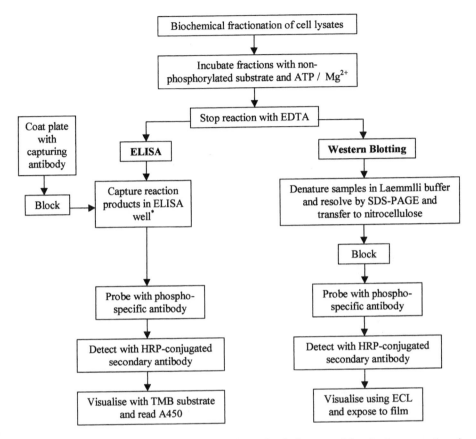

Fig. 3. Flowchart describing the general manipulations used for the immunochemical kinase assay in vitro.

be used to detect the serine 392-phosphorylated p53, in which case an affinity-purified anti-p53 polyclonal, CM5, is used to capture the p53 protein. **Figure 4** shows that the αp53-Pser392 serum can be used to quantify the levels of serine 392-phosphorylated p53 using p53 protein phosphorylated in vitro with CK2.

3.3.1. ELISA Kinase Assay

1. Coat the wells with antibody overnight using 50 μL/well of 1 μg/mL monoclonal antibody DO-1 solution in 0.1 M carbonate buffer using 96-well plates (Falcon).
2. Block the wells for 2 h with PBS containing 3% bovine serum albumin (BSA).
3. Add p53 protein (without or with kinase fractions) in 50 μL of kinase buffer.
4. Capture the p53 protein for 1 h at 4°C.
5. Incubate the captured p53 protein with 50 μL of either αp53-Pser392 or CM5 polyclonal antibodies (each diluted 1:2000 in PBS containing 1% BSA and 50 m*M* NaF).

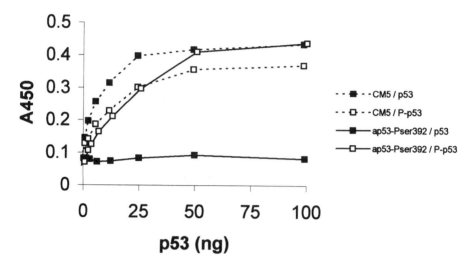

Fig. 4. ELISA to quantify levels of p53 protein phosphorylated in vitro at serine 392. ELISA wells were precoated with DO-1 and used to capture either unmodified p53 protein produced in *E. coli* (solid squares) or p53 phosphorylated by CK2 in vitro (open squares) p53. The captured protein was detected with either CM-5 (dashed lines) or αp53-Pser392 (solid lines) polyclonal antisera (1:2000 dilution) followed by goat anti-rabbit HRP.

6. Incubate the immune complex for 1 h with 50 µl of HRP-conjugated goat antirabbit immunglobulins (diluted 1:1000) in PBS containing 1% BSA and 50 mM NaF.
7. Wash the wells between each step three times with 200 µL 0.1% Tween in PBS, followed one wash with PBS alone.
8. Develop by incubating the immune complexes with 50 µL of TMB substrate (Sigma) for between 30 s and 10 min (empirically determined dependent on strength of signal obtained), reactions are stopped by the addition of 50 µL of 0.5 M H$_2$SO$_4$.
9. Read the absorbance at 450 nM in an ELISA plate reader.

3.4. Development of Nonradioactive Kinase Assays for Use in Chromatography

Conventional assays for the assay of kinase activity measure the transfer of radioactivity from [^{32}P]-labeled ATP into either standard protein substrates or short peptides representing the consensus sequence for phosphorylation by known enzymes *(30)*. Such assays can be used to accurately and rapidly determine enzyme activity in large numbers of samples but disadvantages, apart from the use of radioactivity, are that peptide substrates may not be recognized by enzymes that target the full-length protein. In the case of CK2 for example, this enzyme can be readily assayed on an anionic polypeptide substrate, but the sequence surrounding serine 392 of p53 is not a classic CK2 consensus site and the enzyme must bind amino acids *N*-terminal of this

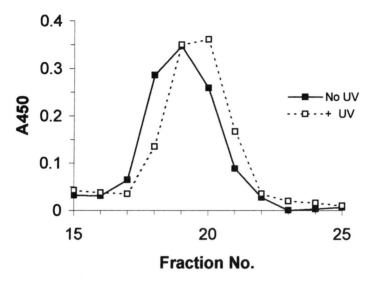

Fig. 5. Nonradioactive kinase assay to quantify p53-serine 392 kinase activity in fractionated cell lysates. A375 cells were exposed to either 0 or 25 J/m^2 UV-C irradiation and lyzed 5 h later. Lysates were fractionated and assayed for enzyme activity as described in the text. UV-C has no significant effect on the activity of CK2, suggesting that the observed changes in p53 phosphorylation at this site in vivo are regulated by a different mechanism, for example, the action of a protein phosphatase.

site in order to phosphorylate serine 392. We have thus used nonradioactive in vitro kinase reactions followed by ELISA or Western blotting to assay biochemically fractionated cell lysates for kinases that target-specific sites on full-length tetrameric p53 protein (*see* **Note 4**). This technique can be used in both the investigation of stress-induced changes in enzyme activity and the purification of potentially novel enzymes. For example, the experiment shown in **Fig. 5** uses this approach to examine the effect of UV irradiation on the activity of the major p53 serine 392-site kinase in cells after column chromatography. A single peak of enzyme activity was found whose specific activity is unchanged after irradiation.

1. Dilute whole cell lysates ten-fold in buffer A, pass the lysate through a 0.22-μM filter and apply 10 mg of protein to a HiTrap Q column.
2. Elute the bound proteins with a linear gradient of KCl in buffer A. All operations are performed at 4°C.
3. Assay 1 μL of each fraction for kinase activity towards p53 by incubating at 30°C for 20 min in a 10μL reaction volume containing: 10% glycerol, 25 mM HEPES, pH 7.6, 0.05 M KCl, 0.5 mg/mL BSA, 1 mM NaF, 1 mM benzamidine, 5 mM MgCl$_2$, 1 mM DTT, 250 μM ATP, and 50 ng p53.

4. Stop the reactions by the addition of 40 µL of ice-cold PBS containing 10 m*M* EDTA and 50 m*M* NaF.
5. Determine phosphate incorporation at serine 392 by ELISA using the αp53-Pser392 antibody as described in the previous section (**Subheading 3.3.1.**).

3.5. Summary

We have demonstrated how phospho-specific antibodies against specific epitopes can be generated and used to study regulable phosphorylation events. These reagents can be used to complement conventional techniques and have many advantages: principally that they are nonradioactive, noninvasive, and highly sensitive reagents for the quantitative study of site-specific phosphorylation. In addition to the uses highlighted in this chapter, immunoblotting or immunoprecipitation with these antibodies may also be used to study phosphorylation in tissue samples from experimental animals or human biopsies. Some phospho-specific antibodies may also be used in immunocytochemical analysis, providing care is taken over the interpretation of the data.

Disadvantages of this methodology lie predominantly in the difficulty in generating highly specific antibodies in the first place. Antibodies generated against certain phospho-peptides may recognize multiple bands on immunoblots of whole cell lysates, possibly because some kinase phosphorylation motifs are conserved between different proteins. Alternatively, the antibody epitope may be blocked by other covalent modifications to amino acids adjacent to the phosphorylated amino acid of interest, resulting in false-negative results. Finally, before going to the expense of generating phospho-specific antibody reagents, it is generally desirable to have evidence from conventional techniques demonstrating that the particular amino acid of interest is modified by phosphorylation in vivo, although, of course, the speculative production of antibodies to candidate sites does provide the opportunity of discovering novel regulatory mechanisms and signaling pathways.

4. Notes

1. The addition of sodium fluoride to rabbit serum containing phospho-specific IgG antibodies minimizes dephosphorylation of the phospho-peptide during affinity purification of the phospho-specific IgG.
2. Although one drawback of the radiolabeling method for determining quantitative changes in protein phosphorylation in stressed cells is that it induces a p53-dependent growth arrest *(5)*, one advantage is that it gives information on phosphate turnover. In contrast, whereas the immunochemical approach is noninvasive and gives information on steady-state levels of phosphorylation, it does not give information on the rate of turnover. Clearly, a combination of both methods would give meaningful information on changes in rates of phosphate turnover if the pathway being studied is not perturbed inherently by radiolabeling.

3. For the specific case of p53, the use of this ELISA method to quantify site-specific phosphorylation of endogenous p53 protein from cell lysates is complicated by the current lack of any suitable antibodies for capturing the native protein that are not affected by covalent modifications within their epitope. For example, the monoclonal antibodies DO-1, Pab421, and ICA-9 are all inhibited by phosphorylation at serine 20 (by an as yet unidentified kinase), Serine 371, serine 376, and serine 378 (by protein kinase C), and serine 392 (by CK2), respectively.

4. Variations of this technique include the use of phosphorylated protein substrates for the assay of phosphatase activity and the use of biotinylated peptide substrates of kinases or phosphatases (i.e., either unmodified or phosphorylated peptides), which can be captured on streptavidin coated ELISA wells.

References

1. Hunter, T. (1995) Protein kinase and phosphatases: the yin and yang of protein phosphorylation and signalling. *Cell* **80**, 225–236
2. Van der Geer, P., Luo, K., Sefton, B. M., and Hunter, T. (1993) Phosphopeptide mapping and phosphoamino acid analysis on cellulose thin-layer plates, in *Protein Phosphorylation* (Hardie, G., ed.), IRL, Oxford, pp. 31–58.
3. Yeargin, J. and Haas, M. (1995) Elevated levels of wild-type p53 induced by radiolabeling of cells leads to apoptosis or sustained growth arrest. *Curr. Biol.* **5**, 423–431.
4. Dover, R., Jayaram, Y., Patel, K., and Chinery, R. (1994) p53 expression in cultured cells following radioisotope labelling. *J. Cell. Sci.* **107**, 1181–1184.
5. Bond, J. A., Webley, K., Wyllie, F. S., Jones, C. J., Craig, A., Hupp, T., and Wynford-Thomas, D. (1999) p53-Dependent growth arrest and altered p53-immunoreactivity following metabolic labelling with ^{32}P ortho-phosphate in human fibroblasts. *Oncogene* **18**, 3788–3792.
6. Czernik, A. J., Girault, J.-A., Nairm, A. C., Chen, J. , Snyder, G., Kebabian, J., and Greengard, P. (1991) Production of phosphorylation state-specific antibodies, in *Methods in Enzymology* Vol. 201, Academic Press, London, pp. 264–283.
7. Ueno, Y., Makino, S., Kitagawa, M., Nishimura, S., Taya, Y., and Hata, T. (1995) Chemical synthesis of phosphopeptides using the arylthio group for protection of phosphate: application to identification of cdc2 kinase phosphorylation sites. *Int. J. Peptide Protein Res.* **46**, 106–112
8. Alberts, A. S., Arias, J., Hagiwara, M., Montminy, M. R., and Feramisco, J. R. (1994) Recombinant cyclic-AMP response element-binding protein (creb) phosphorylated on ser-133 is transcriptionally active upon its introduction into fibroblast nuclei. *J. Biol. Chem.* **269**, 7623–7630
9. Weng, Q. P., Kozlowski, M., Belham, C., Zhang, A., Comb, M., and Avruch, J. (1998) Regulation of the p70 S6 kinase by phosphorylation *in vivo. J. Biol. Chem.* **273**, 16,621–16,629.
10. Kitagawa, M., Higashi, H., Junag, H.-K., Suzuki-Takahashi, I., Ikeda, M., Tamai, K., Kato, J.-Y., Segawa, K., Yoshida, E., Nishimura, S., and Taya, Y. (1996) The

consensus motif for phosphorylation by cyclin D1-cdk4 is different from that for phosphorylation by cyclin A/E-cdk2. *EMBO J.* **15,** 7060–7069.

11. Blaydes, J. P. and Hupp, T. R. (1998) DNA damage triggers DRB-resistant phosphorylation of human p53 at the CK2 site. *Oncogene* **17,** 1045–1052.

12. Lu, H., Taya, Y., Ikeda, M., and Levine, A. J., (1998) Ultraviolet radiation, but not γ radiation or etoposide-induced DNA damage, results in the phosphorylation of the murine p53 protein at serine 389. *Proc. Natl. Acad. Sci. USA* **95,** 6399–6402.

13. Kapoor, M. and Lozano, G. (1998) Functional activation of p53 via phosphorylation following DNA damage by UV but not γ radiation. *Proc. Natl. Acad. Sci. USA* **95,** 2834–2837.

14. Waterman, M. J., Stavridi, E. S,. Waterman, . J. L., and Halazonetis, T. D. (1998) ATM-dependent activation of p53 involves dephosphorylation and association with 14-3-3 proteins. *Nat. Genet.* **19,** 175–178.

15. Shieh, S.-Y., Ikeda, M., Taya, Y., and Prives, C. (1997) DNA damage-induced phosphorylation of p53 alleviates inhibition by MDM2. *Cell* **91,** 325–334.

16. Siliciano, J. D., Canman, C. E., Taya, Y., Sakaguchi, K., Appella, E., and Kastan, M. (1997) DNA damage induces phosphorylation of the amino terminus of p53. *Genes Dev.* **11,** 3471–3481.

17. Harlow, E. and Lane, D. P. (1988) *Antibodies: A Laboratory Manual.* Cold Spring Harbor Laboratory, New York.

18. Atherton, E. and Sheppard, R. C. (1989) *Solid-Phase Peptide Synthesis. A Practical Approach.* IRL Press.

19. Stephen, C. W., Helminen, P., and Lane, D. P. (1995) Characterisation of epitopes on human p53 using phage-displayed peptide libraries: insights into antibody-peptide interactions. *J. Mol. Biol.* **248,** 58–78.

20. Ravera, M. W., Carcamo, J., Brissette, R., Alam-Moghe, A., Dedova, O., Cheng, W., Hsiao, K. C., Klebanov, D., Shen, H., Tang, P., Blume, A., and Mandecki, W. (1998) Identification of an allosteric binding site on the transcription factor p53 using a phage-displayed peptide library. *Oncogene* **16,** 1993–1999.

21. Stephen, C. and Lane, D. P. (1992) Mutant conformation of p53: Precise epitope mapping using a filamentous phage epitope library. *J. Mol. Biol.* **225,** 577–583.

22. Vojtesek, B., Dolezalova, H., Lauerova, L., Svitakova, M., Havlis, P., Kovarik, J., Midgley, C. A., and Lane, D. P. (1995) Conformational changes in p53 analysed using new antibodies to the core DNA binding domain of the protein. *Oncogene* **10,** 389–393.

23. Wade-Evans, A. and Jenkins, J. R. (1985) Precise epitope mapping of the murine transformation-associated protein, p53. *EMBO J.* **4,** 699–706.

24. Hupp, T. R. and Lane, D. P. (1994) Allosteric activation of latent p53 tetramers. *Curr. Biol.* **4,** 865–875.

25. Craig, A. L., Burch, L., Vojtesek, B., Mikutowska, J., Thompson, A., and Hupp, T. R. (1999) Novel phosphorylation sites of human tumour suppressor protein p53 at Ser[20] and Thr[18] that disrupt the binding of MDM2 (mouse double minute 2) protein are modified in human cancers. *Biochem. J.* **342,** 133–141.

26. Craig, A. L., Blaydes, J. P., Burch, L. R., Thompson, A. M., and Hupp, T. R. (1999) Dephosphorylation of p53 at Ser^{20} after exposure to low levels of non-ionizing radiation. *Oncogene* **18,** 6305–6312.
27. Meek, D. W., Simon, S., Kikkawa, U., and Eckhart, W. (1990) The p53 tumour suppressor protein is phosphorylated at serine 389 by casein kinase II. *EMBO J.* **9,** 3252–3260.
28. Hupp T. R., Meek D. W., Midgely, C. A., and Lane D. P. (1992) Regulation of the specific DNA binding function of p53. *Cell* **71,** 875–886.
29. Sakaguchi, K., Sakamoto, H., Lewis, M. S., Anderson, C. W., Erickson, J. W., Appella, E., and Xie, D. (1992) Phosphorylation of serine-392 stabilizes the tetramer formation of tumor-suppressor protein-p53. *Biochemistry* **36,** 10,117–10,124.
30. Wang, Y. and Roach, P. J. (1993) Purification and assay of mammalian protein (serine/threonine) kinases, in *Protein Phosphorylation* (Hardie, G., ed.), IRL, Oxford, pp. 121–142.

15

Peptide Assay of Protein Kinases and Use of Variant Peptides to Determine Recognition Motifs

D. Grahame Hardie

1. Introduction

It is now clear that the major mechanism by which cellular function is switched from one state to another in eukaryotic cells, including the response to cellular stress *(1)*, is via changes in phosphorylation of key proteins. This is usually achieved by modulation of the activity of pre-existing protein kinases (and/or protein phosphatases),rather than by changes in expression at the transcriptional or translational level: most protein kinases are in fact constitutively expressed. Assays of kinase expression levels (e.g., by Western blotting) are therefore not sufficient. In some cases (e.g., MAP kinases), activating phosphorylation events may produce a shift in mobility of the kinase on a Western blot which can be used as an index of activation. However, since nonactivating phosphorylation events can also produce mobility shifts, this method can be unreliable and there remains a requirement to develop direct assays of protein kinase activity. In many cases, phosphorylation of synthetic peptide substrates provides the simplest assay for routine use.

The *Saccharomyces cerevisiae* genome encodes around 113 members of the "eukaryotic" protein kinase superfamily, representing approx 2% of all open reading frames *(2)*. Assuming that the human genome encodes a similar density of protein kinases, there may be around 1500 in our own species. A major challenge for protein kinase researchers will be to identify the downstream targets for all of these protein kinases. One important step in achieving this is the definition of the primary sequence motif that the protein kinase recognizes, and once again synthetic peptides can provide a powerful approach.

From: *Methods in Molecular Biology, vol. 99: Stress Response: Methods and Protocols*
Edited by: S. M. Keyse © Humana Press Inc., Totowa, NJ

The synthetic peptide approach is not successful in all cases. Some protein kinases (e.g., MAP kinase kinases) are highly specific for their target proteins, and these often phosphorylate synthetic peptide substrates poorly, if at all. These kinases may be recognizing some aspect of the three-dimensional structure of their target protein as well as, or even instead of, a primary sequence motif. Alternatively, they may bind to additional determinants on the target protein other than the site at which phosphorylation occurs (e.g., the c-Jun kinase, JNK1 *[3]*). These additional determinants may be remote from the phosphorylation site in the primary sequence, but they may be juxtaposed with it in the three dimensional structure, analogous to the recognition of conformationally sensitive epitopes by some antibodies. Assays of these protein kinases will require purification or expression of the target protein in a form that is readily phosphorylated. Determination of their recognition mechanism may require site-directed mutagenesis of the expressed target protein, three-dimensional structural models of the kinase and the target protein, and/or methods of mapping of regions of protein–protein interaction, such as yeast two-hybrid analysis. Consideration of these methods is beyond the scope of this chapter.

In other cases, particular protein kinases may have multiple targets in vivo, and small synthetic peptides (say, 15 residues) derived from sequences around the phosphorylation site(s) are often phosphorylated with kinetic parameters similar to those obtained using intact target proteins. These kinases appear to be recognizing a short primary sequence motif *(4)*, and synthetic peptides are extremely useful as substrates for kinase assays. A single synthesis can yield enough peptide for tens of thousands of assays, and the assay procedure can be simple and reliable. Studies of phosphorylation of variants of the original peptide can also be used to define the recognition motif. A valuable spin-off of the latter type of study is that one can sometimes obtain optimized peptide substrates that are more selective than the original peptide chosen. A second possible spin-off is that one may be able to design a "pseudosubstrate" peptide that contains all of the recognition determinants but lacks the phosphorylatable residue and that might, therefore, act as a potent and specific inhibitor of the kinase. The precedent here is a 20-residue peptide derived from a naturally occurring heat-stable inhibitor protein of cyclic AMP-dependent protein kinase. It is a very specific "pseudosubstrate" inhibitor of the kinase which has a binding affinity in the low nanomolar range *(5)*.

It is worth noting that the selectivity of protein kinases for particular targets in the intact cell may only partly depend on their intrinsic specificity in solution. It is becoming clear that many protein kinases are localized in the cell, and this undoubtedly affects their selectivity for targets in vivo. Most protein–

tyrosine kinases are localized at the plasma membrane via intrinsic transmembrane domains, or via lipid modifications such as myristoylation, and/or via protein–protein interactions such as the association of SH2 domains with phosphotyrosine motifs on other membrane proteins. Even for an apparently soluble protein kinase such as cyclic AMP dependent protein kinase (PKA), a large family of anchoring proteins (A-kinase anchoring proteins, AKAPs) exist *(6)* that target the kinase to different locations in the cell.

1.1. Synthetic Peptide Kinase Assays

The most widely used method for synthetic peptide kinase assays involves incubation of the kinase with the peptide and [γ-^{32}P]ATP, followed by stopping the reaction by pipetting an aliquot of the mixture onto a square of phosphocellulose paper that is washed with dilute phosphoric acid (*see* **Subheading 3.1.**). As long as the amino acid composition is suitable (*see* ref. *7*), phosphorylated and dephosphorylated peptides bind quantitatively to phosphocellulose paper, whereas unreacted [γ-^{32}P]ATP and free [^{32}P] phosphate are washed off. Binding of peptides to phosphocellulose paper works well for basic, but not acidic, peptides. Quantitative binding of peptides to phosphocellulose can usually be ensured by the addition of two or three basic residues at the N- or C-terminus of the peptide, as long as this does not interfere with the recognition process. Other methods have been developed that are applicable to both basic and acidic peptides, such as the use of thin layer chromatography to separate peptides from ATP *(8)* or the use of ferric adsorbent paper to selectively bind tritium-labeled peptides in their phosphorylated form *(9)*. However, where it is applicable, the phosphocellulose paper method remains the most convenient and widely used.

Consideration of the experience of this laboratory in developing a peptide kinase assay may be helpful. We originally established that ser-79 on rat acetyl-CoA carboxylase (ACC) was a site rapidly phosphorylated by the AMP-activated protein kinase in vitro *(10)*. We synthesized a peptide HMRSSMSGLHLVKRR, containing the sequence from residue 73 to residue 85 on ACC, with two arginines added at the C-terminus to aid binding to phosphocellulose paper. This peptide was a substrate for both PKA and AMPK, but the synthesis of the variant HMRSAMSGLHLVKRR (note the ser→ala change in position 5) eliminated the PKA site without affecting the phosphorylation by AMPK *(11)*. This peptide (the *SAMS* peptide) remains the basis of the AMPK assay that is now used almost universally (*see also* Chapter 6, this volume). The same assay could be used for the budding yeast *(12)* and higher-plant homologs *(13)* of AMPK, even though at the time none of their physiological target proteins were known.

1.2. The "Classical" Approach to Defining Recognition Motifs: Rational Design of Peptide Variants

The "classical" approach to the definition of a kinase recognition motif is to start by mapping and sequencing several sites on protein targets for the kinase and to look for conserved features whose importance can then be tested by systematically designing variant peptides. As an example, our own work on the AMP-activated protein kinase identified six sites phosphorylated on four protein targets. We aligned the sequences of these with those of equivalent sites on homologous proteins from other species where the kinase was also likely to act (14). This analysis showed that the only universally conserved features were hydrophobic residues (M, L, I, F, or V) at P-5 (i.e., five residues N-terminal to the phosphorylated serine) and P+4, and at least one basic residue (R, K, or H) at P-3 or P-4. The importance of these residues was tested using 12 variants of the SAMS peptide. Replacement of the hydrophobic residues at P-5 and P+4, or of the arginine at P-4, with glycine dramatically decreased the V_{max}/K_m for phosphorylation, showing the importance of these residues. We also showed that lysine and histidine could replace the arginine at P-4, although arginine was preferred (14). These results were comfirmed by a second series of peptides based on the parent AMARAASAAALARRR (the AMARA peptide), which contained the serine and the key residues for recognition at P-5, P-3, and P+4, but where other residues (other than the basic tail) were alanine. Experiments with this series of peptides additionally examined the preference for particular hydrophobic residues at the P-5 and P+4 positions and the effects of moving the hydrophobic and basic residues up or down by one or more positions (15). A spin-off of this study was the finding that the AMARA peptide is a much better substrate for the higher-plant homologs than SAMS: The former is now used in the routine assay for the plant kinases.

1.3. The "Modernist" Approach to Defining Recognition Motifs: Random Peptide Libraries

A major disadvantage of the "classical" approach is that it requires prior knowledge of sequences around at least a small number of phosphorylation sites on target proteins. How can the recognition motif for a novel protein kinase be established when one may not know a single protein target? Recently, methods involving "random" peptide libraries have been developed. Two in particular seem worthy of mention:

1. Lam's group (16) utilized a "one bead–one peptide" approach in which peptides were coupled to polystyrene beads. For each cycle of peptide synthesis, the beads were divided into 19 pools (cysteine being avoided because of technical problems), each pool was coupled with a single amino acid, and the pools were

recombined and redivided for the next cycle. Pentapeptide and heptapeptide libraries were phosphorylated using PKA and [γ-^{32}P]ATP, and the mixture of labeled and unlabeled beads were cast into an agarose gel. Labeled beads were detected by autoradiography and picked out, and after a second round of screening to ensure that a single bead was selected, the peptide attached to the labeled bead was sequenced. The method successfully identified the RRXS recognition motif for PKA that had been established previously by the "classical" approach, although in the limited number of positive peptides sequenced (four), the additional preference of PKA for a hydrophobic residue at the P+1 position was not evident.

2. Cantley's group *(17)* utilized an "oriented" peptide library in which certain cycles of synthesis utilized a single amino acid precursor, whereas others contained a mixture. The initial library used had the sequence MAXXXXSXXXXAKKK, where X was any amino acid except the phosphorylatable residues Ser, Thr, and Tyr (Cys and Trp were also omitted because of potential problems of oxidation). The polylysine tail improved the solubility of the peptides and also helped to minimize washout during sequencing. An equivalent library with tyrosine replacing the central serine was also constructed for analysis of protein–tyrosine kinases. Unlike Lam's approach *(16)*, the peptides were used in the soluble phase: The library was phosphorylated using [γ-^{32}P]ATP and a particular kinase, and phosphorylated peptides were then resolved from the bulk of dephosphopeptides by chromatography on a ferric iminodiacetic acid column. The mixture of phosphorylated peptides were then subjected to automated sequencing. The principle of the method is that where the kinase preferred a particular amino acid in one of the eight degenerate positions, recovery of that amino acid was enhanced in that cycle during sequencing. The method successfully predicted that PKA prefers arginine at the P-3 and P-2 positions (although *see* proviso below).

These peptide library approaches have the great advantage that no prior knowledge of the protein substrates for a kinase is required, so they are, at first sight, very attractive. However there are some potential problems. For example, the "classical" approach shows that critical residues for the AMP-activated protein kinase (AMPK) occur at the P-5 and P+4 positions (i.e., nine residues apart) and the particular libraries made by Lam's and Cantley's groups would not have been long enough to detect both of these. Method (2) appears to be excellent where a single position somewhere between P-4 and P+4 is critical, but would be less successful if a second or third position was also essential, because the amount of any peptide containing a unique combination of two or three residues in the degenerate positions ($1/15^2 = 0.4\%$, or $1/15^3 = 0.03\%$ of library respectively) would be too low to be detected by sequencing. This problem can be overcome to some extent by making a second oriented library where the residue found to be most highly preferred with the initial library is fixed, but the other positions remain degenerate. Another potential problem can be seen from results obtained using approach 2 *(17)*, which suggested that PKA prefers basic residues at the P-4 and P-1 positions, as well as P-3 and P-2, and also hydrophobic residues anywhere P+1 and P+4. By contrast, the "classical" approach suggests that if basics are present at the optimal P-3 and P-2 positions and a hydrophobic at the P+1 position, the residues at the P-4, P-1, and P+2 through P+4 are probably irrelevant. These unexpected results of approach 2 probably arise because a basic residue at the P-4 or P-1 positions may be a poor substitute if the preferred P-3/P-2 basics are not present. Similarly, a hydrophobic at one or more of the P+2 through P+4 positions may be a pooor substitute if there is not a hydrophobic at the P+1 position. A third problem with approach 2 is that because synthesis at degenerate positions is performed with a mixture of 15 amino acid precursors, the coupling conditions are a compromise and the library may not be truly random. This could perhaps be circumvented by dividing the library into 15 pools for coupling and then recombining after each step.

Because of the "one bead–one peptide" design of approach 1, this method seems to get around most of these problems, although it has not been widely used as yet. It does make the assumption that having the peptide attached to a polystyrene solid-phase matrix does not affect the recognition motif, but in this respect it is encouraging that the method does pick out the same RRXS motif for PKA as was found using the "classical" approach.

2. Materials

The laboratory should be equipped for general biochemical techniques, including facilities for the handling of radioisotopes and liquid scintillation counting. It is not necessary to have peptide synthesis facilities in-house, as many institutions, as well as commercial companies, now provide a custom peptide synthesis service. With the exception of **Subheading 3.1.**, most of the methods in this chapter are given as lists of guidelines rather than detailed protocols, and it is not appropriate to list materials.

The assay of AMP-activated protein kinase (*see also* Chapter 6, this volume) is described as an example of a typical peptide kinase assay.

1. HEPES-Brij buffer: 50 mM Na HEPES, pH 7.4, 1 mM dithiothreitol (DTT), 0.02% Brij-35: can be kept for a few days at 4°C, otherwise store at –20°C. The Brij-35 is a nonionic detergent that helps to stabilize proteins when they are at low concentrations.
2. 100 mM Unlabeled ATP: dissolve slowly with stirring in HEPES-Brij buffer, keeping the pH just above 7.0 with NaOH solution. ATP solutions must be neutralized *before* the addition of MgCl$_2$, otherwise an insoluble MgATP complex will precipitate during neutralization.
3. [γ-^{32}P]ATP: we use the approx 30 Ci/mmol formulation for protein kinase assays (Amersham, cat. no. PB10132). A laboratory stock solution is prepared in the

following manner: 1 mCi (100 μL) is added to 890 μL of water and 10 μL of 100 mM unlabeled ATP to give a solution (1 mCi/mL, 1 mM), which can be stored at –20°C. This stock solution is then diluted to the desired specific radioactivity with 1 mM unlabeled ATP with 5 μL/mL of 5 M MgCl$_2$ added (final Mg^{2+} concentration 25 mM).

4. 1 mM AMP, dissolved in HEPES-Brij buffer; can be kept for a few days at 4°C, otherwise store in aliquots at –20°C. Other kinases that do not require an allosteric activator will not need this addition.

5. *SAMS* peptide (HMRSAMSGLHLVKRR), 1 mM in HEPES-Brij buffer. Peptides can be kept for a few days at 4°C, otherwise store in aliquots at –20°C. Peptides made by a peptide synthesis service should be high performance liquid chromatography (HPLC) purified and the concentration determined by amino acid analysis. Oxidation of the methionines during prolonged storage at 4°C can affect ability to act as substrates.

6. P81 phosphocellulose paper (Whatman): cut into 1-cm^2 squares.

7. Optiscint "Hisafe" scintillation cocktail (Wallac).

3. Methods

Apart from **Subheading 3.1.**, where the assay of AMP-activated protein kinase is described as an example of a typical peptide kinase assay, detailed protocols are not appropriate for this chapter, and only general guidelines are given.

3.1. Typical Peptide Assay

1. Reaction mixtures are prepared on ice containing the following (*see* **Note 1**):
   5 μL   1 mM [γ-^{32}P]ATP, 25 mM MgCl$_2$ (specific activity 250–500 cpm/pmol)
   5 μL   1 mM AMP in HEPES-Brij buffer
   5 μL   1 mM SAMS or AMARA peptide in HEPES-Brij buffer
   5 μL   HEPES-Brij buffer
   Blank reactions are also performed containing HEPES-Brij buffer in place of peptide.

2. Reactions are initiated by the addition of AMPK (5 μL). In skilled hands, assays can be started and stopped at 15-s intervals.

3. Incubate at 30°C for 10 min, remove 15 μL, and spot onto a P81 paper square (number the squares with a hard pencil before use). After the liquid has soaked in (1–2 s), the square is dropped into a beaker containing 1% (v/v) phosphoric acid.

4. After all incubations have been stopped, the paper squares are stirred gently on a magnetic stirrer for 2–3 mins The phosphoric acid (*caution*: radioactive!) is poured off and a second phosphoric acid wash is performed. The papers are rinsed briefly in water before soaking in acetone. They are then laid out on a paper towel and allowed to dry.

5. Dried filters are counted after immersing in 5 mL of Optiscint "Hisafe" scintillation cocktail.

6. Kinase activity is expressed in units of nanomoles of phosphate incorporated into substrate peptide per minute at 30°C. To determine the specific radioactivity of the ATP, spot 5 μL onto a paper square, dry it, and count in scintillation fluid. Keep

this vial and recount it every time you count some assays using the same ATP. With a 1-m*M* stock solution of [γ-^{32}P]ATP, the counts obtained correspond to 5 nmol of ATP, and radioactive decay is corrected for automatically. Using this method of counting, the counting efficiency is constant and need not be determined.

7. If you use a nonaqueous scintillation cocktail such as Optiscint "Hisafe," the vial of scintillation fluid can usually be reused after removing the paper square. Do not reuse the vial that you used to count the ATP.

3.2. The "Classical" Approach to Determining a Kinase Recognition Motif: General Strategy

1. Determine the phosphorylation sites for the kinase on at least two or three target proteins. A minimum of four or five sites (some of which may be on the same protein) is recommended.
2. Align the amino acid sequences around these sites using the phosphorylated amino acid as reference. Also include sequences from closely related species that are also likely to be targets for the kinase (*see* **Note 2**).
3. Look for conserved residues at certain positions (e.g., all hydrophobic, all basic, all acidic?). Also, note positions where there is no obvious conservation of sequence.
4. Based on hypotheses about the recognition motif derived from **step 3**, design a series of synthetic peptides to test these hypotheses (e.g., replace conserved hydrophobic, basic, or acidic residues with alanine or glycine).
5. Testing of these hypotheses will be an iterative process and may lead to the design of additional synthetic peptides.
6. Once one or more key residues have been identified, it may be worth designing another series of peptides where the key residues are present, but the noncritical, nonconserved residues are all replaced with alanine or glycine (*see* **Note 3**). This is particularly useful if you wish to test the positioning of the key residue, as it obviates the problem of what to do with the residue that you are replacing when you move the key residue.

3.3. Testing Phosphorylation of Variant Peptides

1. Use a peptide assay based on **Subheading 3.1.** to measure the initial rate of phosphorylation.
2. Carry out the assay over a range of peptide concentrations and determine V_{max} and K_m (*see* **Note 4**) by fitting to the Michaelis-Menten equation.
3. If the exact molar concentration of the kinase is known, calculate the turnover number, k_{cat} [$k_{cat} = (V_{max}$ in moles substrate/s) ÷ (total enzyme concentration in mol)].
4. The ratio V_{max}/K_m (or, better, k_{cat}/K_m, sometimes called the selectivity constant) provides a single value that is useful when ranking different peptides for their ability to act as substrates. The higher the value of V_{max}/K_m or k_{cat}/K_m, the better the substrate.
5. One could also consider using a method such as biomolecular interaction analysis (surface plasmon resonance) to directly measure binding affinities and asso-

ciation and dissociation rate constants for binding of peptide substrates to the kinase. However, bear in mind that protein kinases generally work by forming a ternary complex among the kinase, the protein–peptide substrate, and MgATP, and that the presence of MgATP will almost certainly affect the affinity. It is not feasible to measure binding of a substrate peptide in the presence of ATP because the catalytic reaction will occur. It may be worth attempting measuring of binding affinities for "pseudosubstrate" peptides (i.e., variants of substrate peptides in which the phosphorylatable residue is replaced by a nonphosphorylatable residue such as alanine).

3.4. Peptide Library Approaches

The author has no personal experience of these, so it is not appropriate to provide protocols. Detailed protocols are provided in the original papers by Wu et al. *(16)* and Songyang et al. *(17)*.

4. Notes

1. This protocol uses final concentrations in the assay of 200 μM ATP, 5 mM MgCl$_2$, and 200 μM peptide. There is always a temptation to use ATP at very low concentrations in order to maintain a high specific radioactivity. However, the concentration should preferably be at least 5–10 times the apparent K_m for this nucleotide, so that it is almost saturating. Using concentrations of ATP at or below the apparent K_m, the reaction will be nonlinear with respect to time. For most protein–serine kinases, the K_m is in the low micromolar range, and 200 μM ATP is a reasonable compromise. The real substrate of kinases is the Mg/ATP^{2-} complex, and total Mg^{2+} should be maintained at a concentration in the millimolar range (unless free Mg^{2+} inhibits the system). The peptide should also preferably be used at a concentration at least 5–10 times the apparent K_m so that it is almost saturating. Peptides that are good kinase substrates usually have K_m values in the low micromolar range, so that 200 μM peptide is a reasonable compromise. Using these conditions, the reaction rate will usually be linear with time and enzyme concentration if the degree of phosphorylation of the peptide is kept to <10%.
2. Phosphorylation sites that are physiologically important have probably evolved quite early during evolution and will be very likely to be conserved. For example, the three phosphorylation sites for AMP-activated protein kinase on acetyl-CoA carboxylase are conserved between rat and chicken, whereas the single site on HMG-CoA reductase is conserved among vertebrates, insects, sea urchins, and higher plants, although not yeast or bacteria (*see* **ref.** *14*). It is therefore worthwhile including sequences of homologs from related species in the alignment, even though these may not have been directly shown to be targets for the kinase.
3. Peptides containing long stretches of alanine can cause problems because of the development of secondary structure. Consult your peptide synthesis service for advice.
4. Strictly, because protein kinases catalyze a bisubstrate reaction by varying the peptide concentration at a fixed ATP concentration one is only measuring an

apparent K_m. If ATP is saturating, this should, however, equal the true K_m. To measure the true K_m values, one will need to vary the peptide concentration at a series of different concentrations of MgATP (*see* **ref. 16**).

Acknowledgments

Studies in this laboratory were supported by a Programme Grant from the Wellcome Trust and Research Studentships from the UK Medical Research and Biotechnology and Biological Sciences Research Councils (MRC and BBSRC).

References

1. Hardie, D. G. (1994) An emerging role for protein kinases: the response to nutritional and environmental stress. *Cell. Signal.* **6,** 813–821.
2. Hunter, T., and Plowman, G. D. (1997) The protein kinases of budding yeast: six score and more. *Trends Biochem. Sci.* **22,** 18–22.
3. Hibi, M., Lin, A. N., Smeal, T., Minden, A., and Karin, M. (1993) Identification of an Oncoprotein-Responsive and UV-Responsive protein kinase that binds and potentiates the c-Jun activation domain. *Genes Dev.* **7,** 2135–2148.
4. Pinna, L. A. and Ruzzene, M. (1996) How do protein kinases recognize their substrates? *Biochim. Biophys. Acta* **1314,** 191–225.
5. Knighton, D. R., Zheng, J., Ten Eyck, L. F., Xuong, N. H., Taylor, S. S., and Sowadski, J. M. (1991) Structure of a peptide inhibitor bound to the catalytic subunit of cyclic adenosine monophosphate-dependent protein kinase. *Science* **253,** 414–420.
6. Rubin, C. S. (1994) A kinase anchor proteins and the intracellular targeting of signals carried by cyclic AMP. *Biochim. Biophys. Acta.* **1224,** 467–479.
7. Toomik, R., Ekman, P., and Engström, L. (1992) A potential pitfall in protein kinase assay: phosphocellulose paper as an unreliable adsorbent of produced phosphopeptides. *Anal. Biochem.* **204,** 311–314.
8. Lou, Q., Wu, J., and Lam, K. S. (1996) A protein kinase assay system for both acidic and basic peptides. *Anal. Biochem.* **235,** 107–109.
9. Toomik, R., Ekman, P., Eller, M., Jarv, J., Zaitsev, D., Myasoedov, N., Ragnarsson, U., and Engstrom, L. (1993) Protein kinase assay using tritiated peptide substrates and ferric adsorbent paper for phosphopeptide binding. *Anal. Biochem.* **209,** 348–353.
10. Munday, M. R., Campbell, D. G., Carling, D., and Hardie, D. G. (1988) Identification by amino acid sequencing of three major regulatory phosphorylation sites on rat acetyl-CoA carboxylase. *Eur. J. Biochem.* **175,** 331–338.
11. Davies, S. P., Carling, D., and Hardie, D. G. (1989) Tissue distribution of the AMP-activated protein kinase, and lack of activation by cyclic AMP-dependent protein kinase, studied using a specific and sensitive peptide assay. *Eur. J. Biochem.* **186,** 123–128.
12. Wilson, W. A., Hawley, S. A., and Hardie, D. G. (1996) The mechanism of glucose repression/derepression in yeast: SNF1 protein kinase is activated by phos-

phorylation under derepressing conditions, and this correlates with a high AMP:ATP ratio. *Curr. Biol.* **6,** 1426–1434.

13. MacKintosh, R. W., Davies, S. P., Clarke, P. R., Weekes, J., Gillespie, J. G., Gibb, B. J., and Hardie, D. G. (1992) Evidence for a protein kinase cascade in higher plants: 3-hydroxy-3-methylglutaryl-CoA reductase kinase. *Eur. J. Biochem.* **209,** 923–931.

14. Weekes, J., Ball, K. L., Caudwell, F. B., and Hardie, D. G. (1993) Specificity determinants for the AMP-activated protein kinase and its plant homologue analysed using synthetic peptides. *FEBS Lett.* **334,** 335–339.

15. Dale, S., Wilson, W. A., Edelman, A. M., and Hardie, D. G. (1995) Similar substrate recognition motifs for mammalian AMP-activated protein kinase, higher plant HMG-CoA reductase kinase-A, yeast SNF1, and mammalian calmodulin-dependent protein kinase I. *FEBS Lett.* **361,** 191–195.

16. Wu, J., Ma, Q. N., and Lam, K. S. (1994) Identifying substrate motifs of protein kinases by a random library approach. *Biochemistry* **33,** 14,825-14,833.

17. Songyang, Z., Blechner, S., Hoagland, N., Hoekstra, M. F., Piwnica Worms, H., and Cantley, L. C. (1994) Use of an oriented peptide library to determine the optimal substrates of protein kinases. *Curr. Biol.* **4,** 973–982.

III

THE ANALYSIS OF STRESS-INDUCED GENE EXPRESSION

16

Assaying NF-κB and AP-1 DNA-Binding and Transcriptional Activity

Judith M. Mueller and Heike L. Pahl

1. Introduction

Nuclear factor-κB (NF-κB) and activator protein 1 (AP-1) are well-characterized ubiquitously expressed transcription factors that play important roles in the response to cellular stress situations. In unstimulated resting cells, NF-κB and AP-1 are inactive. These transcription factors are induced by a great variety of stimuli and conditions that represent internal or external stress situations for the cells. These include pathological stimuli, such as viruses, bacteria, oxidative stress, hypoxia, and inflammatory mediators as well as internal cellular stress (e.g., endoplasmic reticulum overload) (1–7).

NF-κB is a multisubunit transcription factor of higher eukaryotes. In vertebrates, the NF-κB/Rel family comprises five cloned DNA-binding subunits, that can homo- and heterodimerise in various combinations (1,7). They are called p50, p52, p65 (RelA), c-Rel, and Rel-B (6). Often, NF-κB is composed of p50/p65 (RelA) heterodimers. NF-κB activity is regulated by interaction with specific inhibitory proteins, called IκB's. To date five IκB proteins are known: IκB-α, -β, -γ, -δ, and -ε (6). Upon cell stimulation IκB becomes phosphorylated and subsequently ubiquitinylated. Finally, it is degraded by the 26S proteasome. IκB proteolysis releases NF-κB, which translocates to the nucleus, where it binds to its target sequences. This rapidly initiates transcription of "defense genes" involved in inflammatory, immune and acute-phase responses (1,7, see also, Chapter 10, this volume).

AP-1 is a dimer composed of two DNA-binding subunits belonging to the *fos* and *jun* multigene family of transcription factors (reviewed in **refs. 8,9**). The products of the proto-oncogenes *c-fos* and *c-jun* are two well-studied members of these families. AP-1 is predominantly activated by *de novo* synthesis of

From: *Methods in Molecular Biology, vol. 99: Stress Response: Methods and Protocols*
Edited by: S. M. Keyse © Humana Press Inc., Totowa, NJ

its subunits, which is controlled by pre-existing transcription factors *(10)*. An exception to this are c-Jun homodimers, which pre-exist in resting cells. Stress situations such as ultraviolet (UV) radiation, growth factors, or hypoxia have been reported to induce *c-fos* and *c-jun* genes *(11–13)*.

Two methods will be described in this chapter that allow the detection of NF-κB or AP-1 activation in cellular extracts. Using these assays, the involvement of these two transcription factors in cellular stress responses can be studied. A short description and the advantages of each method are given in the following paragraphs. The first technique is detection of DNA-binding in EMSAs (electrophoretic mobility shift assays) *(14)*. Cells simply have to be treated with the substance or stimulus under investigation. Cell extracts are made and used in binding reactions with radioactive DNA probes. The differential mobility of protein–DNA complexes vs DNA alone is then resolved in native polyacrylamide gels (*see* **Fig. 1** for details). This extremely sensitive method requires only a small number of cells. However, a radioactive DNA–probe is required and results have to be carefully evaluated.

The second technique, the so-called reporter gene assay, gives more information than an EMSA. Rather than assaying simple DNA binding, this method measures transactivation of a target gene in transient transfections *(15–17)*. A plasmid carrying binding sites for the transcription factor in front of a reporter gene is introduced into cells. Activation of the transcriptional activity of the 5' binding factor will then increase the amount of mRNA of the reporter gene, and subsequently, the amount of protein made. Transactivation can then be measured using the activity of the gene product as a readout. In this protocol, luciferase-generated light units are measured. Compared to EMSAs, reporter gene assays take longer (3 d vs 1 d). In addition, reporter gene assays have to be repeated several times to obtain statistically reliable data. It is important to note that reporter gene assays are sensitive to several confounding factors and have to be carefully standardized.

To test NF-κB and AP-1 activation, it is advisable to perform EMSAs first. If EMSAs show activation of NF-κB or AP-1, reporter gene assays should follow.

2. Materials

2.1. Tissue Culture

1. Phosphate-buffered saline (PBS) pH 7.4: 10 mM Na$_2$HPO$_4$, 5 mM NaH$_2$PO$_4$, 5.4 mM KCl, 137 mM NaCl; sterilize by autoclaving and store at 4 °C.
2. Rubber policeman or cell scraper.
3. Tissue culture materials: Dulbecco's modified Eagle's medium (DMEM) containing 1 g/L glucose, fetal calf serum (FCS), penicillin/streptomycin solution (10,000 U/mL penicillin, 10,000 μg/mL streptomycin), sterile plasticware for propagation of cells and 60-mm dishes; sterile tissue culture material, such as pipets, and so forth.

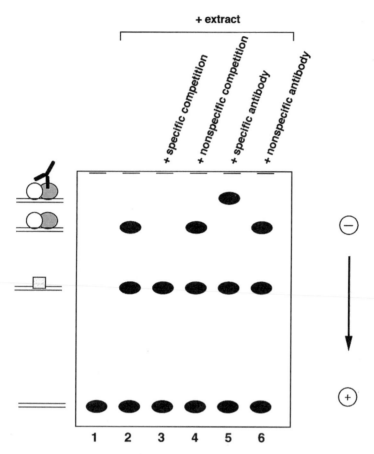

Fig. 1. Schematic representation of an EMSA autoradiogram. On the left side, the complexes are schematically indicated at their respective position in the gel. On the right side, the direction of current flow in shown. In lane 1 only the labeled oligonucleotide is present. The addition of extract results in two bands with reduced mobility. The lower band represents nonspecific binding, the upper band shows specific binding to the oligonucleotide. Specific competition with a self-oligonucleotide results in competition and loss of the upper band, whereas the lower band remains intact. Both bands are unchanged, when an unrelated oligo is used (lane 4). A supershift is seen in lane 5 after addition of an specific antibody. An unrelated antibody does not interfere with protein DNA binding (lane 6). Antibodies that destroy protein–DNA complexes give a pattern that resembles that seen in lane 3.

2.2. EMSAs

1. Cell extract buffers; can be mixed and stored at 4°C for several months; add protease inhibitors immediately before use (0.5 m*M* Pefa-Block® [Boehringer Mannheim] and 0.2 U/mL Aprotinin). NF-κB TOTEX buffer: 20 m*M* HEPES, pH

7.9, 0.35 M NaCl, 1 mM MgCl$_2$, 0.5 mM EDTA, 1% (v/v) Nonidet P-40, 20% (w/v) glycerol, and 5 mM dithiothreitol (DTT). AP-1 TOTEX buffer: 10 mM Tris-HCl, pH 7.5, 30 mM sodium pyrophosphate, 5 mM Na$_3$VO$_4$, 2 mM iodoacetic acid, 50 mM NaCl, 50 mM NaF, 5 mM ZnCl$_2$, 1% (w/v) Triton X-100.

2. Specific and nonspecific oligonucleotides; specific NF-κB and AP-1 DNA probes (available from Promega); factor binding sites are underlined.

   κB: 5'-AGTTGAG*GGGACTTTCC*CAGGC-3'
           3'-TCAACTCCCCTGAAAGGGTCCG-5'

   AP-1: 5'-CGCTTGA*TGAGTCA*GCCGGAA-3'
           3'-GCGAACTACTCAGTCGGCCTT-5'

3. As the nonspecific oligonucleotide use an unrelated oligonucleotide (e.g., binding site for the Sp-1 transcription factor).
4. TBE 5X: 445 mM Tris, 445 mM borate, 5 mM EDTA.
5. Glass plates (25 cm × 20 cm); comb and spacers (approx 1 mm thick) available from Gibco-BRL (Rockville, MD); glass plate coating (e.g., Gel Slick [FMC Bioproducts, Rockland, ME]) or BlueSLick (Serva, Heidelberg, Germany).
6. 30% Acrylamide/bisacrylamide solution (29.2% acrylamide; 0.8% bisacrylamide).
7. N,N,N',N'-tetramethylethylenediamine (TEMED); 10% ammonium persulfate (APS) (from a 10% solution [w/v] in H$_2$O, solution should not be older than 1 wk or store in aliquots at –20°C).
8. Buffer for binding reaction *(18)*. 5X binding buffer: 100 mM HEPES (NaOH) pH 7.9, 300 mM KCl, 20% Ficoll (w/v), 10 mM DTT, 2 mM Pefa-Bloc. (Store in aliquots at –20°C.)
9. Nonspecific competitor: Poly(dI-dC) dissolve at 1 mg/mL. (Store in aliquots at –20°C.)
10. 50 mM MgCl$_2$ solution; sterilize by autoclaving
11. 10 mg/mL bovine serum albumin (BSA) solution (fraction V) in H$_2$O. (Store at –20°C.)
12. Whatman paper 3 MM.
13. Gel dryer.
14. Exposure cassette and X-ray films (Kodak X AR 5) or Beta-imager.
15. For labeling of oligonucleotides: radioactivity room; [γ-^{32}P]ATP; T4 polynucleotide kinase and its buffer; Microspin™ S-200 HR columns (Pharmacia, Uppsala, Sweden).

2.3. Reporter Assays

1. DNA of reporter plasmids (pure).
2. Carrier DNA: Sonicated salmon sperm DNA or pUC 18 plasmid.
3. Transfection solutions: sterilize by filtration through a 0.22-µm filter. (Store in aliquots at –80°C.) 250 mM CaCl$_2$ solution. 2X HEPES-buffered saline (HBS): 55 mM HEPES, pH 7.1 (NaOH), 1.5 mM Na$_2$HPO$_4$, 10 mM KCl, 273 mM NaCl, 12 mM glucose.
4. Lysis buffer L: 25 mM Gly-Gly, pH 7.8, 15 mM MgSO$_4$, 4 mM EGTA, 1% (v/v) Triton X-100, 1 mM DTT.

5. Assay buffer: 25 m*M* Gly-Gly, pH 7.8, 15 m*M* MgSO$_4$, 4 m*M* EGTA, 15 mM potassium phosphate, pH 7.6, 1 m*M* DTT, 2 m*M* ATP. Lysis and assay buffer are made freshly when required using stock solutions. (Store 1 *M* DTT and 50 m*M* ATP stocks at –20°C.)

6. 0.3 mg/mL luciferin in H$_2$O (sodium salt from Sigma [St. Louis, MO]; potassium salt from Promega); light sensitive. (Store at –20°C.) Alternatively, commercially available buffers can be used (e.g., reporter lysis buffer from Promega).

7. Luminometer (e.g., Micro Lumat LB96P [EG&G Berthold]) and a flat-bottomed pigmented 96-well microplate for measuring (e.g., Microlite™ from Dynex Tech.).

3. Methods

In principle, any higher eukaryotic cell line can be investigated. Here, as an example, we will describe the methods for 293 cells, a human embryonic kidney cell line (ECACC 85120602). Different cell lines may be required depending on the stimulus under investigation *(19)*. The 293 cells are cultured in DMEM, supplemented with 10% fetal calf serum and 1% (v/v) penicillin–streptomycin.

3.1. Electrophoretic Mobility Shift Assays for Detection of NF-κB and AP-1 DNA-Binding Activity

The fastest and most convenient method to monitor DNA binding of activated transcription factors such as NF-κB or AP-1 are EMSAs (*see* **Fig. 1** for principle). Because radioactive oligonucleotides are used in EMSAs, general safety regulation for working with radioactive materials should be followed.

Oligonucleotides are labeled at 37°C for 30–60 min using T4 polynucleotide kinase (PNK). Standard reactions contain 50 ng of an oligonucleotide, 50 µCi [γ-^{32}P]ATP in 20 µL reaction of PNK-buffer and 1 U PNK. Unincorporated free ATP is removed, for example, by using Microspin™ columns according to the manufacture's recommendation. One microliter of the labeled oligo is subjected to Cerenkov counting in a β–counter. The specific activity of the radioactive oligonucleotide should be at least 10^6 cpm/µg. Lower specific activities result in loss of sensitivity of EMSAs, such that DNA-binding activity cannot be detected.

3.1.1. Preparation of Total Cell Extracts

1. Plate 293 cells to a density of 5 × 10^5 per 60-mm dish in 3 mL of medium in the afternoon.

2. The next day, treat cells as desired (usually 5–120 min).

3. Aspirate medium and try to remove all traces (*see* **Note 1**).

4. Scrape off in 1 mL PBS per plate with a rubber policeman.

5. Collect cells in Eppendorf tubes and pellet 5 s by centrifugation in a microcentrifuge at 13,000*g*.

6. Lyse the cells by vortexing using 100 μL ice-cold TOTEX buffer per 10^6 cells (*see* **Note 2**). Please note that different buffers are used for analysis of NF-κB or AP-1, respectively (*see* **Subheading 2.2.1.** and **Note 3**).
7. Incubate on ice for 20–30 min.
8. Centrifuge lysates for 10 min in a cooled microcentrifuge at 14,000*g*.
9. Transfer the supernatant into new Eppendorf tubes.
10. Measure protein concentration using the Bradford assay (e.g., Bio-Rad Protein Microassay). Take 1–5 μL of lysate and the respective amount of Totex buffer as a negative control (blank).
11. Extracts can then be brought to equal protein concentration with cold extract buffer.
12. Use extracts either directly or freeze aliquots in liquid nitrogen and store at –80°C (*see* **Note 4**).

3.1.2. Gel Casting

1. Clean glass plates carefully (no detergent should be present).
2. Coat smaller glass plates using nontoxic substances (e.g., Gel Slick or BlueSLick).
3. Assemble glass plates with the spacer.
4. Gels can be poured horizontally, which prevents leaking.
5. Cast 4% polyacrylamide (PAA) gel with 0.5X final TBE concentration by combining 8 mL of a 30% acrylamide/bisacrylamide solution with 6 mL of 5X TBE and 46 mL H_2O. Start polymerization by adding 30 μL of TEMED and 300 μL of 10% APS.
6. Pour gel between the glass plates and insert comb. Polymerization takes around 20 min.

3.1.3. Binding Reaction

1. Place gel in its chambers with 0.5X TBE as running buffer.
2. Wash slots with running buffer using a syringe.
3. Prerun gel for 15 min at 180 V. Then start the binding reaction (*see* **Note 5**).
4. Thaw 5X binding buffer, poly(dI-dC), BSA, and extracts on ice.
5. Mix 10 μg of cell extract with 4 μL of 5X binding buffer, 10 μg of BSA, and 1 μg of poly(dI-dC) in a final volume of 19 μL (*see* **Note 6**). For AP-1 shifts, 5 mM MgCl$_2$ has to be present in the binding reaction (2 μL of the 50 mM MgCl$_2$ solution).
6. Add 1 μL of the oligo corresponding to 20,000 cpm.
7. Incubate for 20 min at room temperature (RT) (20–25°C).

3.1.4. Gel Loading, Running, Drying, and Exposure

1. Wash slots with running buffer using a syringe.
2. Load binding reactions in the wells (*see* **Note 7**).
3. Run gels at 200 V for approx 1.5 h (milliamps decrease over time; *see* **Note 8**).
4. Transfer the gel onto a Whatman paper and cover with Saran wrap.

5. Place a second Whatman underneath to avoid contamination of the gel dryer (*see* **Note 9**).
6. Dry gel at 80°C for 40 min (*see* **Note 10**).
7. Remove Saran Wrap.
8. Expose gel to an X-ray film for several hours or overnight (*see* **Notes 11** and **12**).

3.1.5. Use of Competing Oligos to Identify Specific Binding

Specificity of binding is tested using competitive binding to an excess of unlabeled oligos. Complexes that disappear in a reaction with the unlabeled oligo, that contains the binding site of interest (self), but stay constant with an unrelated unlabeled oligo (nonself), are specific for this oligo.

1. Add unlabeled oligo in a 50-fold molar excess in **step 5** of **Subheading 3.1.3.** (*see* **Note 13**).
2. Incubate 10 min at RT.
3. Proceed with **step 6** of the binding reaction (**Subheading 3.1.3.**).

3.1.6. Use of Antibodies to Identify Specific Binding

In order to identify the protein(s) in protein–DNA complexes, antibodies are used. Specific immunoglobulins either inhibit the formation of the specific protein–DNA complex or cause its further retardation in EMSAs (referred to as "supershift"). A partial inhibition or supershift with one antibody alone suggests that a protein–DNA complex contains a mixture of dimer combinations, which are not all immunoreactive. In this case, the possibility has to be ruled out that too little antibody or too much protein extract was used. Antibodies for almost all NF-κB and AP-1 family members are commercially available (e.g., from Santa Cruz Bio-Technology Inc. [Santa Cruz, CA]). The concentration of the antibody should be high enough to add a few microliters without altering the binding conditions too much.

1. Add 1–2 μL antibodies (amount has to be tested in titration experiments) in **step 5** of **Subheading 3.1.3.** (*see* **Note 14**).
2. Incubate either 25 min at RT or 60 min on ice.
3. Proceed with **step 6** of the binding reaction (**Subheading 3.1.3.**).
4. Use equal amounts of an nonspecific antibody as a control.

3.2. Reporter Assays for Detection of NF-κB and AP-1 Transcriptional Activity

A major difference between this technique and the EMSA is that the production of a reporter gene protein is relatively slow, whereas EMSAs may detect transcription factor activation within minutes following stimulation. One disadvantage, therefore, is that cells have to be treated for a much longer period of time with the potentially toxic compounds under investigation. An obvious advantage of transactivation assays is that they demonstrate the potential of a treatment to either induce or specifically prevent gene transcription.

3.2.1. Reporter Plasmids

Generally, reporter plasmids consist of NF-κB or AP-1 binding sites cloned in front of a minimal promoter driving the reporter gene. Firefly luciferase is often used as a reporter gene because of its easy and fast detection after transfection *(16)*.

To detect AP-1-dependent transactivation the reporter construct –73/+63 Col-Luc and the control construct –60/+63 Col-Luc are useful *(3,4,20)*. The expression vector –73/+63 Col-Luc contains one AP-1 motif upstream of a human collagenase promoter. The –60/+63 Col-Luc promoter contains the minimal human collagenase promoter lacking the AP-1 binding-site. To analyze NF-κB activation, the plasmid 6XκB-TK Luc can be used. It contains three repeats of the human immunodeficiency virus type 1 (HIV-1) tandem NF-κB sites in front of a minimal thymidine kinase (TK) promoter *(4)*.

3.2.2. Cell Lines and Transfection

1. Plate 293 cells to a density of 0.5×10^5 per 60-mm dish in 3 mL of medium in the afternoon.
2. The next morning prepare the transfection solution (*see* **Note 15**): Bring 1 μg of reporter plasmid and 3 μg of carrier DNA to 135 μL with sterile H_2O in a sterile Eppendorf tube (*see* **Note 16**).
3. Mix with 15 μL 2.5 M $CaCl_2$.
4. While vortexing, add 150 μL 2X HBS dropwise for 30 s (*see* **Note 17**).
5. Incubate 15 min at room temperature.
6. Apply $Ca_3(PO_4)_2 \cdot DNA$ precipitate dropwise to the cells.
7. Distribute evenly by slightly tilting the dish.
8. Eight hours after transfection, the cells are washed carefully with PBS and 3 mL of fresh medium is added (*see* **Note 18**).
9. Treat cells with the stimulus under investigation (usually 6–14 h) (*see* **Note 19**). Include unstimulated and control cells, for example, with solvent alone.

3.2.3. Preparation of Cell Lysates

1. Cells are harvested 24–36 h after transfection.
2. Aspirate medium and try to remove all traces (*see* **Note 1**).
3. Cells are scraped from the dish in 300 μL lysis buffer L and transferred to Eppendorf tubes.
4. Clear by centrifugation for 3 min in a microcentrifuge at 14,000g.
5. Transfer supernatant into a new Eppendorf tube.

3.2.4. Detection of Luciferase Activity

1. Into one well of a special pigmented 96-well microplate, pipet 150 μL assay buffer and 50 μL cellular extract from **step 5** of **Subheading 3.2.3.**
2. Light emission generated by luciferase is then measured in a luminometer (Microlumat LB 96 P, EG&G Berthold), programmed to inject 100 μL of luciferin (0.3 mg/mL) per well (*see* **Note 20**).

3.2.5. Controls and Standards

In transactivation assays, specificity of a stimulatory or inhibitory effect must be demonstrated. For this purpose, control reporter constructs are used that are either regulated by unrelated transcription factors or have mutations in the binding sites for the transcription factors under investigation. The transcriptional activity of these constructs should not change with treatment. A good control, for example, is the minimal TK promoter alone in front of the luciferase gene. All other steps of the reporter gene assay are done in the same way.

Because transfection efficiency varies greatly between individual samples, it is absolutely necessary to standardize transfection efficiency. This can be done by including a second reporter gene in the transfection. Nearly all companies offer control plasmids together with buffers to analyze them after transfection. For example, pSV-β-Galactosidase (Promega) or pβgal-Control (Clontech) use β-galactosidase as reporter gene, pSEAP2-Control (Clontech) uses secreted alkaline phosphatase, and pRL-TK uses the *Renilla* luciferase (Promega).

To study the effect of various cellular stress situations in more complex genomic settings, the use of Northern blotting, quantitative polymerase chain reaction methods or immunodetection of gene products may provide more biological information. For instance, if a particular treatment is found to induce or inhibit NF-κB, other genes known to be controlled by NF-κB can be studied.

4. Notes

1. Other cells are washed once with ice-cold PBS, but 293 cells detach very easily.
2. For fractionation of cells into cytoplasmic and nuclear extracts, we obtained good results using the simple and fast method by Schreiber et al. *(21)*.
3. If both transcription factors are being analyzed, divide the cell-PBS suspension after **step 4** of **Subheading 3.1.1.**
4. Best results are obtained in EMSAs when freshly prepared extracts are used, but extracts snap-frozen in liquid nitrogen may also produce good results even after repeated freezing and thawing.
5. It is advisable to test the quality of an extract and the specificity of an effect by testing the DNA-binding activity of other unrelated transcription factors by simply using distinct [^{32}P]-labeled DNA probes in EMSAs (several are available from Promega [e.g., OCT1]).

6. For optimal salt conditions extract volume should be 2–5 µL. An equal volume of extract buffer has to be added in each reaction to obtain comparable conditions. If less than 2 µL of extract is used, adapt salt to a final concentration of 120 mM with KCl or add extract buffer.

7. We usually use a Hamilton syringe to load the binding reactions onto the gel, but a P20 pipet works as well. Let the binding reaction sink smoothly in the slots and take care not to mix it with the running buffer. Do not add any running dye.

8. We always run our EMSA gels for the same distance in order to be able to compare the positions of the bands. This can be done by loading 0.025% Bromphenol Blue in 40% glycerol in a free lane and stopping the gel run after the same migration distance (e.g., 10 cm).

9. Take care to dispose of all radioactive materials in the radioactive waste. Anode buffer can be radioactive and should be checked regularly for contamination. The second Whatman paper used during gel drying will also be radioactive.

10. If the gel is not dry after 40 min, bands tend to become fuzzy. Saran Wrap can be removed easily, as soon as the gel is dry.

11. Radioactive protein–DNA complexes can be analyzed and quantitated by β-imaging (Molecular Dynamics imager).

12. Background binding of AP-1 can be lowered by starving the cells overnight in medium containing 0% or 0.5% FCS.

13. The amount of oligo recovered after purification from unincorporated ATP can be roughly estimated as being 90% of the amount used in the labeling reaction.

14. In EMSAs using antibodies (supershift assays), it is advisable to reduce the amount of extract used, as the antibody–protein ratio is critical.

15. If using other cell lines, it must be shown that the transfection method itself does not activate transcription factor activity.

16. Usually, no cotransfection with NF-κB or AP-1 subunits is necessary, because enough of the endogeneous binding factors are present.

17. The slow dropwise addition of 2X HBS solution is critical for the DNA precipitate and transfection efficiency. All precipitates should be handled in the same way.

18. The cell-culture medium can exert a powerful influence on the results obtained and this should be analyzed separately. For AP-1, it may be useful to lower the amount of serum in the medium. For some substances, it may be important to use MEM (modified Eagle's medium) instead of DMEM, because it contains no iron.

19. The optimal time of exposure has to be determined (e.g., by harvesting the cells 4, 6, 8, 12, and 20 h after the treatment).

20. Instead of pipetting the assay buffer, a ready mix containing luciferin and all buffer components (e.g., from Promega) can be injected by the luminometer.

References

1. Baeuerle, P. A. and Henkel, T. (1994) Function and Activation of NF-κB in the immune system. *Annu. Rev. Immunol.* **12,** 141–179.
2. Schreck, R., Rieber, P., and Baeuerle, P. A. (1991) Reactive oxygen intermediates as apparently widely used messengers in the activation of NF-κB transcription factor and HIV-1. *EMBO J.* **10,** 2247–2258.

3. Rupec, R. A. and Baeuerle, P. A. (1995) The genomic response of tumor cells to hypoxia and reoxygenation: differential activation of transcription factors AP-1 and NF-κB. *Eur. J .Biochem.* **234,** 632–640.

4. Meyer, M., Schreck, R., and Baeuerle, P. A. (1993) H₂O₂ and antioxidants have opposite effects on activation of NF-κB and AP-1 in intact cells: AP-1 as secondary antioxidant-responsive factor. *EMBO J.* **12,** 2005–2015.

5. Pahl, H. L. and Baeuerle, P. A. (1995) A novel signal transduction pathway from the nucleus to the ER. *EMBO J.* **14,** 2876–2883.

6. Piette, J., Piret, B., Bonizzi, G., Schoonbroodt, S., Merville, M. P., Legrand-Poels, S., and Bours, V. (1997) Multiple redox regulation in NF-κB transcription factor activation. *Biol. Chem.* **378,** 1237–1245.

7. Sha, W. C. (1998) Regulation of immune responses by NF-κB/Rel transcription factors. *J. Exp. Med.* **187,** 143-146.

8. Karin, M., Liu, Z., and Zandi, E. (1997) AP-1 function and regulation. *Curr. Opin. Cell Biol.* **9,** 240–246.

9. Rahmsdorf, H. J. (1994) The FOS and JUN families of transcription factors. (Angel, P. E. and Herrlich, P. A., eds.), CRC Press, Boca Raton, FL.

10. Angel, P. and Karin, M. (1991) The role of Jun, Fos, and the AP-1 complex in cell-proliferation and transformation. *Biochem. Biophys. Acta* **1072,** 129–157.

11. Dérijard, B., Hibi, M., Wu, I.-H., Barrett, T., Su, B., Deng, T., Karin, M., and Davis, R. J. (1994) JNK1: a protein kinase stimulated by UV light and Ha-Ras that binds and phosphorylates the c-Jun activation domain. *Cell* **76,** 1025–1037.

12. Sachsenmaier, C., Radler-Pohl, A., Zinck, R., Nordheim, A., Herrlich, P., and Rahmsdorf, H. J. (1994) Involvement of growth factor receptors in the mammalian UVC response. *Cell* **78,** 963–972.

13. Meyer, M., Schreck, R., and Baeuerle, P. A. (1993) H₂O₂ and antioxidants have opposite effects on activation of NF-κB and AP-1 in intact cells: AP-1 as a secondary antioxidant-responsive factor. *EMBO J.* **12,** 2005–2015.

14. Garner, M. M. and Revzin, A. (1981) A gel-electrophoresis method for quantifying the binding of proteins to specific DNA regions: Application to components of the E. coli Lactose operon regulatory system. *Nucl. Acids Res.* **9,** 3047–3060.

15. Wigler, M., Pellicer, A., Silverstein, A., and Axel, R. (1978) Biochemical transfer of single-copy eucaryotic genes using total cellular DNA as donor. *Cell* **14,** 725–731.

16. Welsh, S. and Kay, S. A. (1997) Reporter gene expression for monitoring gene transfer. *Curr. Opin. Biotechnol.* **8,** 617–622.

17. Saez, E., No, D., West, A., and Evans, R. M. (1997) Inducible gene expression in mammalian cells and transgenic mice. *Curr. Opin. Biotechnol.* **8,** 608–616.

18. Meyer, R., Hatada, H.-P., Hohmann, H.-P., Haiker, M., Bartsch, C., Rötlisberger, U., Lahm, H.-W., Schlaeger, E. J., van Loon, A. P. G. M., and Scheidereit, C. (1991) Cloning of the DNA-binding subunit of human nuclear factorκB: the level of its mRNA is strongly regulated by phorbol ester or tumor necrosis factor α. *Proc. Natl. Acad. Sci. USA* **88,** 966–970.

19. Mueller, J. M., Rupec, R. A., and Baeuerle, P. A. (1997) Study of gene regulation by NF-κB and AP-1 in response to reactive oxygen intermediates. *Methods* **11,** 301–312.

20. Deng, T. and Karin, M. (1993) JunB differs from c-Jun in its DNA-binding and dimerization domains, and represses c-Jun by formation of inactive heterodimers. *Genes Dev.* **7,** 479–490.

21. Schreiber, E., Matthias, P., Müller, M. M., and Schaffner, W. (1989) Rapid detection of octamer binding proteins with "mini extracts" prepared from a small number of cells. *Nucleic Acids Res* . **17,** 6419.

17

Analysis of the Mammalian Heat-Shock Response

Inducible Gene Expression and Heat-Shock Factor Activity

Anu Mathew, Yanhong Shi, Caroline Jolly, and Richard I. Morimoto

1. Introduction

The evolutionarily conserved heat-shock response has been extensively studied as a model for transcriptional regulation. In eukaryotic cells, the regulation of heat-shock gene expression is mediated by a family of related proteins, the heat-shock transcription factors (HSFs) *(1–8)*. Smaller eukaryotes such as yeast and *Drosophila melanogaster* usually express single members of the HSF family *(1–3)*, while larger eukaryotes express multiple HSFs. At least four HSF family members have been identified in vertebrate systems *(4–8)*. Multiple HSFs may have arisen to allow expression of heat shock proteins under different conditions, such that divergent signaling pathways converge to result in the production of a common class of proteins, the heat-shock proteins.

Activation of HSF proteins in response to stress involves a conversion from an inert form to a transcriptionally active state *(9–12)*. The activation process is a multistep event in which induction of DNA binding can be uncoupled from the emergence of transcriptional activity. In the yeast *S. Pombe (13)*, and in most other eukaryotic systems, HSF exists in an inert monomeric (yHSF, vertebrate HSF1) or dimeric (vertebrate HSF2, HSF3) form that trimerizes upon activation, with subsequent association with DNA, and transcriptional activation *(14–18)*. The HSF in *S. cerevisiae (1,2)*, and *K. lactis (19)* bypasses the regulation of DNA binding, being constitutively trimeric and bound to DNA and acquiring transcriptional activity upon exposure to stress. Treatment of mammalian cells with anti-inflammatory drugs has also been shown to result in a DNA-binding competent, but transcriptionally inert form of HSF1 *(20–22)*. Posttranslational modification, particularly inducible hyperphosphorylation,

From: *Methods in Molecular Biology, vol. 99: Stress Response: Methods and Protocols*
Edited by: S. M. Keyse © Humana Press Inc., Totowa, NJ

which is not observed in this anti-inflammatory drug-induced form *(22,23)*, may also be involved in the regulation of HSF activity.

Various conditions induce activation of HSFs. Classical stresses, including heat-shock, exposure to heavy metals, amino acid analog incorporation, and oxidative stress, primarily activate HSF1 in most vertebrate tissues *(10–12,24)*. The avian HSF3 exhibits similar properties but seems to be activated under conditions of severe stress, such as exposure to temperatures above 45°C *(18,25)*. The highly conserved HSF2 is activated in a number of differentiation associated systems *(17,26–30)*. HSF2 activity is also induced under conditions that impair the proper function of the ubiquitin-proteasome pathway and, hence, is sensitive to the protein degradative capacity of the cell *(31)*.

The aim of this chapter is to outline strategies that have been successfully used to analyze the different regulatory stages of vertebrate HSF activities and to analyze the downstream events of HSF activation (i.e., the regulation of heat-shock gene expression). The first section will discuss protocols for HSF protein analyses using various biochemical assays, as well as a method to visualize their cellular distribution. The second section describes protocols to determine the transcriptional activity of HSF under various conditions by monitoring expression of endogenous HSF target genes and using appropriate HSF reporter constructs.

2. Materials

2.1. Equipment

1. Cell-culture glass slides with two chambers (Lab-Tek).
2. Coplin-jars.
3. Cover slips: circle 22 mm, 22 × 40 mm, 22 × 50 mm, 18 × 18 mm.
4. Dot-blot apparatus.
5. Fast protein liquid chromatography (FPLC) system (Pharmacia).
6. Fluorescence microscopy facilities (confocal laser scanning microscope, epifluorescence microscope).
7. Fractionation device (if available).
8. Gradient maker.
9. Hoefer DE 102 series tube gel electrophoresis apparatus.
10. Peristaltic pump.
11. Plasticene.
12. Rubber cement.
13. Superdex 200 HR column.
14. Equipment for sodium dodecyl sulfate-polyacrylamide gel electrophoresis (SDS-PAGE), DNA sequencing, and agarose gel analysis.
15. Thermocycler.
16. TLC plates.

2.2. Reagents

2.2.1. Common Reagents

1. Acrylamide/bis-acrylamide stock solutions: (38 g:2 g, or 28.38 g:1.62 g, per 100 mL, as required). Store in a dark bottle (*see* **Note 1**).
2. Ammonium acetate ($NH_4C_2H_3O$), 8 *M*.
3. Ammonium persulfate (APS) stock solution 10% (store at 4°C).
4. Chloroform-isoamyl alcohol (24:1 v/v).
5. [γ-^{32}P]dATP (7000 Ci/mmole) (*see* **Note 2**).
6. dNTP solutions (Pharmacia).
7. Dithiothreitol (DTT) stock solution, 1 *M* (store in aliquots at -20°C).
8. Ethylenediaminetetraacetate (EDTA) 0.5 *M*, pH 8.0.
9. Formamide (Fluka BioChemika and UltraPure from Sigma [St. Louis, MO]) (*see* **Note 3**).
10. 1X Laemmli sample buffer: 2% SDS, 10% glycerol, 60 m*M* Tris-HCl, pH 6.8, 0.25% Bromophenol Blue, and reducing agents (1–10% β-mercaptoethanol and/or 0.1 *M* DTT). Reducing agents are omitted for nonreducing gels.
11. Phosphate-buffered saline (PBS) 10X, pH 7.4: 26.7 m*M* Kcl, 1.38 *M* NaCl, 11.5 m*M* KH_2PO_4, 80.6 m*M* Na_2HPO_4.
12. Phenol, saturated with TE buffer.
13. Phenol–chloroform (1:1, v/v).
14. Poly (dI-dC)-(dI-dC) (Pharmacia) dissolved in TE at 5 µg/µL and stored at –20°C.
15. Proteinase K (1 mg/mL) stock solution (store in aliquots at –20°C).
16. Sequencing gel loading buffer: formamide 80–90%, 0.5X TBE, 0.25% Bromophenol Blue, 0.25% xylene cyanol (*see* **Note 3**).
17. Sodium acetate ($C_2H_3O_2Na$), 3 *M*, pH 5.2.
18. SDS stock solution 20% (store at room temperature).
19. SSC 20X stock solution: 3 *M* NaCl and 0.3 *M* sodium citrate.
20. T4 polynucleotide kinase.
21. TBE 5X stock solution: 445 m*M* boric acid, 446 m*M* Tris base, 10 m*M* EDTA.
22. TE buffer: 10 m*M* Tris-HCl, pH 8.0, 1 m*M* EDTA.
23. Tnana 10X stock solution: 67 m*M* Tris-HCl, pH 7.5, 10 m*M* EDTA, 33 m*M* sodium acetate.
24. Yeast tRNA 10 mg/mL (store at –20°C).

2.2.2. Reagents for Preparation of Cell Extracts

1. Buffer C: 20 m*M* *N*-2-hydroxyethylpiperazine-*N'*-2-ethanesulfonic acid (HEPES), pH 7.9, 25 % (v/v) glycerol, 0.42 *M* NaCl, 1.5 m*M* $MgCl_2$, 0.2 m*M* EDTA. Immediately prior to use, add 0.5 m*M* each of phenylmethylsulfonyl fluoride (PMSF) and DTT.
2. RIPA buffer: 10 m*M* Tris-HCl, pH 7.4, 150 m*M* NaCl, 1% sodium deoxycholate, 1% Triton-X 100. Immediately prior to use, add protease inhibitors: 1 m*M* PMSF, 2 µg/mL leupeptin A, 2 µg/mL pepstatin.

2.2.3. Reagents for Immunoblot and Immunoprecipitation Protocols

1. Blocking solution: low fat milk powder 2.5% (w/v) in 1X PBS.
2. Modified secondary antibodies to visualize immunoblots, diluted in blocking solution.
3. Immunoblot washing solution: 0.1–0.2% polyoxyethylenesorbitan monolaurate (Tween 20) in 1X PBS.
4. Primary antibodies to HSFs, diluted in 1X PBS, 1% BSA, 0.02% NaN$_3$
5. Protein A- or protein G-conjugated beads (prepared in advance and stored in 1X PBS/0.1% NaN$_3$ at 4°C).

2.2.4. Reagents for Immunofluorescence (IF)

1. Anti-fading solution: 90% glycerol, 2.33% DABCO (Sigma), 20 mM Tris-HCl, pH 8.0 (store at 4°C in the dark).
2. DNA counterstaining solution: propidium iodide (PI) (from a 1 mg/mL stock-solution, –20°C) at 100 ng/mL diluted in the anti-fading solution (store at 4°C in the dark).
3. Fixative: 4% formaldehyde, 1X PBS (prepared fresh from 37% formaldehyde solution).
4. IF blocking solution: 10% fetal bovine serum (FBS), 0.3% Triton, 1X PBS (store at 4°C).
5. IF washing solution: 2% FBS, 0.3% Triton X-100, 1X PBS (prewarm the solution and the coplin-jar at 45°C).
6. Secondary antibodies modified by linkage to fluorochromes.

2.2.5. Reagents for the Determination of HSF Oligomeric States

1. Elution buffer for gel filtration: 1% glycerol, 20 mM Tris-HCl, pH 7.9, 200 mM KCl, 1.5 mM MgCl$_2$.
2. Ethylene glycol bis (succinimidylsuccinate) (EGS) dissolved in dimethyl sulfoxide (DMSO).
3. Glycerol buffers (10% and 40%): 20 mM HEPES, pH 7.9, 100 mM NaCl, 5 mM MgCl$_2$, 0.5 M EDTA, 1 mM DTT, 10 or 40% glycerol.
4. Trichloroacetic acid (TCA) 10% (w/v).

2.2.6. Reagents for Electrophoretic Mobility Shift Assay

1. Sample dye: 0.2% Bromophenol Blue, 0.2% xylene cyanol, 50% glycerol.
2. Binding buffer 2X stock: 20 mM Tris-HCl, pH 7.8, 100 mM NaCl, 1 mM EDTA, and 10% glycerol.

2.2.7. Reagents for In Vivo Genomic Footprinting

1. Dilution solution: 17.5 mM MgCl$_2$, 42.3 mM DTT, 125 µg/mL BSA. Should be freshly prepared and kept on ice.
2. Dimethyl sulfate (DMS) (*see* **Note 4**).
3. Double-stranded linker oligonucleotides.

4. Gene-specific primers 1, 2, and 3.
5. Labeling mix: 5X *Taq* buffer diluted to 1X final with double-distilled water, plus 2 mM dNTPs, and 1–10 pmoles end-labeled primer 3 per reaction. Chill on ice. Immediately before use, add 2.5 U *Taq* polymerase per reaction. Use 5 μL per reaction.
6. Ligation mix: 10 mM MgCl$_2$, 20 mM DTT, 3 mM rATP (Pharmacia), 50 μg/mL BSA, 50 mM Tris-HCl, pH 7.7, 100 pmoles linker per reaction, three Weiss units of T4 DNA ligase per reaction. (Note that the linkers are in 250 mM Tris-HCl.).
7. 5X magnesium-free sequenase buffer: 200 mM Tris-HCl, pH 7.7, 250 mM NaCl.
8. Mg/DTT/dNTP solution: 20 mM MgCl$_2$, 20 mM DTT, 0.1 mM of each dNTP (should be freshly prepared and kept on ice).
9. 1 M Piperidine (Fisher) diluted 1:10 with ddH$_2$O.
10. Sequenase.
11. Solution I: 60 mM Tris-HCl, pH 8.2, 60 mM KCl, 15 mM NaCl, 0.5 mM spermidine, 0.15 mM spermine, 0.5 mM EDTA, 0.3 M sucrose (store at 4°C).
12. Solution II: same as solution I, but with 1% Nonidet P-40 (store at 4°C).
13. Solution III: same as solution I, but lacking sucrose (store at 4°C).
14. Nuclear lysis buffer: 0.5 M EDTA, 1% Sarcosyl, 500 μg/mL RNAse A.
15. *Taq* DNA polymerase (Perkin Elmer).
16. 5X *Taq* polymerase buffer: 200 mM NaCl, 50 mM Tris-HCl, pH 8.9, 50 mM MgCl$_2$, 0.05% (w/v) gelatin (store at –20°C).
17. Taq stop solution : 260 mM sodium acetate, 10 mM Tris-HCl, pH 7.5, 4 mM EDTA.

2.2.8. Reagents for In Vitro Footprinting

1. 5 mM CaCl$_2$.
2. 10 mg/mL DNase I stock solution.
3. DNase I stop solution: 1% SDS, 200 mM NaCl, 20 mM EDTA, 100 μg/mL yeast tRNA.
4. 4 mM Fe(NH$_4$)$_2$(SO$_4$)$_2$.
5. Glycogen.
6. 1.5 mM Methidiumpropyl-EDTA (MPE) stock solution (light and temperature sensitive, store at –80°C until needed).
7. 10 mM MgCl$_2$.
8. 1X PCR buffer: 10 mM Tris-HCl, pH 8.8, 50 mM KCl, 6 mM MgCl$_2$, 1 mM DTT.
9. Stop solution for cleavage reaction : mix 2 μL of 100 mM thiourea, 1 μL of 250 mM EDTA, and 2 μL of 3 M sodium acetate, per reaction.
10. T7 and T3 promoter primers.
11. 2X transcription buffer: 24 mM HEPES, pH 7.9, 120 mM KCl, 24 % glycerol, 16 mM MgCl$_2$, 2 mM DTT, 1 mM EDTA.

2.2.9. Reagents for Run-On Analysis

1. 50X Denhardt's solution: 1% Ficoll, 1% polyvinylpyrrolidone, and 1% bovine serum albumin (BSA).

2. DNase I: Worthington RNase-free DNase (DPRF) at 10,000 U/mL in 50% glycerol (store at –20°C).
3. Hybridization buffer: 50% formamide, 6X SSC, 10X Denhardt's solution, 0.2% SDS (*see* **Note 3**).
4. Lysis buffer: 10 mM NaCl, 3 mM MgCl$_2$, 10 mM Tris-HCl, pH 7.4, 0.5% Nonidet P-40.
5. Nuclei storage buffer: 40% glycerol, 50 mM Tris-HCl, pH 8.5, 5 mM MgCl$_2$, 0.1 mM EDTA.
6. Proteinase K stock solution 50 mg/mL (store at –20°C).
7. 2X reaction cocktail: 50 µL [^{32}P]UTP (500 µCi), 250 µL 4X reaction mix, 125 µL 8 X tri-phosphate mix (*see* **step 11**), 75 µL ddH$_2$O (enough for 10 reactions at 50 µL per reaction).
8. Nuclei run on stop buffer: 2% SDS, 7 M urea, 0.35 M NaCl, 1 mM EDTA, 10 mM Tris-HCl, pH 8.0.
9. 4X reaction mix: 100 mM HEPES, pH 7.5, 10 mM MgCl$_2$, 10 mM DTT, 300 mM KCl, 20% glycerol (store at 4°C).
10. 50% TCA (w/v).
11. 8X tri-phosphate mix: 2.8 mM ATP, 2.8 mM GTP, 2.8 mM CTP, 3.2 µM UTP (store at –20°C).

2.2.10. Reagents for Fluorescence In Situ Hybridization (FISH)

1. Avidin-FITC (Sigma).
2. BioPrime DNA labeling kit (Gibco-BRL).
3. Human placental DNA (Cot I™ DNA, Gibco-BRL).
4. Detergent solution: 0.5% saponin (Sigma), 0.5% Triton X-100, 1X PBS.
5. FISH blocking solution: 3% BSA, 0.1% Tween-20. 4X SSC (store at 4°C).
6. FISH detection solution: 1% BSA, 0.1% Tween-20, 4X SSC (store at 4°C).
7. FISH washing solution: 4X SSC, 0.1% Tween-20 (prewarm to 45°C).
8. Glycerol 20% in 1X PBS (store at 4°C).
9. Hybridization mixture: 20% dextran sulfate, 4X SSC (store at 4°C).
10. Liquid nitrogen.
11. Post-hybridization washing solution: 60% formamide (Fluka BioChemika, *see* **Note 3**), 2X SSC (adjust pH to 7.0, and store at 4°C; prewarm to 45°C before use).
12. Salmon sperm DNA (Sigma).

2.2.11. Reagents for Reporter Assays

1. 4 mM Acetyl-Coenzyme A (store at –80°C).
2. [^{14}C] chloramphenicol (54 mCi/mmol).
3. Chloroform–methanol (19:1 v/v).
4. Ethyl acetate (ice cold).

2.2.12. Reagents for mRNA Analysis

1. DEPC (diethylpyrocarbonate) treated ddH$_2$O: add 0.1% DEPC to ddH$_2$O, stir for 30 min to overnight, and autoclave for 30–60 min to inactivate the DEPC. Plastic-ware can also be treated with DEPC to inactivate RNase activity (*see* **Note 5**).

2. Guanidine solution (solution A): 4 M guanidine thiocyanate, 10 mM Tris-HCl, pH 7.5, 0.5% sarkosyl, 0.1 M β-mercaptoethanol.
3. 5X hybridization buffer for primer extension: 1.25 M KCl, 10 mM Tris-HCl, pH 8.0, 1 mM EDTA.
4. Primer extension buffer: 10 mM MgCl$_2$, 5 mM DTT, 20 mM Tris-HCl, pH 8.0, 10 μg/mL actinomycin D, 0.5 mM each dNTP (dATP, dTTP, dCTP, dGTP).
5. MMLV RT (Moloney murine leukemia virus reverse transcriptase, Gibco-BRL).
6. 10X hybridization buffer for S1 nuclease protection: 4 M NaCl, 0.4 M Pipes, pH 6.8, 0.02 M EDTA.
7. S1 digestion buffer: 66 mM sodium acetate, 0.3 M NaCl, 4 mM ZnSO$_4$.
8. S1 nuclease.

2.2.13. Reagents for Heat Shock Protein Analyses

1. Isofocusing overlay: 3.6 g ultrapure urea, 100 μL 40% ampholines, 0.1% Bromophenol Blue, and ddH$_2$O to a final volume of 10 mL (store at –20°C).
2. 95% Laemmli sample buffer to which 5% β-mercaptoethanol is added immediately before use.
3. Loading gel: 1% agarose in 95% sample buffer (w/v) (for two gels, mix 0.1% agarose in 9.5 mL Laemmli sample buffer).
4. Lower reservoir buffer: 0.01 M H$_3$PO$_4$.
5. Tube gel composition: 1.38 g ultrapure urea, 0.5 mL 10 % Nonidet P-40, 125 μL 40 % ampholines, acrylamide/bisacrylamide solution (28.38:1.62), 0.49 mL ddH$_2$O, and immediately before pouring gels, add 5 μL TEMED and 4 μL 10% APS (makes 10 tube-gels).
6. Upper reservoir buffer: 0.02 M NaOH.

3. Methods

3.1. Immunological Detection and Analysis of HSF Proteins

3.1.1. Sample Preparation for HSF Analyses

Cell extracts may be prepared from tissue or cultured cell systems by any of a number of techniques, depending on the intended use of the extracts. High-salt buffer C extractions *(32)* maintain the integrity of HSF activities and may be used for experiments such as electrophoretic mobility shift analyses and enzymatic reporter assays. Detergent lysis, such as with RIPA buffer *(33)*, is suitable for immunoanalyses although not designed for long-term storage of active HSF.

1. Prior to extract preparation, spin suspended cells for 1–2 min at 750g in a table-top centrifuge (e.g., Sorvall Technospin) and remove the medium. Wash the cells with cold 1X PBS, transfer to Eppendorf tubes, and spin at maximum speed in a refrigerated microcentrifuge for 5 s (*see* **Note 6**). Remove the PBS and use the cell pellets immediately, or flash-freeze in a dry-ice bath for storage. Frozen cell pellets may be stored indefinitely at –80°C. For isolation of adherent cells,

remove the tissue culture medium, wash the cells with cold 1X PBS, add 1 mL of 1X PBS, and harvest with a cell scraper. Transfer the cells to Eppendorf tubes for processing as above.

2a. Flash-freeze and thaw cells once prior to buffer C extraction. Add cold buffer C equivalent to 3–5 cell pellet volumes and resuspend the cells by gentle pipetting. Spin the lysates at 108,900g in a Beckman TL100 tabletop ultra-centrifuge, or at maximum speed in a microcentrifuge at 4°C for 15 min (*see* **Note 7**). Transfer the supernatants to a fresh eppendorf tube and use immediately, or store indefinitely at –80°C. Determine the protein concentration of extracts and use equal amounts of protein for analyses. Alternatively, cell number equivalents may be used.

2b. For preparation of RIPA extracts, lyse cells in 0.3–1 mL RIPA buffer, using enough buffer to prevent the lysate from being viscous. Resuspend the pellets by repeated pipetting and leave on ice for 5 min. Clear the lysates by microcentrifugation at 4°C for 15 min at maximum speed. Transfer the supernatants to fresh tubes. The amount of extract to use for specific assays may be based on either protein concentrations or equivalent cell numbers.

3.1.2. Immunoblot Analyses

Generally, 10 µg of cell protein is sufficient for detection of HSF in whole-cell extracts when using an acrylamide mini-gel apparatus.

1. Prepare cell extracts by addition of Laemmli sample buffer, and boil for 5 min prior to use. Resolve these samples by SDS-PAGE in 6–10% polyacrylamide gels.
2. Following electrophoresis, soak the gels in transfer buffer and transfer proteins to nitrocellulose filters, using an electroblotting device. The filters can then be dried (the dry blot can be stored at this point) or used immediately.
3. Incubate the filters in blocking solution for a minimum of 1 h. Wash the filters three times with immunoblot washing solution and incubate with shaking in diluted primary antibody for 1 h at room temperature.
4. Wash three times and incubate with a dilution of the appropriate secondary anti-sera in blocking solution for approx 45 min, at room temperature.
5. Three more washes removes unbound secondary antibody. Visualize the bound antibodies by ECL or other techniques, as per the manufacturers' instructions (*see* **Note 8**). The HSF proteins should appear as approximately 70 kDa in size. The constitutively phosphorylated inactive HSF1 will appear as a distinct band, whereas the inducibly hyperphosphorylated form (*24*) appears as a slower migrating band or series of bands (*34*) (**Fig. 1**).

3.1.3. Immunoprecipitation

Cells used for immunoprecipitation analyses using HSF-specific antisera should be actively growing and dense for optimal protein yield. RIPA extracts are suitable for immunoprecipitation analyses (*33*).

1 2 3 4 5

HSF1 →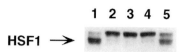

Fig. 1. Immunoblot detection of HSF1 in control and heat-shock-treated NIH-3T3 cells. Buffer C extracts (10 µg) of cells incubated at 42°C for 0, 0.5, 1, 2, and 4 h (lanes 1–5, respectively) were resolved by SDS-PAGE and HSF1 was detected by immunoblot analysis using antisera raised against murine HSF1. The immunoreactive bands were visualised by ECL. The inducible phosphorylation of HSF1 results in retarded migration of the protein (lanes 2–4) and parallels the acquisition of DNA-binding activity at 42°C. Reversion to the control state occurs upon prolonged treatment at 42°C (lane 5).

1. Add specific antisera to cell lysates and rotate to mix for approx 1 h at 4°C. The amount of antibody to be used and the incubation volume must be determined empirically. The efficacies of the antibodies used for HSFs vary, and both polyclonal and monoclonal antibodies have been used successfully (*see* **Note 9**). Prepare parallel samples using preimmune antisera, or other nonspecific antibodies, to determine the specificity of the reaction.
4. Add 30–50 µL of a 1:1 slurry of PBS/Protein A-, or Protein G-conjugated beads, depending on the species of antibodies used (*see* **Note 10**), and mix for 1 h at 4°C.
5. Wash the beads five times with 1 mL ice cold RIPA/0.1% SDS. Resuspend the beads by inverting and flicking the tube, and pellet by microcentrifugation at maximum speed for 15 s. Leave 50–100 µL of buffer behind when removing the washes, to prevent inadvertent removal of beads.
6. After the last wash, remove all but 100 µL of the wash buffer. Remove the remaining wash buffer with a bent 22-gauge needle attached to a 1-mL syringe or a Hamilton syringe (preferably one with a slightly bent needle tip). This should draw up buffer while leaving the beads behind.
7. Add Laemmli sample buffer, mix, and boil for 5 min. Pulse spin out the beads and the sample (supernatant) is ready to load for SDS-PAGE.
8. Visualize the immunoprecipitated proteins by immunoblotting. Alternatively, if extracts are prepared from metabolically labeled cells (pulse or steady-state labeled), the immunoprecipitates may be visualized by autoradiography (*see* **Note 11**).

3.1.4. Immunofluorescence

Immunofluorescence uses antibody fluorochrome conjugates to detect the cellular distribution of proteins of interest *(35)*. This technique allows one to visualize the expression of endogenous or transfected HSFs in intact cells *(36,37)*. The cells to be analyzed are fixed and incubated with specific antibodies to the HSF proteins. This is followed by incubation with modified secondary antibodies against the primary antibody species, which are conjugated to fluorophores with characteristic excitation and emission wavelengths (*see* **Note**

12). The antibody staining pattern can be visualized by microscopy (**Fig. 2**). Adherent cells, such as HeLa and human fibroblasts, are ideal for the experiments described. Cells may be grown directly on two-chamber glass slides (LabTek), as in the following steps. However, cells may alternatively be grown on sterile glass cover slips placed in tissue culture plates, and the ensuing protocol modified appropriately. Treatments that activate HSF are carried out directly in the two-chamber glass slides. For heat-shock treatment, the slides may be sealed with parafilm and immersed in a water bath at the desired temperature. Unless otherwise stated, all steps are performed in a coplin-jar with shaking. Prepare all the required solutions before starting. Note that the volume necessary for a circular coplin-jar is 40 mL, and 80 mL for a square one.

1. Remove the slides from the plastic chambers and place them in a coplin-jar containing 1X PBS and wash for 5 min.
2. Fix the cells for 10 min in fixative. Quench the reaction by a 5-min wash with 0.1 M Tris-HCl (pH 6.8). Wash briefly with 1X PBS.
3. To prevent non-specific binding of the antibodies, add 200 μL of IF blocking solution per slide and cover with a 22 × 50-mm cover slip (*see* **Note 13**). Place all the slides in a closed box containing a wet paper towel or Kim wipes (to maintain a moist atmosphere and prevent evaporation of the blocking solution) and place the box in an incubator at 37°C for 45 min.
4. Meanwhile, prepare the primary antibody, diluting it 1:300 in IF washing solution. At this dilution, three antibodies, a rat monoclonal anti-HSF1 and two rabbit polyclonals, anti-HSF1 and anti-HSF2, produced good results.
5. Remove the IF blocking solution and add 200 μL of diluted antibody per slide and cover with a 22 × 50-mm cover slip. Incubate in the moisture box at 37°C for 2 h. Remove the coverslips and wash three times, for 5 min each time, in prewarmed IF washing solution.
6. Dilute the appropriate secondary antibodies coupled to FITC or other fluorophore 1:100 in the IF washing solution (*see* **Note 14**). After the last wash, add 200 μL of diluted antibody per slide with a 22 × 50-mm cover slip, and incubate in the moisture box at 37°C for 1 h.
7. Remove the coverslips and give them three 5-min washes in the prewarmed IF washing solution. At this point, the coplin-jar should be wrapped in aluminum foil to protect the slides from light.
8. After the last wash, add 30 μL of the DNA counterstaining solution per slide, and cover with a 22 × 40-mm cover slip (*see* **Note 15**). Seal the coverslips on the slides with nail polish. The slides can be stored at 4°C for months.
9. In order to visualize the antibody staining pattern, a regular epifluorescence or confocal laser scanning microscope may be used. Appropriate filter sets for each fluorochrome should be chosen when using an epifluorescence microscope. If a confocal laser scanning microscope is used, the appropriate lasers should be chosen to excite the fluorophores. It is recommended to attenuate the lasers as much

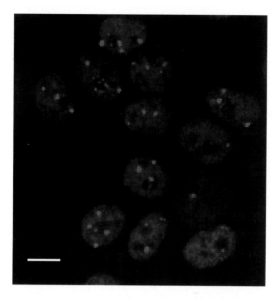

Fig. 2. Determination of HSF1 cellular localization by immunofluorescence analysis. Polyclonal anti-HSF1 antisera *(16)* were used to detect HSF1 in HeLa cells which had been treated for 1 h at 42°C. In heat-shock-treated human cells, the transcription factor is concentrated in the nucleus in several granules of varying size, in addition to a diffuse nucleoplasmic staining. Bar: 5 μm.

as possible to limit the fading of the signals. If possible, acquire both green and red images simultaneously if two antibodies are being used, to limit the exposure of the preparation to the excitation wavelength. The images can be acquired using a conventional camera mounted on the microscope, or a CCD camera may be used to acquire digital images. The use of a cooled CCD camera is recommended when imaging fluorescent signals of low intensity.

3.2. Determination of HSF Oligomeric States

The difference in oligomeric states of the inert and active forms of vertebrate HSFs allows one to employ techniques to separate the two forms based on their sizes and hydrodynamic properties. Glycerol gradient fractionation has been used successfully for this purpose *(17,38,39)*, as has gel filtration *(17,18,38,39)*. Electrophoretic analysis of crosslinked cell extracts also provides information on the oligomeric state of HSF under various conditions *(16,17,39,40)*.

3.2.1. Glycerol Gradient Fractionation

In this procedure, cellular extracts are fractionated through a gradient of glycerol concentrations by centrifugation. The sedimentation properties of a protein are dependent primarily on its size and its sedimentation coefficient (S). The fractions obtained are analyzed by immunoblotting and the sedimentation profile of the protein being analyzed is compared to that of a mixture of standards with known S values.

1. Using a gradient maker, prepare linear glycerol gradients in ultracentrifuge tubes using 2.5 mL each of 10% and 40% glycerol buffers. Chill the gradients for up to 4 h at 4°C.
2. Load whole-cell extracts (100–500 µg in 200 µL) onto the gradients. Load a mixture of standards with known S values (alcohol dehydrogenase = 7.4S, BSA = 4.3S, cytochrome c = 1.9S) on to a separate gradient. Adjust the salt and glycerol concentrations of the loaded samples to resemble the composition of the 10% glycerol buffer.
3. Centrifuge at 4°C in a precooled Beckman SW50.1 rotor for 36 h at 192,000g in a Beckman ultracentrifuge. Remove the gradients carefully from the rotor buckets and collect 250-µL fractions using a peristaltic pump system, preferably from top to bottom (*see* **Note 16**).
4. If 500-µg aliquots of extract are used, 10–12 µL from each fraction may be used directly for SDS-PAGE analyses. If small amounts of extracts are used, precipitate the total protein content of each fraction by the addition of an equal volume of cold 10% TCA, incubate on ice for 10 min, and centrifuge. Remove and discard the supernatants, wash the pellets with ice cold acetone, and allow to air-dry at room temperature.
5. Resuspend the pelletted protein in Laemmli sample buffer with the addition of a few microliters of 1 M Tris-HCl (pH 8.0) (to neutralize the acidity of the precipitate). Use for SDS-PAGE and immunoblot analyses. Fractions from the standard mix are used for SDS-PAGE and the proteins visualized by staining with Coomassie blue.

3.2.2. Gel Filtration

Gel filtration through a sizing column allows relatively precise fractionation of the different oligomeric forms of HSF proteins as well as determination of their Stokes radii. An estimation of the apparent size of the fractionated HSFs can be obtained by comparison to the fractionation profile of protein standards with known molecular weights and Stokes radii (*17,39*).

1. Apply whole cell extracts (0.5–1 mg per 500 µL) to a Superdex 200 HR column with a fast-protein liquid chromatography system (Pharmacia) and elute fractions (0.5 mL each) at 0.25–0.3 mL/min with elution buffer. Analyze aliquots of the fractions as described earlier for glycerol gradient fractionation.
2. Use a mixture of protein standards, such as thyroglobulin (669 kDa, 85.0 Å), ferritin (440 kDa, 61.0 Å), aldolase (158 kDa, 48.1 Å), and albumin (67 kDa, 35.5 Å).

3.2.3. Crosslinking Experiments

The dimeric and trimeric species of HSF are formed by noncovalent association of HSF monomers. Crosslinking creates covalent links between the components of oligomeric complexes that will then resolve electrophoretically according to size. The concentration of crosslinking agent is critical and because many crosslinking agents decay with time, it is often advisable to prepare fresh stocks. Excess crosslinking agent will result in covalent-bond formation between proteins that do not normally associate as stable complexes. To overcome problems resulting from variation of crosslinking efficiency, utilize a range of crosslinker concentrations. Typical concentrations used for HSF analyses with the crosslinking agent ethylene glycobis(succinimidylsuccinate) (EGS) *(41)* are 0.5, 1.0, and 2.0 m*M* *(16,17)*.

1. Dissolve EGS in dimethylsulfoxide (DMSO) and dilute to various concentrations (each as a 10X stock). To separate aliquots of extract, add the different dilutions of crosslinker such that the same volume (maximum of one-tenth of the extract volume) of DMSO is added to each sample.
2. Allow the crosslinking reaction to proceed at 25°C for a determined period of time (usually 30 min). Quench the reaction by addition of concentrated glycine solution to 75 m*M*.
3. Add Laemmli sample buffer to the samples and analyze by SDS-PAGE (5% resolving gel) and immunoblotting for specific HSF proteins. Trimeric, dimeric, and monomeric HSF species can be easily distinguished, as they run approximately according to oligomeric size.

3.3. Detection and Study of HSF Binding to DNA

3.3.1. Electrophoretic Mobility Shift Assay

The electrophoretic mobility shift assay (EMSA) *(42,43)* allows detection of the HSF DNA-binding activities in an extract, or preparation of recombinant HSF. A radioactively labeled oligonucleotide probe containing consensus HSF binding sites or heat-shock elements (HSEs) is incubated with the test sample. The HSF-bound probe is separated from a free probe by native polyacrylamide gel electrophoresis *(44)* (**Fig. 3**).

1. Obtain HSE-containing oligonucleotides. Probes commonly used for mammalian HSF analyses include an HSE-containing region from the human *hsp70* promoter or an idealized HSE-containing oligonucleotide (*see* **Note 17**). Generally, one strand of a double-stranded oligonucleotide is labeled and then annealed to an excess of the complementary strand. In the case of the idealized oligonucleotide, however, the double-stranded species may be labeled.
2. Oligonucleotides are 5' end-labeled by T4 polynucleotide kinase, using [γ-^{32}P]ATP as the source of label. Prepare a 50 µL sample containing 1X kinase buffer (sup-

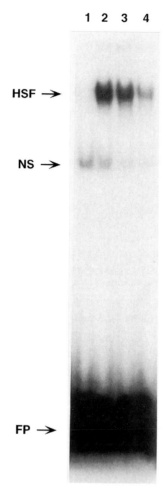

Fig. 3. Analysis of HSF activity by EMSA. The HSF DNA-binding activities in extracts of human Peer cells exposed to 42°C for 0, 1, 2, and 4 h (lanes 1–4, respectively) were assayed by EMSA using a labeled probe containing idealized HSF binding sites (*see* **Note 17**). NS denotes a nonspecific binding activity and FP indicates the position of free unbound probe.

plied with the kinase, or use Pharmacia One-Phor-All buffer), 100 ng oligonucleotide, 1µl [γ-^{32}P]ATP, and 10 U of T4 kinase, added in that order. Incubate at 37°C for 30 min.

3. Add 2 µL 0.5 *M* EDTA and 48 µL TE. This mixture may be extracted with phenol-chloroform, but this is optional. Separate the labeled oligonucleotide from unincorporated [^{32}P]ATP by centrifuging through a Sephadex G-50 spin column. The final concentration of the labeled oligonucleotide is approx 1 ng/µL.

4. Mix the labeled strand (1 ng/µL) with a four- to fivefold excess of the unlabeled strand to which it is to be annealed (5 ng/µL), at a 1:1 volume ratio, for a final concentration of approximately 1 ng/µL double-stranded oligonucleotide. Heat to 85°C for 5 min in a heating block and allow the mixture to cool slowly to room temperature (approx 4 h). Use a diluted aliquot of the labeled probe to determine its specific activity (should be between 2–5 × 10⁵ cpm/ng). Store the labeled oligonucleotide at –20°C until use.

5. For the binding reaction, mix the labeled probe (0.1 ng, 10–50,000 cpm) and poly (dI-dC)-(dI-dC) (0.5 µg) in 1X binding buffer in a final volume of 25 µL minus the volume of extract to be added. The addition of BSA (10 µg) is optional. Add extract (10 µg) to start the reaction and mix by pipetting (*see* **Note 18**). Incubate at 25°C for 20 min.

6. Add 2.5 µL of sample dye to stop the reaction. Load the samples onto a prerun 4% (38:2 acrylamide-bisacrylamide) native polyacrylamide gel. Buffer systems commonly used for HSF EMSA gels are 0.5X TBE and 1X Tnana.

7. Run the gel at approx 11 V/cm (120–160 V for 2.5–3 h for a 15 cm gel) until the bromophenol blue dye is two-thirds of the distance from the bottom of the wells (the free oligonucleotides will migrate below this dye) (*see* **Note 19**). Tnana buffer must be recirculated during electrophoresis.

8. Dry the gel on 3MM Whatman paper, and visualize the results by autoradiography, or systems for detection and quantitation of radioactive signals (e.g., PhosphorImager analysis).

3.3.2. HSF Footprinting Analysis

Whereas electrophoretic mobility shift assays detect the presence of HSE-binding activities, footprinting assays delineate the regions of HSE-containing sequences occupied by an HSF activity, or activities. In vivo genomic footprinting provides a powerful tool for studying the in vivo occupancy by both basal transcription factors and HSFs of their corresponding DNA binding elements. This approach has allowed definition of the role of HSF-HSE interactions in the regulation of heat-shock gene expression *(17,29,45–47)*. Preparations of recombinant HSF1 and HSF2 have been used for in vitro footprinting experiments using the *hsp70* promoter sequence, to further define their binding properties and to confirm the observations obtained from *in vivo* footprinting assays *(48,49)*. Such studies demonstrated that the heat-shock-activated HSF (HSF1) occupies the *hsp70* promoter more extensively than does HSF2, such that each factor has a characteristic "footprint." The differences in DNA occupancy were also shown to be the result of, at least partially, varying extents of cooperativity. HSF1 exhibits cooperative binding to DNA, whereas HSF2 does not *(17,48,49)*.

3.3.3. In Vivo Genomic Footprinting

A ligation-mediated polymerase chain reaction (LMPCR) protocol for in vivo genomic footprinting is described *(50)*. Genomic DNA is first purified and treated with dimethylsulfate (DMS), which methylates guanine (G) residues at the N7 position and makes them susceptible to subsequent cleavage by piperidine. The methylation and cleavage of DNA at G residues creates a DNA ladder. However, G residues located at, or flanking, sites of protein contact exhibit differential sensitivity to modification, being either protected from or hypersensitive to the modification. This will result in a cleavage pattern distinct from that obtained with protein-free, or naked, DNA. The mixture of cleaved DNA products obtained in either case is denatured, and used as templates for DNA polymerase, using a primer (primer 1) specific to the gene of interest. The products obtained will have blunt ends and can be ligated to a common linker DNA (*see* **Note 20**). This creates a population of DNA fragments with identical ends that may be amplified by the polymerase chain reaction (PCR), using one component of the linker DNA (the 25mer, in our experiments) and a second gene specific primer, primer 2, located between the linker DNA and primer 1 (*see* **Note 21**). The amplified DNA products are then indirectly end-labeled with a third gene-specific primer, primer 3, located between the linker DNA and primer 2 (*see* **Note 21**). Analysis of the radiolabeled DNA products on DNA sequencing gels allows visualization of the DNA methylation and cleavage pattern.

3.3.3.1. Isolation of Genomic DNA

This protocol may be used for in vivo (described below) and in vitro DMS-treated (*see* **Note 22**) genomic DNA from cultured cells for use in conjunction with ligation-mediated PCR. Use a minimum of 2×10^7 cells/procedure.

1. For cells in suspension, spin down cells and resuspend the cell pellet in 1 mL of fresh medium. Transfer to a 15-mL Corex tube, place on ice and in a fume hood, add 2 μL DMS, pipetting repeatedly to dissolve. Incubate at 25°C for 4 min. Add 10 mL cold 1X PBS to stop the reaction and spin down the cells at 4°C for 5 min at 900*g* in a Sorvall centrifuge, using an SA600 rotor. Wash the cell pellet with 10 mL cold 1X PBS and centrifuge as above (*see* **Note 4**).
2. For adherent cells, remove the medium and add fresh medium to a 10-cm plate of cells. Add 2 μL DMS/mL medium and pipette to dissolve. Incubate for 4 min at 20°C. Wash the plates twice with 25 mL cold 1X PBS, harvest the cells into 5-mL cold 1X PBS, and transfer to a 15-mL Corex tube. Spin down the cells as described for suspension cells (*see* **Note 4**).
3. Resuspend the cell pellets in 1.5 mL of solution I. Add 1.5 mL of solution II, mix well, and incubate on ice for 5 min. Spin down the resultant nuclei at 1300*g*, for 5 min at 4°C, and resuspend the pellet in 3 mL of solution III. Spin down the nuclei as earlier.
4. Resuspend the pellet in 500 μL of 0.5 *M* EDTA. Once in solution, add 500 μL of nuclear lysis buffer, and incubate at 37°C for 3 h.
5. Add proteinase K to a concentration of 250 μg/mL and incubate at 37°C overnight.

6. Add an equal volume of phenol, mix by hand, and spin at 14,500g for 10 min in a Sorvall centrifuge, at 4°C. Because of the high EDTA concentration, the aqueous layer will be on the bottom, so remove and discard the upper phenol phase and leave the interface and lower phase. Repeat this once.

7. Perform a phenol-chloroform extraction. This time, however, the aqueous layer will be on top. Transfer the aqueous layer and the material in the interface into dialysis tubing (cutoff of 12–14 kDa), and dialyze overnight against 3 liters of TE buffer with one change of buffer.

8. Transfer the dialysate to a 15-mL Corex tube and add one-tenth the volume of 3 M sodium acetate. Precipitate the genomic DNA with 2 vol of ethanol at –20°C for 30 min, and centrifuge at 14,500g for 20 min at 4°C.

9. Discard the supernatant and leave the tube inverted for 30 min to dry the pellet. Add 1 mL TE and leave overnight at 37°C to dissolve the DNA. Determine the DNA concentration and store at –20°C.

10. Perform an overnight restriction digestion of 50 µg of DNA (final volume of 400 µL) with an enzyme that does <u>not</u> cut within the region of DNA to be examined (*Eco*RI for the human *hsp70* gene).

11. Perform two phenol-chloroform extractions, followed by a chloroform-isoamyl alcohol (24:1) extraction, in each case mixing by hand, and microcentrifuging for 5 min at maximum speed.

12. Add 45 µL of 3 M sodium acetate and precipitate the DNA with 900 µL of ethanol at –20°C for 30 min, followed by a 20-min microcentrifugation at maximum speed.

13. Wash the pellet with 80% ethanol and dissolve in 100 µL of 1 M piperidine. Incubate at 90°C for 30 min. Chill the tube on ice and pulse spin to collect everything at the bottom.

14. Dry the sample in a speed-vac with heat, resuspend in 100 µL ddH$_2$O, and dry down again. Repeat this once more. Finally, resuspend the pellet in 100 µL of ddH$_2$O, and add 11 µL of 3 M sodium acetate and 250 µL ethanol. Incubate at –20°C for 30 min and microcentrifuge to precipitate the DNA. Dissolve the pellet in 20 µL ddH$_2$O and measure the DNA concentration.

3.3.3.2. Genomic Footprinting by LMPCR

1. Creation of Blunt-Ended Elongation Products: Mix 6 µg of the cleaved DNA, 0.3 pmole of primer 1, and 3 µL of 5X magnesium-free Sequenase reaction buffer, in a final volume of 15 µL. Incubate at 95°C for 2 min, then 60°C for 30 min. Transfer to ice, and pulse spin at 4°C (to remove condensation). Keep the DNA on ice for all subsequent steps, unless otherwise specified.

2. Add 7.5 µL of ice cold Mg/DTT/dNTP solution, pipetting to mix. Add 1.5 µL of ice cold, freshly diluted Sequenase (1:4 in TE), and mix gently with the pipet. Incubate at 48.5°C for 5 min and 60°C for 5 min.

3. Add 6 µL of 310 mM Tris-HCl (pH 7.7) at room temperature, pipet to mix, and immediately transfer to 67°C for 10 min. Transfer to ice and pulse spin at 4°C.

4. Ligation of Linker DNA: Add 20 µL of ice cold dilution solution, pipetting to mix. Add 20 µL of ice cold ligation mix containing the linker DNA. Transfer to 18–19°C and incubate overnight.

5. Heat-inactivate the ligase by transferring to 70°C for 10 min and pulse spin to remove condensation before adding 8.4 µL 3 *M* sodium acetate, 1 µL of 10 mg/mL tRNA, and 220 µL ethanol. Incubate at –20°C for at least 2 h.

6. PCR-Mediated Amplification: Centrifuge the samples for 15 min at 4°C. Wash the pellet with 75% ethanol and resuspend in 70 µL ddH$_2$O (let it sit for 15–30 min at room temperature to allow the DNA to dissolve).

7. Add 20 µL of 5X Taq buffer, 20 nmoles of each dNTP, 5 U of *Taq* polymerase, and 10 pmoles each of primer 2 and of the 25mer (part of the linker DNA, *see* **Note 20**), pipetting to mix. Denature at 94°C for 2 min, and perform PCR, with 1 min at 94°C, 2 min at 66°C, and 3 min at 76°C for 15 cycles. Transfer the PCR products to ice.

8. Add 5 µL of ice cold labeling mix containing 1 to 10 pmol of end-labeled primer 3 (labeled as described for probe preparation for in vitro footprinting, below). Incubate at 94°C for 2 min, 69°C for 2 min, and 76°C for 10 min. Add 295 µL of *Taq* stop solution and 10 µg of tRNA (can be freshly added to the stop solution and added together) to stop the reaction.

9. Precipitate the DNA with 2.5 vol of ethanol, at –20°C for at least 2 h. Microcentrifuge at maximum speed for 15 min. Wash the pellet with 75% ethanol, and finally resuspend in 12 µL of sequence loading buffer, load an aliquot on to a 6% sequencing gel, which is dried and analyzed by autoradiography.

3.3.4. In Vitro Footprinting

Several in vitro footprinting techniques have been used for HSF studies *(51,52)*, two of which, DNase I footprinting and MPE footprinting, are described here. For the first protocol, deoxyribonuclease (DNase) I digestion is used to generate DNA ladders from protein bound and naked DNA *(51)*. Regions of DNA in contact with protein will be protected from hydrolysis and comparison of the digestion patterns obtained allows mapping of the specific binding sites of proteins on DNA and the approximate boundaries of interaction along the DNA phosphate backbone. The second protocol presented uses methidiumpropyl-EDTA (MPE), which binds to the minor groove of DNA and generates hydroxyl radicals *(52–54)*. Because of its small size, MPE can penetrate regions of protein-DNA interaction inaccessible to DNase I and provides information about the tightest interactions between protein and DNA. The pattern of cleavage obtained is also dictated by the structure of DNA in the minor groove and thus gives us more information on the region studied *(54)*. Both techniques utilize labeled probes generated from plasmids containing heat-shock promoter sequences. A protocol for preparation of an *hsp70* promoter-specific probe is outlined first.

3.3.4.1. Preparation of Labeled Probes

1. Radioactive probes of high specific activity are generated by PCR, using labeled primers flanking the region of interest. In this example T7 and T3 promoter spe-

cific primers are used to prepare single stranded end-labeled probes corresponding to both DNA strands, with a plasmid containing the human *hsp70* promoter
sequence as template. Label primer with 20 units of T4 polynucleotide kinase
and 1 mCi of [γ-^{32}P]ATP per μg of primer, in 1 X kinase buffer, for 30 min at
37°C. Remove unincorporated label as described above for preparation of EMSA
probes. The labeled primer may be stored at –20°C in an appropriately shielded
container until use. In our system, the T7 primer is used for footprint analysis of
the coding strand and the T3 primer for analysis of the noncoding strand.

2. For the coding strand, the PCR reaction is performed in 50 μL with 0.25 μ*M* of
   the labeled T7 primer and unlabeled T3 primer, 1 U of *Taq* polymerase, 1X PCR
   buffer, 200 μ*M* dNTPs, and 1 ng of template (or 5 μL of a 1:1000 dilution of a
   mini-preparation of plasmid DNA). PCR with 94°C for 1 min, 50°C for 1 min,
   and 72°C for 1 min, for 30 cycles, followed by incubation at 72°C for 5 min.
   Carry out a similar reaction for the noncoding strand using labeled T3 primer and
   excess unlabeled T7 primer.

3. Precipitate the labeled probes with addition of 0.25 vol of 8 *M* ammonium acetate,
   5 μg of glycogen, and 1 vol of isopropanol, incubation for 10 min at 25°C, and
   microcentrifugation at maximum speed for 10 min.

4. Resuspend the recovered DNA in 400 μL TE. Check the integrity and purity of
   the DNA by electrophoresis on a 10% (19:1) polyacrylamide gel. The concentration of the labeled DNA can be accurately estimated by a direct spectrophotometric analysis of the entire sample (400 μL) at 260 nm, (*see* **Note 2**).

3.3.4.2. DNASE I FOOTPRINTING

1. Mix single-stranded end-labeled probe (1×10^{-10} *M*), competitor DNA (100 ng)
   poly (dI-dC)·(dI-dC), and 50 μL of 2X transcription buffer in a final volume of
   100 μL at room temperature. It is best to prepare one large reaction mix and then
   aliquot 100 μL into each sample tube.

2. Add HSF1 or HSF2 protein (amounts required for complete protection varies
   from 0.5–10 n*M*) and incubate at 23°C for 20 min. While this binding reaction
   proceeds, thaw a 10 mg/mL stock of DNase I (–20°C freezer, 50-μL aliquots),
   dilute 1:100 in ice cold water, and keep on ice at all times.

3. Add 100 μL each of 10 m*M* MgCl$_2$ and 5 m*M* CaCl$_2$ immediately before DNase
   I treatment. Add 4 μL DNase I to final concentration of 2 μg/mL. After 1 min
   digestion, add 200 μL of DNase I stop solution. Because there are normally multiple samples, four to five samples should be processed at a time for precise tim-
   ing of digestion.

4. Extract once with phenol-chloroform extraction and precipitate with 2 vol of 95%
   ethanol, at –20°C overnight, or in a dry ice bath for 15 min. Microcentrifuge for
   15 minutes and allow the pellet to air-dry.

5. Resuspend the sample in 4–5 μL of sequencing gel loading buffer and load onto
   a prerun 6% sequencing gel, to analyze by autoradiography.

3.3.4.3. MPE FOOTPRINTING

1. Establish a binding reaction as in DNase I footprinting.

2. Right before use, mix 2.5 µL of a 1.5 mM stock solution (kept frozen at −80°C) of MPE mixed with 4 µL of 4 mM Fe(NH$_4$)$_2$(SO$_4$)$_2$ and immediately dilute to 100 µL with cold ddH$_2$O.
3. Add 1 µL of the MPE-Fe(II) solution to the HSF binding reaction. After 3 min, add 1 µL of 100 mM DTT and allow the cleavage reaction to proceed for an additional 2 min.
4. To stop the cleavage reaction, add 5 µL of stop solution and 60 µL of 95% ethanol per 25 µL binding reaction. The samples are then processing as in **steps 4** and **5** under **Subheading 3.3.4.2.** for DNase I footprinting.

3.4. The Analysis of HSF-Mediated Gene Expression

Under conditions in which HSF activation is observed, it is necessary to determine the functional status of the HSF (i.e., whether there is an increase in its transcriptional activity). Many techniques have been used to study HSF transcriptional activity and heat-shock gene expression. Nuclear run-on transcription analysis is one such technique that allows direct determination of the transcription rate of endogenous heat-shock genes. The change in heat-shock gene transcriptional state may be visualized *in situ* by use of modified probes specific to heat shock genes of interest, as described in the second procedure (FISH). Transcriptional activity may be monitored by measuring the enzymatic activities of proteins encoded by transfected reporter genes (reporter assays) or by assessing reporter gene expression at the mRNA level. The RNA-analysis protocols described (primer extension and S1 nuclease protection assays) are also used for quantitative assessment of endogenous heat-shock gene expression. The expression of endogenous heat-shock proteins may also be monitored by pulse-labeling procedures and immunoblot analyses.

3.4.1. Nuclear Run-on Transcription

In vitro transcription in isolated nuclei is used to measure the rate of transcription of a specific gene or genes. RNA transcription initiated in vivo is completed in vitro with incorporation of a radiolabeled nucleotide. There is no reinitiation of transcription; thus, this technique provides information about which of the genes of interest were actively being transcribed at the time that the nuclei were isolated (55–58). The total RNA (including the labeled RNA) is isolated and hybridized to DNA probes blotted on nitrocellulose filters. We have used this analysis to measure the rate of transcription of heat-shock genes under various stress conditions (17,29,41,48,47,59,60) (**Fig. 4**). The method to be described was adapted from protocols previously used in our laboratory (59) and optimized for convenience, especially in the steps involving use of radioactivity. Once the cells are added to a tube, all of the subsequent steps up to the hybridizations are performed in the same tube. The samples are maintained on ice throughout the procedure, unless otherwise specified.

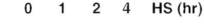

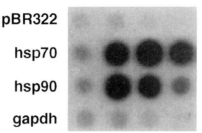

Fig. 4. Analysis of heat-shock gene transcription. Human Peer cells were heat shocked at 42°C for 0, 1, 2, and 4 h. The transcription rates of *hsp*70 and *hsp*90α genes in these cells were measured by run-on transcription analysis. The plasmid vector pBR322 was included as a control for nonspecific hybridization, and the *gapdh* gene used as a normalization control for transcription.

3.4.1.1. ISOLATION OF NUCLEI

1. Wash 6×10^6 cells or more (one or more 100×20-mm plates of HeLa cells, for example) in 1X PBS and harvest as described above.
2. Lyse the cells by the addition of several volumes of lysis buffer (about 200 µL for 6×10^6 cells) and repeated pipetting to disperse the pellet. Spin briefly (3 s) in a microcentrifuge to pellet the nuclei.
3. Estimate the nuclear pellet volume (usually around 50–200 µL) and resuspend in an equal volume of nuclei storage buffer. Pipetting the nuclei should be done with a wide-bore pipet tip (made by cutting off the end of the pipet tip with a clean razor blade) to avoid breaking them. Store at –80°C until needed.

3.4.1.2. DNA DOT BLOT PREPARATION

1. The amount of DNA we use is typically 1 µg per dot. The procedure described is for preparation of 10 reactions using the Minifold™ filtration manifold from Schleicher and Schuell. Sonicate 10 µg DNA in 750 µL ddH$_2$O for 15 s to linearize DNA.
2. Add 30 µL of 10 N NaOH to each sample, and chill on ice to denature the DNA. Neutralize with one volume of 2 M ammonium acetate (780 µL).
3. Filter 1 µg (150 µL in our case) of the DNA solution on to a nitrocellulose membrane on top of one piece of 3MM Whatman paper using a dot blot apparatus. Cut the nitrocellulose membrane into strips each containing one set of DNA samples to be tested (*see* **Note 23**) and bake in a vacuum oven for 2 h at 80°C. The DNA blots can be stored at room temperature.

3.4.1.3. TRANSCRIPTION REACTIONS

1. Freshly prepare the 2X reaction cocktail that contains radiolabeled nucleotide. Thaw the nuclei on ice and add 50 µL of nuclei (using a wide-bore pipet tip) to 50 µL

of the 2X reaction cocktail. Incubate at room temperature for 20 min. Stop the reaction by adding 2 µL of DNase I and incubating at 37°C for 10 min.

2. Add 300 µL stop buffer, 300 µg proteinase K (final concentration of 1 mg/mL), and 100 µg tRNA to each reaction. Homogenize by pipetting and incubate at 40–50°C for 2 h.
3. Precipitate by adding ice cold TCA to each reaction to a final concentration of 10% and incubating on ice for 20 min.
4. Spin 15 min to pellet the nucleic acids. Wash the pellet with cold absolute ethanol to remove any trace of TCA. Air-dry the pellet and resuspend in 50 µL of TE containing 0.5% SDS. Incubate at 65°C for 15–30 min to dissolve the RNA.

3.4.1.4. HYBRIDIZATION REACTIONS

1. Prepare hybridization buffer. Prehybridize filter-bound DNA in 2.5 mL hybridization buffer at 42°C for at least 6 h. It is convenient to hybridize the nitrocellulose strip in 12 × 75-mm culture tubes with caps.
2. Add 50 µL radiolabeled RNA to the prehybridization buffer. Hybridize at 42°C for at least 72 h.
3. Wash the DNA blots once with 6 X SSC and 0.2% SDS at room temperature for 10 min, then twice with 2X SSC and 0.2% SDS, followed by two washes in 0.2X SSC and 0.2% SDS, all at 65°C for 10–30 min each wash. Place the DNA blot strips on an old X-ray film as backing, cover with plastic wrap, and analyze by autoradiography.

3.4.2. Fluorescence In Situ Hybridization

In situ RNA hybridization has been used to assess transcriptional activity of specific genes in intact cells (*61*). This procedure has been used to study the transcriptional activity of HSF and the relative distribution of heat shock gene transcription sites (*37*) (**Fig. 5**). The nuclear transcripts of the three HSF1-regulated genes, *hsp70*, *hsp90α*, and *hsp90β*, have been detected by FISH and this approach used to assess the *in situ* transcriptional activity of HSF1. This protocol has been optimized to ensure high efficiency for RNA detection and good preservation of both cellular morphology and nuclear texture (*62*). As described above for immunofluorescence, HeLa and human fibroblast cells may be used, at a confluency of 70–80%. Unless otherwise stated, all steps are performed in a coplin-jar and require shaking.

3.4.2.1. PREPARATION OF LABELED PROBES

1. Label probes with biotin by random priming, using the BioPrime Labeling kit from Gibco-BRL. Alternatively, probes can be labeled with digoxygenin using standard nick-translation procedures. Precipitate labeled probe (100 ng) and 20 µg of salmon sperm DNA together by addition of sodium acetate (0.1 *M* final concentration) and 2 volumes of 100% ethanol. When genomic probes are used, it is necessary to add *Cot*I DNA before precipitation to allow suppression

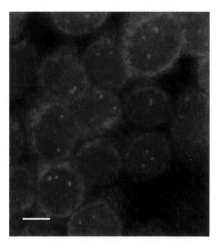

Fig. 5. Detection of hsp70 Transcripts *in situ*. HeLa cells treated for 1 h at 42°C were analyzed by FISH using a probe specific to *hsp70*. The detected nuclear transcripts appear as foci corresponding to the allelic sites of transcription. Diffuse nuclear and cytoplasmic staining represent diffusely distributed *hsp70* transcripts. Bar: 5 μm.

of repeated sequences (to reduce background fluorescence and enhance the specific signals) *(63)*.

2. After centrifugation to precipitate the DNA, wash the pellet once with 70% ethanol, spin again, and dry to remove all traces of ethanol (air-dry or speed-vac). Resuspend the pellet in 5 μL of Sigma UltraPure formamide.

3. Add 5 μL of hybridization mixture and mix well. Just before use, denature the probe for 5 min in a heating block at 75°C and place immediately on ice. In the case of genomic probes, a 45-min incubation in a 37°C water bath is necessary after denaturation to allow preannealing of the repeated sequences with the *Cot*I DNA *(63)*.

3.4.2.2. Cell Treatment and Hybridization

1. The cells to be used for this procedure are first washed, fixed and quenched as described above for the immunofluorescence protocols. After the 0.1 *M* Tris-HCl treatment, rinse the slides briefly with 1X PBS, followed by three 5-min washes with the detergent solution.

2. Rinse the slides with 1X PBS, and immerse them in the 20% glycerol/PBS solution for at least 20 min.

3. Permeabilize the cells by dipping/thawing the slides three times in liquid nitrogen (let them stand for 3 s in liquid nitrogen each time). Slides can be reimmersed in 20% glycerol/PBS between each dipping if necessary.

4. After the last thawing step, rinse the slides in 1X PBS. Dehydrate the cells through sequential incubations in 70%, 90%, and 100% ethanol baths for 5 min each. After the last ethanol bath, allow the slides to dry completely.

5. Add 10 µL of the labeled probe on each slide and cover with an 18 × 18-mm cover slip. To avoid dehydration of the slide during incubation, the cover slips should be sealed with rubber cement. Place the slides in a moisture box, and incubate overnight in a 37°C incubator.
6. Remove the rubber cement. Immerse the slides in a bath of prewarmed 60% formamide/2 X SSC solution and remove the cover slips carefully. Wash three times for 5 min in the same solution at 45°C followed by three 5-min washes with 2X SSC at room temperature.
7. Add 200 µL of FISH blocking solution per slide with a 22 × 50-mm cover slip and incubate for 45 min in the moisture box at 37°C. Meanwhile, dilute avidin-FITC (Sigma) (1:200) in the FISH detection solution.
8. Remove the coverslip and add 200 µL of the diluted avidin solution per slide. Cover with a 22 × 50-mm cover slip, and incubate in the moisture box for 45 min at 37°C.
9. Remove the cover slip and wash three times for 5 min in the prewarmed FISH washing solution. Add 30 µL of DNA counterstaining solution per slide, and cover with a 22 × 40-mm coverslip. Seal the cover slips on the slides with nail polish. Microscopy is used to visualize the images as described for the immunofluorescence protocol (*see* **Subheading 3.1.4.**).

3.4.3. Enzymatic Reporter Assays

A number of reporter genes suitable for mammalian cell transfection are available, such as those encoding chloramphenicol acetyl transferase (CAT) *(64)*, firefly luciferase *(65,66)*, and β-galactosidase *(67)*. Heat-shock gene-promoter-regulated versions of these reporter genes have been constructed and used successfully. When using such reporter systems, it is wise to cotransfect a second reporter controlled by a constitutive promoter to serve as an internal control for transfection. Various combinations of reporter and control plasmids may be used. We will describe a system in which the activity of the enzyme CAT (which catalyzes the transfer of an acyl group from acetyl-CoA to chloramphenicol) gene is used as a measure of HSF activity. To examine endogenous HSF activity, we use pHBCAT as the reporter plasmid which has the CAT gene driven by a promoter containing heat-shock elements *(30)*. To study the transcriptional capability of HSF uncoupled from its DNA-binding activity, regions of HSF may be fused to the GAL4 DNA-binding domain and cotransfected with the G5BCAT reporter plasmid that has the CAT gene under the control of a promoter containing five GAL4 DNA binding sites *(60,68)*. The control plasmid for transfection efficiency is that encoding the firefly luciferase gene or the β-galactosidase gene under the control of strong constitutive promoters. Following transfection and treatment of the cells to activate HSF, cell lysates are prepared and incubated with acetyl-CoA and radiolabeled chloramphenicol. The products of the CAT activity, acetylated chloramphenicol, are separated from the nonacetylated substrate by thin-layer chromatography (TLC).

1. Transfect 10-cm plates of cells, such as COS cells, with 1 μg reporter plasmid, HSF expression systems (such as fusion constructs, if needed), 1 μg internal control plasmid for transfection efficiency, and carrier DNA needed to bring the total DNA amount to 20 μg (the DNA amount is optimized for the calcium phosphate precipitation transfection method, but other transfection protocols have also been used successfully).
2. Harvest cells within 48 h posttransfection and freeze cell pellets quickly on dry ice. Prepare cell lysates in buffer C and determine the protein concentration.
3. Combine 1 μL of 25 μCi/mL [^{14}C] chloramphenicol, 20 μL of 4 mM acetyl CoA, and 32.5 μL of 1 M Tris-HCl (pH 7.4). Add 12.5 μg cell extract and ddH$_2$O to a final volume of 150 μL. Mix and incubate at 37°C for 1 h (*see* **Note 24**).
4. After incubation, add 1 mL ice-cold ethyl acetate to the reaction and vortex for a few seconds. Microcentrifuge for 1 min at 4°C and transfer the top layer to a new tube, avoiding the interface. Dry down the supernatant in a speed-vac for 20–30 min and resuspend the pellet in 15 μL ethyl acetate.
5. Spot the samples with even spacing on a TLC plate (silica gel) and develop in a chromatography tank containing a chloroform-methanol mixture at a ratio of 95:5. The tank should have been preequilibrated at least 2 h prior to developing. Run the sample until the solvent is close to the top of the plate.
6. Remove the TLC plate from the tank, air-dry, and expose to X-ray film at room temperature.

3.4.4. RNA Analyses

Reporter systems may also be analyzed at the level of mRNA expression. High-quality RNA preparations from cells to be analyzed must be obtained, and an RNA isolation procedure suitable for our analyses is first described *(69)*. Two protocols for analysis of specific mRNA species are subsequently described.

3.4.4.1. ISOLATION OF TOTAL RNA

The critical factor in RNA isolation is to take all necessary precautions to minimize RNA degradation by the highly active and ubiquitous RNase enzymes. All solutions to be used for these procedures should be prepared in water treated with diethyl pyrocarbonate (DEPC), which inactivates RNase activities. Solutions that can be autoclaved should also be autoclaved. It is advisable to use as much sterile disposable plasticware as possible and to bake any essential glassware at 180°C. Gloves should be worn at all times to prevent RNase contamination from skin. Finally, speed is of the essence to minimize the time of exposure of RNA to any residual RNase activity.

1. Harvest cells, wash in cold 1X PBS, and lyse in solution A, which contains a high concentration of guanidine thiocyanate (a denaturant which inactivates RNase activities released upon lysis). Use 100 μL solution A per 10^7 cells.
2. Sequentially add 0.1 vol of 2 M sodium acetate, 1 vol of TE-saturated phenol, and 0.2 vol of chloroform-isoamyl alcohol. Mix by inversion after adding each reagent. Shake vigorously for 10 s and incubate on ice for 15 min. Microcentrifuge at maximum speed for 15 min at 4°C.
3. Remove the aqueous phase to a new tube and mix with 1 vol of isopropanol. Store at –20°C for at least 1 h. Microcentrifuge at maximum speed for 15 min at 4°C to pellet the RNA.
4. Add 100 μL of solution A to the pellet and incubate at 65°C for 30 min to dissolve. Add 1 vol of isopropanol and store at –20°C for at least 1 h, followed by microcentrifugation at maximum speed for 15 min at 4°C.
5. Wash the pelleted RNA in 0.5 mL of 80% ethanol and dissolve in DEPC-treated water, incubating at 65°C to aid solution. Quantitate the RNA concentration by measuring the absorbance of a diluted aliquot at 260 nm.

3.4.4.2. PRIMER EXTENSION ANALYSIS OF RNA

Primer extension analysis is useful for mapping the transcription start site of a gene of interest and assessment of mRNA expression levels *(70,71)*. It is a more sensitive measure of heat-shock-induced reporter gene expression than the enzymatic reporter assays. An end-labeled primer specific to the gene of interest is hybridized to total RNA and radiolabeled cDNA synthesized by the reverse transcriptase. The number of cDNA molecules synthesized will reflect the number of mRNA molecules present in a given preparation of RNA. The products of this reaction are resolved by electrophoresis and the specific mRNA levels may be quantified and normalized to an internal control for transfection efficiency. Unlike in the enzymatic reporter assay described earlier, when using CAT reporter systems for primer extension analysis, it is best to cotransfect the constitutively expressed RSV-CAT as the internal control. The same CAT gene-specific primer may be used for simultaneous reverse transcription of CAT genes driven by promoter containing the HSF (or GAL4) binding sites and that driven by the RSV promoter. The primer is designed in such a way that the different sized products will resolve upon electrophoresis. This protocol is also used to measure the mRNA levels of endogenous heat-shock genes using heat-shock gene-specific primers *(68,71)* (**Fig. 6**).

1. Select a gene specific primer and label 50 ng of this oligonucleotide with 1 μL of [γ-^{32}P]ATP and 1 μL of T4 polynucleotide kinase in 1X kinase buffer at 37°C for 30 min. Bring the volume to 100 μL and remove the unincorporated radioactivity as described for EMSA probe preparation (*see* **Subheading 3.3.1.**). The labeled primer may be stored at –20°C in an appropriately shielded container until use.

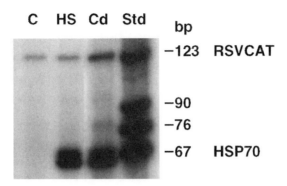

Fig. 6. Primer extension analysis of *hsp70* gene expression. RNA isolated from control, untreated mouse NIH-3T3 cells, or cells exposed to either heat shock at 42°C for 2 h (HS), or 40 μM cadmium sulfate for 8 h (Cd) was used for reverse transcription primer extension analysis. The products were resolved by gel electrophoresis and the levels of the endogenous *hsp70* gene transcripts were measured. Transfected RSVCAT was used as a normalization control. A mixture of oligonucleotides of defined sizes was used as standards (Std).

2. Add 10 µg purified RNA to 10 µl 5 X hybridization buffer with 1 ng of [^{32}P]-labeled primer and ddH$_2$O to a total volume of 50 µL. Incubate at 85°C for 5 min to denature the template and allow to anneal at 56°C for 3 to 4 h.
3. Add 115 µL primer extension buffer and 100 U MMLV reverse transcriptase to each sample. Incubate at 42°C for 1 h.
4. Precipitate the cDNA with 500 µL of ethanol by centrifugation at 4°C for 15 min. Wash the DNA pellet with 70% ethanol and allow to air-dry. Resuspend the pellet in 10 µL formamide loading solution for electrophoretic separation on a sequencing gel.

3.2.4.3. S1 Nuclease Protection Assay

This RNA analysis protocol *(72)* uses an end-labeled single-stranded DNA probe complementary to the RNA of interest and may be used to determine the boundaries of the transcriptional unit. An end-labeled DNA probe is first hybridized to RNA, then S1 nuclease is added to digest all single-stranded, unhybridized probe. The products are resolved by electrophoresis and quantified. When vast probe excess is used, this method allows quantitation of the relative amount of a specific RNA species. This protocol can be used to measure endogenous heat-shock gene expression as well as analysis of reporter gene expression. However, when using reporter systems, controls for normalization of transfection cannot be determined simultaneously, as in primer extension analyses. Probes to endogenous housekeeping genes may be used to normalize for the amount of total RNA used.

1. For 5' mapping, digest the DNA template to be labeled from the corresponding plasmid DNA. Dephosphorylate the 5' termini of DNA with 40 U/mL of calf intestinal phosphatase (CIP) in 50 mM Tris-HCl (pH 9.0), 1 mM MgCl$_2$, 0.1 mM ZnCl$_2$, and 1 mM spermidine, at 37°C for 1 h.
2. Stop the reaction with an equal volume of 10 mM Tris-HCl (pH 8.0), 0.1 M NaCl, 1 mM EDTA, and 0.5% SDS. Incubate at 68°C for 15 min and extract with 1 vol of phenol-chloroform.
3. Radio-label and purify the DNA fragment with [γ-^{32}P]ATP as described for primer extension analysis, except that 1 µg of DNA is used.
4. Digest the 5' end-labeled DNA with a suitable restriction enzyme to define the 3' end of the probe. The probe is isolated on an alkaline agarose gel (*see* **Note 25**).
5. Templates for mapping the 3'-terminus of a message are prepared by selective end-labeling with the Klenow fragment of *Escherichia coli* DNA polymerase I and an appropriate [α-^{32}P] dNTP.
6. Ethanol precipitate 100 µg of total RNA and an excess of end-labeled probe (in the range of 5×10^4 Cerenkov counts), with yeast tRNA as carrier to facilitate RNA precipitation. Resuspend the pellet in 10 µL ddH$_2$O.
7. Add 80 µL formamide and 10 µL of 10X hybridization buffer. Vortex and incubate at 65–75°C for 15 min. Transfer the sample to 53°C, taking care not to allow the temperature to drop below 53°C, and incubate for more than 4 h.
8. Add 300 µL ice-cold S1 buffer and place the tubes on ice. Add 200–400 U of S1 nuclease and incubate at 37°C for 60 min.
9. Phenol extract the samples. Add 1 mL absolute ethanol to precipitate the nucleic acids. Allow the pellets to air-dry and resuspend in 10 µL ddH$_2$O. Add sequence loading dye to an aliquot of each sample and resolve on a sequencing gel for analysis of products.

3.5. Heat Shock Protein Analyses

The expression of heat-shock proteins as a result of HSF activation can be monitored by analysis of the protein synthetic profile of cells. The increased expression of major heat-shock proteins may be seen indirectly by examination of extracts from pulse labeled cells, because a heat-shock response results in a characteristic profile of protein synthesis by one-dimensional, and more convincingly, by two dimensional gel electrophoreses. Cell extracts may also be used for immunoblot analysis to directly monitor the expression level of specific heat-shock proteins. For some HSF-activating treatments, such as short periods of heat shock, it may be necessary to allow cells to recover (30 min or more, for example) following the activating treatment, to allow accumulation of detectable amounts of heat-shock proteins. Protocols for pulse-labeling of cells and for two-dimensional gel electrophoresis are now described.

3.5.1. Pulse-Labeling of Cells

1. Determine which labeled amino acid is to be incorporated and obtain medium lacking this amino acid. Wash cells to be labeled following defined treatments,

with prewarmed 1X PBS or the amino-acid-deficient medium. A short incubation in the deficient culture medium (to ensure depletion of the amino acid to be incorporated) prior to addition of the labeled amino acid is optional *(73)*.

2. For analyses of chaperone expression, pulse-labeling cells with 50 µCi/mL Tran [^{35}S] label (ICN) (a mixture of [^{35}S]-labeled methionine and cysteine), in Met- and Cys-deficient medium produces satisfactory results, even without pre-incubation to deplete residual methionine and cysteine. Add the labeled amino acids directly to the Met- and Cys-minus medium and incubate 30–60 min at 37°C in a 5% CO_2 atmosphere (*see* **Note 26**).
3. Remove the labeling medium and wash the cells with cold 1X PBS. Use the cell pellets for whole-cell-extract preparation and resolve by one- or two-dimensional gel analyses and visualize by autoradiography. Typical heat-shock protein expression profiles produce an elevated expression of 70- and 90-kDa proteins because of increased expression of Hsp70 and Hsp90 in one-dimensional gels (**Fig. 7**), and a characteristic pattern of protein expression is obtained for two-dimensional gels.

3.5.2. Two-Dimensional Gel Electrophoresis

This procedure allows the separation of proteins based on two criteria. In the first dimension, proteins are resolved in tube gels according to their charge, and then separated on the basis of size in the second dimension by SDS-PAGE in slab gels *(74)*. The ensuing protocol for tube gel preparation has been developed for the Hoefer DE 102 series tube-gel electrophoresis apparatus. The tube gels are cast in 200-µL Clay Adams Micropipets Accu-fill 90, which have an internal diameter of 1.5 mm and therefore require 1.5-mm slab gels for the second dimension (we use 13 × 13-cm SDS-PAGE gels).

1. Make a 10-cm mark on each glass tube. Using a small rasp file or punch tool, widen the tip of a 1-cc syringe so that the tubes screw very tightly into the tip of the syringe. It is useful to have on hand a second prepared syringe, as the acrylamide polymerizes quickly and the syringe will stretch after several uses, making it useless.
2. Flatten and mold a piece of plasticene into a 4–6 × 1-in. strip. Place the plasticene on top of a piece of parafilm and align with the bottom edge of a test tube rack around which a large rubber band will fit (to support the tubes while they polymerize).
3. Prepare the tube gel solution. With a pipet attached to the 1-cc syringe, fill to the 10-cm mark, remove the tube from the solution, and pull the gel up approx 0.5 cm above the mark (to avoid loss due to dripping). Carefully slide the tube between the support and the rubber band, push the gel level back to the 10 cm mark, and push squarely into the plasticene to plug the end before removing the syringe (**Fig. 8**). When all of the tube gels have been poured, overlay each with 5–10 µL ddH$_2$O. Allow 2–4 h for polymerization.
4. Remove the tube gels from the plasticene and, using a sharp object, remove any plasticene adhering to, or plugging, the bottom of the tubes. Select only good

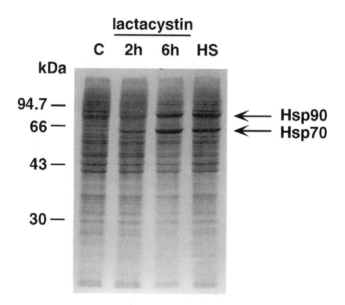

Fig. 7. Analysis of heat-shock protein expression by pulse-labeling of cells. Human K562 cells were treated with the proteosome inhibitor lactacystin (which activates HSF2) for 2 and 6 h, or exposed to heat shock at 42°C for 30 min followed by a 30-min recovery (HS). The treated cells and untreated control K562 cells (C) were allowed to incorporate [^{35}S]-labeled methionine for 30 min. Cell extracts were prepared and analyzed by SDS-PAGE and autoradiography. The positions of Hsp70 and Hsp90 are indicated.

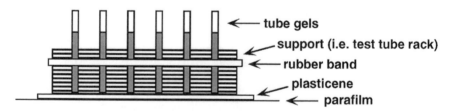

Fig. 8. Isofocusing tube-gel apparatus. This is used for the first dimension in the two-dimensional gel electrophoresis system.

tubes to use, avoiding any that have leaked or contain air bubbles in the acrylamide.

5. Wipe each tube with a wet Kim-wipe and "flick" the tube to remove the ddH$_2$O overlay. Carefully push the tubes through pinholes in the rubber inserts of the electrophoresis apparatus to approx 1.5–2 in. from the top of the apparatus.

6. Squirt lower reservoir buffer into the space in the bottom of each tube to eliminate air bubbles. Fill the lower reservoir chamber approximately two-thirds full with buffer. Place the isofocusing unit into the lower buffer chamber. Check once again for air bubbles.

7. Fill any remaining rubber inserts with unusable poured gels or empty tubes with their upper ends above the surface of the upper buffer to prevent leakage of upper buffer to the lower chamber. Add upper reservoir buffer to the upper chamber. Mark the upper buffer level to monitor for possible leakage into the lower chamber.

8. Load the samples onto the tube gel. For 10-cm tube gels, the volume of sample cannot exceed 25 μL. Overlay each with 2–4 μL isofocusing overlay. Run the gels at 300 V constant voltage for 14.5 h, then increase to 800 V for 2.5 h (*see* **Note 27**).

9. Remove the tube gel. For each tube, flick off buffer from the top and wipe dry. Fit tightly onto a prepared 10 cc syringe whose plunger has been preset at the 10-cc mark. Extrude each tube gel into a separate tube on ice, containing 5 mL of 1X Laemmli sample buffer to which 5% β-mercaptoethanol had been freshly added. The tubes can be frozen at –20°C.

10. Prepare the second dimension SDS-PAGE gel in advance. The stacking gel should be poured without a comb, until 0.3–0.5 cm from the top of the plate and overlay with ddH$_2$O Alternatively, a comb that will create a well wide enough to accommodate the tube gel when placed horizontally may be used.

11. Prepare loading gel in 10-mL aliquots (enough for two gels each). Place these aliquots in a boiling water bath and leave them there until used. Add 500 μL β-mercaptoethanol immediately before use. Because large amounts of β-mercaptoethanol are used, it is best to carry out the subsequent steps in a ventilated hood.

12. Place all the materials required for the subsequent steps in the ventilated hood, along with an additional layer of bench paper and a generous supply of paper towels.

13. Pour the contents of the tube containing the tube gel onto a firm plastic piece with a thin edge suitable for loading the tube gel onto the SDS polyacrylamide gel. Using blunt-ended forceps, align the tube gel straight along the thin edge (pour off excess sample buffer into a waste container).

14. Remove a tube of loading gel from the boiling water and add 500 μL β-mercaptoethanol. Using a Pasteur pipet, mix well and add a layer of loading gel to the top of the stacking gel (*see* **Note 28**).

15. After all tube gels have been loaded, let them sit for a few minutes to ensure that the loading gel has solidified completely. Carefully add reservoir buffer to the electrophoresis unit and run the gels at low constant voltage (80 V in our system) through the stacking gel. The voltage can subsequently be increased. Run the gels until the dye front reaches the bottom of the plates. Run the gels longer for better separation of higher-molecular-weight proteins.

16. The gels may be analyzed by immunoblotting or staining for total proteins. Alternatively, if radiolabeled cell extracts are used, the gels may be visualized by autoradiography.

4. Notes

1. Acrylamide and bis-acrylamide are neurotoxins. Wear gloves at all times when handling unpolymerized solutions and polymerize all waste solutions before disposal.
2. Take appropriate precautions and follow institutional and other safety regulations when handling and disposing of radioactive materials.
3. Handle with care, as formamide is toxic and a teratogen.
4. Toxic, treat all DMS waste with 5 *N* NaOH prior to disposal.
5. DEPC is toxic, handle with care and use in a well-ventilated fume hood.
6. Avoid long centrifugations, as this may cause HSF activation.
7. Ultracentrifugation produces higher-quality extracts, because the higher *g* force generated ensures complete pelleting of insoluble cellular components.
8. If a high background or a low intensity of signal is a problem, alter the stringencies of the washes and lengths of incubations. Also, changing the ratio of dilution of the primary and secondary antisera may be of help.
9. Polyclonal antisera, by definition, recognize multiple epitopes of the target protein, whereas monoclonal antibodies recognize single epitopes. The availability of these epitopes in the native folded protein will contribute to the efficacy of a given antiserum in immunoprecipition studies.
10. The choice of matrix for immobilization of antibodies will depend on the species in which the antisera were raised and the classes and subclasses of immunoglobulin obtained. Of the common species of antibodies used, rabbit antibodies bind well to Protein-A, whereas rat and mouse antibodies bind less efficiently. The affinities of the latter two species of antisera for Protein-G vary with subtype. Commercially available rat monoclonal antisera against HSF1 and HSF2 have been used successfully, using a Protein-G matrix, or using a rabbit anti-rat linker antibody and Protein-A–Sepharose matrix.
11. If a high background is a problem, try switching beads to a new tube before the last wash, and change the stringency (salt and detergent concentrations), number, and length of washes.
12. Fluorescein-isothiocyanate (FITC) is a green fluorophore with maximum excitation and emission wavelengths of 490 and 520 nm, respectively. The red fluorophore tetramethylrhodamin (TRITC) can be used instead of FITC. Its maximum excitation and emission wavelengths are 540 and 570 nm, respectively. Propidium iodide (PI) is a red-emitting fluorophore whose maximum excitation and emission wavelengths are 536 and 623 nm, respectively.
13. Slides should never be allowed to dry. To avoid drying, treat each slide separately when removing cover slips and adding solutions to them.
14. All the steps where fluorescent molecules are used should be protected from direct light.
15. As PI is a red-color fluorochrome, it may be useful to use another DNA counterstain such as DAPI (blue fluorescence). In this case, it is necessary to ensure that the epifluorescence microscope has appropriate filter sets for DAPI (or an ultraviolet laser if a confocal microscope is used). DAPI is used at a final concentra-

tion of 250 ng/mL in the same antifading solution. Its maximum excitation and emission wavelengths are 350 and 470 nm, respectively.

16. A Buchler Auto Densi-Flow IIC gradient collector may be used to collect fractions from the top. Alternatively, a peristaltic pump may be used to collect fractions from the bottom of the gradient, by attaching the tubing to a capillary tube which is inserted to the bottom of the gradient. Though the latter method is more disruptive to the gradient, it produces good results.

17. The sequence of both strands of an oligonucleotide containing three intact HSF binding sites (nGAnn) from the human *hsp70* promoter is as follows *(44)*:

       5 ' GATCTCGGCTGGAATATTCCCGACCTGGCAGCCGA 3 '
          3 ' AGCCGACCTTATAAGGGCTGGACCGTCGGCTCTAG 5 '

    The sequence of a self-complementary oligonucleotide containing four ideal HSE (nGAAn) sequences when annealed is as follows *(16)*:

       5 ' CTAGAAGCTTCTAGAAGCTTCTAG 3 '
       3 ' GATCTTCGAAGATCTTCGAAGATC 5 '

18. This binding reaction may be altered to perform antibody supershift assays to allow one to distinguish which HSF(s) is activated by a particular treatment. In this case, the incubation of extract with the oligonucleotide is preceded by a 20-min incubation of extract with dilutions of HSF-specific antisera. Ternary complexes that form among the HSF, antibody, and oligonucleotide will appear as a slower migrating species (supershift) compared to samples from which antisera were excluded. Another variation of the gel shift assay is the inclusion of excess unlabeled oligonucleotide to determine the specificity of the HSF-HSE interaction.

19. The free probe should not run off the gel because it is essential to determine that the amount of probe used is in excess, such that the intensity of the free probe bands should not change between samples.

20. The linker we have used is *(50)*:

       5 ' GCGGTGACCCGGGAGATCTGAATTC 3 '   25mer, 60% GC
                  3 ' CTAGACTTAAG 5 '   11mer, 36% GC

    The exact sequence, GC content, or length of the linker is not critical. What are important are as follows: (1) The short oligomer should be able to bind to the long oligomer under ligation conditions, but not under Taq conditions; (2) The long oligomer should be suitable for PCR amplification with primer 2; and (3) The double-stranded linker is a blunt-end ligatable structure. Gel purify the 25mer before annealing and combine with the 11mer in 250 mM Tris-HCl (pH 7.7) to a final concentration of 2 pmol/µL. Heat to 95°C for 5 min. Transfer to 70°C and gradually cool over a period of 1 h to room temperature. Leave at room temperature for 1 h then gradually cool over a period of 1 h to 4°C. Leave at 4°C for 24 h and then store at -20°C. Thaw and keep the linkers on ice during use.

21. Primer 2 must be 3' to primer 1. If they overlap, the overlap should be less than 12 bases. Primer 3 must overlap with primer 2. It can completely overlap primer 2 and extend a few extra bases 3' of it, or it can overlap primer 2 for about 15 bases. In general, the T_m of the primers should increase from primer 1—> 2—> 3.

22. For in vitro DMS treatment, naked DNA is purified from cells, 40 µg of which is digested (with EcoRI in our case) and purified by phenol-chloroform-isoamyl alcohol extraction and ethanol precipitation. Purified DNA is resuspended in 6 µL of ddH$_2$O. In a fume hood, add 200 µL of DMS buffer (50 mM Na-Cacodyoate, 10 mM MgCl$_2$, and 1 mM EDTA), vortex and chill on ice. Add 1 µL DMS, vortex, and incubate at 25°C for 7 min (*see* **Note 4**). Stop the reaction by addition of 50 µL of DMS stop solution (1.5 M sodium acetate, pH 7.0, 1.0 M β-mercaptoethanol) and 750 µL of cold ethanol. Vortex and chill on dry-ice for 30 min. Subsequent piperidine treatment and purification are as described for DMS treatment of genomic DNA.

23. Plasmids encoding heat-shock protein family members may be used. Controls should also be included on each strip and containing genes for messages that are constitutively expressed, such as glutaraldehyde 3-phosphate dehydrogenase, and a vector control for nonspecific hybridization.

24. The incubation time and amount of extract used can be altered to obtain signals in the appropriate range. Take care not to allow the reaction to proceed until the nonacetylated substrate is no longer in excess.

25. Labeled probe is isolated on an alkaline agarose gel to denature it and obtain single-stranded species.

26. Discard all materials and media used for this procedure as radioactive waste. The incubator in which the labeling is performed should contain a fresh open plate of activated charcoal to absorb any volatile radioactivity released into the atmosphere. This charcoal should be discarded as radioactive waste after each labeling procedure.

27. The isofocusing dyes will separate within a few minutes of applying current. Check each gel for separation to make sure there is no problem, such as air bubbles, present.

28. It is important to move quickly once the loading gel is removed from the boiling water because the gel will solidify in a few minutes.

References

1. Sorger, P. K. and Pelham, H. R. (1988) Yeast heat shock factor is an essential DNA-binding protein that exhibits temperature-dependent phosphorylation. *Cell* **54,** 855–864.
2. Wiederrecht, G., Seto, D., and Parker, C. S. (1988) Isolation of the gene encoding the *S. cerevisiae* heat shock transcription factor. *Cell* **54,** 841–853.
3. Clos, J., Westwood, J. T., Becker, P. B., Wilson, S., Lambert, K., and Wu, C. (1990) Molecular cloning and expression of a hexameric *Drosophila* heat shock factor subject to negative regulation. *Cell* **63,** 1085–1097.
4. Rabindran, S. K., Giorgi, G., Clos, J., and Wu, C. (1991) Molecular cloning and expression of a human heat shock factor, HSF1. *Proc. Natl. Acad. Sci. USA* **88,** 6906–6910.
5. Sarge, K. D., Zimarino, V., Holm, K., Wu, C., and Morimoto, R. I. (1991) Cloning and characterization of two mouse heat shock factors with distinct inducible and constitutive DNA-binding ability. *Genes Dev.* **5,** 1902–1911.

6. Schuetz, T. J., Gallo, G. J., Sheldon, L., Tempst, P., and Kingston, R. E. (1991) Isolation of a cDNA for HSF2: evidence for two heat shock factor genes in humans. *Proc. Natl. Acad. Sci. USA.* **88**, 6911–6915.
7. Nakai, A. and Morimoto, R. I. (1993) Characterization of a novel chicken heat shock transcription factor, heat shock factor 3, suggests a new regulatory pathway. *Mol. Cell. Biol.* **13**, 1983–1997.
8. Nakai, A., Kawazoe, Y., Tanabe, M., Nagata, K., and Morimoto, R. I. (1995) The DNA-binding properties of two heat shock factors, HSF1 and HSF3, are induced in the avian erythroblast cell line HD6. *Mol. Cell. Biol.* **15**, 5268–5278.
9. Lis, J. T. and Wu, C. (1993) Protein traffic on the heat shock promoter: parking, stalling and trucking along. *Cell* **74**, 1-20.
10. Morimoto, R. I., Jurivich, D. A., Kroeger, P. E., Mathur, S. K., Murphy, S. P., Nakai, A., Sarge, K., Abravaya, K., and Sistonen, L. T. (1994) Regulation of heat shock gene transcription by a family of heat shock factors, in *The Biology of Heat Shock Proteins and Molecular Chaperones* (Morimoto, R. I., Tissieres, A., and Georgopoulos, C., eds.), Cold Spring Harbor Laboratory, Cold Spring Harbor, NY, pp. 417–455.
11. Wu, C. (1995) Heat shock transcription factors: structure and regulation. *Annu. Rev. Cell. Dev. Biol.* **11**, 441–469.
12. Morimoto, R. I., Kroeger, P. E., and Cotto, J. J. (1996) The transcriptional regulation of heat shock genes: a plethora of heat shock factors and regulatory conditions, in *Stress-Inducible Cellular Responses.* (Feige, U., Morimoto, R. I., Yahara, I., and Polla, B., eds.), Birkauser Verlag, Basel, pp. 139–163.
13. Gallo, G. J., Schuetz, T. J., and Kingston, R. E. (1991) Regulation of heat shock factor in *Schizosaccharomyces pombe* more closely resembles regulation in mammals than in *Saccharomyces cerevisiae. Mol. Cell. Biol.* **11**, 281–288.
14. Baler, R., Dahl, G., and Voellmy, R. (1993) Activation of human heat shock genes is accompanied by oligomerization, modification, and rapid translocation of heat shock transcription factor HSF1. *Mol. Cell. Biol.* **13**, 2486–2496.
15. Rabindran, S. K., Haroun, R. I., Clos, J., Wisniewski, J., and Wu, C. (1993) Regulation of heat shock factor trimer formation: role of a conserved leucine zipper. *Science* **259**, 230–234.
16. Sarge, K. D., Murphy, S. P., and Morimoto, R. I. (1993) Activation of heat shock gene transcription by heat shock factor 1 involves oligomerization, acquisition of DNA-binding activity, and nuclear localization and can occur in the absence of stress. *Mol. Cell. Biol.* **13**, 1392–1407.
17. Sistonen, L., Sarge, K. D., and Morimoto, R. I. (1994) Human heat shock factors 1 and 2 are differentially activated and can synergistically induce hsp70 gene transcription. *Mol. Cell. Biol.* **14**, 2087–2099.
18. Nakai, A., Kawazoe, Y., Tanabe, M., Nagata, K., and Morimoto, R. I. (1995) The DNA-binding properties of two heat shock factors, HSF1 and HSF3, are induced in the avian erythroblast cell line HD6. *Mol. Cell. Biol.* **15**, 5268–5278.
19. Jakobsen, B. K. and Pelham, H. R. (1991) A conserved heptapeptide restrains the activity of the yeast heat shock transcription factor. *EMBO J.* **10**, 369–375.

20. Jurivich, D. A., Sistonen, L., Kroes, R. A., and Morimoto, R. I. (1992) Effect of sodium salicylate on the human heat shock response. *Science* **255**, 1243–1245.

21. Giardina, C. and Lis, J. T. (1995) Sodium salicylate and yeast heat shock gene transcription. *J. Biol. Chem.* **270**, 10,369–10,372.

22. Cotto, J. J., Kline, M., and Morimoto, R. I. (1996) Activation of heat shock factor 1 DNA binding precedes stress-induced serine phosphorylation. *J. Biol. Chem.* **271**, 3355–3358.

23. Lee, B. S., Chen, J., Angelidis, C., Jurivich, D. A., and Morimoto, R. I. (1995) Pharmacological modulation of heat shock factor 1 by antiinflammatory drugs results in protection against stress-induced cellular damage. *Proc. Natl. Acad. Sci. USA.* **92**, 7207–7211.

24. Morimoto, R. I., Tissieres, A., and Georgopoulos, C. (1990) The stress response, function of the proteins, and perspectives, in *Stress Proteins in Biology and Medicine* (Morimoto, R. I., Tissieres, A., and Georgopoulos, C. eds.), Cold Spring Harbor Laboratory, Cold Spring Harbor, NY, pp. 1–36.

25. Tanabe, M., Nakai, A., Kawazoe, Y., and Nagata, K. (1997) Different thresholds in the responses of two heat shock transcription factors, HSF1 and HSF3. *J. Biol. Chem.* **272**, 15,389–15,395.

26. Mezger, V., Rallu, M., Morimoto, R. I., Morange, M., and Renard, J. P. (1994) Heat shock factor 2-like activity in mouse blastocytes. *Dev. Biol.* **166**, 819–822.

27. Rallu M., Loones, M., Lallemand, Y., Morimoto, R., Morange, M., and Mezger, V. (1997) Function and regulation of heat shock factor 2 during mouse embryogenesis. *Proc. Natl. Acad. Sci. USA* **94**, 2392–2397.

28. Sarge, K. D., Park-Sarge, O. K., Kirby, J. O., Mayo, K. E., and Morimoto, R. I. (1994) Expression of heat shock factor 2 in mouse testis: potential role as a regulator of heat-shock protein gene expression during spermatogenesis. *Biol. Reprod.* **50**, 1334–1343.

29. Sistonen, L., Sarge, K. D., Phillips, B., Abravaya, K., and Morimoto, R. I. (1992) Activation of heat shock factor 2 during hemin-induced differentiation of human erythroleukemia cells. *Mol. Cell. Biol.* **12**, 4104–4111.

30. Theodorakis, N. G., Zand, D. J., Kotzbauer, P. T., Williams, G. T., and Morimoto, R. I. (1989) Hemin-induced transcriptional activation of the HSP70 gene during erythroid maturation in K562 cells is due to a heat shock factor-mediated stress response. *Mol. Cell. Biol.* **9**, 3166–3173.

31. Mathew, A., Mathur, S. K., and Morimoto, R. I. (1998) Heat shock response and protein degradation: regulation of HSF2 by the ubiquitin-proteasome pathway. *Mol. Cell. Biol.* **18**, 5091–5098.

32. Dignam, J. D., Lebovitz, R. M., and Roeder, R. G. (1983) Accurate transcription initiation by RNA polymerase II in a soluble extract from isolated mammalian nuclei. *Nucleic Acids Res.* **11**, 1475–1489.

33. Harlow, E. and Lane , D. (1988) Antibodies: A laboratory Manual. Cold Spring Harbor Laboratory. Cold Spring Harbor, NY.

34. Larson, J. S., Schuetz, T. J., and Kingston, R. E. (1988) Activation *in vitro* of sequence-specific DNA binding by a human regulatory factor. *Nature* **335**, 372–375.

35. Coons, A. H., Creech, H. J., and Jones, R. N. (1941) Immunological properties of an antibody containing fluorescent group. *Proc. Soc. Exp. Biol. Med*. **47,** 200–202.

36. Cotto J, Fox S, and Morimoto R (1997) HSF1 granules: a novel stress-induced nuclear compartment of human cells. *J. Cell Sci*. **110,** 2925–2934.

37. Jolly, C., Morimoto, R., Robert-Nicoud, M., Vourc'h, C. (1997) HSF1 transcription factor concentrates in nuclear foci during heat shock: relationship with transcription sites. *J. Cell Sci.* **110,** 2935–2941.

38. Siegel, L. M., and Monty, K. J. (1966) Determination of molecular weights and frictional ratios of proteins in impure systems by use of gel filtration and density gradient centrifugation: Application to crude preparations of sulfite and hydroxylamine reductases. *Biochim. Biophys. Acta.* **112,** 346–362.

39. Westwood, J. T. and Wu, C. (1993) Activation of *Drosophila* heat shock factor: conformational change associated with a monomer-to-trimer transition. *Mol. Cell. Biol.* **13,** 3481–3486.

40. Perisic, O., Xiao, H., and Lis, J. T. (1989) Stable binding of *Drosophila* heat shock factor to head-to-head and tail-to-tail repeats of a conserved 5 bp recognition unit. *Cell* **59,** 797–806.

41. Abdella, P. M., Smith, P. K., and Royer, G. P. (1979) A new cleavable reagent for cross-linking and reversible immobilization of proteins. *Biochem. Biophys. Res. Commun.* **87,** 734–742.

42. Fried, M. and Crothers, D. M. (1981) Equilibria and kinetics of lac repressor-operator interactions by polyacrylamide gel electrophoresis. *Nucleic Acids Res.* **9,** 6505–6525.

43. Garner, M. M. and Revzin, A. (1981) A gel electrophoresis method for quantifying the binding of proteins to specific DNA regions: application to components of the Escherichia coli lactose operon regulatory system. *Nucleic Acids Res.* **9,** 3047–3060.

44. Mosser, D. D., Theodorakis, N. G., and Morimoto, R. I. (1988) Coordinate changes in heat shock element-binding activity and HSP70 gene transcription rates in human cells. *Mol. Cell. Biol*. **8,** 4736–4744.

45. Abravaya, K., Phillips, B., and Morimoto, R. I. (1991) Attenuation of the heat shock response in HeLa cells is mediated by the release of bound heat shock transcription factor and is modulated by changes in growth and in heat shock temperatures. *Genes Dev.* **5,** 2117–2127.

46. Abravaya, K., Phillips, B., and Morimoto, R. I. (1991) Heat shock-induced interactions of heat shock transcription factor and the human hsp70 promoter examined by *in vivo* footprinting. *Mol. Cell. Biol.* **11,** 586–592.

47. Phillips, B., Abravaya, K., and Morimoto, R. I. (1991) Analysis of the specificity and mechanism of transcriptional activation of the hsp70 gene during infection by DNA viruses. *J. Virol*. **65,** 5680–5692.

48. Kroeger, P. E., Sarge, K. D., and Morimoto, R. I. (1993) Mouse heat shock transcription factors 1 and 2 prefer a trimeric binding site but interact differently with the HSP70 heat shock element. *Mol. Cell. Biol*. **13,** 3370–3383.

49. Kroeger, P. E. and Morimoto, R. I. (1994) Selection of new HSF1 and HSF2 DNA-binding sites reveals difference in trimer cooperativity. *Mol. Cell. Biol*. **14,** 7592–7603.

50. Mueller, P. R. and Wold, B. (1989) *In vivo* footprinting of a muscle specific enhancer by ligation mediated PCR. *Science* **246**, 780–786.
51. Dynan, W. S. (1987) DNase I footprinting as an assay for mammalian gene regulatory proteins, in *Genetic Engineering: Principles and Methods* (Setlow, J. ed.), Plenum Press, NY. Vol. 9. pp. 75–87.
52. Hertzberg, R. P. and Dervan, P. B. (1984) Cleavage of DNA with methidiumpropyl-EDTA-iron(II): reaction conditions and product analyses. *Biochemistry* **23**, 3934–3945.
53. O'Halloran, T. V., Frantz, B., Shin, M. K., Ralston, D. M., and Wright, J. G. (1989) The MerR heavy metal receptor mediates positive activation in a topologically novel transcription complex. *Cell* **56**, 119–129.
54. Tullius, T. D., Dombroski, B. A., Churchill, M. E., and Kam, L. (1987) Hydroxyl radical footprinting: a high-resolution method for mapping protein-DNA contacts. *Methods Enzymol.* **155**, 537–558.
55. Reeder, R. H. and Roeder, R. G. (1972) Ribosomal RNA synthesis in isolated nuclei. *J. Mol. Biol.* **67**, 433–441.
56. Marzluff, W. F. Jr, Murphy, E. C. Jr., and Huang, R. C. (1974) Transcription of the genes for 5S ribosomal RNA and transfer RNA in isolated mouse myeloma cell nuclei. *Biochemistry* **13**, 3689–3696.
57. Udvardy, A. and Seifart, K. H. (1976) Transcription of specific genes in isolated nuclei from HeLa cells *in vitro. Eur. J. Biochem.* **62**, 353–363.
58. Groudine, M., Peretz, M., and Weintraub, H. (1981) Transcriptional regulation of hemoglobin switching in chicken embryos. *Mol. Cell. Biol* . **1**, 281–288.
59. Banerji, S. S., Theodorakis, N. G., and Morimoto, R. I. (1984) Heat shock-induced translational control of HSP70 and globin synthesis in chicken reticulocytes. *Mol. Cell. Biol.* **4**, 2437–2448.
60. Shi, Y., Mosser, D. D., and Morimoto, R. I. (1998) Molecular chaperones as HSF1-specific transcriptional repressors. *Genes Dev.* **12**, 654–666.
61. Lawrence, J. B., Singer, R. H., and Marselle, L. M. (1989) Highly localized tracks of specific transcripts within interphase nuclei visualized by *in situ* hybridization. *Cell* **57**, 493–502.
62. Jolly, C., Mongelard, C., Robert-Nicoud, M., and Vourc'h, C. (1997). Optimization of nuclear transcripts detection by FISH and combination with fluorescence immunocytochemical detection of transcription factors. *J. Histochem. Cytochem.* **45**, 1585–1592.
63. Lichter, P., Cremer, T., Borden, J., Manuelidis, L., and Ward, D. C. (1988) Delineation of individual chromosomes in metaphase and interphase cells by *in situ* suppression hybridization using recombinant DNA libraries. *Hum. Genet* . **80**, 224–234.
64. Gorman, C. M., Moffat, L. F., and Howard, B. H. (1982) Recombinant genomes which express chloramphenicol acetyltransferase in mammalian cells. *Mol. Cell. Biol.* **2**, 1044–1051.
65. Gould, S. J. and Subramani, S. (1988) Firefly luciferase as a tool in molecular and cell biology. *Anal. Biochem.* **175**, 5–13.

66. Brasier, A. R., Tate, J. E., and Habener, J. F. (1989) Optimized use of the firefly luciferase assay as a reporter gene in mammalian cell lines. *Biotechniques* **7,** 1116–1122.
67. Herbomel, P., Bourachot, B., and Yaniv, M. (1984) Two distinct enhancers with different cell specificities coexist in the regulatory region of polyoma. *Cell* **39 (Pt 2),** 653–662.
68. Shi, Y., Kroeger, P., and Morimoto, R. I. (1995) The Carboxyl-terminal transactivation domain of heat shock factor 1 is negatively regulated and stress responsive. *Mol. Cell. Biol*. **15,** 4309–4318.
69. Chomczynski, P. and Sacchi, N. (1987) Single-step method of RNA isolation by acid guanidinium thiocyanate-phenol-chloroform extraction. *Anal. Biochem.* **162,** 156–159.
70. Dynan, W. S. and Tjian, R. (1985) Control of eukaryotic messenger RNA synthesis by sequence-specific DNA-binding proteins. *Nature* **316,** 774–778.
71. Wu, B., Hunt, C., and Morimoto, R. (1985) Structure and expression of the human gene encoding major heat shock protein HSP70. *Mol. Cell. Biol*. **5,** 330–341.
72. Berk, A. J. and Sharp, P. A. (1978) Structure of the adenovirus 2 early mRNAs. *Cell* **14,** 695–711.
73. Young, P. R., Hazuda, D. J., and Simon, P. L. (1988) Human interleukin 1 beta is not secreted from hamster fibroblasts when expressed constitutively from a transfected cDNA. *J. Cell Biol*. **107,** 447–456.
74. O'Farrell, P. H. (1975) High resolution two-dimensional electrophoresis of proteins. *J. Biol. Chem.* **250,** 4007–4021.

18

Approaches to Define the Involvement of Reactive Oxygen Species and Iron in Ultraviolet-A Inducible Gene Expression

Charareh Pourzand, Olivier Reelfs, and Rex M. Tyrrell

1. Introduction

Oxidative stress is implicated in a wide range of human diseases as well as the ageing process and major efforts have been concentrated on development of markers of this state in cells. The discovery that the expression of the gene which encodes the heme catabolic enzyme, heme-oxygenase-1 (*HO-1*), is strongly induced by oxidizing agents such as ultraviolet-A (UVA, 320–400nm) radiation, has provided a powerful indicator of cellular redox state (*see also* Chapter 23). This phenomenon, which was originally observed in primary human cultured skin fibroblasts, was later shown to occur in most human cell types, although not in keratinocytes (*1*). The expression of the *HO-1* gene is clearly dependent on cellular reducing equivalents. It is induced by several oxidants, including hydrogen peroxide (H_2O_2), and glutathione (GSH) depletion strongly enhances both basal levels and oxidant-induced expression of the gene (*2,3*). The enhanced expression of the *HO-1* gene by UVA and other agents can be entirely accounted for by a very strong enhancement in transcription rate (*4*) and given the fact that the mRNA for human *HO-1* is relatively stable (half-life 2.7 h, **ref.** *[4]*), the measurement of mRNA accumulation by the Northern blot technique is evidently the simplest way of monitoring expression. This approach is currently used in several laboratories as a positive control when testing other genes suspected of being oxidant inducible. A second redox-regulated gene is *CL-100*, whose transcription is also strongly acti-

From: *Methods in Molecular Biology, vol. 99: Stress Response: Methods and Protocols*
Edited by: S. M. Keyse © Humana Press Inc., Totowa, NJ

vated by UVA and oxidants *(5)*. *CL-100* is interesting since it encodes a dual-specificity (tyrosine/threonine) protein phosphatase that specifically inactivates cellular mitogen-activated protein (MAP) kinases. It is therefore implicated in the modulation of signal transduction events involved in the cellular stress response *(3,5)*.

Although it is known that the UVA radiation component of sunlight exerts its biological effects primarily by oxidative pathways *(3)*, and that the induction of *HO-1* by UVA is a general response to oxidative stress, it was important to determine the effector species involved in UVA activation of the *HO-1* gene. To answer this question, several strategies were developed in this laboratory that led to the identification of certain reactive oxygen species (ROS) involved in the activation of *HO-1* by UVA (e.g., **ref. 6**). Later studies based on this approach provided further clues as to how other oxidant-inducible genes could be activated. In this chapter, we provide an overview of approaches that can identify the oxidizing species involved in initiation of oxidant-regulated gene expression and discuss the usefulness and limitations of such techniques.

2. Materials

2.1. Equipment

1. Tissue culture facilities and 5% CO_2 incubator (37°C).
2. Millipore MilliQ purification system (Millipore, Bedford, MA).
3. Autoclave.
4. Refrigerator (+4°C) and freezer (–20°C).
5. Water bath.
6. pH Meter.
7. Vortex mixer.
8. Spectrophotometer with thermostatic control.
9. Light box.
10. Radiometer.
11. Rubber policeman.
12. Coulter counter.
13. Scintillation counter.
14. Glass gel plates (20 × 20 cm), 1.5-mm spacers, 12- to 16-well combs.
15. Electrophoresis apparatus and power pack.
16. Cold-room facilities.
17. Gel dryer and vacuum pump.
18. X-ray films, exposure cassettes, intensifying screens, X-ray developing apparatus.

2.2. Reagents and Stock Solutions

Water used to prepare the reagents should be freshly drawn from a deionization purification system, such as a Millipore MilliQ cartridge (*see* **Note 1**).

1. Phosphate-buffered saline (PBS): Use only recently made (a few days old) autoclaved solutions (*see* **Note 2**).
2. Buthionine [*S,R*]-sulfoximine (BSO): 10 mM stock solution in PBS, filter-sterilize (0.2 µm membrane), aliquot, and store at –20°C.
3. NADPH: 5 mM in 0.5% sodium bicarbonate (NaHCO$_3$), aliquot and store at –20°C.
4. 0.1 M Phosphate buffer pH 7.0–7.2: for 250 mL stock solution, add 9 mL of 1 M KH$_2$PO$_4$, and 16 mL of 1 M K$_2$HPO$_4$ to 225 mL Millipore water (*see* **Note 3**).
5. 0.1 M Phosphate buffer pH 7.0–7.2/1 mM EDTA: Add 0.5 mL of EDTA 0.5 M, pH 8.0, to 250 mL phosphate buffer prepared as in **step 4** (*see* **Note 3**).
6. DTNB [5,5'-dithiobis(2-nitrobenzoic acid)]: 1.5 mg/mL stock solution in 0.5% NaHCO$_3$. Dissolve at 37°C for 1 h (in the dark, *see* **Note 3**).
7. Trichloroacetic acid (TCA): 10% (w/v) stock solution in Millipore water. Store at room temperature.
8. 5% TCA/2 mM EDTA: For 10 mL solution, add 5 mL of 10% TCA and 40 µL of 0.5 M EDTA, pH 8.0, to 4.96 mL Millipore water (*see* **Note 3**).
9. *N*-Ethylmaleimide (NEM): 1 M stock solution in absolute ethanol, aliquot, and store at –20°C.
10. MOPS [3-(*N*-morpholino)propanesulfonic acid]: 1 M stock solution in Millipore water. 0.2 µm filter-sterilize and keep at room temperature.
11. KOH: 3 M stock solution in Millipore water.
12. 0.3 M MOPS/2 M KOH: For 10 mL solution, mix 6.66 mL of 3 M KOH stock solution and 3 mL 1 M MOPS solution with 0.34 mL Millipore water (*see* **Note 3**).
13. GSSG: 1 M stock solution in PBS (*see* **Note 3**).
14. *N*-acetyl, L-cysteine (NAC): 2 M stock solution in PBS, pH 7.4–7.7. Resuspend the powder in about half of the required final volume of PBS. Then, while agitating the solution, adjust the pH by dropwise addition of 1–10 M NaOH. Make up to the required final volume with PBS and 0.2-µm filter-sterilize (*see* **Note 3**).
15. GSH-ester (e.g., monoethylester): 1 M stock solution in PBS (*see* **Note 3**).
16. Cysteamine: 0.65 M stock solution in PBS (*see* **Note 3**).
17. Deuterium oxide (D$_2$O): D$_2$O stock solution (99.9% of deuterium atoms) based PBS: this is prepared by the dissolving of one PBS tablet (Oxoid, Basingstoke, England) in 100 mL of 99.9% D$_2$O. Filter and store the solution at room temperature.
18. Sodium azide (NaN$_3$): 1 M stock solution in PBS. Filter and keep at 4°C in the dark.
19. L-Histidine: 0.25 M stock solution in PBS. Filter and keep at 4°C.
20. Mannitol: 1 M stock solution in PBS. Filter and keep at 4°C.
21. Dimethyl sulfoxide (DMSO): Commercially available solution (99.5%).
22. Rose Bengal: 1 mM stock solution in DMSO, store at –20°C in the dark.
23. H$_2$O$_2$: Commercially available as a stabilized 30% (w/w) H$_2$O$_2$ solution, store at 4°C. Prepare intermediate dilutions freshly in water, and final dilution in serum-free medium.
24. ATZ [3-amino-1,2,4-triazole]: 1–2 M stock solution in PBS (*see* **Note 4**). Filter-sterilize at 4°C in the dark.
25. Iron citrate (Fe-citrate): Prepare 100 mM stock solutions of ferric chloride (FeCl$_3$) and sodium citrate (Na$_3$-citrate) in Millipore H$_2$O. Filter sterilize and store at

room temperature. On the day of the treatment, prepare Fe-citrate by mixing these solutions in a 1:1 ratio.

26. Hemin: 10–100 mM stock solution in DMSO (*see* **Note 5**). Aliquots are stored at –20°C, wrapped in aluminium foil to avoid direct exposure to light.

27. Desferrioxamine (Desferal): 150 mM stock solution in Millipore H$_2$O. Aliquots are stored at –20°C.

28. Salicylaldehyde isonicotinoyl hydrazone (SIH): 500 mM stock solution in DMSO. Aliquots are stored at –20°C.

29. HEPES pH 7.5: 100 mM stock solution in Millipore water, filter-sterilize, and store at 4°C.

30. MgCl$_2$: 1 M stock solution in Millipore water, autoclave, and store at room temperature.

31. KCl: 1 M stock solution, autoclave, and store at room temperature.

32. Glycerol: 87% stock solution, autoclave, and store at room temperature.

33. Nonidet P-40 (NP-40): 10% (w/v) stock solution in Millipore water, store at room temperature.

34. Phenylmethyl sulfonyl fluoride (PMSF): 100 mM stock solution in isopropanol, aliquot, and store at –20°C (*see* **Note 6**).

35. 1X Munro lysis buffer (10 mM HEPES, 3 mM MgCl$_2$, 40 mM KCl, 5% glycerol, 0.3% NP-40, and 1 mM PMSF): For 1 mL 1X Munro buffer, add 100 μL of 100 mM HEPES buffer, 3 μL of 1 M MgCl$_2$, 40 μL of 1 M KCl, and 57.5 μL of 87% glycerol to 760 μL of Millipore water. Aliquot 960 μL per Eppendorf tube and store at –20°C. On the day of experiment and just prior to lysis, to one aliquot of 1X Munroe buffer (960 μL), add 30 μL of 10% NP-40 and 10 μL of 100 mM PMSF. Vortex and add to cell pellets for lysis.

36. 5X Munro buffer: 50 mM HEPES, 15 mM MgCl$_2$, 200 mM KCl, 25% glycerol. To prepare 1 mL 5X Munro buffer, add 500 μL of 100 mM HEPES buffer, 15 μL of 1 M MgCl$_2$, 200 μL of 1 M KCl to 287.5 μL of 87% glycerol. Aliquot and store at –20°C.

37. Iron responsive element (IRE) plasmid DNA template: pGEM-3Zf(+/–) plasmid containing double-stranded H-ferritin IRE motif (kind gift from L.Kuhn, Swiss Institute for Experimental Cancer Research, Switzerland). Plasmid DNA template should be linearized with *Bam*HI prior to RNA (T7-transcript) preparation.

38. T7 RNA-polymerase: 20 U/μL (Promega, cat. no. P207B). Store at –20°C.

39. 5X TSB (Transcription optimized buffer): 5X buffer for T7 RNA-polymerase, contains 200 mM Tris-HCl, pH 8.0, 30 mM MgCl$_2$, 50 mM NaCl, 10 mM spermidine (Promega, cat. no. P118B). Aliquot and store at –20°C.

40. Ultrapure NTP set (ATP, GTP, UTP): 100 mM stock solutions (Pharmacia Biotech, cat. no. 27-2050-01). Aliquot and store at –20°C. Prior to use, dilute each aliquot (1:9) in Millipore water.

41. Dithiothreitol (DTT): 100 mM stock solution (Pharmacia Biotech., cat. no. P117B), aliquot and store at –20°C (*see* **Note 7**).

42. Ribonuclease inhibitor (RNAsin): 40 U/μL (Promega, cat. no. N2111), store at –20°C.

43. [α-^{32}P]CTP: 800 ci/mmol (Amersham, cat. no. PB20382), store at –20°C.

44. Glycogen: 20 mg/mL (Boehringer Mannheim), aliquot and store at -20°C.

45. Ammonium acetate: 7.5 M stock solution in Millipore water. Autoclave and store at room temperature.
46. NaOH: 0.1 M stock solution in Millipore water, store at room temperature.
47. Ultrapure Acrylamide/bis-acrylamide solution: Readysol DNA/PAGE 40% stock solution (Pharmacia Biotech., cat. no. 17-1308-01), store at 4°C.
48. Ammonium persulfate (APS): 10% stock solution in Millipore water. Filter-sterilize and store at 4°C (*see* **Note 8**).
49. Ultrapure TEMED: Pharmacia Biotech (cat. no. 17-1312-01), store at room temperature.
50. Heparin: 50 mg/mL stock solution in Millipore water. Aliquot and store at 4°C.
51. IRP/IRE Loading dye: 30 mM Tris-HCl, pH 7.5, 40% sucrose, 0.2% Bromophenol blue, 0.2 μM filter-sterilize, aliquot, and store at −20°C.
52. 10X TBE: for 1000 mL, weigh 108 g of Tris base and 55 g of boric acid, add 40 mL of 0.5 M EDTA and then adjust the volume with millipore water to 1000 mL. Autoclave and store at room temperature.
53. 0.3X TBE: Dilute 10X TBE in millipore water prior to use (*see* **Note 3**).
54. Gel fixing solution: 10% methanol/10% acetic acid mixture in Millipore water.

3. Methods

3.1. Glutathione as a Diagnostic of Redox-Regulated Gene Expression

There is a direct correlation between the levels of sensitization to oxidizing agents (e.g., UVA) and cellular GSH content. GSH as an ubiquitous thiol-containing tripeptide is present in most mammalian cells at high concentrations (often in the range 3–5 mM) and is, therefore, a major component of the machinery for the maintenance of the cellular redox state. Lowering the cellular GSH levels will reduce scavenging of both intermediates generated during normal metabolism and those generated as a result of exogenous oxidizing insult.

3.1.1. Glutathione Depletion

The first step in identification of redox regulated genes is usually to monitor whether lowering cellular GSH levels enhances the expression of the gene under study. The compound buthionine [*S,R*]-sulfoximine *(7)* is an extremely potent inhibitor of γ-glutamylcysteine synthetase, which catalyzes the production of γ-glutamylcysteine (an intermediate that arises at an earlier stage of glutathione biosynthesis). This drug is capable of significantly reducing the level of GSH in cells within a few hours of treatment. The effective concentration of BSO varies from cell type to cell type; for example, treatment of human fibroblasts with 5 μM BSO for 18 h in normal culture medium reduces the levels of GSH to undetectable levels *(8)*, however a human lymphoblastoid cell line (TK6) requires 10 times this concentration of BSO in order to achieve the same degree of GSH depletion.

3.1.1.1. BSO Treatment

Treatment with BSO at 0.1–1000 μ*M* final concentration in medium for 18 h at 37°C (*see* **Note 9**).

3.1.1.2. Glutathione Measurement

To determine the effective concentration of BSO, the level of GSH (reduced glutathione) and GSSG should be measured. Here we describe the spectrophotometric method adapted from Tietze *(9)*, which, as a simple and fast methodology, enabled us to determine the cellular glutathione levels following BSO treatment of the primary fibroblast cell line, FEK4. In this method, glutathione is conveniently assayed by an enzymatic recycling procedure in which it is sequentially oxidized by DTNB and reduced by NADPH in the presence of glutathione reductase. The formation of the resulting chromogenic product (2-nitro-5-thiobenzoic acid) is monitored spectrophotometrically at 412 nm (for 6–10 min) and the glutathione present in cells is evaluated by comparison of the resulting values with a standard curve. This assay detects both glutathione (GSH) and oxidized GSSG.

1. After treatment of cell monolayers with a range of BSO concentrations, remove the growth medium and store at –20°C for subsequent determination of extracellular glutathione concentrations.
2. Rinse cell monolayers with PBS, trypsinize, resuspend in cold PBS, count cells and pellet them by centrifugation (200*g*) at 4°C.
3. Extract the cell pellets with a freshly prepared mixture of 5% TCA/2 m*M* EDTA (*see* **Subheading 2.2.**), in order to give 1 mL of extract per 2×10^6 cells.
4. After centrifugation (1100*g*) at 4°C, divide each supernatant into two aliquots in order to measure both GSH and the oxidized form, GSSG, using the spectrophotometric method adapted from Tietze *(9)*.
5. For GSSG determination, add *N*-ethylmaleimide (NEM, 50 m*M* final concentration) to the samples, and then incubate for 1 h at 25°C. Next, remove free NEM from the samples by 10 extractions with ether, prior to GSSG reductase recycling assay using DTNB. The rapid and complete reaction of NEM with GSH to form a stable complex *(10)* prevents participation of the reduced form in the enzymatic assay as well as its possible oxidation to GSSG. Ether extraction is used to ensure removal of the untreated sulfhydryl reagent, which is an inhibitor of glutathione reductase.
6. For GSSG reductase recycling assay, mix 1 mL of phospate-buffer/EDTA (*see* **Subheading 2.2.**) with 20 μL DTNB (1.5 mg/mL in 0.5% NaHCO$_3$), 50 μL NADPH (5m*M*), and 100 μL samples (or blanks, PBS or medium) in an Eppendorf tube and transfer to a thermostatically controlled cuvet at 25°C. To the prewarmed mixture add 100 μL GSSG reductase (18 U/mL in phosphate buffer) and then monitor the change in absorbance in the spectrophotometer for 6–10 min at 412 nm (25°C).

7. In order to calculate the level of intracellular glutathione equivalents, a standard curve is determined along with the sample measurements. For this purpose, GSSG stock solution is diluted in 5% TCA/2 mM EDTA in a range of 0.01–10 μM and is processed as in **step 6**.

3.1.2. GSH Supplementation

Although modulation by GSH depletion is a strong indication that a given gene is redox regulated, GSH supplementation in the form of *N*-acetyl, L-cysteine (NAC), GSH-ester or cysteamine is more complex to interpret. NAC is readily taken up by cells and is hydrolyzed to cysteine so that it will provide thiol groups for scavenging. However it is also a substrate for γ-glutamylcysteine synthetase and enters the GSH biosynthesis pathway. GSH itself, as well as being a scavenger, acts as the unique hydrogen donor for the crucial antioxidant enzyme glutathione peroxidase. Thus a modulating effect of NAC could indicate the involvement of one of several cytoplasmic oxidizing intermediates or lipid peroxides. The latter is also true for GSH-ester and cysteamine, which also serve as extracellular sources of GSH. Nevertheless, experiments with these thiol precursors can provide useful general information. Concentrations of the order of 5 mM and higher would seem a reasonable choice, given the normal endogenous GSH levels. However in subconfluent primary skin fibroblasts, FEK4, concentrations higher than 2 mM NAC, cysteamine, or GSH–ester were toxic to the cells (*unpublished observation*, this laboratory). One explanation is that GSH levels may be much higher in subconfluent cells than confluent cells and therefore added GSH would become cytotoxic. This explanation is based on the observation made by Mbemba et al. *(11)*, who showed that in human WI-38 fibroblasts, GSH levels vary with cell growth and are higher (up to threefold) in subconfluent cells than in confluent non-dividing cells. Finally the lack of an effect or even an enhancing effect upon addition of NAC, cysteamine or GSH–ester does not necessarily mean that the biological endpoint under study is not dependent on cellular reducing equivalents. For example, NAC treatment decreases *HO-1* activation in fibroblasts *(12,13)*, but it has no effect on UVA-induced activation of NF-κB binding to its recognition sequence (O. Reelfs, *unpublished data*).

3.1.2.1. NAC, GSH–ester, and Cysteamine Treatments

NAC, GSH–ester, and cysteamine: 0.5–10 mM final concentration in medium for 30 min to 18 h at 37°C (*see* **Notes 9–11**).

3.1.3. Methods Suitable for Analysis of Redox-Regulated Gene Expression

Following treatments (e.g., GSH depletion or supplementation) and verification of the efficiency of the treatments (e.g., monitoring the modulation of intracellular glutathione equivalents using the method of Tietze), cells can be processed in a number of ways. These include Northern blotting, RNase protection or Western blotting. Alternatively, in the case of redox-regulated transcription factors (e.g., NF-κB), DNA band-shift assay could be used for the analysis of the effects of the given treatment (*see also* Chapter 16).

In our laboratory, Northern analysis following treatment of human cultured skin fibroblasts with BSO (50 μM for 18 h), showed an increase in the basal level of *HO-1* mRNA of approx fivefold (2). This provides strong evidence that the expression of this gene is sensitive to reactive oxygen intermediates generated during normal metabolism. In the low-UVA-dose range, GSH depletion also increased the peak levels of *HO-1* mRNA accumulation by about fivefold, thus providing strong support that it is the thiol-scavengable intermediates generated by UVA radiation that trigger the signaling pathways that activate the gene. Similarly, using a DNA band-shift assay, we observed a five- to sixfold increase in UVA-mediated NF-κB-binding activity following treatment of fibroblasts FEK4 with BSO (5 μM for 18 h). This is also a strong indication that UVA-mediated NF-κB binding activity is redox regulated (O. Reelfs, *unpublished data*).

3.2. Evidence for the Involvement of Singlet Oxygen

A common approach for detecting $^1O_2$ involvement in reactions is the use of scavengers such as DABCO, β-carotene, diphenylisobenzofuran, sodium azide and L-histidine. The addition of these compounds should inhibit a reaction that is dependent on $^1O_2$. If when added at high concentrations, none of these compounds inhibits the biological effect under study, it is probable that $^1O_2$ is not involved in the process. However if the given effect is quenched by these scavengers, this does not provide absolute proof of $^1O_2$ involvement, as all these compounds also react with the hydroxyl radical (·OH) as evaluated by the in vitro rate constants (14–16). Furthermore DABCO, diphenylisobenzofuran and β-carotene also react with $RO·_2$ radicals (17). Another approach is to replace water by deuterium oxide (D_2O) in experimental solutions since the lifetime of $^1O_2$ is enhanced 10 to 15 times under such conditions (*see* **ref. 16**). Enhancement of a biological reaction in this way may be taken as strong evidence for $^1O_2$ involvement, assuming that isotope effects are negligible. A combination of the deuterium oxide and inhibitor approach is really essential for providing reasonable evidence that $^1O_2$ is involved in a reaction. In a cultured human fibroblast cell line, FEK4, both the cytotoxic effect of UVA on cell populations and the UVA activation of the *HO-1* gene were enhanced more than twofold by D_2O. Moreover, sodium azide and L-histidine, which are scavengers of both $^1O_2$ and ·OH reduced the dose-dependent accumulation of *HO-1* mRNA following UVA irradiation of FEK4 cells. Taken together, these results suggested that $^1O_2$ is a critical species in the UVA induction of the human heme oxygenase gene *(6)*. In order to broaden such approaches to investigate other oxidant-inducible genes, several parameters have to be taken in account. For example, sodium azide, although an ROS scavenger, is also a strong inhibitor of RNA polymerase in cells, so this must be controlled if the biological endpoint is dependent on transcription. The quenching effect of L-histidine also provides valuable information about the superoxide anion. L-histidine has a negligible interaction with the superoxide anion, so quenching by this compound indicates that the involvement of the anion is unlikely.

3.2.1. D₂O/Inhibitor Treatments

1. D₂O treatment: Treatment for 15 min at 37°C. Prior to this, remove growth medium from the cells and rinse twice gently with normal PBS. PBS should be carefully decanted and removed, yet without leaving the cells to dry. Then, in order to remove as much liquid from the cells as possible, cover the cells with D₂O-based PBS (e.g., 2 mL per 10-cm plate), swirl, and aspirate. Finally, cover the cells with PBS-based D₂O (5 mL per 10-cm plate) and incubate for 15 min at 37°C (*see* **Note 12**).
2. NaN₃ treatment: Treatment up to 100 mM (final concentration) in PBS for 15–30 min at 37°C (*see* **Notes 9** and **12**).
3. L-Histidine treatment: Treatment with up to 10 mM (final concentration) in PBS for 15 min at 37°C (*see* **Notes 9** and **12**).

Following these treatments, cells may be processed by methods outlined in **Subheading 3.1.3.**

3.2.2. In Vitro Generation of $^1O_2$

As corroborative evidence of singlet oxygen involvement in the biological effect under study, the effect should also be seen to occur after appropriate generation of this intermediate in vitro. For example either exogenous porphyrin or rose bengal (RB) when irradiated with red light can generate $^1O_2$ and activate the *HO-1* gene. However interpretation of results obtained using these compounds is complicated by the fact that they can generate radicals (type I reactions) in addition to $^1O_2$ (type II reactions). To avoid this problem, it is recommended that the RB/visible light reaction is carried out in the presence of D₂O, in which type I reactions are presumably unaffected *(18)*. Here, we describe the latter approach, which has been successfully used in our laboratory to confirm the effectiveness of $^1O_2$ in the induction of the *HO-1* gene *(6)*.

3.2.2.1. RB/D₂O TREATMENT

1. Pretreatment for 30 min at 37°C. Prior to this, remove growth medium from the cells and rinse gently twice with normal PBS. PBS should be carefully decanted and removed, without leaving the cells to dry. Then, in order to remove as much

liquid from the cells as possible, cover the cells with D_2O-based PBS (e.g., 2 mL per 10-cm plate), swirl, and aspirate. Finally, cover the cells with D_2O-based PBS (5 mL per 10-cm plate) containing 1 μM RB and incubate for 30 min in the dark at 37°C (*see* **Note 12**).

2. After the pretreatment, irradiate cells with a range of doses (0–1500 J/m²) of broad-spectrum (400–700 nm) visible light by placing the culture dishes on a fluorescent light box.
3. After irradiation, aspirate the D_2O-based PBS/RB, wash the cells with normal PBS, and add back the original medium for the desired time.
4. Cells may be processed by methods outlined in **Subheading 3.1.3.**

3.2.2.2. Other Available Approaches

As an alternative to the above methods, irradiation of the sensitizer such as RB attached to beads or to membrane in close vicinity to the target cells but with an air interface may provide a way to generate pure $^1O_2$ *(19)*. Another method is to generate $^1O_2$ by the degradation of an unstable endoperoxide (*see* **ref. 20**). Special care must be taken to manipulating this compound, because of its explosive nature. However, this technique has been used to show that collagenase, another UVA activated gene, is induced by pure singlet oxygen *(21)*.

3.3. Evidence for the Involvement of Hydroxyl Radical

Hydroxyl radical may be generated in fairly pure form using H_2O_2 in combination with a ferric-EDTA complex to drive a Fenton reaction *(22)*. A similar process is thought to occur in vivo to generate the radical. The short-lived ·OH is the most reactive oxygen radical known, with a highly positive reduction potential. Multiple studies have shown that some putative antioxidants are ·OH scavengers when assayed in vitro. In contrast, in in vivo systems, the fact that almost any molecule can react quickly with ·OH, makes the use of such scavengers less feasible. In general, huge concentrations of the scavengers need to be added to the system so that they compete with biological molecules for any ·OH generated. The sugar alcohol, mannitol, is often used as an ·OH scavenger in laboratory experiments, but its rate constant for reaction with ·OH is equal to or less than that of many biomolecules (*see* **ref. 16**). Other established ·OH scavengers in biological systems are DMSO, ethanol, methanol, thiourea, acetates, and formates. A possible quenching effect by these scavengers is only an indication of possible ·OH involvement because all these compounds have the potential to react with other radical species (e.g. thiourea reacts with superoxide anion, H_2O_2, HOCl, and ONOO⁻ in addition to ·OH).

3.3.1. Examples of ·OH Scavenger Treatments

1. Mannitol: Treatment with a range of millimolar to molar (final concentration) in PBS for 15 min at 37°C (*see* **Notes 9** and **12**).

2. DMSO: Treatment with 0.5–4% (final concentration) in PBS for 15 min at 37°C (*see* **Notes 9** and **12**).

Following these treatments, cells may be processed by methods outlined in **Subheading 3.1.3.**

3.3.2. Limitations of ·OH Scavengers

Although a consistent lack of effect of a set of such scavengers is fairly strong evidence that ·OH scavengers are not involved, it is conceivable that these scavengers may fail to inhibit ·OH simply because they are not present at an adequate concentration in a given cellular compartment. In vivo, the ·OH generated via reaction of H_2O_2 with iron that is bound to a biological molecule (Fenton reaction) may preferentially attack the molecule to which the metal is bound rather than the scavenger, because the local concentration of that molecule is overwhelmingly greater than that of added scavenger.

In investigations of in vivo Fenton-mediated generation of ·OH, the evidence for the involvement of the two other participants in the reaction (i.e., H_2O_2 and iron) will provide complementary information, especially because the effects seen as a result of using ·OH scavengers are rarely conclusive.

3.4. Evidence for the Involvement of Hydrogen Peroxide

Hydrogen peroxide is a by-product of several biochemical pathways, notably the dismutation of superoxide by the superoxide dismutases (SODs). It may also be generated intracellularly by interaction of UVA with biomolecules such as NADH and NADPH, resulting in the formation of the superoxide anion, which is then dismutated by SOD to yield H_2O_2. Because rates of H_2O_2 production by cells and organelles are often only in the nanomoles per minute range (except for activated phagocytes), the intracellular detection of H_2O_2 generation is rather difficult. A popular assay employs the H_2O_2-dependent conversion of the nonfluorescent compound dichlorofluorescein diacetate (DCFH–DA) to the fluorescent 2',7' dichlorofluorescein (*see also* Chapter 4). DCFH–DA is taken up by cells and tissues, usually undergoing deacetylation by esterase enzymes. Oxidation of DCFH within cells generates dichlorofluorescein, which can be easily visualized (strong emission at 525 nm using excitation at 488 nm). In addition to peroxidase/H_2O_2, several species cause DCFH oxidation, which include RO_2·, RO·, OH·, HOCl, and $ONOO^-$, but not superoxide. Therefore it appears that this fluorescent imaging is an assay of generalized oxidative stress rather than of production of any particular oxidizing species and therefore, it is not a conclusive diagnostic test for H_2O_2.

The modulation of cellular levels of peroxide by catalase, the enzyme responsible for H_2O_2 breakdown, can provide useful information concerning the involvement of H_2O_2 in a biological reaction. Catalase may either be added exogenously or overexpressing plasmids may be used to enhance intracellular levels of the enzyme *(23)*. In the case of former, it is assumed that extracellular catalase can still function to lower the overall H_2O_2 concentration, because the peroxide can diffuse through membrane (*see* **ref. 23**). A further approach is to enhance the level of endogenous H_2O_2 formation in cells by using the known catalase enzyme inhibitor 3-amino-1,2,4-triazole (ATZ). In this case, the concentration of ATZ has to be adjusted for a given cell line in order to abolish at least 90% of the catalase activity (e.g., *ref. 24*). Peroxide itself is necessary for this compound to function because it acts at the level of compound 1, which is generated as a short-lived intermediate in the catalysis of peroxide by catalase. The extent of inactivation of catalase by ATZ can also be used as a tool to calculate the level of H_2O_2 production in isolated cells or organs. If inhibition of catalase by ATZ potentiates the level of the biological effect under study, then one can conclude that H_2O_2 is involved in the oxidative process. However even if H_2O_2 is proven to be involved using this inhibitor, ·OH may not necessarily be involved. If ATZ treatment does not modulate the biological effect, it is unlikely that catalase is the major pathway of detoxification of H_2O_2 that is generated endogenously following the oxidizing insult. However, glutathione peroxidase, which is another major H_2O_2-detoxifying enzyme pathway present in cells, could be involved. As corroborative evidence of H_2O_2 involvement in a given effect, the effect should also be seen to occur after direct treatment of cells with a bolus of H_2O_2 or by treament of the cells with an H_2O_2-generating system such as glucose/glucose oxidase. For example, the generation of H_2O_2 in this way in a murine cell line has been used to provide evidence that the peroxide is involved in the activation of iron regulatory protein-1 *(25)*.

3.4.1. H_2O_2/ATZ Treatments

1. Exogenous H_2O_2 treatment: Treatment in serum-free medium for 15–30 min at 37°C. Prior to treatment, cell monolayers are rinsed twice thoroughly with prewarmed serum-free medium (5–10 mL per 10-cm plate) in order to remove any trace of catalase from serum.
2. ATZ: Treatment with final concentrations up to 50 mM for 90 min in serum-free medium at 37°C, shaded from direct light. Prior to this, cell monolayers are rinsed twice with prewarmed serum-free medium (*see* **Notes 9** and **12**).

Following these treatments, cells may be processed by methods outlined in **Subheading 3.1.3.**

3.5. Evidence for the Involvement of Iron

Iron acts as a catalyst in reactions between ROS and biomolecules and is commonly considered as the major cellular catalyst in the Fenton reaction.

However, iron also acts as a catalyst for other oxidative reactions, including the lipid peroxidation chain reaction. Furthermore, $^1O_2$ may also be generated by the Fenton reaction *(26)*. Therefore, it is clear that an effect of iron levels can be used only as a general indicator of oxidative reactions.

3.5.1. Modulation of Intracellular Iron Levels

When iron is implicated in an oxidative event leading to gene activation, iron chelators should reduce the effect and iron loading should enhance it. The choice of iron chelators is crucial, because some chelators such as *o*-phenanthroline also strongy chelate other metals. The use of a set of iron chelators that localize to different cellular compartments is advisable. For example, we have used three chelators individually to test for a role of iron depletion in particular effects. First we used desferrioxamine (Desferal), a highly specific iron scavenger that does not efficiently enter cell membranes but rather binds iron as it is transported out of the cell. A second useful compound is pyridoxal isonicotinoyl hydrazone (PIH, 100–500 μM), which has a 50/50 lipid/water partition coefficient, and, finally, salicylaldehyde isonicotinoyl hydrazone (SIH, 100–500 μM), which is extremely lipophilic and can readily enter cells. In practice, all three compounds appear to be effective chelators of free intracellular iron when left in contact with the cells for 18 h. Unfortunately, SIH can undergo redox cycling which can complicate the interpretation of the results.

Cells may also be loaded with iron in various ways. The most common iron-loading treatments imply either overnight incubation of the cells with Fe-citrate (10–100 μM) or hemin treatment (5–100 μM) for 2–4 h. Most of the iron that enters the cells will be stored within a few hours in the iron storage protein ferritin. Therefore, if the endpoint of the study relies on the effect of iron, the effect should be analyzed for a suitable period after addition of iron, in order to ascertain that a lack of effect is not the result of storage of iron in ferritin molecules.

3.5.1.1. IRON/IRON CHELATOR TREATMENTS

1. Desferal: Treatment with 100 μM (final concentration) for 18 h or 1 mM for 2 h in medium at 37°C (*see* **Note 9**).
2. SIH: Treatment with 100–500 μM (final concentration) for 18 h in medium at 37°C (*see* **Note 9**).
3. Fe-citrate: Treatment with 100 μM (final concentration) in medium for 18 h at 37·C (*see* **Note 9**).
4. Hemin: Treatment with 5-100 μM (final concentration) in medium for 2–4 h at 37°C (*see* **Note 9**).

Following these treatments, cells may be processed by methods outlined in **Subheading 3.1.3.**

3.5.2. Estimation of Intracellular Levels of Iron

It is crucial to determine the concentration of drug necessary for modulation of the iron level in the cells following iron-chelation or iron-loading treatments. One of the most sensitive ways of estimating the intracellular levels of free iron in the cells relies on the measurement of the level of activation of iron regulatory protein (IRP) in the cytoplasmic extract of treated cells, as this protein binds to specific elements within mRNA molecules (IREs) in response to low iron levels. This results in either an inhibition of ferritin synthesis or an increase in the stability of the transferrin receptor mRNA, both of which will lead to increased levels of free intracellular iron. Therefore, extremely sensitive estimates of free iron can be made by monitoring changes in IRP/RNA binding by a bandshift assay that uses a radiolabeled probe containing the specific IRE from H-chain ferritin. Cytoplasmic extracts of cells to be examined are mixed with the RNA probe and run on acrylamide gels to separate the bound complex. Levels of the complex estimated by autoradiography will provide information about the availability of intracellular free iron *(27,28)*. Alternatively the level of aconitase activity in the cytosolic fractions of cells, devoid of mitochondria, could provide an estimate of intracellular fee iron, because in response to high iron, IRP-1 protein becomes inactive as an IRE-binding protein and instead functions as a cytosolic aconitase. Nevertheless, IRP-1 protein itself may be subjected to structural modifications following oxidative treatment. Indeed, It has been shown that treatment of murine cells with hydrogen peroxide can activate the IRP binding to IRE, and the latter is unrelated to the level of iron in the cells *(29)*. Interpretation of IRP activation following oxidative treatment should therefore be approached cautiously. Another approach to define the level of intracellular free iron is a method developed by Cabantchick and coworkers *(30)* that is based on total dequenching of the sector of calcein (a divalent fluorescent chelator) fluorescence that is quenched by bound iron. In this method, cells loaded with calcein are transferred to a spectrofluorimeter and the level of intracellular calcein-bound iron is determined by the increase in fluorescence produced by the addition of the highly permeable iron chelator SIH. To establish the relationship between fluorescence change and intracellular chelatable iron concentrations, calibrations should be undertaken using various concentrations of ferrous ammonium sulfate (0.1–1 μM) in the presence of ionophore A23187 (10 μM) and the corresponding change in fluorescence monitored. We have used the latter technique to follow the increased levels of free, chelatable iron following UVA irradiation of primary human FEK4 skin fibroblasts, *(31)* as well as in cells overexpressing HO1 following hemin treatment (E. Kvam, V. Hejmadi, S. Ryter, Ch. Pourzand, and R. M. Tyrrell, *unpublished data*). Here, we describe one of the above-mentioned techniques to estimate the intracellular levels of iron following iron-loading or iron-chelation treatments.

3.5.2.1. IRP/IRE Bandshift Assay

1. Preparation of cytoplasmic extracts: Following treatment, aspirate media and wash the plates two times with cold PBS (e.g., 10 mL per 10-cm plate).
2. Next, scrape the cells with rubber policeman in cold PBS (e.g., 3 mL per 10-cm plate) and centrifuge for 5 min at $200g$ (4°C). Resuspend cells in 1 mL cold PBS. Use 50 µL of cell suspension for counting.
3. Following a second spin at $200g$ (4°C), lyse cells in 1X Munroe buffer/NP40/ PMSF (usually 960 µL lysis buffer for 3.6×10^7 cells). Centrifuge the lysate for 10 min at $1100g$ (4°C) to pellet nuclei.
4. Transfer supernatant (cytoplasmic extract) into a new tube. Aliquot lysates and store at –70°C (*see* **Notes 13** and **14**).
5. Preparation of T7-transcripts (IRE probe, *see* **Note 15**): Mix in an eppendorf tube, 6 µL of buffer 5X TSB, 1.5 µL of 10 mM AGU-TP, 3 µL of 100 mM DTT, 3 µL of [α-^{32}P]CTP (60 µCi), 8 µL of H-ferritin IRE DNA stock (2 µg, *see* **Note 16**), 4.5 µL of Millipore water, 1 µL of RNAsin (40 U) and 2 µL of T7 RNA-polymerase (40 U). Incubate for 1–2 h at 38.5°C.
6. Stop reaction by adding 1 µL 0.5 M EDTA, pH 8.0. Next, precipitate the probe with 2 volumes of absolute ethanol, half volume of 7.5 M ammonium acetate, and 1 µL of glycogen stock as carrier. Let stand 5 min at room temperature.
7. Centrifuge 15 min at $9,000g$; take 1% aliquot of supernatant for counting. Remove rest of supernatant, wash pellet 1X with 0.5 mL 70% EtOH (*see* **Note 17**).
8. Recentrifuge 1 min at $9,000g$. Remove supernatant carefully and dissolve moist pellet (*see* **Note 18**) in 0.1–0.2 mL Millipore water. Take 1–2 µL aliquot for counting and calculate incorporation rate (*see* **Note 19**).
9. Preparation of polyacrylamide gel: First treat gel plates (20×20-cm glass plates) for at least 4 h with 0.1 N NAOH. Next, clean the plates as well as three 1.5-mm spacers and one 12 well comb with detergent, rinse with hot tap water, rinse with Millipore water, and then with 70% EtOH, dry.
10. Assemble the glass plates and spacers. Prepare a 6% acrylamide mixture in 0.3X TBE (for 60 mL, add 9 mL of 40% ultrapure acrylamide/bis-acrylamide, 1.8 mL of 10X TBE, 48.5 mL of Millipore water, 720 µL of 10% APS, and 18 µL of TEMED).
11. Cast gel and insert the comb. Following polymerisation (20 min at room temperature) store the gel in the cold room until use.
12. IRE/IRP binding reaction: Dilute the protein extracts in 1X Munro buffer (without NP40/PMSF) to a final concentration of 1 µg/µL.
13. Dilute the IRE probe in Millipore water to 1.3×10^5 cpm/µL.
14. Prepare a premixture per sample by adding 13 µL Millipore water, 4 µL of 5X Munro, and 1 µL of diluted IRE probe.
15. Start the reaction by adding 2 µL of diluted extract to each premixture and leave 10 min at room temperature to allow the formation of the IRP/IRE complexes. Next, add 2 µL of 50-mg/mL heparin, and leave for 10 min at room temperature (*see* **Note 20**).
16. Stop the reaction by adding 4 µL of loading dye.

17. Electrophoresis of IRP/IRE complexes and autoradiography: Remove the third spacer (in the bottom of the gel), assemble gel and electrophoresis apparatus in the cold room, fill with cold 0.3X TBE, remove comb, and rinse slots with running buffer.
18. Prerun the gels for about 10 min at 200 V (14 V/cm = 200 V in 20 × 20-cm gels).
19. Load the reaction mixtures. Run at 200 V for 2 h (4°C).
20. Separate the plates, leaving the gel on one of the glass plates. Soak the gel for 15 min in fixing solution. Dry 1–2 h on the "hot" setting on vacuum gel dryer or overnight at the "low" setting.
21. Expose for autoradiography (*see* **Note 21**).

3.6. Evidence for the Involvement of the Superoxide Anion

In investigating the role of superoxide in biological systems, a common approach relies on the use of the superoxide scavenger, Tiron (1,2-dihydroxybenzene-3,5-disulfonate). However, this scavenger is relatively nonspecific because it reacts with ·OH, with a greater rate constant than for reaction with superoxide anion *(32,33)*. Furthermore, like many diphenols, it can be oxidized by many systems, including peroxidases and $RO·_2$ radicals, so confirmatory evidence is required to show that superoxide is responsible for the biological effect under study *(34)*. The ability of known superoxide generators such as Pyrogallol, vitamin K_3, and menadione to enhance given effects could provide further evidence for the involvement of the superoxide anion. Indeed the latter approach has been used in several recent studies dealing with the role of superoxide in mitochondrial injury during oxidant-induced apoptosis (e.g., **refs.** *35* and *36*). In vitro superoxide generating systems such as the xanthine/ xanthine oxidase combination have also been used to provide evidence that superoxide anion is involved in tumor promotion *(37)*. In such studies, the production of superoxide anion in cells has to be confirmed following treatment by methods such as the monitoring of the reduction of ferricytochrome *c* or nitroblutetrazolium (NBT) by the superoxide generated. In order to confirm that the observed reduction is superoxide mediated, inhibition by SOD is recommended because many other substances have the ability to reduce cytochrome *c* (especially ascorbate and thiols) and therefore can interfere with superoxide determination (*see* **ref.** *16*). The notion that detection of the superoxide overlaps considerably with the methods available for assaying SOD enzyme activities, provides a basis for the use of SOD inhibitors to enhance the level of superoxide anion.

SOD eliminates the generation of superoxide either during normal cellular metabolism or during oxidative injury. The use of SOD inhibitors such as diethyldithiocarbamate in a given system will therefore enhance the level of the superoxide anion and will provide the easiest approach to define the involvement of this intermediate in the cells. However, in common with many enzyme inhibitors, this compound is relatively nonspecific. Another approach would be to use SOD-mimetic agents that are known to quench effects as a result of the superoxide anion. One such compound is Cu(II) diisopropyl salicylate (CuDips). However, the chemical is not very soluble and may not always reach appropriate target sites so that negative effects are not conclusive. Alternatively, overexpression of SOD via transfection of cells with inducible vectors may be used to provide evidence for superoxide involvement *(38)*. Finally, a possible effect of L-histidine, which reacts only at very low rates with the superoxide anion, may be taken as evidence that this species is not involved (e.g., *HO-1* activation by UVA *[6]*).

4. Notes

1. This is to avoid traces of metal ions which could affect the outcome of the experiments.
2. Old (weeks or months old) PBS solutions tend to accumulate significant amounts of trace elements, notably iron *(18)*.
3. It is recommended that this be freshly prepared.
4. ATZ at concentrations higher than 2 *M* will not dissolve easily. Extensive vortexing is required.
5. Insoluble above 100 m*M*.
6. PMSF tends to precipitate at –20°C. Therefore prior to use vortex thoroughly to bring back the component into solution. Furthermore, because PMSF is only stable for 3–4 h at room temperature, use a fresh aliquot each time.
7. Suitable for single use.
8. APS in aqueous solution is only stable for 7 d at 4°C.
9. The efficiency and toxicity of the treatment is to be checked for each cell system.
10. Addition of these compounds to the medium may result in alteration of the pH; this should then be adjusted back to its original value before proceeding.
11. Cell treatment with GSH–esters should preferably be done in serum-free medium, as serum may contain esterases.
12. The oxidative treatment is carried out in the presence of the compound.
13. It is highly recommended to take an aliquot for protein determination by the Bio-Rad Bradford Assay at this stage to normalize the band-shift assays for identical protein content.
14. Do not reuse frozen aliquots.
15. Use freshly autoclaved Eppendorf tubes, gilson tips, MilliQ water, and so forth in order to avoid RNAse contamination.
16. More DNA always gives better synthesis, because promoter is limiting.
17. Do not vortex; just invert tube gently.
18. It is not necessary to dry pellet.
19. Probe can be stored at 4°C for up to 14 d.
20. Heparin displaces nonspecifically bound proteins off the IRE.
21. Expose either for 1 h at –70°C with sensitive Amersham Hyperfilms and intensifying screen or overnight (at –70°C) with insensitive Fuji films (without intensifying screen).

Acknowledgments

Many studies described herein were supported by core grants from the Association for Cancer Research (UK), the Department of Health (contract no. 121/6378), the European Union Fourth Framework Environment program financed by the Swiss Office of Education and Science under contract OFES 95.0509 with additional support from the League Against Cancer of Central Switzerland, the Neuchateloise League Against Cancer and the Swiss National Science Foundation. We would like also to thank the colleagues who have contributed to the studies described herein, including Sharmila Basu-Modak, Stefan Ryter, Egil Kvam, Glenn Vile, Stephen Keyse, Vidya Hejmadi, Jonathan Brown, and Richard Watkin.

References

1. Tyrrell, R. M. (1996) UV activation of mammalian stress proteins in *Stress-Inducible Cellular Responses*. (Feige, U., Morimoto, R. I., Yahara, I., and Polla, B., eds.) Birkhauser Verlag, Basel, Switzerland, pp. 255–271.
2. Lautier, D., Luscher, P., and Tyrrell, R. M. (1992) Endogenous glutathione levels modulate both constitutive and UVA radiation/hydrogen peroxide inducible expression of the human heme oxygenase gene. *Carcinogenesis* **13,** 227–232.
3. Tyrrell, R. M. (1996) Activation of mammalian gene expression by the UV component of sunlight-from models to reality. *BioEssays* **18,** 139–148.
4. Keyse, S. M., Applegate, L. A., Tromvoukis, Y., and Tyrrell, R. M. (1990) Oxidant stress leads to transcriptional activation of the human heme oxygenase gene in cultured skin fibroblasts. *Mol. Cell. Biol.* **10,** 4967–4969.
5. Keyse, S. M. (1995) An emerging family of dual specificity MAP-kinase phosphatases. *Biochim. Biophys. Acta* **1265,** 152–160.
6. Basu-Modak, S. and Tyrrell, R. M. (1993) Singlet oxygen: a primary effector in the ultraviolet A/near-visible light induction of the human heme oxygenase gene. *Cancer Res.* **53,** 4505–4510.
7. Meister, A. and Anderson, M. E. (1983) Glutathione. *Annu. Rev. Biochem.* **52,** 711–760.
8. Tyrrell, R. M. and Pidoux, M. (1988) Correlation between endogenous glutathione content and sensitivity of cultured human skin cells to radiation at defined wavelengths in the solar ultraviolet range. *Photochem. Photobiol.* **47,** 405–412.
9. Tietze, F. (1969) Enzymatic methods for quantitative determination of nanogram amounts of total and oxidised glutathione: applications to mammalian blood and other tissues. *Anal. Biochem.* **27,** 502–521.
10. Gunthenberg, H. and Rost, J. (1966) The true oxidized glutathione content of red blood cells obtained by new enzymatic and paper chromatographic methods. *Anal. Biochem.* **15,** 205–210.
11. Mbemba, F., Houbion, A., Raes, M., and Remacle, J. (1985) Subcellular localization and modification with ageing of glutathione, glutathione peroxidase and glutathione reductase activities in human fibroblasts. *Biochim. Biophys. Acta* **838,** 211–220.

12. Caltabiano, M. M., Koestler, T. P., Poste, G., and Grieg, R. G. (1986) Induction of 32- and 34-kDa stress proteins by sodium arsenite, heavy metals, and thiol-reactive agents. *J. Biol. Chem.* **261**, 13,381–13,386.

13. Tyrrell, R. M. and Basu-Modak, S. (1994) Transient enhancement of heme oxygenase 1 mRNA accumulation: a marker of oxidative stress to eukaryotic cells. *Methods Enzymol.* **234**, 224–235.

14. Wilkinson, F. and Brummer, J. (1981) Rate constants for the decay and reactions of the lowest electronically excited singlet state of molecular oxygen in solution. *J. Phys. Chem. Ref. Data* **10**, 809–999.

15. Farhataziz, B. and Ross A. B. (1977) Selected specific rates of reactions of transients from water in aqueous solution. III. Hydroxyl radical and perhydroxyl radical and their radical ions. *Natl. Stand. Ref. Data Ser. (U. S. Natl. Bur. Stand.)* **59**, 1–113.

16. Halliwell, B. and Gutteridge, J. M. C. (1999) *Free Radicals in Biology and Medicine*, Clarendon Press, Oxford.

17. Packer, J. E., Mahood, J. S., Mora-Arellano, V. O., Slater, T. F., Willson, R. L., and Wolfenden, B. S. (1981) Free radicals and $^1O_2$ scavengers: reaction of a peroxy radical with β-carotene, diphenylfuran and DABCO. *Biochem. Biophys. Res. Commun.* **98**, 901–906.

18. Halliwell, B. and Gutteridge, J. M. C. (1989) in *Free Radicals in Biology and Medicine*, Clarendon Press, Oxford.

19. Perez, M. D. (1985) in *Handbook of Methods for Oxygen Radical Research* (Greenwald, R. A., ed.), CRC Press, Boca Raton, FL, pp.111–113.

20. Foote, C. S. and Clennan, E. L. (1995) Properties and reactions of singlet dioxygen, in *Active Oxygen in Chemistry*, Vol. 2 (Foote, C. S., Valentine, J. S., Greenberg, A., and Liebman, J. F., eds.), Blackie Academic and Professional, Glasgow, UK, pp. 105–140.

21. Scharffeter, K., Wlaschek, M., Hogg, A., Bolsen, K., Schothorst, A., Goerz, G., Krieg, T., and Plewig, G. (1991) UVA irradiation induces collagenase in human dermal fibroblasts in vitro and in vivo. *Arch. Dermatol. Res.* **283**, 506–511.

22. Gutteridge, J. M. C. (1985) Superoxide dismutase inhibits the superoxide-driven Fenton reaction at two different levels. Implications for a wider protective role. *FEBS Lett.* **185**, 19–23.

23. Tyrrell, R. M. (1997) Approaches to define pathways of redox regulation of a eukaryotic gene: The heme oxygenase 1 example, in *Methods: A Companion to Methods in Enzymology 11,* (Demple, B., ed.), Academic Press, UK, pp. 313–318.

24. Moysan, A., Marquis, I., Gaboriou, F., Santus, R., Dubertret, L., and Morliere, P. (1993) Ultraviolet A-induced lipid peroxidation and antioxidant defense systems in cultured human skin fibroblasts. *J. Invest. Dermatol.* **100**, 692–698.

25. Pantopoulos, K., Mueller, S., Atzberger, A., Ansorge, W., Stremmel, W., and Hentze, M. W. (1997) Differences in the regulation of iron regulatory protein-1 by extra- and intracellular oxidative stress. *J. Biol. Chem.* **272**, 9802–9808.

26. Khan, A. U. and Kasha, M. (1994) Singlet molecular oxygen in the Haber-Weiss reaction. *Proc. Natl. Acad. Sci. USA* **91**, 12,365–12,367.

27. Henderson, B. R., Menotti, E., Bonnard, C., and Kuhn, L. C. (1994) Optimal sequence and structures of iron responsive elements, selection of RNA stem-loops with high affinity for iron regulatory factor. *J. Biol. Chem.* **268,** 27,327–27,334.
28. Pourzand, C., Reelfs, O., Kvam, E., and Tyrrell, R. M. (1999) The iron regulatory protein can determine the effectiveness of 5-aminolevulininc acid in inducing protoporphyrin IX in human primary skin fibroblasts. *J. Invest. Dermatol.* **112,** 419–425.
29. Pantopoulos K. and Hentze, M. W. (1995) Rapid responses to oxidative stress mediated by iron regulatory protein. *EMBO J.* **14,** 2917–2924.
30. Breuer, W., Epsztejn, S., and Cabantchik, Z. I. (1996) Dynamics of the cytosolic chelatable iron pool of K562 cells. *FEBS Lett.* **382,** 304–308.
31. Pourzand, C., Watkin, R. D., Brown, J. E., and Tyrrell, R. M. (1999) Ultraviolet A radiation induces immediate release of iron in human primary skin fibroblasts: The role of ferritin. *Proc. Natl. Acad. Sci. USA* **96,** 6751–6756.
32. Greenstock, C. L. and Miller, R. W. (1975) The oxidation of tiron by superoxide anion. Kinetics of the reaction in aqueous solution in chloroplasts. *Biochim. Biophys. Acta* **396,** 11–16.
33. Bors, W., Saran, M., and Michel, C. (1979) Pulse-radiolytic investigations of catechols and catecholamines. II. Reactions of Tiron with oxygen radical species. *Biochim. Biophys. Acta* **582,** 537–542.
34. Kahn, V. (1989) Tiron as a substrate for horseradish peroxidase. *Phytochemistry* **28,** 41.
35. Virag, L., Salzman, A. L., and Szabo, C. (1998) Poly(ADP-Ribose) synthetase activation mediates mitochondrial injury during oxidant-induced cell death. *J. Immunol.* **161,** 3753–3759.
36. Godar, D. E. (1999) Ultraviolet-A1 radiation triggers two different final apoptotic pathways. *J. Invest. Dermatol.* **112,** 3–12.
37. Zimmerman, R. and Cerutti, P. (1984) Active oxygen acts as a promoter of transformation in mouse embryo C3H/10T1/2/C18 fibroblasts. *Proc. Natl. Acad. Sci. USA* **81,** 2085–2087.
38. Schmidt, K. N. Amstad, P., Cerutti, P., and Bauerle, P. A. (1995) The roles of hydrogen peroxide and superoxide as messengers in the activation of transcription factor NF-κB. *Chem. Biol.* **2,** 13–22.

19

The Human Immunodeficiency Virus LTR-Promoter Region as a Reporter of Stress-Induced Gene Expression

Michael W. Bate, Sushma R. Jassal, and David W. Brighty

1. Introduction

Human immunodeficiency virus type-1 (HIV-1) replication and proviral gene expression are exquisitely responsive to factors that induce cellular stress. Oxidants, ultraviolet (UV) light, osmotic stress, heat shock and pro-inflammatory cytokines all promote proviral gene expression and enhance viral proliferation *(1–6)*. In quiescent or unstimulated cells, the HIV-1 promoter exhibits a low basal level of transcriptional activity that can be activated 12- to 150-fold following exposure to a mitogenic or stress-inducing stimulus *(3,6)*. This dramatic induction of transcription is achieved by integrating a complex network of cellular signal-transduction pathways, with a variety of highly responsive transcription factors that bind to and modulate the transcriptional activity of the viral promoter. Therefore, not only is the viral promoter highly responsive to diverse physiological stimuli, but this transcriptional activity is correlated directly with the activation status of specific kinase-regulated signal transduction pathways *(6,7)*.

The robust transcriptional activation of the HIV-1 promoter in response to mitogenic or stress-inducing stimuli has generated considerable experimental interest in studies that explore the factors promoting viral replication. Moreover, the compact nature of the HIV-1 promoter, coupled with its ability to retain regulated transcriptional activity when removed from its natural context and placed into heterologous chimeric promoter–reporter constructs, has provided a versatile tool for analysis of the physiological and molecular mechanisms that underlie the transcriptional activation of highly inducible stress-responsive genes. Here, we discuss the salient features of the HIV-1 pro-

From: *Methods in Molecular Biology, vol. 99: Stress Response: Methods and Protocols*
Edited by: S. M. Keyse © Humana Press Inc., Totowa, NJ

moter, the transcription factors that contribute directly to stress-induced promoter activation, and the stimuli and signal-transduction pathways that modulate promoter activity. Finally, we discuss the HIV-1 promoter–reporter constructs and assays that have been used as indicators of stress-induced gene expression.

1.1. The HIV-1 Promoter Region

Situated within the proviral 5' long terminal repeat (LTR), the HIV-1 promoter exhibits a classical modular structure (**Fig. 1**). Three functionally important promoter regions have been defined: (1) a core promoter region; (2) an enhancer region; and (3) Tar, an RNA element, distal to the transcript initiation site, which tethers the viral Tat trans-activator to the promoter (*see* **Subheading 1.5.**). The core promoter region is required for recruitment and assembly of the basal transcription apparatus, and this region includes a canonical TATA element that binds TBP and associated factors, thereby ensuring efficient and accurate initiation of transcription *(8–10)*. Also required are three high-affinity binding sites for the constitutively active cellular transcription factor SP1, which have been mapped immediately 5' of the TATA element *(8)*. Deletion of any of these core promoter elements severely impairs promoter function, prevents induction in response to multiple stimuli, and renders the promoter unresponsive to Tat *(8–10)*. In addition, sites recognised by the cellular transcription factors LBP-1, YY1 and USF map within the core promoter region close to, or overlapping with, the site of transcript initiation *(11–14)*. Although USF can stimulate core promoter activity, YY1 and LBP-1 act synergistically to inhibit HIV gene expression, and mutagenesis of the sites recognised by these latter factors can relieve this repressive effect *(13)*. Thus, competitive binding of these factors to the core promoter region may differentially modulate promoter activity.

Within the promoter–enhancer region, a plethora of cis-acting DNA sequence elements have been mapped that have the capability to bind a variety of trans-acting cellular factors that regulate transcriptional activity. Potential or confirmed binding sites for the transcription factors AP1, COUP, a number of homodimeric and heterodimeric nuclear hormone receptors, NFAT, USF, LEF-1, and C/EBP have all been mapped within the enhancer region (**Fig. 1**). Although many of these transcription factors bind to LTR-derived probes in vitro, activate transcription of LTR-reporter constructs in cell-based assays, and activate or modulate transcription in vitro *(15)*, the contribution of these factors to viral replication has not been fully explored. Nevertheless, LTR-luciferase reporter assays have been used to demonstrate that a cellular transcription factor such as C/EBP contributes directly to stress-induced promoter activation *(16)* and that binding of such factors is required for full promoter

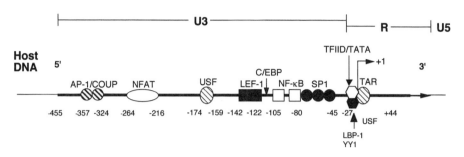

Fig. 1. Schematic representation of the HIV-1 LTR. The HIV-1 LTR region is depicted and the positions of the various transcription factors that bind to the LTR are shown relative to the transcription initiation site. For details, see the text.

activity in response to lipopolysaccharide (LPS) and phorbol myristate acetate (PMA) treatment of monocytic/macrophage cell lines *(16)*.

1.2. NF-κB-Dependent Transcriptional Activation

A major player in regulating proviral gene expression and enhancing LTR-driven transcription is the transcription factor, NF-κB. NF-κB is a principal co-ordinator of the cellular response to immunological, toxic, or inflammatory stress and is a potent activator of HIV-1 gene expression *(1,2,6,17,18)*. Comprised of p50 and RelA heterodimers NF-κB is expressed constitutively but in an inactive form that is sequestered within the cytoplasm through binding to the NF-κB inhibitor IκB *(see also* Chapter 10). Upon stimulation by any one of a variety of signals, IκB proteins are degraded releasing NF-κB and allowing the functional transcription factor to relocate to the nucleus where it binds to DNA in a sequence-specific manner and promotes expression of NF-κB-responsive genes *(17,18; see also* Chapter 16). Activation of HIV-1 gene expression by NF-κB has been the focus of considerable attention and tandem NF-κB sites have been mapped at -80 to -105 immediately adjacent to the SP1 sites of the core promoter region *(6)*. Upon activation, NF-κB binds to these sites and greatly enhances LTR-driven transcription *(1,2,6,8)*. Mutation of either one or both NF-κB sites acutely impairs enhancer function and prevents induction of proviral or LTR-reporter gene expression in response to multiple stimuli *(1,6,8)*. In particular, it has been demonstrated that activation and binding of NF-κB at the HIV enhancer–promoter is required for, or can support, enhanced LTR-mediated transcription in response to cytokines such as tissue necrosis factor-α (TNF-α) and interleukin-1 (IL-1), H_2O_2, phorbol 12-myristate 13-acetate (PMA), lectins, UV irradiation, bacterial lipopolysaccharide, and calcium ionophores *(1,2,5,6)*. A common theme that connects these seemingly disparate stimuli is the production and activity of reactive oxygen intermediates (ROI). Importantly, antioxidants can inhibit the activation of NF-κB in response to these stimuli; indicating that ROI may play a central role in regulating the transcriptional activity of NF-κB *(1,17)*.

1.3. NF-κB-Independent Stress-Responsive Promoter Activation

Although NF-κB exerts a powerful influence on transcription from the LTR it is not essential for promoter function *(19,20)*. Studies with LTR-reporter constructs have uncovered a striking example of stress-regulated promoter activation that underscores both the complex and dynamic nature of transcriptional activation from the HIV-1 LTR and the role played by stress-activated protein kinases in promoter activation. Transcription from the HIV-1 LTR is potently activated in response to a UV stimulus (**Fig. 2**) *(3,4,21)*. UV induction of transcription is dependent on an active p38/RK mitogen-activated protein (MAP) kinase pathway, and is highly sensitive to the p38/RK MAP kinase inhibitor SB203580 *(5; see also* Chapter 13). UV rapidly activates p38/RK MAP kinase and this activation precedes UV-stimulated transcription from the LTR *(5)*. Addition of SB203580 to UV-treated cells inhibited p38/RK activity and also abolished UV-induced LTR-driven transcription *(5)*. These results clearly establish the stress-activated p38/RK MAP kinase pathway as the principal route of signal transduction in UV-induced proviral gene expression. Importantly however, although NF-κB is clearly activated following UV irradiation and can stimulate transcription from the HIV-1 LTR, the NF-κB sites are not essential to UV-induced proviral gene expression. Indeed, it has been demonstrated that the NF-κB sites can be deleted from the HIV LTR and that as long as the LTR-reporter constructs are stably integrated into the host cell genome they remain highly responsive to UV irradiation (**Fig. 3**) *(22,23)*. Moreover, the inhibitory effect of SB203580 on UV-stimulated transcription from the LTR does not require the NF-κB sites (**Fig. 4**) *(5;* Bate and Brighty, unpublished results), indicating that NF-κB is not the target of the p38/RK pathway and that additional LTR-associated events are required for the response to UV-induced cellular stress.

In a series of definitive experiments the groups of Valerie and Rosenberg *(3,21,23,24)* demonstrated that, although the NF-κB sites are dispensable for UV-induced activation of the LTR, sequences within the core promoter region are essential for transcriptional activation in response to UV-induced cellular stress. Moreover, these authors suggested that decondensation of chromatin may contribute directly to activation of integrated HIV-1 LTR-reporter constructs by increasing the accessibility of the core promoter region to the basal transcription apparatus *(24)*. If this is the case, a degree of promoter specificity must be invoked because a variety of unrelated and stably integrated promoter–reporter constructs do not respond to UV-irradiation *(24;* our unpublished results).

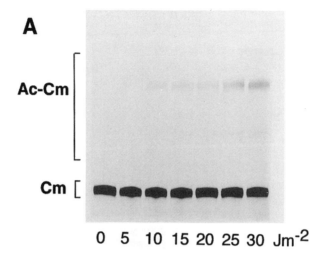

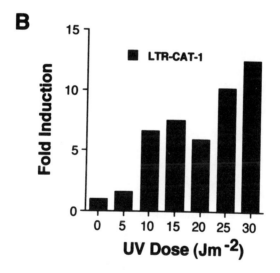

Fig. 2. Transcriptional activation of the HIV-1 LTR in response to UV. HeLa cells were stably transfected with pLTR-CAT and exposed to UV irradiation (0–30 J/m²). (A) Autoradiograph of TLC plate showing the CAT activity expressed from the stably transfected cells following UV irradiation. The positions of both acetylated (Ac-Cm) and nonacetylated chloramphenicol (Cm) are indicated. (B) The CAT activity for each stimulus was quantitated and the fold activation, relative to untreated cells, is shown. CAT activity was assayed as described in **Subheadings 2.** and **3.**

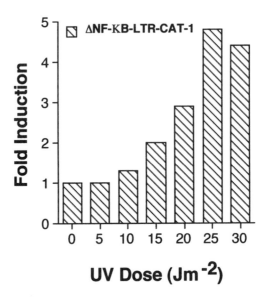

UV Dose (Jm⁻²)

Fig. 3. UV-induced transcriptional activation of an integrated HIV-1 LTR occurs by an NF-κB-independent pathway. HeLa cells were stably transfected with pLTRΔNF-KB-CAT, an LTR-reporter construct in which the NF-κB binding sites have been deleted, and exposed to UV irradiation (0–30 J/m^2). The CAT activity for each stimulus was quantitated and the fold activation, relative to untreated cells, is shown. CAT activity was assayed as described in **Subheadings 2.** and **3.**

1.4. Integrated HIV-1 LTR-Reporter Genes are Responsive to Changes in Chromatin Structure

Using a range of LTR-reporter constructs and also using HIV-1-infected cells, a number of groups have demonstrated that the HIV-1 promoter is highly responsive to factors that induce changes in local DNA topology. Transcriptionally silent HIV-1 promoter reporter constructs can be activated both in vivo and in vitro by histone deacetylase inhibitors *(15,25)*. Indeed, an integrated and chromatin repressed promoter, which was unresponsive to NF-κB, could be transcriptionally activated following treatment of cells with trichostatin A (TSA) in a manner that supported synergistic enhancement by NF-κB *(26)*. These results have important implications for the reactivation of latent virus in infected cells, and a model for HIV-1 promoter reactivation in response to cellular stress and cytokine stimulation, that is consistent with all of the available data, can be developed if we invoke a two-step process for LTR activation. In this model, NF-κB or other enhancer-binding transcription factors bind to the integrated HIV LTR but fail to promote transcription unless an additional dominant stress-responsive activating event has occurred. This additional event may

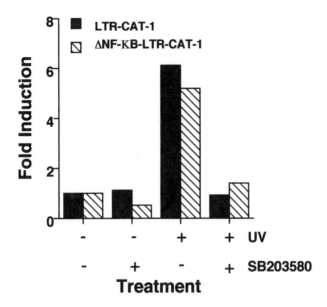

Fig. 4. The p38/RK MAP kinase inhibitor SB 203580 prevents transcriptional activation of the HIV-1 LTR and LTR-NF-κB deletion mutants in response to UV. HeLa cells stably transfected with pLTR-CAT and pLTRΔNF-KB-CAT were exposed to UV irradiation (0–30 J/m^2) in the presence or absence of the MAP kinase inhibitor SB 203580. The CAT activity for each stimulus was quantitated and the fold activation, relative to untreated cells, is shown. CAT activity was assayed as described in **Subheadings 2.** and **3.**

involve a stress-responsive alteration in chromatin structure possibly achieved by the displacement or repositioning of inhibitory nucleosome structures within the HIV-1 promoter region. Importantly, transiently transfected reporter constructs would not be responsive to this aspect of stress-responsive gene expression because they are not assembled into chromatin.

1.5. Tat-Mediated Transactivation

No discussion of the HIV-1 LTR as a reporter of cellular stress would be complete without mention of the viral Tat transactivator *(27)*. Tat is a potent activator of transcription and enhances proviral gene expression by binding to an RNA stem loop structure, designated Tar, that is found at the 5' end of all proviral transcripts (reviewed in **ref.** *27*). In binding to Tar, Tat recruits a cellular kinase, Tat associated kinase (TAK), to the HIV-1 promoter *(28)*. TAK hyperphosphorylates the carboxyl-terminal domain of RNA polymerase II, which enhances transcription complex processivity, prevents premature termination of transcription, and greatly stimulates the production of full-length transcripts *(28)*. Importantly, Tat contributes directly to the production and maintenance of cellular stress by perturbing cellular gene expression and promoting apoptosis *(29,30)*. These effects of Tat may hasten progression to acquired immunodeficiency syndrome (AIDS).

1.6. Activation of the HIV-1 Promoter as a Model for Stress-Induced Gene Expression

HIV-1 LTR-containing reporter genes and their mutant-LTR derivatives provide researchers with versatile reagents that can be used to probe the stress-responsive pathways that activate viral and cellular gene expression. The sensitive stress-induced activation of LTR-reporter genes allows a simple, yet effective, readout of cellular processes that are otherwise difficult to assay. By determining the promoter regions and transcription factors that are responsive to a given stimulus, coupled with the selective use of inhibitors or genetically transdominant mutant proteins that are specific for a given signal-transduction pathway, it is possible to derive testable models to account for the cell's response to diverse physiological insults. Such studies are likely to produce information that will enhance our understanding of the cellular response to stress and the factors that contribute to viral replication and pathogenesis.

1.7. Generation of LTR-Reporter Constructs

A variety of reporter genes can be employed as markers of LTR-driven transcription. These include bacterial chloramphenicol acetyl transferase (CAT) *(3,21,23)*, bacterial β-galactosidase *(22)*, firefly luciferase *(31,32)*, and the green fluorescent protein obtained from the jellyfish *Aquorea victoria (33,34)*. Each reporter gene has particular properties that suit different experimental protocols, but none is universally ideal and the specific attributes of each reporter should be considered in view of the procedures planned.

Typical LTR-reporter vectors place the HIV-1 LTR within a multiple-cloning site upstream of one of the reporter genes described above (**Fig. 5**). The reporter gene is oriented so that it can be expressed from the LTR. An intron is often included downstream of the reporter gene, and the transcription unit should be terminated by inclusion of a polyadenylation signal; these pre-mRNA processing signals improve expression and ensure efficient mRNA accumulation within eukaryotic systems. These elements are particularly important when the gene is of bacterial origin (e.g., CAT and β-galactosidase). Vectors should also encode antibiotic resistance markers and origins of replication for growth and manipulation of the vectors in bacterial hosts. Also, a drug-resistance marker may be included for selection of the reporter vector in mammalian cells to aid production of stably transfected cell lines. Suitable LTR-reporter plasmids can be constructed from a variety of traditional promoter–probe vectors

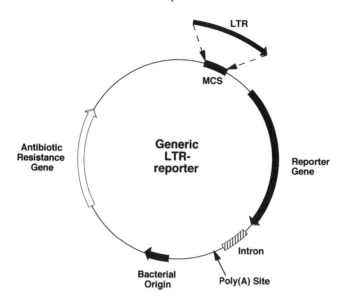

Fig. 5. Schematic representation of a generic LTR-reporter construct. The design of HIV-1 LTR-reporter vectors is discussed in detail in the the text.

and reporter genes that are available from commercial sources (Promega, Pharmacia, and Clontech).

The CAT gene is commonly used in LTR-reporter constructs because the assays are simple, robust, and relatively inexpensive (*see* **Subheading 3.5.**). The CAT assay has been extensively accredited, the results are widely accepted, and a hard-copy visual result can be obtained following autoradiography. However, CAT assays are limited by their sensitivity and cannot be used to follow stress-responsive gene expression in single cells. β-galactosidase has also been used extensively in LTR-reporter constructs *(22)*; the assays for β-galactosidase activity are straightforward and require no unusual items of equipment. However, a significant advantage of β-galactosidase is the range of substrates available: *O*-nitrophenyl β-D-galactopyranoside (ONPG), 5-bromo-4-chloro-3-indolyl β-D-galactoside (X-gal), chlorophenol red β-D-galactopyranoside (CPRG) and 4-methyl-umbelliferyl-β-D-galactoside (MUG). In each case the enzyme hydrolyzes the substrate to release a product that can be directly quantified. Assays using OPNG are the least sensitive, and can be measured by simple photometric analysis, whereas assays using CPRG and MUG exhibit several orders of magnitude greater sensitivity and are detected by colorimetry and spectrofluorimetry, respectively. The oxidized indoxyl product of X-gal is poorly soluble in aqueous solution and, consequently, has been used effectively to identify individual cells expressing LTR–

β-gal reporter constructs following histochemical staining of tissues samples from transgenic mice *(22)*.

Luciferase is found in the firefly species *Photinus pyralis*. The enzyme catalyzes the conversion of luciferin in the presence of ATP and oxygen to oxyluciferin, resulting in the emission of light that is proportional to the concentration of luciferase. The assay requires a luminometer (an expensive item), but the technique has the advantage of being extremely sensitive and noninvasive. The exquisite sensitivity of the LTR–luciferase system has enabled researchers to assay LTR activity where only a small proportion of a cell population is transcriptionally active *(31)*. Moreover, improved imaging technologies have permitted the study of temporal regulation of HIV-1 LTR-driven gene expression in single cells *(32)*.

Green fluorescent protein (GFP) has also been used to quantify LTR-driven gene expression *(33)*, but the detection methods, which employ the latest technology in cytofluorimetry, fluorescence microscopy and flow cytometry, are intrinsically expensive. Nevertheless, GFP allows analysis of the transcriptional activity of the LTR in viable metabolically active single cells, provides the opportunity to track expressing cells in infected tissue and animals, and allows fluorescence assisted cell sorting (FACS) analysis of transcriptionally active cell populations.

2. Materials

2.1. Materials for Transient and Stable Transfections

All reagents used for procedures involving tissue culture should be sterilized either by autoclaving or filtration through a 20-µm filter.

1. Phosphate-buffered saline (PBS): 10 mM Na_2HPO_4, 5 mM NaH_2PO_4, 5.4 mM KCl, and 137 mM NaCl, pH 7.4.
2. Tissue culture medium: Dulbecco's modified Eagle medium (DMEM) containing 4500 g/L glucose and 2 mM L-glutamine, 10% fetal bovine serum (FBS), 100 µg/mL penicillin/100 U/mL streptomycin (all available from Gibco-BRL [Gaithersburg, MD], Life Technologies [Bethesda, MD]). For selection of stably transfected cell lines the medium should be supplemented with 400 µg/mL G418 (Life Technologies).
3. Trypsin 2.5% (w/v) (Life Technologies).
4. 2.5 M $CaCl_2$ solution containing 183.7g $CaCl_2$ dihydride (Sigma [St. Louis, MO]; tissue culture grade) dissolved in water and adjusted a final volume of 500 mL.
5. 2X HEPES-buffered saline solution (HeBS): 16.4 g/L NaCl, 11.9 g/L HEPES acid, and 0.21 g/L Na_2HPO_4, pH 7.05.
6. 5 µg Plasmid DNA. To aid plasmid integration for stable transfections, the plasmid DNA may be linearized by digestion with a restriction enzyme that cuts plasmid backbone out with the transcription unit or selectable marker.

7. HeLa cells subcultured into 35-mm dishes at a cell density of $0.5-1 \times 10^5$ cells/ 30-mm dish for stable and transient transfections. 75-cm^2 flasks for the selection and maintenance of stable cell lines. Preparation of the DNA/CaCl$_2$ solution requires the use of a mechanical pipettor (e.g., Drummond [Broomall, PA] Pipet-aid), Pasteur pipet and vortex (e.g., Vortex-genie 2, Scientific Industries). A tissue culture incubator (5% CO$_2$) at 37°C is required for cell growth (e.g., LEEC MkII Proportional Temperature Controller).

2.2. Materials for UV Irradiation of Transient and Stable Cell Lines

1. Transfected cells in 35-mm dishes at 60–75% confluence.
2. 0–30 J/m^2 UVC (254 nm) produced by a Stratalinker model 2400 (Stratagene, La Jolla, CA).
3. Tissue culture medium containing DMEM, 10% FBS, penicillin/streptomycin, and G418. Pasteur pipets, PBS, a mechanical pipettor, and 37°C (5% CO$_2$) incubator.

2.3. Materials for CAT Assays

1. 35-mm dishes placed on ice containing transfected cells exposed to UV irradiation or other inducing agent.
2. An aspirator, ice-cold PBS, spatula, Gilson pipet, and Eppendorf tubes.
3. Ice-cold hypotonic lysis buffer: 25 mM Tris-HCl, pH 7.5, 2 mM MgCl$_2$, and 0.5% Triton X-100. Alternatively, a commercially available buffer such as Reporter Lysis Buffer® (Promega, Madison, WI) may be used.
4. Techne (Duxford, UK) Dri-block DB-3D at 65°C and microcentrifuge (e.g., Eppendorf centrifuge 5415 C).
5. Bio-Rad (Hercules, CA) Protein Assay reagent, plastic cuvets (Clinicon) and a spectrophotometer set at 595 nm (e.g., Ultrospec 2000 UV/visible light spectrophotometer; Pharmacia Biotech, Uppsala, Sweden).
6. A master mix containing acetyl-Co A (3.5 μg/μL; Sigma), distilled water, and [^{14}C]-labeled chloramphenicol (111 mCi/mmol; ICN Biomedicals [Wycombe, UK]). A plastic tray, paper towels, and radioactive waste bags are used at this stage to reduce the risk of radioactive contamination.
7. Ethyl acetate and a Speedyvac (Savant, Heckville, NY) set on a moderate heat.
8. A thin-layer chromatography (TLC) tank containing two rectangular sheets of Whatman paper (3 MM) and a mixture of chloroform and methanol (95:5 v/v). The tank and the reagents are prepared in a fume cupboard.
9. A TLC plate (Whatman LK6DF).
10. Quantitative analysis is carried out using the Bio-Rad PhosphorImager and Molecular Analyst program.

2.4. β-*Galactosidase Assay*

All reagents, unless indicated, are molecular biology grade and are available from Sigma or Aldrich (Milwaukee, WI).

1. PBS, pH 7.4.
2. Lysis buffer: 50 mM potassium phosphate, pH 7.5, 1 mM MgCl$_2$, 0.2% Triton X-100.

3. Assay buffer: 50 mM potassium phosphate, pH7.5, 1 mM MgCl$_2$, 1.65 mM red β-D-galactopyranoside (CPRG; Boehringer Mannheim, Mannheim, Germany).
4. Eppendorf tubes and disposable cuvets.
5. Spectrophotometer

2.5. Histological Staining for β-Galactosidase

1. Fixer solution: 0.1 M sodium phosphate buffer, pH 7.3, 2% formaldehyde, 0.2% glutaraldehyde, 5 mM EGTA, 2 mM MgCl$_2$.
2. Standard wash: PBS, 2 mM MgCl$_2$.
3. Detergent wash: 0.1 M sodium phosphate buffer, pH 7.3, 2 mM MgCl$_2$ 0.1% sodium deoxycholate, 0.02% NP-40, 0.05% BSA.
4. Stain solution: detergent wash, 0.085% NaCl, 5 mM K$_3$Fe(CN)$_6$, 5 mM K$_4$Fe(CN)$_6$ 0.024% spermidine (6 mg/mL stock, dissolve and filter), 0.1% X-gal (*see* **Note 1**; available from Biogene Ltd).
5. Incubator set at 37°C.
6 Inverted microscope (e.g., Olympus IX50 or equivalent).

3. Methods

3.1. Transient and Stable Transfections and Construction of HIV-1 LTR-CAT Plasmids

We have routinely used the calcium phosphate coprecipitation *(35)* method to transfect cells with plasmid DNA, both for stable and transient transfections. Other techniques commonly employed to introduce DNA into cells such as DEAE–dextran transfection, liposome-mediated transfection, and electroporation may also be used. If the reporter construct encodes a selectable drug-resistance marker, or, alternatively, if the reporter construct is cotransfected with a plasmid encoding drug resistance, then stable cell lines can be obtained by drug selection using G418 or hygromycin.

3.2. Transient Transfection

Plasmid constructs encoding the bacterial chloramphenicol acetyl transferase or β-galactosidase gene under control of the HIV-1 LTR have been used in our assays.

1. Seed 0.5–1 × 10^5 HeLa cells into 35-mm dishes containing 2 mL DMEM medium, and 10% FBS and incubate overnight at 37°C (5% CO$_2$) or until the cells become 40–60% confluent.
2. When the cells are sufficiently confluent, dissolve 5µg plasmid DNA in 450 µL sterile water and 50 µL 2.5 M CaCl$_2$.
3. Add 500 µL 2X HEPES-buffered saline (HeBS) to a 15-mL conical tube. Use a mechanical pipet aid to bubble air through the solution to mix the 2X HeBS while adding DNA/CaCl$_2$ dropwise with a Pasteur pipet.

4. Mix the solution for 5 s with a vortex, and then incubate the solution for 20 min at room temperature. Incubation allows a DNA–calcium phosphate coprecipitate to form that can be added to the cell monolayer.

5. Add the DNA–calcium phosphate coprecipitate to the cells and agitate the dishes gently to mix the precipitate with the medium.

6. After 4–16 h, remove the precipitate and wash the cells twice with PBS. Add fresh tissue culture medium to the cells and return the dishes to the incubator. Then 16–20 h after transfection the cells may be exposed to UV irradiation or other stimulus (*see* **Subheading 3.4.**).

3.3. Stable Transfections

1. Linearize the DNA to prepare the plasmid for stable transfection.
2. Repeat **steps 1–6** of **Subheading 3.2.**
3. Forty-eight hours after transfection replace the tissue culture medium with medium containing the selective drug (usually G418 or hygromycin). When the cells become confluent, split and transfer them to 75-cm^2 tissue culture flasks at a concentration of 5×10^4 cells/mL. Every 2–3 d the medium should be removed, the cells washed with PBS to remove cell debris, and fresh medium containing selective drug added (*see* **Note 2**). When using G418, stable cell lines can be selected over 3–4 wk.

3.4. UV Irradiation of Transient and Stable Cell Lines

1. Before UV irradiation, transfer 1×10^5 cells to 35-mm dishes and incubate at 37°C (5% CO_2) until the cells reach 70–80% confluence (*see* **Note 3**).
2. Remove the medium with an aspirator and wash the cells with PBS.
3. Remove the PBS and place the dishes without the lids into a Stratalinker model 2400 (Stratagene). Expose the cells to 10–30J/m^2 UVC (254 nm).
4. Following UV exposure, return the original medium to the cells, and incubate overnight at 37°C.

3.5. CAT Assay

The CAT gene was adapted for use in mammalian systems by Gorman et al. *(36)*. In this technique, the response of a promoter to a stimulus can be measured by expression of the CAT gene and the capacity of its protein product to acetylate the two hydroxyl groups of [^{14}C]-labeled chloramphenicol. The cells are lysed in hypotonic buffer containing 25 mM Tris-HCl at pH 7.5, 2 mM MgCl$_2$, and 0.5% (v/v) Triton X-100. The buffer maintains a physiological pH for the assay.

1. At various time points after UV induction or other stimulus, remove the medium from the cells using an aspirator and wash the cells twice with ice-cold PBS.
2. Remove the PBS and add 200 µL lysis buffer to each dish. With the dishes on ice, scrape the lysing cells vigorously with a spatula. Using a Gilson pipet, transfer

the cell lysates to a clean eppendorf tube and pellet the insoluble material by centrifugation at 14,000g in a microfuge at 4°C for 5 min.

3. The aqueous phase from each sample is transferred to a fresh microfuge tube and placed in a Techne Dri-block for 10 min at 65°C to inactivate endogenous deacetylase activity that might interfere with the assay.

4. Denatured and insoluble protein is again removed by centrifugation in an microfuge for 10 min at 14,000g and 4°C.

5. Transfer the supernatant containing CAT activity to clean Eppendorf tubes (*see* **Note 4**).

6. At this point the protein concentration of each sample can be determined using the Bio-Rad Protein Assay, as described by the manufacturer. In brief: Remove 5 μL of supernatant from each tube and add 795 μL distilled water followed by 200 μL Bio-Rad Protein Assay reagent. Vortex each tube for 10 s and allow the tubes to stand at room temperature for at least 5 min. The optical density of each lysate can be measured using a spectrophotometer set at 595 nm. When the optical densities have been measured, the protein concentration of each lysate can be determined and normalized for each sample to be assayed (*see* **Note 5**).

7. Prepare a master mix containing 14 μL acetyl-Co A (3.5 μg/μL), 10 μL distilled water, and 3 μL [^{14}C]-labeled chloramphenicol (111 mCi/mmol) for each reaction.

8. Equilibrate the lysate concentrations with lysis buffer; the final volume should be 100 μL for each reaction.

9. Add 27 μL of the master mix to each tube, mixing the contents by vortex. Transfer the tubes to a 37°C water bath and incubate from 1 to 24 h according to the linear range of the assay.

10. Add 550 μL of ethyl acetate to each reaction to extract [^{14}C]-labeled chloramphenicol and its acetylated product. Ethyl acetate effectively separates the organic components of the reaction from the aqueous fraction.

11. Vortex each tube for 10 s to mix the reaction with ethyl acetate.

12. Centrifuge the reactions for 10 min at 14,000g.

13. With a Gilson pipet, remove the ethyl acetate from each tube and transfer the solvent to a clean Eppendorf tube.

14. Allow the ethyl acetate to evaporate in a Speedyvac centrifuge set to a moderate heat. Evaporation should take 1–1.5 h to allow the radioactive chloramphenicol compounds to crystallize.

15. Dissolve the crystallized pellets in 30 μL ethyl acetate. Vortex briefly and centrifuge for 2 min at 14,000g. Load the reactions onto a TLC plate (Whatman).

16. During evaporation of the ethyl acetate (*15*), 95:5 v/v chloroform and methanol should be mixed and added to a TLC tank. Two pieces of Whatman paper should be inserted and fastened to each side of the tank, the lid replaced, and sealed with parafilm. The tank should be allowed to equilibrate and a uniform layer of solvent should be adsorbed by the paper, which will prevent uneven migration of the chloramphenicol compounds. The TLC plate is placed into the equilibrated tank and allowed to develop.

17. After 50 min, remove the TLC plate from the tank and allow it to dry in a fume cupboard.

18. Cover the plate with Saran Wrap and quantify CAT activity by PhosphorImage analysis using the Bio-Rad Molecular Analyst program. The percentage of acetylated [^{14}C]-labeled chloramphenicol is calculated as follows: % acetylated = [counts in acetylated species/(counts in acetylated species + counts in nonacetylated chloramphenicol)] × 100 (*see* **Note 6**). The relative fold induction of each reaction is determined as a factor of the basal level of CAT activity calculated using the value obtained from a lysate prepared from unirradiated (control) cells.

19. Alternatively, the developed TLC plate can be exposed to X-ray film for autoradiography. An overnight exposure is usually sufficient and the signal from the radiolabeled spots can be quantified using densitometry.

3.6. β-Galactosidase Assay

This assay is adapted from a method employed to determine β-galactosidase activity in transgenic flies *(37)*.

1. The 35-mm tissue culture dishes contain confluent cell monolayers expressing β-galactosidase from the HIV-1 LTR and are treated with inducing agents or untreated control cells.
2. Wash cells with PBS.
3. Add 100–200 µL lysis buffer to cover the cells.
4. Scrape cells from the plate using a rubber policeman or cell scraper.
5. Transfer cell lysate to an Eppendorf tube and centrifuge at 14,000g for 5 min at 4°C to remove insoluble material. Recover supernatant and transfer to a clean Eppendorf tube.
6. Remove 5µl of supernatant from each tube and perform a Bio-Rad Protein Assay as described by the manufacturer (*see* **Subheading 3.5., step 6**). Normalize the protein concentration of the samples by addition of lysis buffer.
7. Transfer 50 µL of each extract to a clean test tube and add 950 µL of assay buffer (*see* **Note 7**).
8. Incubate the samples at 37°C for 1–4 h for color to develop.
9. Determine absorbance of each sample at 574 nm against a β-galactosidase negative control. Activity can be calculated as optical density (OD) U/h/µg cell extract.

3.7. Histological Staining for β-Galactosidase

The following protocol adapted from Glaser et al. *(38)* describes the histological staining of confluent cell monolayers plated into 35-mm dishes or grown on cover slips, and have been treated with a stress-inducing agent (such as UV irradiation) 4–20 h previously. The protocol is also useful for monitoring infection of LTR-β-galactosidase expressing cells with HIV-1 and the procedure can be adapted for staining of histological sections (*see* **Note 8**).

1. Remove medium from cells and wash cell monolayer with PBS.
2. Add 1 mL of fixer to each dish and incubate at 4°C for 20 min.
3. Remove fixing solution and add 1 mL standard wash and incubate at 4°C for 10 min; repeat this process twice more.

4. Aspirate off standard wash and add 1 mL detergent wash, incubate for 30 min at 4°C, and repeat detergent wash twice more.
5. Remove detergent wash and add 1 mL stain solution incubate in the dark at 37°C for 6 h (*see* **Note 9**).
6. Staining can be stopped by removing the stain solution and washing with 1 mL PBS.
7. Remove the PBS and replace with 1 mL 70% glycerol for long-term storage. The stained cells should be stored at 4°C in the dark.
8. Cells expressing β-galactosidase from the HIV-1-LTR can be examined by light microscopy (X40).

4. Notes

1. X-gal stock solution is prepared at a concentration of 25 mg/mL in dimethyl formamide (DMF) in a glass container (DMF will damage some plastics). The stock solution should be stored at –20°C in the dark. X-gal is added to the stain solution immediately before use.
2. The concentration of G418 required for selection of stable cell lines varies with the cells used and the cell density; for example, resistant HeLa cells can be selected using 400 μg/mL G418, but Jurkat T-cells or THP-1 monocytic cells require G418 concentrations of 800 or 1000 μg/mL, respectively. The optimum concentration for each cell type should be determined empirically. Following isolation of stable cell lines, G418 may be omitted from the medium or retained at a low level of 200 μg/mL for maintenance of the selected cell populations without loss of expression from the integrated LTR-reporter construct.
3. Drugs such as SB 203580 and *N*'-acetyl-cysteine, both of which affect the levels of viral transcription, can be introduced to the medium 1 h before an inducing stimulus.
4. The cell extracts may be frozen at this stage for analysis at a later date, but avoid repeated freeze–thaw cycles, as this may decrease the stability of the CAT enzyme.
5. A minimum of 10 μg of protein should be taken from each lysate to ensure that the CAT assay works effectively. The suggested amounts of cell lysate work well for stable cell lines carrying an integrated LTR-reporter gene, however, increased volumes of lysate may be required for transiently transfected cell populations.
6. Samples in which more than 40% of the [^{14}C]-chloramphenicol has been converted to the acetylated form should be considered to be outwith the linear range of the assay. These samples should be diluted with buffer and reassayed.
7. The assay is time dependent and consequently the reaction should be started by adding substrate to consecutive samples at a constant time interval to allow accurate reading of the sample activity.
8. Histological sections are prepared from frozen tissues which are mounted in two drops of Gurr gum on a cork mounting disk and frozen using a cryostat spray. The gum-embedded tissues are cut by a freezing cryostat (5–8 μ) and mounted onto slides (coated in poly-L-lysine); the slides are left to dry for 30 min at room temperature before fixing. Multiple slides are placed into a slide rack that is immersed into the solutions for the histological staining protocol.
9. The cells may be stained overnight to detect low level expression.

References

1. Schreck, R. P., Rieber, P., and Baeuerle, P. (1991) Reactive oxygen intermediates as apparently widely used messengers in the activation of the NF-κB transcription factor of HIV-1. *EMBO J.* **10,** 2247–2258.
2. Westendorp, M. O., Shatrov, V. A., Schulze-Osthoff, K., Frank, R., Kraft, M., Los, M., Krammer, P. H., Dröge, W., and Lehmann, V. (1995) HIV-1 Tat potentiates TNF-induced NF-κB activation and cytotoxicity by altering the cellular redox state. *EMBO J.* **14,** 546–554.
3. Valerie, K., Delers, A., Bruck, C., Thiriart, C., Rosenberg, H., Debouck, C., and Rosenberg, M. (1988) Activation of human immunodeficiency virus type 1 by DNA damage in human cells. *Nature* **333,** 78–81.
4. Stein, B., Krämer, M., Rahmsdorf, H. J., Ponta, H., and Herrlich, P. (1989) UV-induced transcription from the human immunodeficiency virus type 1 (HIV-1) long terminal repeat and UV-induced secretion of an extracellular factor that induces HIV-1 transcription in nonirradiated Cells. *J. Virol.* **63,** 4540–4544.
5. Kumar, S., Orsini, M. J., Lee, J. C., McDonnell, P. C., Debouck, C., and Young, P. R. (1996) Activation of the HIV-1 Long Terminal Repeat by Cytokines and Environmental Stress Requires an Active CSBP/p38 MAP Kinase. *J. Biol. Chem.* **271,** 30,864–30,869.
6. Nabel, G. and Baltimore, D. (1987) An inducible transcription factor activates expression of human immunodeficiency virus in T cells. *Nature* **326,** 711–713.
7. Rabbi, M. F., al-Harthi, L., Saifuddin, M., and Roebuck, K. A. (1998) The cAMP protein kinase A and protein kinase C-beta pathways synergistically interact to activate HIV-1 transcription in latently infected cells of monocyte/macrophage lineage. *Virology* **245,** 257–269.
8. Berkhout, B. and Jeang, K-T. (1992) Functional roles for the TATA promoter and enhancers in basal and Tat-induced expression of the human immunodeficiency virus type 1 long terminal repeat. *J. Virol.* **66,** 139–149.
9. Garcia, J. A., Wu, F. K., Mitsuyasu, R., and Gaynor, R. B. (1987) Interactions of cellular proteins involved in the transcriptional regulation of the human immunodeficiency virus. *EMBO J.* **6,** 3761–3770.
10. Garcia, J. A., Harrich, D., Soultanakis, E., Wu, F., Mitsuyasu, R., and Gaynor, R. B. (1989) Human immunodeficiency virus type 1 LTR TATA and TAR region sequences required for transcriptional regulation. *EMBO J.* **8,** 765–778.
11. Kato, H., Horikoshi, M., and Roeder, R. G. (1991) Repression of HIV-1 transcription by a cellular protein. *Science* **251,** 1476–1479.
12. Yoon, J-B., Li, G., and Roeder, R. G. (1994) Characterization of a family of related cellular transcription factors which can modulate human immunodeficiency virus type 1 transcription *in vitro. Mol. Cell. Biol.* **14,** 1776–1785.
13. Romerio, F., Gabriel, M. N., and Margolis, D. M. (1997) Repression of human immunodeficiency virus type1 through the novel co-operation of human factors YY1 and LSF. *J. Virol.* **71,** 9375–9382.
14. Du, H., Roy, A. L., and Roeder, R. G. (1993) Human transcription factor USF stimulates transcription through the initiator elements of the HIV-1 and the Ad-ML promoters. *EMBO J.* **12,** 501–511.

15. Sheridan, P. L., Mayall, T. P., Verdin, E., and Jones, K. A. (1997) Histone acetyltransferases regulate HIV-1 enhancer activity *in vitro. Genes Dev.* **11,** 3327–3340.
16. Henderson, A. J., Zou, X., and Calame, K. L. (1995) C/EBP proteins activate transcription from the human immunodeficiency virus type1 long terminal repeat in macrophages/monocytes. *J. Virol.* **69,** 5337–5344.
17. Verma, I. M., Stevenson, J. K., Schwarz, E. M., Van Antwerp, D., and Miyamoto, S. (1995) Rel/NF-kB/IkB family: intimate tales of association and dissociation. *Genes Dev.* **9,** 2723–2735.
18. Baeuerle, P. A. (1998) IκB-NF-κB structures: at the interface of inflammation control. *Cell* **95,** 729-731.
19. Zhang, L., Huang, Y., Yuan, H., Chen, B. K., Ip, J., and Ho, D. D. (1996) Identification of replication-competent, pathogenic human immunodeficiency virus type 1 with a duplication in the Tcf-1 region but lacking NF-κB binding sites. *J. Virol.* **71,** 1651–1656.
20. Chen, B. K., Feinberg, M. B., and Baltimore, D. (1997) The κB sites in the human immunodeficiency virus type 1 long terminal repeat enhance virus replication yet are not absolutely required for viral growth. *J. Virol.* **71,** 5495–5504.
21. Sadaie, M. R., Tschachler, E., Valerie, K., Rosenberg, M., Felber, B. K., Pavlakis, G. N., Klotman, M. E., and Wong-Staal, F. (1990) Activation of *tat*-defective human immunodeficiency virus by ultraviolet light. *New Biologist* **2,** 479–486.
22. Zider, A., Mashhour, B., Fergelot, P., Grimber, G., Vernet, M., Hazan, U, Couton, D., Briand, P., and Cavard, C. (1993) Dispensable role of the NF-κB sites in the UV-induction of the HIV-1 LTR in transgenic mice. *Nucleic Acids Res.* **21,** 79–86.
23. Valerie, K., Singhal, A., Kirkham, J. C., Laster, W. S., and Rosenberg, M. (1995) Activation of human immunodeficiency virus gene expression by ultraviolet light in stably transfected human cells does not require the enhancer elements. *Biochemistry* **34,** 15,760–15,767.
24. Valerie, K. and Rosenberg, M. (1990) Chromatin structure implicated in activation of HIV-1 gene expression by ultraviolet light. *New Biologist* **2,** 712–718.
25. Van Lint, C., Emiliani, S., Ott, M., and Verdin, E. (1996) Transcriptional activation and chromatin remodelling of the HIV-1 promoter in response to histone acetylation. *EMBO J.* **15,** 1112–1120.
26. El kharroubi, A., Piras, G., Zensen, R., and Martin, M. A. (1998) Transcriptional activation of the integrated chromatin-associated human immunodeficiency virus type 1 promoter. *Mol. Cell. Biol.* **18,** 2535–2544.
27. Cullen, B. R. (1990) The HIV-1 Tat protein: an RNA sequence-specific processivity factor. *Cell* **63,** 655–657.
28. Jones, K. A. (1997) Taking a new TAK on Tat transactivation. *Genes Dev.* **11,** 2593–2599.
29. Westendorp, M. O., Rainer, F., Ochsenbauer, C., Stricker, K., Dhein, J., Walczak, H., Debatin, K-M., and Krammer, P. H. (1995) Sensitisation of T cells to CD95-mediated apoptosis by HIV-1 Tat and gp120. *Nature* **375,** 497–500.

30. Li, C. J., Friedman, D. J., Wang, C., Metelev, V., and Pardee, A. B. (1995) Induction of apoptosis in uninfected lymphocytes by HIV-1 tat protein. *Science* **268**, 429–431.

31. White, M. R., Masuko, M., Amet, L., Elliot, G., Braddock, M., Kingsman, A. J., and Kingsman, S. M. (1995) Real-time analysis of the transcriptional regulation of HIV and hCMV promoters in single mammalian cells. *J. Cell Science* **108**, 441–455.

32. Recio, J. A. and Aranda, A. (1997) Activation of the HIV-1 long terminal repeat by nerve growth factor. *J. Biol. Chem.* **272**, 26,807–26,810.

33. Dorsky, D. I., Wells, M., and Harringtonm R. D. (1996) Detection of HIV-1 infection with a green fluorescent protein reporter system. *J. Acquired Immune Defici. Syndr. Human Retrovirol.* **13**, 308–313.

34. Gervaix, A., West, D., Leoni, L. M., Richman, D. D., Wong-Staal, F., and Corbeil, J. (1997) A new reporter cell line to monitor HIV infection and drug susceptibility *in vitro. Proc. Natl. Acad. Sci. USA* **94**, 4653–4658.

35. Graham, F. L. and van der Eb, A. J. (1973) A new technique for the assay of infectivity of human adenovirus 5 DNA. *Virology* **52**, 456.

36. Gorman, C. M., Moffat, L. F., and Howard, B. H. (1982) Recombinant genomes which express chloramphenicol acetyltransferase in mammalian cells. *Mol. Cell. Biol.* **2**, 1044–1051.

37. Simon, J. A. and Lis J. T. (1987) A germline transformation analysis reveals flexibility in the organization of heat shock consensus elements. *Nucleic Acids Res.* **7**, 2971–2988.

38. Glaser, R. L., Wolfner, M. F., and Lis, J. T. (1986) Spatial and temporal pattern of hsp26 expression during normal development. *EMBO J.* **4**, 747–754.

20

SAGE

The Serial Analysis of Gene Expression

Jill Powell

1. Introduction

The characteristics of an organism or tissue are determined by the genes expressed within it. The determination of the genomic sequence of higher organisms, including humans, is now a real and attainable goal, as seen by the progress of the Human Genome Mapping Project. This information is needed in order to define the underlying genetic changes that give rise to a variety of normal, developmental, and disease states.

Methods to study expression changes in one specific gene at a time have traditionally used Northern blot and mRNA dot-blot analyses to determine mRNA abundance relative to a control transcript. It is important, however, to study how genes work together as a complete system rather than in isolation. Technologies that yield a comprehensive analysis of gene expression within any cell type would provide important tools for elucidating complex, global patterns of gene expression. Examples of such methods include differential display (*see* Chapter 22) and subtractive hybridization. Both are limited to identifying genes expressed at relatively high abundance and/or showing large expression differences between samples. In addition, the number of transcripts that can be analyzed is limited *(1,2)*. Microarray technologies have recently been described *(3)*. These may eventually provide comprehensive transcript profiles but are currently restricted to known genes and rely on the development and use of specialised and costly equipment.

The serial analysis of gene expression (SAGE), originally described by Velculescu et al. *(4)*, is a technique that facilitates the analysis of global gene expression. Furthermore, SAGE results in the generation of a quantitative tran-

From: *Methods in Molecular Biology, vol. 99: Stress Response: Methods and Protocols*
Edited by: S. M. Keyse © Humana Press Inc., Totowa, NJ

script profile, an ability lacking in other analysis technologies. This quantitative and simultaneous analysis of a large number of transcripts generates expression profiles of mRNA samples for direct comparison.

SAGE generates short sequence tags of 9–10 base pairs (bp) which are positionally located within the cDNA molecule from which they are derived. This allows specific detection of that cDNA from a large number of different transcripts. The tags are generated as dimers, or ditags, and are ligated together end to end forming concatamers which are then cloned. Sequencing of these clones allows over 30 individual tags to be read from each lane of an automated sequencing gel. The abundance of a particular tag detected after sequencing many clones relates directly to the expression level of the gene from which it is derived. This serial analysis of many thousands of gene specific tags allows the accumulation of information from genes expressed in the tissue of interest and gives rise to an expression profile for that tissue.

The choice of starting material is an important consideration upon embarking on a SAGE analysis. SAGE can be used to determine differences in gene expression in different tissue states (e.g., normal vs tumor) in a single individual, even where the availability of material is limited. However, in order to directly compare two transcript profiles, the source RNA has to be matched in order to take into account variations, for example, between individual tissue samples. Because of these considerations cell lines provide an ideal source of material for SAGE analysis. For example, a comparison of identical cell lines with and without the expression of a transfected gene provides a common background upon which to analyze the changes in gene expression which result from expression of that protein. As SAGE technology improves, it is hoped that even smaller amounts of starting material will be used to make libraries, therefore making the technique available for comparisons between tissues of very limited availability. Published uses of SAGE take into account such considerations and demonstrate the value of each in the construction of gene-expression patterns and, importantly, the identification of previously uncharacterized transcripts.

There is a huge range of SAGE applications. These include comprehensive global expression analyses such as the yeast expression map [transcriptome *(5)*]. Because of the availability of positional information for the yeast genome, the integration of this information with the transcriptome was possible and allowed the generation of chromosomal expression maps identifying physical regions of transcriptional activity. In humans, Zhang et al. *(6)* performed detailed expression analyses between carefully matched tissues of different disease states in order to study gene expression in gastrointestinal tumors. These results showed extensive similarity between the samples chosen, but more than 500 transcripts were expressed at significantly different levels, underlining the differences between normal and neoplastic cells and leading to the identification of genes for use as diagnostic or prognostic markers.

Cell lines were used to study the effect on gene expression of p53 induction in rat *(7)* and humans *(8)*. In the latter case, many genes found to be markedly upregulated in response to p53 induction were predicted to code for proteins that could generate or respond to oxidative stress. Collectively, these gene-expression changes led to the proposal of a novel pathway through which p53 results in apoptosis. This latter application illustrates the utility of the SAGE technique in the analysis of stress-induced changes in gene expression. In this regard, SAGE is one of the most powerful technologies currently available for the molecular analysis of the mammalian stress response.

1.1. Preparation of SAGE Libraries

The SAGE protocol is outlined in **Fig. 1**. The starting material consists of double-stranded cDNA constructed from an mRNA sample from the tissue of interest using a biotinylated oligo-dT primer. The cDNA is cleaved using a frequent-cutting (four base-pair recognition sequence) enzyme termed the "Anchoring enzyme," commonly *Nla*III. The cDNA is then bound to streptavidin beads, thus isolating the 3' ends of each transcript. This ensures that tags will be generated from a defined position in the cDNA transcript; this being the site of the anchoring enzyme restriction site closest to the 3' end.

The bead-bound cDNA is then split into two fractions and each is ligated to one of two linkers each containing polymerase chain reaction (PCR) primers and the five base-pair recognition site for a type IIS restriction enzyme, the "Tagging enzyme," commonly *Bsm*FI. Type IIS restriction enzymes recognize their target sequences, but then cleave up to 20 bp 3' from this site *(9,10)*. Each fraction is digested with the tagging enzyme and blunt-ended. As described in **ref. 4**, this results in *Bsm*FI cleaving a minimum of 12 bp 3' of its recognition sequence. This 12 bp consists of 3 bp of the anchoring enzyme recognition site and 9 bp that are specific to the cDNA; this 9 bp is referred to as the SAGE tag. The two resulting blunt-ended pools are then ligated to one another to make "ditags" and amplified using primer sequences in the attached linkers. As the ditags are generated and ligated prior to amplification any bias from PCR can be eliminated by discarding ditags that occur more than once. This is because any two tags, even from the most abundant species, would not be expected to come together and form the same ditag more than once.

The PCR primers and linkers are then removed by digestion with the anchoring enzyme and the resulting ditags purified, concatenated, and cloned. The sequence of one such clone can, therefore, result in the identification of 30–40 SAGE tags. The anchoring enzyme recognition site separates each ditag and serves as a punctuation mark to orient and define each ditag.

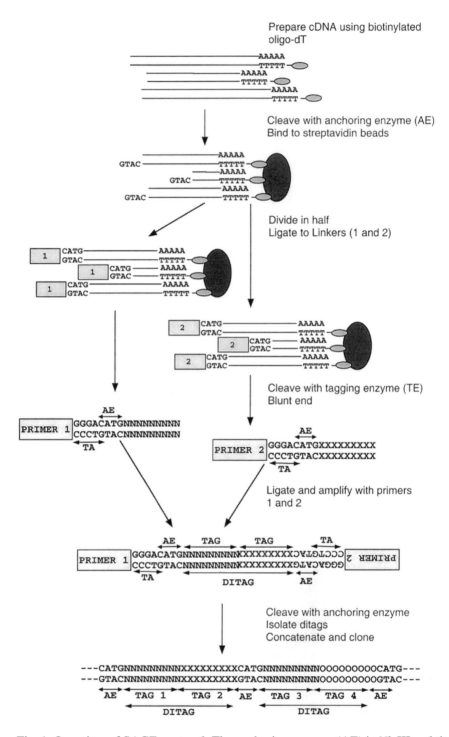

Fig. 1. Overview of SAGE protocol. The anchoring enzyme (AE) is *Nla*III and the tagging enzyme (TE) is *Bsm*FI.

This results in the generation of many thousands of SAGE clones that represent a SAGE library for the tissue of origin. The gene expression information is contained in the abundance of a particular tag; in order to reveal this information, the SAGE library has to be sequenced and the resulting data analysed.

The amount of sequencing required for a SAGE project is dependent on the purpose for which it is being used. In order to analyze the complete set of genes expressed from the yeast genome, or transcriptome, Velculescu et al. *(5)* generated over 60,000 SAGE tags. Their results showed that the number of unique genes identified for this organism reached a plateau at approximately 60,000 tags. The generation of further SAGE tags would, therefore, yield few additional unique genes expressed in the transcriptome. This is a very powerful study, resulting in the identification of nearly all of the estimated 6000 yeast genes. SAGE analyses of rat embryo fibroblast cells *(7)* determined that at 60,000 total tags analyzed, 15,000 genes were represented and that new genes were still being added as more tags were analyzed. This is in line with the estimated 10,000 to 50,000 expressed genes in a given mammalian cell population. The study conducted by Zhang et al. *(6)* achieved a detailed analysis of human gene expression differences between normal and cancer cells by analyzing over 300,000 tags giving expression level information on at least 45,000 different genes.

These are very large studies aimed at gathering information at all levels of expression. For quicker, smaller-scale studies, SAGE provides a means of identifying expression changes in the most abundantly expressed transcripts. This requires the analysis of approximately 10,000–20,000 tags.

1.2. Analysis of SAGE Tags

The clones are amplified by PCR and size selected to maximize the tag yield per lane on an automated sequencer, 600–800 bp yielding 30–40 tags per clone. The results of automated sequencing are then transferred directly to the SAGE software and used to build project files for each SAGE library constructed. The software allows input of SAGE library parameters such as tagging and anchoring enzymes, the tag and ditag lengths and automatically reads the anchoring enzyme punctuation between each tag. The software can then analyze the collection of tags in various ways. These include the determination of the most abundant transcripts and direct comparison of SAGE libraries. This allows the determination of the fold differences in specific tags between different libraries in order to reveal differences in expression patterns.

The reproducibility of the method can be assessed as the accumulation of the tags progresses by comparing the transcript levels for genes encoding ribosomal proteins and housekeeping genes expected to be expressed at comparable levels between samples. Internal controls can also be incorporated into the experimental design by assessing the expression level of a particular gene by other means and comparing this with the SAGE results.

Finally, the software allows the tag sequences to be compared to sequences in Genbank (taking into account the position and orientation of the 9-bp tags) in order to identify matches corresponding to known genes or expressed sequence tags (EST) sequences. Where tags do not find a match, such uncharacterized transcripts can be further analyzed in order to "rescue" cDNA clones with more sequence information using the tag plus a tagging enzyme site (13–15 bp oligo as described in **ref. *4***). Changes in specific genes can then be confirmed by more conventional means such as Northern blots.

The criteria for selection of candidate differentially expressed genes will depend on the individual project. One can imagine that exposing cells to highly toxic chemicals and drugs might induce gene expression differences of 10-to 50-fold over that normally seen. In contrast, a study of gene-expression under normal physiological conditions may not reveal such dramatic changes and this has to be taken into consideration when selecting candidates. The SAGE software has facilities for determining the statistical significance of an observed expression difference; these are described later.

2. Materials

2.1. General Reagents

1. 10 M Ammonium acetate.
2. 20 µg/µL Glycogen.
3. LoTE: 3 mM Tris-HCl, pH 7.5, 0.2 mM EDTA, pH 7.5, in dH$_2$O, store at 4°C.
4. PC8 (*Caution:* care must be taken when using phenol, as it is caustic and poisonous): 480 mL phenol (warm to 65°C), 320 mL of 0.5 M Tris-HCl, pH 8.0, 640 mL chloroform.
5. Binding and washing buffer: 10 mM Tris-HCl (pH 7.5), 1 mM EDTA, 2.0 M NaCl, store at room temperature.
6. Ethanol.
7. 20% Novex (San Diego, CA) gels (cat. no. EC6315).
8. Apparatus for running polyacrylamide and agarose gels.

2.2. Preparation of Starting Materials: RNA and cDNA (see Note 1)

1. Total RNA: ToTALLY RNA—Total RNA Isolation Kit (Ambion, Austin, TX, , cat. no. 1910).
2. mRNA: MicroPoly (A) Pure—mRNA Isolation Kit (Ambion, cat. no. 1918).
3. cDNA synthesis system: Gibco-BRL (Gaithersburg, MD) (cat. no. 18267-013).
4. Biotinylated oligo-dT: 5' [biotin]T$_{18}$ (*see* **Note 2**).
5. Streptavidin (Sigma, St. Louis, MO, cat. no. S-4762).

2.3. Preparation of SAGE Libraries

1. Anchoring enzyme used here: *Nla*III (10 U/μL, plus buffer 4 and 100X BSA (NEB, cat. no. 125S) (*see* **Note 3**).
2. Dynabead M-280 streptavidin slurry (Dynal, cat. no. 112.05, 10 mg/mL).
3. Dynal MPC–E: Magnetic particle concentrator for microtubes of Eppendorf type, 1.5 mL (Dynal, cat. no. 120.04).
4. PCR primers for amplification of ditags (*see* **Note 4**):
   Primer 1: 5' GGA TTT GCT GGT GCA GTA CA 3'
   Primer 2: 5' CTG CTC GAA TTC AAG CTT CT 3'
5. 40% Polyacrylamide (19:1 acrylamide:bis, Bio-Rad, Hercules, CA, cat. no. 161-0144).
6. 50X Tris Acetate buffer (Quality Biological, cat. no. 330-008-161).
7. 10% Ammonium persulfate (Sigma, cat. no. A-9164).
8. TEMED (Amresco, cat. no. 0761-50Ml).
9. DNA size markers: 20-bp ladder, 1-kb ladder.
10. Spin X microcentrifuge tubes (Costar, Cambridge, MA, cat. no. 8160).

2.4. Cloning Concatamers and Sequencing

1. pZero cloning kit (Invitrogen, San Diego, CA, cat. no. K2500-01).
2. *Sph*I restriction enzyme (NEB, cat. no. 182S).
3. Purification of PCR reactions: Microcon 100 columns (Amicon,, Danvers, MA, cat. no. 42413), QIAquick 8 PCR purification kits (Qiagen, Chatsworth, MD, cat. no. 28142/28144).

2.5. Analysis of SAGE Data

This is carried out using specialist SAGE software, freely available to academic users; details are available on the SAGE web page (*see* **Note 5**).

3. Methods
3.1. General Methods
3.1.1. Preparation of 12% PAGE Gels

40% Polyacrylamide (19:1 acrylamide:bis)	10.5 mL
dH$_2$O	23.5 mL
50X Tris acetate buffer	700 μL
10% Ammonium persulfate (APS)	350 μL
TEMED	30 μL

Mix and add to vertical gel apparatus. Insert the comb and leave for at least 30 min at room temperature to polymerize.

3.1.2. Preparation of PC8

1. Warm 480 mL phenol to 65°C in a waterbath then add, in the following order, 320 mL 0.5 *M* Tris-HCl (pH 8.0), and 640 mL chloroform.

2. Shake gently to mix and store at 4°C.
3. After 2–3 h mix again.
4. Finally, after another 2–3 h, aspirate off the aqueous layer, aliquot, and store at –20°C.

3.1.3. PC8 Extraction

1. Add an equal volume of PC8 to sample and mix by vortexing for several seconds.
2. Spin for 2 min at full speed in microcentrifuge.
3. Transfer aqueous (top) layer to a new microcentrifuge tube.

3.1.4. Ethidium Bromide Dot Quantitation

1. Use any solution of pure DNA to prepare the following standards: 0 ng/µL, 1 ng/µL, 2.5 ng/µL, 5 ng/µL, 7.5 ng/µL, 10 ng/µL, 20 ng/µL.
2. Use 1 µL of sample DNA to make 1/5, 1/25, and 1/125 dilutions in LoTE.
3. Add 4 µL of each standard or 4 µL of each diluted sample to 4 µL 1 µg/mL ethidium bromide and mix well.
4. Place a sheet of plastic wrap on a UV transilluminator and spot each 8 µL sample onto the plastic wrap.
5. Photograph under UV light and estimate the DNA concentration by comparing the intensity of the sample to the standards.

3.2. Preparation of Starting Materials: RNA and cDNA

3.2.1. Testing Biotinylation of Biotin-Oligo-dT (see **Note 6**)

1. Add several hundred nanograms of biotin-oligo-dT to 1 µg streptavidin.
2. Incubate for 3–5 min at room temperature.
3. Run out the sample on a 20% Novex gel using unbound oligo as a standard. If the oligo is well biotinylated, the entire sample containing streptavidin should be shifted to a higher molecular weight when compared with the unbound material.

Alternatively, increasing amounts of oligo (from several hundred nanograms to several micrograms) can be incubated with and without separate aliquots of 100 µL of Dynal streptavidin beads. After 15 min, the beads are separated from the supernatant using a magnet. The supernatant is removed, and DNA quantitation is determined using a spectrophotometer to read ultraviolet light absorbance at 260nm (OD_{260}). At low amounts of oligo, when bead binding capacity is not saturated, the ratio of unbound oligo to the total oligo will indicate the percentage of oligo that is not biotinylated.

3.2.2. Preparation of Total and mRNA

Prepare as per manufacturer's instructions or according to preferred method; 5–10µg of mRNA will be needed for cDNA synthesis.

3.2.3. Preparation of cDNA

Prepare as per manufacturer's instructions or according to preferred method using typically 2.5 µg biotinylated oligo-dT.

3.3. Preparation of SAGE Libraries

In order to comply with the conditions imposed by the licenser of SAGE technology, Genzyme Molecular Oncology (One Mountain Road, P.O. Box 9322, Framingham, MA 01701-9322), the precise experimental details required to perform SAGE cannot be published here. However, this information is freely available to academic users and can be obtained on the Internet through the SAGE World Wide Web page (*see* Note 5). In this way the terms of the license are adhered to and users receive the very latest protocols, as the web page is constantly updated.

An outline of the steps of the Detailed Protocol (as depicted in **Fig. 1**) is now given, together with a modification that significantly enhances the efficiency of SAGE *(11)*.

1. Cleave biotinylated cDNA with an anchoring enzyme—*Nla*III is shown.
2. Bind biotinylated cDNA to streptavidin-coated magnetic beads.
3. Ligate linkers to bead-bound cDNA.

A variety of linkers can be used at this point in SAGE. Linkers must contain the appropriate anchoring enzyme overhang, a restriction site for a type IIS enzyme (tagging enzyme), and a priming site for PCR amplification. Linkers must be kinased and annealed to their complementary linker before use (this can be performed chemically at the time of oligo synthesis or enzymatically as described in the Detailed Protocol).

4. Release cDNA tags using the tagging enzyme—*Bsm*FI is shown.
5. Blunt end released cDNA tags.
6. Ligate tags to form ditags.
7. PCR amplification of ditags.

Amplify ditags using primers 1 and 2. PCR products are analyzed on a 12% polyacrylamide gel. Amplified ditags should be 102 bp. A background band of equal or lower intensity occurs around 80 bp (linker–linker artefacts). All other background bands should be of substantially lower intensity (*see* Note 7). **Figure 2** shows results from a typical ditag PCR experiment. After PCR conditions have been optimized, large-scale PCR (100–200, 50-µL reactions) can be performed.

3.3.1. Isolation of Ditags

The latest version (1.0c) of the Detailed Protocol includes a gel purification of the 102 bp ditag band away from contaminating linker–linker dimers in order to address the following problem. Restriction enzyme digestion with the tagging enzyme yields 26-bp ditags and 40-bp linkers (derived from those that flank the ditags in the 102-bp PCR product and those from the 80-bp linker–linker artifact molecules.) Both these molecules have the same sticky ends and

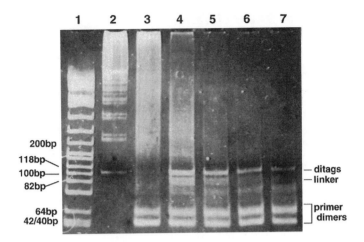

Fig. 2. Results of ditag PCR amplification. 12% polyacrylamide gel showing amplified ditags at 102bp, linker dimer background band at 80 bp, and primer dimers. Lanes: 1—FX174/*Hin*FI marker; 2—100-bp ladder marker; 3—Negative control (no 102-bp band after 35 cycles); 4—1/10 dilution of ligation reaction; 5—1/50 dilution; 6—1/100 dilution; 7—1/200 dilution (102-bp band still visible).

linker molecules will ligate to ditags (**Fig. 3A**). In order to overcome this problem, the ditags must be purified away from the linkers before ligation to form concatamers. If this is not done, the linkers, which do not have the correct ends for ligation into the cloning vector, will swamp the ditags and produce short concatamers and molecules that will not clone. These effectively poison the concatamer ligation reaction.

I have recently developed an enhanced ditag purification method to address this problem (**Fig. 3B**) and this has been compared with the method in the Detailed Protocol provided by Genzyme *(11)*. This enhanced method is described as follows. For this modification, PCR primers 1 and 2 must have been synthesized with a 5' biotinylated base. The principle of the modification is that linkers are generated as biotinylated molecules during the bulk PCR reaction by the use of these biotinylated primers. The linkers and ditags are then digested apart, using the anchoring enzyme, this leaves biotinylated linkers and nonbiotinylated ditags. The unwanted linkers can then be removed by binding to streptavidin beads. Hence, no attempt to remove linkers from the ditags is necessary until after isolation of the 24- to 26-bp ditags. In avoiding extra gel purifications, the ditag yield is maximized, so that after removal of the linkers, a high yield of pure ditags is recovered. The use of this modification results in the rapid generation of high ditag yields and clones with large average insert sizes (**Table 1**).

A

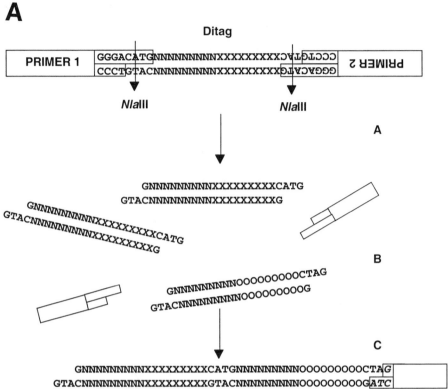

Fig. 3A. Contaminating linker problem **(A)** Restriction enzyme digestion with the tagging enzyme yields 26-bp ditags and 40-bp linkers (derived from those that flank the ditags in the 102-bp PCR product and those from the 80-bp linker-linker artefact molecules.) Both of these molecules have the same sticky ends. **(B)** If linkers are allowed to persist in the ditag sample, when concatenated, they do not have the correct ends for ligation into the cloning vector. **(C)** Thus the presence of linkers will produce short concatamers and molecules which will not clone. They effectively poison the concatamer ligation reaction.

1. Perform PCR of ditags as described in the Detailed Protocol but using biotinylated primers 1 and 2. Once conditions have been optimized, perform bulk PCR reactions. These must also use biotinylated PCR primers 1 and 2.
2. Pool PCR reactions into microfuge tubes (approx 450 µL each).
3. Extract with PC8 and transfer 330 µL of the aqueous supernatant to 1.5-mL microcentrifuge tubes.
4. Ethanol precipitate by adding the following to the 330 µL sample. 100 µL ammonium acetate, 3 µL glycogen, and 1000 µL 100% ethanol. Mix end over end several times and vortex.
5. Spin for 10 min at full speed in an Eppendorf microcentrifuge at room temperature, discard supernatant, and wash the pellet twice with 70% ethanol.

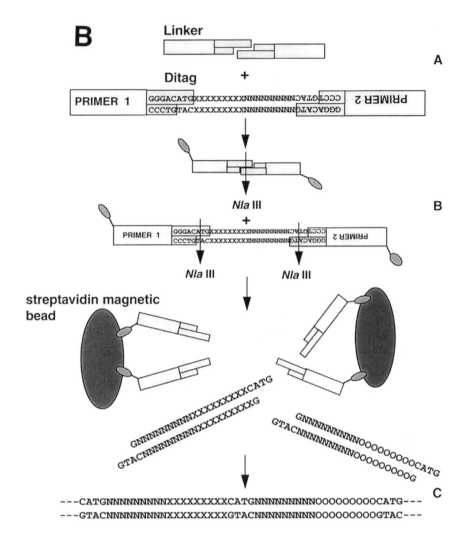

Fig. 3B. Diagram showing how contaminating linkers are removed by modification of the basic SAGE method *(11)*. (**A**) Shown here are the products resulting from Methods, 3.3 step 6. These are an unwanted 80-bp linker–linker artifact and the desired product, a 102-bp linker–ditag–linker molecule. (**B**) Bulk PCR reactions are performed using biotinylated primers 1 and 2. Subsequent *Nla* III digestion releases linkers and ditags as shown. If allowed to remain and participate in the following step, concatamer ligation, the linkers will ligate to ditag molecules, as they have compatible sticky ends derived from the anchoring enzyme. Any molecule to which a linker has ligated will not concatenate further and will not clone because of incompatible ends, the reaction is effectively poisoned. The unwanted biotinylated linkers are removed by binding to streptavidin magnetic beads. (**C**) The remaining ditags are free from contaminating linkers and can ligate to form long clonable concatamers.

6. Resuspend each pellet in LoTE, pool samples, and adjust volume to 320 μL with LoTE.
7. Divide the sample into four tubes (80 μL each).
8. Dot quantitate the DNA sample. The total amount of DNA should be between 60 and 100 μg.
9. Digest the PCR products with *Nla*III by adding the following to each sample tube:

   PCR products in LoTE   80 μL
   NEB buffer 4 (10X)   10 μL
   BSA (100X)   1 μL
   *Nla*III (10 U/μL)   10 μL

   Incubate for 1 h at 37°C.
10. Extract with an equal volume of PC8, spin and transfer 200 μL of the aqueous phase to a fresh Eppendorf tube.
11. Ethanol precipitate by adding the following to the 200-μL sample:

    10 *M* ammonium acetate   67 μL
    Glycogen   3 μL
    100% Ethanol   733 μL

    Vortex and place in a dry-ice/ethanol bath for 10 min.
12. Spin at maximum speed in a microcentrifuge at 4°C for 15 min.
13. Wash once with 75% ethanol, resuspend each tube in 50 μL LoTE, and pool samples (200 μL total).
14. Dot quantitate the sample. Total DNA at this stage should be between 60 and 100 μg.
15. Prepare two, 10-well 12% polyacrylamide gels.
16. Add 35 μL loading buffer to the above sample and divide among 18 wells (nine lanes per gel) leaving one lane on each gel for a 20-bp marker. Run at 160 V for 2 h.
17. Cut out the 24- to 26-bp band from all 18 lanes and place three cutout bands in each of 6 × 0.5-mL microcentrifuge tubes.
18. Pierce the bottom of each tube, place it in a 1.5-mL Eppendorf tube and spin at maximum speed for 2 min in an Eppendorf microcentrifuge. This breaks up the gel fragments by forcing them through the hole in the 0.5-mL tube.
19. Discard the 0.5-mL tubes and add 300 μL LoTE to the resulting gel slurry in each tube, vortex, and place at 37°C for 15 min.
20. Transfer the contents of each tube to one Spin X microcentrifuge tube and spin each Spin X column in a microcentrifuge for 5 min at full speed.
21. Transfer the eluate to a 1.5-mL tube and ethanol precipitate the sample as described in **step 11** using dry ice (use nine tubes, each containing 200-μL sample).
22. Wash twice with 75% ethanol and dry the pellets in a Speedvac. Resuspend each DNA sample in 10 μL LoTE (pool samples and make up to 100 μL total with LoTE).
23. Dot quantitate. Total DNA at this stage is usually 4–5μg, but this does contain contaminating linkers.

3.3.1.1. Removal of Contaminating Biotinylated Linkers

1. Add the ditag sample to 100 μL 2X binding and washing buffer and divide in half (in order not to exceed the binding capacity of the beads).
2. Add 100 μL streptavidin beads, which have been prewashed with 1X binding and washing buffer, to each sample.
3. Mix and leave the samples at room temperature for 15 min with intermittent gentle agitation to allow binding of the beads to the biotinylated linkers.
4. Immobilize the complex formed between the streptavidin beads and contaminating biotinylated linkers using a magnet and carefully remove the supernatant.
5. Wash the immobilized beads once with 1X binding and washing buffer and once with LoTE, reserving the supernatant after each wash.
6. Ethanol precipitate all three supernatants on dry ice as described in **Subheading 3.3.1., step 11**, combine the pellets, and resuspend in 7.5 μL LoTE. After dot quantitation (remove a 1-μL aliquot), the yield of pure ditags at this stage is 1–2 μg.

The purified ditag sample is then used in concatamer and clone formation resuming the Detailed Protocol as described in **Subheading 3.3.2.**

A direct comparison of the method described above and that described in the Detailed Protocol was carried out *(11)*. For each method, the starting material consisted of 50 bulk PCR reactions of 100 μL, generated using biotinylated PCR primers. Concatamers were generated and cloned. Several parameters were then determined for the SAGE libraries resulting from each method. The results are shown in **Table 1**.

For optimum performance of SAGE, maximum information is required from each clone sequenced in order to minimize the sequencing load per experiment. The yield of ditags generated is critical to the outcome, with several hundred nanograms of material being required for successful cloning of large concatenated inserts. **Table 1** shows that the method described here gave a greater yield of ditags. In addition, the average clone size was longer, and the number of tags per clone was 43% greater, which increases the efficiency of the SAGE method. These results indicate that the modification described here represents a rapid and effective means of improving the efficiency of the SAGE method. Further SAGE libraries have been constructed in order to assess the reproducibility of the method. The results from these libraries are also presented in **Table 1** and confirm the reproducibility of the modified method in improving average insert size.

3.3.2. Ligation of Ditags to Form Concatamers

The length of the ligation reaction depends on the quantity and purity of the ditags obtained. Several hundred nanograms of pure ditags without contaminating linkers require between 30 min to 2 h at 16°C. However, where less DNA has been obtained, or there is significant linker contamination, then

Table 1
Results of the Protocol Comparison

Experiment	Protocol	Ditag yield*a* (µg)	Average clone insert size*b* (bp)	Number of tags per clone*c*
1	Velculescu et al.	400	500	21
2	Powell	800	620	30
3	Powell	2000	670	34
4	Powell	1500	740	39
5	Powell	1100	700	36

Ditags were generated using each protocol as described. After concatamer formation and cloning, 100 clones with inserts were analyzed from the SAGE library resulting from each method. Experiment 1 describes results when SAGE was carried out using the Detailed Protocol. Experiment 2 shows the results obtained using the modification described here. Experiments 3–5 describe results from three further SAGE libraries constructed using the modification described here, showing that the technique gives a reproducible increase in clone size resulting from the effective removal of the contaminating linkers.

*a*Ditag yield describes the amount of ditags produced using each method, from a starting material of 50 identical bulk PCR reactions.

*b*The average clone insert size is an important parameter, as large clone inserts are essential to the efficiency of SAGE. A single automated sequencing run can yield 600–1000 bp of readable sequence. Insert sizes approaching this range are therefore desirable.

*c*A clone insert consists of 226 bp of vector plus concatenated ditags each of which is 26 bp. Each ditag represents two tags. Therefore: a clone of 616 bp equates to 30 SAGE tags—226 bp of vector sequence plus 15 × 26 bp ditags.

longer incubations will be required. After ligation, the entire concatenated sample is visualized using polyacrylamide gel electrophoresis. Concatamers will form a smear on the gel with a size range from about 100 bp to several thousand base pairs (**Fig. 4**). The region of interest is cut out from the gel, usually approx 600–2000 bp and subcloned into an appropriate vector in order to isolate clones for sequencing. Kenzelmann and Muhlemann describe an improved method of running these gels in order to maximize the cloning efficiency of SAGE *(12)*.

3.4. Cloning and Sequencing Concatamers

1. Concatamers can be cloned and sequenced in a vector of choice. An *Sph*I cleaved pZERO vector is currently used (*see* **Note 8**).
2. Transform the cells of choice by electroporation and plate 1/10 of transformed bacteria onto each of ten, 10-cm antibiotic-containing plates.
3. Save all 10 plates for each concatamer ligation reaction, because if the insert size appears appropriate, many clones will be required for sequencing.
4. Amplify the cloned inserts by PCR and check the insert sizes by agarose gel electrophoresis.

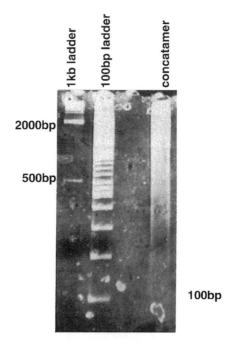

Fig. 4. Ditag concatamers. Eight percent polyacrylamide gel showing concatenated ditags ranging in size from 200–300 bp to >>2500 bp.

5. Purify the remainder of the PCR reactions that contain concatamers comprised of at least 15 ditags (>616 bp [226 bp vector + 26 bp per ditag × 15 ditags]) according to the sequencing method of choice (*see* **Note 9**).
6. Sequencing can be performed manually or on automated sequencers (*see* **Note 10**).

3.5. Analysis of SAGE Data

The SAGE software is used to create project files for the analysis of clone sequences after entering the enzymes used and the tag and ditag length choices. The sequences can be entered directly as ABI.seq files; the SAGE software then identifies the anchoring enzyme sites between ditags (**Fig. 5A**) and automatically records the tags and duplicate ditags found (**Fig. 5B**).

After the sequences of many clones have been entered the different project files of SAGE, tags can be analyzed to determine the expression profile of the starting tissue. **Figure 5C** shows a tag library sorted for tag abundance in one project file; in **Fig. 5D** two project files have been compared with each other to show differences in tag abundance between the two. In order to identify which genes the tags represent, a SAGE tag project is created from the sequences in Genbank in order to find a tag in the correct position in the transcript, depend-

A

37PBA5.Seq

```
GGAGNNNNNNNNTTNNAATCTCAGCTTGCTCAAGCTTGGTACCG
AGCTCGGATCCCTAGTAACGGCCGCCAGTGTGCTGGAATTCTGC
AGATATCCATCACACTGGCGGCCGCTCGAGCATGATCCGGCGCC
TTATTGATGGTACATGACATCATCGATGATGTACGGGGACATGT
CCCTATTAAGCTTTTTATTGAACATGTTCTCTATTGGGATGGCC
AGGCCATGAATGAAAAGGTTGAACTCCTGCATGTGATTTCACTT
CTGACCAGCACCATGCGCAAGCTGGTAGGCTTTGCCCACATGAA
GTAAACTGGTCCCTAACTGCATGAGGCAGACGGTGGAGCAGAAT
GGTCATGTAAAGCCTGTAGATGTACGGGGACATGTCCCCGTACA
TCTTGCCAGCATTCATGAAGAAGATAGAAATGTACGGGGACATG
TGTTCCCCAAATTGAGTTTAGGCATGCCCGACGTGCCACAAGGC
GGGCCATGTGCTGCCCTGTCTCGGGCACTGTCATGGCTTTTTAG
AATACCGTGGGTCATGGAGAGGAAGGGCATTTTTTCTACATGGT
TTAAGTTAACTTAATAGGGACATGGAGGTGGTGCGGCTCGTGTC
GTCCATGTGGCAACCTTTGTTCCGGACGGGCATGCCCATCGTCC
TTTGGAACACAAGCCATGCATCTAGAGGGCCCAATTCGCCTATA
GTGAGTCGTATTACAATTCACTTGGCCGNNNNN
```

Fig. 5A. Results output from SAGE analysis software. (A) SAGE clone sequence showing tagging enzyme site punctuation (in bold) between 24- and 26-bp ditags.

ing on which anchoring and tagging enzymes are used. **Fig. 5E** shows the results of comparing a SAGE tag project with a Genbank project and the genes or ESTs corresponding to the specific tag.

A variety of statistical methods are currently being used in order to determine whether an expression difference is significant given the other parameters of the SAGE tag project. The SAGE software incorporates facilities to perform significance calculations. This calculates a relative likelihood that a difference would be seen by chance for an individual tag. By performing a similar analysis on an entire project, the expected differences can be modeled and used to convert relative likelihoods to approximate absolute likelihoods. To minimize the number of assumptions and to take into account the large number of comparisons being made during a SAGE analysis, the software has facilities to perform Monte Carlo simulations in order to determine statistical significance. The null hypothesis is that the level, kind, and distribution of transcripts is the same for the two populations being compared. For each transcript, 100,000 simulations are performed to determine the relative likelihood because of chance alone of obtaining a difference in expression equal to or greater than the observed difference, given the null hypothesis. This likelihood can be converted to an absolute probability value by simulating 40 experiments in which a representative number of transcripts is identified and com-

B

Date Analyzed: 01-20-1999

Project Name: test

Enzyme: NlaIII-CATG

Tag length: 9bp

Maximum DiTag length: 24bp

Total Tags: 42

Project Duplicate Dimers: 0

Input File: 37PBA5.SEQ

Start position: 1   Stop position: 1000

Sequence Length: 737

- 1)   ATCCGGCGCCTTATTGATGGTA -   54938 - 201937
- 2)   ACATCATCGATGATGTACGGGGA -   19767 - 218546
- 3)   TCCCTATTAAGCTTTTTATTGAA -   218941 - 250049
- 4)   TTCTCTATTGGGATGGCCAGGC -   253392 - 155285
- 5)   AATGAAAAGGTTGAACTCCTG -   14339 - 75966
- 6)   TGATTTCACTTCTGACCAGCAC -   233426 - 190382
- 7)   CGCAAGCTGGTAGGCTTTGCCCA -   102559 - 239875
- 8)   AAGTAAACTGGTCCCTAACTG -   11272 - 77611
- 9)   AGGCAGACGGTGGAGCAGAATGGT-   42119 - 21471
- 10) TAAAGCCTGTAGATGTACGGGGA -   197215 - 218546
- 11) TCCCCGTACATCTTGCCAGCATT -   218546 - 14826
- 12) AAGAAGATAGAAATGTACGGGGA -   8333 - 218546
- 13) TGTTCCCCAAATTGAGTTTAGG -   245077 - 94238
- 14) CCCGACGTGCCACAAGGCGGGC -   88175 - 153184
- 15) TGCTGCCCTGTCTCGGGCACTGT -   237144 - 19350
- 16) GCTTTTTAGAATACCGTGGGT -   163827 - 21612
- 17) GAGAGGAAGGGCATTTTTTCTA -   139907 - 204801
- 18) GTTTAAGTTAACTTAATAGGGA -   195632 - 218941
- 19) GAGGTGGTGCGGCTCGTGTCGTC -   142255 - 137287
- 20) TGGCAACCTTTGTTCCGGACGGG -   238616 - 88923
- 21) CCCATCGTCCTTTGGAACACAAGC -   86894 - 163568

Total Dimers: 21

Short Dimers: 0

Long Dimers: 0

Good Tags: 42

C

Tag Abundance Report

Total tags after excluding tags = 20337

Count	Percent	Tag Sequence	Tag BaseFour Number
562	2.7634	CTGGCCCTC	125278
371	1.8242	TACCATCAA	201937
275	1.3522	CTAAGACTT	115232
269	1.3227	CCCATCGTC	86894
234	1.1506	CCTCCAGCT	95528
136	0.6687	GCGACCGTC	156014
117	0.5753	CTCATAAGG	119563
110	0.5408	AAAACATTC	318
110	0.5408	TTGGTCCTC	256862
102	0.5015	TTCATACAC	250642
100	0.4917	CCCGTCCGG	88923
98	0.4818	AGGGCTTCC	43510
95	0.4671	TGCACGTTT	233920
89	0.4376	GTGAAACCC	188438
82	0.4032	CAAGCATCC	67894
78	0.3835	CCCAAGCTA	86173
76	0.3737	GCCGGGTGG	154299
73	0.3589	ATGGCTGGT	59884
72	0.354	CAAACCATC	65870
70	0.3442	CACCTAATT	71440
68	0.3343	TGTGTTGAG	244707
64	0.3146	AAAAAAAAA	1
62	0.3048	AAGGTGGAG	11171
59	0.2901	GCAGGGCCT	150168
59	0.2901	TCAGATCTT	215264
58	0.2851	AGCACCTCC	37238
57	0.2802	AGCCCTACA	38341

.
.
.
.
.

.etc.

Fig. 5B. *(opposite page)* Results of analysis of a single clone by SAGE software, detection of ditag (dimer) and tag sequences. **(C)** SAGE tag abundance report (first page only) showing breakdown of tag abundance levels for a particular project.

pared. The distribution of transcripts used for these simulations is derived from the average level of expression observed in the original samples. The distribu-

D

Library # 1 = 15774 tags from C:\Program files\SAGE300\test.sum

Library # 2 = 16415 tags from C:\Program Files\SAGE300\test2.sum

Total tags in all projects = 32189

Output = Observed tags, NonNormalized

Lib # 1	Lib # 2	Total	Tag Sequence
562	646	1208	CTGGCCCTC
371	160	531	TACCATCAA
275	166	441	CTAAGACTT
269	321	590	CCCATCGTC
234	283	517	CCTCCAGCT
136	68	204	GCGACCGTC
117	53	170	CTCATAAGG
110	29	139	AAAACATTC
110	123	233	TTGGTCCTC
102	127	229	TTCATACAC
100	77	177	CCCGTCCGG
98	92	190	AGGGCTTCC
95	92	187	TGCACGTTT
89	77	166	GTGAAACCC
82	42	124	CAAGCATCC
78	97	175	CCCAAGCTA
76	93	169	GCCGGGTGG
73	68	141	ATGGCTGGT
72	80	152	CAAACCATC
70	79	149	CACCTAATT
68	71	139	TGTGTTGAG
64	70	134	AAAAAAAAA
62	50	112	AAGGTGGAG

.
.
.
.
.

.etc.

tion of the p-chance scores obtained in the 40 simulated experiments (false positives) are then compared to those obtained experimentally. Based on this comparison, a maximum value for p-chance that gives the desired significance level (i.e., false positive rate) can be selected.

E

FileName = c:\sage\test\test.bag   Total Tags = 16489
Anchoring Enzyme = *Nla*III - CATG   DataBase Link
Tag Length = 9bp   Database = c:.dum
DiTag Length = 26bp

Count	Percent	Tag Sequence	Tag BaseFour Number
500	3.0323 %	CTGGCCCTC	- 125278

Noted Tags = 5   Collected Tags = 3
GGCACC, Class A, X52003, H.sapiens pS2 protein gene.
GGCACC, Class A, X00474, Human pS2 mRNA induced by estrogen
GGCACC, Class A, M12075, Human estrogen receptor mRNA

258	1.5646 %	CCCATCGTC	- 86894

Noted Tags = 22   Collected Tags = 13
CTAGAA, Class A, X15759, Human mitochondrial mRNA for cytochrome c oxidase
CTAGAA, Class C, U12690, Human Hsa2 mitochondrion cytochrome oxidase subuni
CTAGAA, Class C, U12691, Human Hsa3 mitochondrion cytochrome oxidase subuni

236	1.4312 %	CCTCCAGCT	- 95528

Noted Tags = 14   Collected Tags = 8
ACAAAA, Class A, X12882, Human mRNA for cytokeratin 8.
ACAAAA, Class A, X98614, H.sapiens mRNA for cytokeratin.

128	.7762 %	TACCATCAA	- 201937

Noted Tags = 7   Collected Tags = 5
TAAAGT, Class A, X53778, H.sapiens hng mRNA for uracil DNA glycosylase.
TAAAGT, Class A, J02642, Human glyceraldehyde 3-phosphate dehydrogenase
TAAAGT, Class A, M36164, Human glyceraldehyde-3-phosphate dehydrogenase

79	.4791 %	CCCAAGCTA	- 86173

Noted Tags = 8   Collected Tags = 5
GCCACG, Class A, X54079, Human mRNA for heat shock protein HSP27.
GCCACG, Class A, Z23090, H.sapiens mRNA for 28 kDa heat shock protein
·
·
·
·
.etc.

Fig. 5D. *(opposite page)* Comparison of tag abundance levels between two projects. This has been sorted according to Library #1 and differences in tag abundance can be seen. (**E**) Comparison of sequence tags against Genbank. Here, a SAGE tag abundance report has been prepared using the sequences from Genbank (in this case, primate sequences only) and compared with a SAGE tag library. This matches tag sequence with the entries in Genbank in order to identify the genes corresponding to the sequenced tags.

The SAGE software files can also be exported into MS Excel and/or MS Access in order to further analyze the data.

4. Notes

1. In order to generate good SAGE libraries with large clone inserts, it is vital to have the best starting material possible. Many standard protocols and a number of commercially available kits can be used to prepare the RNA and cDNA samples. I have found the ones described here to be reliable.
2. High-quality linkers are crucial to several steps in the SAGE method. Linker oligos and the biotinylated oligo-dT should be obtained in gel-purified form from the oligo-synthesis supplier.
3. *Nla*III has been reported to be quite unstable. It is recommended that it be kept at –70°C for only 3 mo and that test digests are performed just prior to cutting SAGE material to ensure that the enzyme is still active.
4. The enhanced SAGE method requires these PCR primers to be biotinylated at the 5' end. This modification does not affect the PCR conditions or yields but will enable ditags to be efficiently purified as described later.
5. The SAGE Detailed Protocol, Velculescu, V. E., Zhang, L., Zhou, W., Vogelstein, B. and Kinzler, K. W. and SAGE software can be obtained via the SAGE web page: **http://www.sagenet.org/**
6. The gel-purified biotinylated oligo-dT needs to be fully biotinylated to ensure efficient capture of the cDNA, this can be tested as described in these methods or, at the time of oligo synthesis, by high-performance liquid chromatography.
7. This is the first point of analysis after many manipulations and many workers report the lack of a 102-bp band of sufficient intensity. The main points to stress are as follows:
   • Good quality oligos and highly biotinylated oligo-dT primer must be used in order to ensure good quality cDNA and yield after binding to the Dynabeads.
   • The amount of input cDNA. Increasing this may improve the PCR reactions but, ultimately, quality is better than quantity.
   • Quality of linkers and linker pair preparation.
8. The efficiency of cloning deteriorates rapidly upon storage of the cut vector. In order to maximize cloning efficiency, I recommend using the vector the same day that it is digested without freezing.
9. For ABI 310 and 377 automated sequencers, the following methods of PCR purification have worked well:
   • Microcon 100 columns (Amicon, cat. no. 42413).
   • QIAquick 8 PCR purification kits (QIAGEN, cat. no. 28142/28144).
10. Even for small-scale SAGE projects, the amount of sequencing is significant. (The generation of 10,000 tags would require the sequencing of over 300 clones of the size described in this method.) Automated sequencing of large-insert SAGE clones to minimize the sequencing load is therefore recommended.

References

1. Liang, P. and Pardee, A. B. (1992) Differential display of eukaryotic messenger RNA by means of the polymerase chain reaction. *Science* **257,** 967–971.

2. Bertioli, D. J., Schlichter, U. H., Adams, M. J., Burrows, P. R., Steinbiss, H. H., and Antoniw, J. F. (1995) An analysis of differential display shows a strong bias towards high copy number mRNAs. *Nucleic Acids Res.* **23,** 4520–4523.
3. Bains W. (1996) Virtually sequenced: the next genomic generation. *Nat. Biotech.* **14,** 711–713.
4. Velculescu, V. E., Zhang, L., Vogelstein, B., and Kinzler, K. W. (1995) Serial analysis of gene expression. *Science* **270,** 484–487.
5. Velculescu, V. E., Zhang, L., Zhou, W., Vogelstein, J., Basrai, M. A., Bassett Jr, D. E., Heiter, K. W., Vogelstein, B., and Kinzler, K. W. (1997) Characterisation of the yeast transcriptome. *Cell* **88,** 243–251.
6. Zhang, l., Zhou, W., Velculescu, V. E., Kern, S. E., Hruban, R. H., Hamilton, S. R., Vogelstein, B., and Kinzler, K. W. (1997) Gene expression profiles in normal and cancer cells. *Science* **276,** 1268–1272.
7. Madden, S. L., Galella, E. A., Zhu, J., Bertelsen, A. H., and Beaudry, G. A. (1997) SAGE transcript profiles for p53-dependent growth regulation. *Oncogene* **15,** 1079–1085.
8. Polyak, K., Xia, Y., Zweier, J. L., Kinzler, W. K., and Vogelstein, B. (1997) A model for p53-induced apoptosis. *Nature* **389,** 300–305.
9. Szybalski, W. (1985) Universal restriction endonucleases: designing novel cleavage specificities by combining adapter oligonucleotide and enzyme moieties. *Gene* **40,** 169–173.
10. Szybalski, W., Kim, S. C., Hasan, N., and Polhajska, A. J. (1991) Class-IIS restriction enzymes - a review. *Gene* **100,** 13–26.
11. Powell J. (1998) Enhanced concatamer cloning—a modification to the SAGE (serial analysis of gene expression) technique. *Nucleic Acids Res.* **26,** 3445–3446.
12. Kenzelmann, K. and Muhlemann, K. (1999) Substantially enhanced cloning efficiency of SAGE (serial analysis of gene expression) by adding a heating step to the original protocol. *Nucleic Acids Res.* **27,** 917,918.

21

Analysis of Differential Gene Expression Using the SABRE Enrichment Protocol

Daniel J. Lavery, Philippe Fonjallaz, Fabienne Fleury-Olela, and Ueli Schibler

1. Introduction

An important component of the stress response, as well as many other biological systems, is the change in relative concentration of specific mRNAs following induction *(1)*. Conventional techniques of differential screening may be sufficient for the detection and identification of some of these differentially expressed mRNAs, especially those of relatively high abundance. However, mRNAs displaying more subtle changes in their expression may be difficult to detect. This is particularly true for rare mRNAs showing relatively modest differences (10-fold or less) or for mRNAs expressed within only a subpopulation of cells in a sample.

Detection of subtle differences in mRNA accumulation between two populations would be aided if these differences could be specifically amplified. Here, we describe a protocol that accomplishes this amplification, termed Selective Amplification via Biotin- and Restriction-mediated Enrichment (SABRE) *(2)*. SABRE is a polymerase chain reaction (PCR)-based selection procedure for specific enrichment of differentially expressed mRNA species, including rare mRNAs with modest differences in abundance. This protocol is relatively rapid and efficient, owing to an accelerated hybridization procedure and simple manipulations between rounds of selection. In this chapter, we present the SABRE selection protocol in detail, as well as guidelines for the procedures leading up to and following the selection. As an example, we demonstrate that SABRE is able to detect differences in the accumulation of several liver mRNAs in a comparison of wild-type and mutant "knock-out"

From: *Methods in Molecular Biology, vol. 99: Stress Response: Methods and Protocols*
Edited by: S. M. Keyse © Humana Press Inc., Totowa, NJ

strains of mice. One of these differentially expressed mRNA species was found to encode the mouse 70-kDa heat-shock protein, hsp70.

1.1. Competitive Hybridization

Many protocols are currently in use that permit the overall comparison of two mRNA populations, in order to detect differences between them. Examples include differential screening of cDNA libraries with probes derived from two different mRNA populations, and differential display-reverse transcription PCR (DD-RT PCR) (*see also* Chapter 22, this volume). However, these protocols rely upon detection of overexpressed species against a large background of common, equally expressed species. Thus, more subtle differences in expression may be difficult to detect. More recently, methods have been developed that, in fact, enrich these differences between the two populations, thereby increasing the probability of their detection. Several protocols, including SABRE, perform this enrichment via a procedure referred to here as competitive hybridization (*see* **Note 1**). The starting materials for competitive hybridization are two populations of double-stranded DNA. One population, termed the tester, is suspected of containing species present at a higher concentration than in the other population, termed the driver, although most species are found equally in both. For example, the tester population might represent cDNA from heat-shocked cells, whereas the driver population would represent cDNA from control, nonshocked cells. The DNA strands in the two populations have been marked (e.g., during PCR amplification) so that each strand can be distinguished as having come from the driver or tester DNA sample. The two DNA populations are mixed together, denatured, and allowed to reassociate into double-stranded molecules.

Three different double-stranded DNA hybrids can result from this hybridization: tester homohybrids, in which both strands are derived from the tester population; driver homohybrids, in which both strands are derived from the driver population; and driver-tester heterohybrids, in which one strand derives from each population. A species should reform hybrids at a relative ratio of t^2 : $2dt$: d^2 for tester homohybrids, driver-tester heterohybrids, and driver homohybrids, respectively, where t is the fraction of all strands contributed by the tester population and d is the fraction of all strands contributed by the driver population. Thus, following hybridization of equal amounts of tester and driver DNA, a species present equally in both tester and driver ($t = \frac{1}{2}$; $d = \frac{1}{2}$) will be present at a ratio of $\frac{1}{4}$: $\frac{1}{2}$: $\frac{1}{4}$ in tester homohybrids, driver-tester heterohybrids, and driver homohybrids, respectively (**Fig. 1**).

In contrast, a species that has a higher concentration in the tester population will be more likely to form tester homohybrids than will a species expressed equally in the two populations. This is because, for this species, the probability

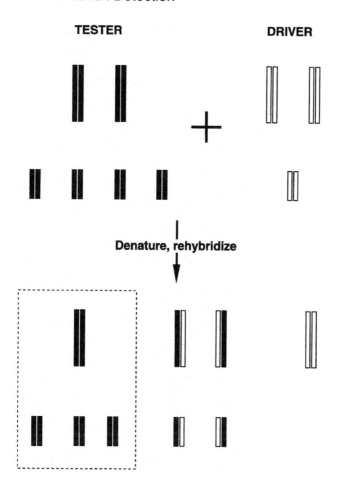

Fig. 1. Competitive hybridization for enrichment of overexpressed species. Two double-stranded DNA populations, tester and driver, are comprised of two DNA species each: one (*large bars*) present at equal abundance in the two populations; the other (*small bars*) overexpressed by four-fold in the tester population relative to the driver population. If the two DNA populations are combined, denatured, and allowed to reassociate at random, three populations of double-stranded molecules of each species can be formed: driver homohybrids, in which both strands are derived from the driver population (both bars open, bottom right); tester homohybrids, in which both strands are derived from the tester population (both bars filled, bottom left); and driver-tester heterohybrids, in which one strand is derived from each population (one bar open, one bar filled; bottom middle). The dashed line delineates the tester homohybrid population, in which the ratio of the overexpressed species to the equally expressed species (3 to 1) has increased as compared to the ratio in the starting tester population (4 to 2). Thus isolation of the tester homohybrid population will increase the relative ratio of overexpressed species.

is greater that one of its DNA strands in the tester population will hybridize with a strand also derived from the tester population. Specifically, if the species is expressed N-fold more in the tester than in the driver [$t = N (d)$], its relative distribution in tester homohybrids, driver-tester heterohybrids, and driver homohybrids will be $(Nd)^2 : 2 Nd^2 : d^2$, respectively. Therefore, the population of double-stranded molecules of this species will have a bias towards tester homohybrids, as compared to double-stranded molecules of the commonly expressed species (**Fig. 1**). This can be best visualized if one imagines a species present in the tester population but completely absent from the driver population; by necessity, all resulting double-stranded DNA molecules of this species will be in the form of tester homohybrids.

Thus, the ratio of tester-overexpressed species to commonly expressed species is higher in the tester homohybrid population than in the starting tester population. This enrichment of tester-overexpressed species can be increased further by using an excess of driver DNA over tester DNA in the hybridization. Such an excess of driver DNA "drives" proportionately more of the commonly expressed species in the tester DNA into heterohybrids, thereby increasing the enrichment of the tester-overexpressed species in the tester homohybrid population (*see* **Note 2**).

The theoretical ratio by which the concentration of the tester-overexpressed species is increased in the tester homohybrids compared with the starting tester population can be calculated as:

$$(1 + R)/(1 + [R/N])$$

where R is the ratio of driver to tester DNA used in the hybridization and N is the fold difference in abundance of this species between the tester and driver populations (*see* **Note 3** for the derivation of this ratio). The value of N can vary from zero for species not found in the tester population, to 1 for common species, and up to infinity for species present only in the tester population.

One observation drawn from this ratio is that the tester homohybrid population will be enriched in the concentration of all species overexpressed in the starting tester population, including those with relatively modest concentration differences (**Fig. 1**). Thus, for a species x whose concentration in the tester population is 10-fold higher than in the driver population ($N = 10$), after hybridization with a 20-fold excess of driver DNA its abundance in tester homohybrids relative to that of a commonly expressed species will be sevenfold greater than in the starting tester population [ratio = $(1 + 20)/(1 + [20/10]) = 7$]. If the tester homohybrids are purified away from the other species, the concentration of the species overexpressed by 10-fold in the starting tester population will now be 70-fold more abundant in the tester than in the driver.

If the protocol permits subsequent rounds of selection, this difference can become even greater. At the beginning of a second selection round, N for species x has passed from 10 to 70. Following hybridization with a 20-fold excess of driver DNA and purification of tester homohybrids, the ratio of x to the common species in tester homohybrids at the end of the second round will be increased 16.3-fold [ratio = (1 + 20)/(1 + [20/70]) = 16.3], to 1143. Thus two rounds of selection have increased the relative concentration of the overexpressed species by more than 100-fold.

1.2. Selection of Tester Homohybrids

In order to enrich overexpressed species, the tester homohybrids must be specifically purified away from the driver homohybrid and driver-tester heterohybrid molecules after hybridization. This can be technically difficult, given that when driver DNA is used in excess, the tester homohybrid population will represent a small fraction of the total pool of double-stranded molecules. For example, in a hybridization reaction containing a 20-fold excess of driver DNA over tester DNA, only 0.23% $[(1/21)^2]$ of the double-stranded molecules will be tester homohybrids. Thus, the purification procedure must be stringent enough to keep contamination by other DNA species low.

The SABRE protocol uses two stringent criteria for tester homohybrid selection: streptavidin-biotin affinity and restriction enzyme site recognition. These are incorporated into the selection process via differences in the PCR primers used to amplify the tester and driver DNA samples for hybridization (**Fig. 2A**). DNA samples from the two populations to be compared are digested with a restriction enzyme, *Sau*3AI, which cuts approximately once every 250 bases. This generates DNA restriction fragments which are more easily manipulated and more faithfully amplified by the PCR reaction. Linker DNA sequences are ligated onto each end of the restriction fragments through complementarity of the end sequences. The ligated linker-restriction fragment products are purified and used as libraries to generate double-stranded tester or driver DNA for hybridization and selection. Tester DNA is amplified using a PCR primer which is biotinylated at the 5' end and contains in its sequence the recognition site (GGATCC) of the restriction enzyme *Bam*HI (*see* **Note 4**). Conversely, the driver PCR primer is unbiotinylated and contains no restriction enzyme site, owing to sequence differences in the 5' ends of the oligos. Otherwise, the sequence of both PCR primers 3' to the restriction site is identical (**Fig. 2A**). Hybridization of these tester and driver DNA molecules results in the formation of double-stranded molecules depicted in **Fig. 2B**. Tester homohybrids are biotinylated at each 5' end and contain a reconstituted restriction enzyme site; driver homohybrids are unbiotinylated, with no restriction enzyme site; and driver-tester heterohybrids contain one biotin group and are

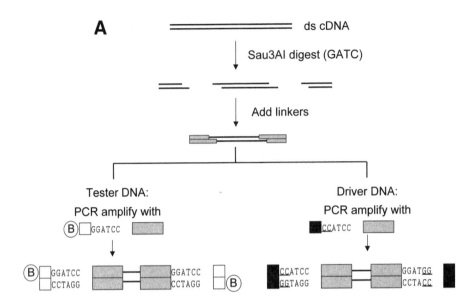

Fig. 2. (**A**) *Generation of SABRE PCR Libraries*. Double-stranded cDNA is digested with the frequent cutting enzyme *Sau*3AI, which recognizes the sequence 5'-GATC-3', cleaving before the G residue. These fragments are then ligated to double-stranded DNA linkers (shaded boxes) with 5' GATC overhangs which complement those of the *Sau*3AI digested cDNA fragments. These ligated products are purified and stored as the master library stock for SABRE selection. PCR amplification on this master library stock is performed to generate tester and driver DNA samples for hybridization. To generate tester DNA, PCR is performed using an oligonucleotide primer which shares similarity to the linker DNA sequence (*shaded box*), and thus can prime DNA synthesis. However, the oligo is also biotinylated at its 5' end (circled B) and differs in its 5' sequence (*open box*), including the presence of the recognition sequence for the restriction enzyme *Bam*HI (5'-GGATCC-3', cutting between the G residues). Likewise, to generate driver DNA, an oligonucleotide is used with sequence similarity to the linker DNA sequence (*shaded box*), but with different sequences at the 5' end (*filled box*). In the case of the driver DNA primer, there is no biotin group at the 5' end, and the *Bam*HI restriction enzyme site has been mutated (<u>CC</u>ATCC, mutations underlined). Note that the same DNA library can be used to make either "tester" DNA or "driver" DNA, depending upon the primer used for PCR amplification. (**B**) *SABRE Selection*. Tester and driver DNA prepared as in **Fig. 2A** are combined, denatured, and allowed to rehybridize. During rehybridization, the three duplex species described in **Fig. 1** and in the text are formed. Note that the tester homohybrids have biotin groups at either 5' end and have reconstituted the double-stranded *Bam*HI restriction enzyme site. Driver homohybrids are unbiotinylated and contain no *Bam*HI restriction site. In driver-tester heterohybrids, sequence differences in the driver and tester primers result in the formation of heteroduplexes with unpaired terminal sequences. These short

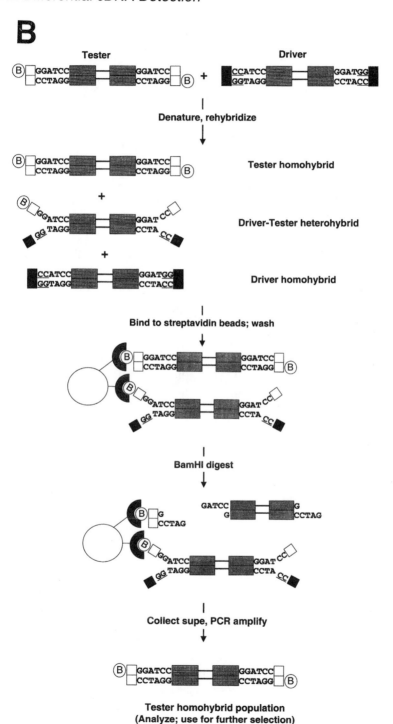

B

Tester homohybrid

Driver-Tester heterohybrid

Driver homohybrid

Tester homohybrid population
(Analyze; use for further selection)

mismatched at either end, such that the restriction enzyme site is not reconstituted. After reannealing, the hybrids are digested with S1 nuclease to remove single-stranded DNA. This treatment not only digests non-hybridized nucleic acids but also unpaired terminal sequences of most heterohybrids (*see* **Fig. 2B**). As a consequence, tester homohybrids constitute the major proportion of biotinylated DNA duplexes after S1 nuclease digestion. These tester homohybrids are first purified from nonbiotinylated molecules by capture with streptavidin-coated magnetic beads. Because of the high affinity of streptavidin for biotin, the tester homohybrids and remaining driver-tester heterohybrids are bound quantitatively to the magnetic beads, which are recovered and stringently washed using a magnetic attractor. In this way, the nonbiotinylated driver homohybrids and S1 nuclease-trimmed driver-tester heterohybrids are largely eliminated.

The bead-bound DNA molecules are next incubated with the restriction enzyme whose site is found in the tester PCR primer. Tester homohybrids are released from the beads, whereas the heterohybrids, in which the duplex restriction site is not reconstituted, remain bound. The homohybrid-containing supernatant is collected, and can be immediately reamplified by PCR with the same PCR primers. These selection products can be analyzed for detection of enriched species or can be used directly for a subsequent round of selection without any further preparation. Thus, the purification of tester homohybrids by SABRE is simple, efficient, and rapid, with minimal preparation time between selection rounds.

---

single-stranded sequences, including the biotinylated 5' terminal nucleotide, are removed from most driver-tester heterohybrids during S1 nuclease digestion. Heterohybrids which do survive the S1 nuclease digestion do not contain a *Bam*HI recognition sequence and hence are not cleaved by this endonuclease.

Following digestion with single-stranded nuclease S1 to eliminate unhybridized species, the hybridization products are incubated with paramagnetic beads (large circles) coated with streptavidin (filled "C" shape). Biotin-containing molecules are retained on the magnetic beads due to the strong interaction of biotin and streptavidin; thus, driver homohybrids are excluded, and removed in stringent wash steps. The DNA-containing beads are next incubated with the restriction enzyme *Bam*HI, which recognizes and cleaves the site GGATCC in the tester homohybrids, releasing these molecules from the beads. The heterohybrids (whose terminal sequences survived S1 nuclease digestion) remain bound to the beads, as their primers do not contain a reconstituted *Bam*HI site. The supernatant containing the released tester homohybrids is collected, and the DNA species are amplified by PCR using the same primers as before, which can hybridize to the sequences 3' to the restriction cleavage site. The PCR products can next be analyzed to detect differentially expressed species, or can be used in a subsequent round of SABRE selection.

1.3. Accelerated Hybridization with Thermal Cycler PERT

As with other competitive hybridization protocols, SABRE selection will only function for double-stranded molecules, as single-stranded species are lost in the selection procedure. This requires that the hybridization reaction proceed as nearly as possible to completion, to avoid the depletion of rare species. Given the complexity of eukaryotic cDNA populations, this can be difficult to achieve: Standard hybridization protocols may call for incubations of 1 wk or more at high temperatures in order to achieve significant hybridization of rare species. This long incubation is not only inconvenient but can also lead to degradation of the sample. To avoid this, the SABRE protocol has incorporated a modification of the phenol emulsion reassociation technique (PERT) *(3)* developed by Miller and Riblet, in which hybridization is performed overnight in a thermal cycler apparatus. Under these conditions, up to 1000-fold higher DNA reassociation rates are achieved as compared to aqueous hybridization protocols *(4)*.

1.4. Autohybridization Control

Because the source of DNA for tester and driver populations is often limited, SABRE uses PCR-based libraries of tester and driver DNA. This permits the amplification of large amounts of DNA from relatively small amounts of starting material. However, as for any PCR-based procedure, certain sequences may be more efficiently amplified by PCR than others, leading to an artificial increase in their relative concentration and, thus, an increase in false-positive species. To help offset this, the SABRE protocol includes a control hybridization, in which the DNA from the control population (cDNA from non-heat-shocked cells, for example) is PCR amplified with both the driver and tester PCR primers, generating both driver and tester DNA. These driver and tester DNAs are then hybridized (non-heat-shock cDNA tester + non-heat-shock cDNA driver) and selected identically to the experimental hybridization (heat shock cDNA tester + non-heat-shock cDNA driver). Because the control DNA is hybridized to itself in this control hybridization, the population at the end of the selection round should, in theory, be identical to the starting population, except for changes introduced by the manipulations. By comparing the experimental and control populations after selection, species with an increased concentration in both populations can be excluded from further analysis. In addition, this control-selected population, not the original driver population, is used as the source for driver DNA in the next round; as this population is enriched in false positives, it will offset the increased concentration of these species in the next hybridization.

1.5. Isolation of Enriched Species

The products of each round of selection are monitored on a denaturing polyacrylamide sequencing gel. In our experience, discrete bands can begin to be distinctly and specifically enriched in the experimental selection products following the second or third round of selection. However, further selection rounds will lead to the appearance of additional enriched species too rare to be significantly enriched earlier in the selection. Individual results will, of course, depend on the degree of difference between the two populations, as well as the efficiency of selection. If the enriched bands appear in an area of the gel lane with few other major species, they can be cut from the gel with a razor blade, reamplified from the gel slice, and cloned for further analysis *(5)*. However, to detect enriched species that might be masked by abundant bands, the entire experimental selected population can be used as a hybridization probe to screen a bacteriophage or plasmid library. The control selected population is used as a hybridization probe in duplicate screening to confirm that the detected plaque or colony is enriched in the experimental selected population. In cDNA expression studies, positive clones can be confirmed by reverse Northern analysis, using as probes the RNA used to generate the original tester and driver populations *(5)*.

2. Materials (*see* Note 5)

2.1. PCR-Based Library Construction

1. Double-stranded DNA of tester or driver samples (usually cDNA); ideally, 2–4 μg, though less can be used (*see* **Note 6**).
2. DNA linker oligonucleotides, annealed, resuspended at 2 μg/μL in 10 m*M* Tris, pH 7.5, 0.1 m*M* EDTA, 50 m*M* NaCl (*see* **Note 7**).
   L1: 5'- GGT CCA TCC AAC C-3'
   L2: 5'-phosphate-GAT CGG TTG GAT GGA CCG T-3'
3. Restriction enzyme *Mbo*I or *Sau*3AI and its digestion buffer, 20 U/reaction.
4. Sodium dodecyl sulfate (SDS) 20%.
5. EDTA, 0.5 *M* pH 8.0.
6. Chloropane: 25:24:1 mixture (v/v) of buffer-saturated phenol, chloroform, and iso-amyl alcohol (*see* **Note 8**).
7. CIA: 24:1 mixture of chloroform and iso-amyl alcohol.
8. Ethanol 100% and ethanol, 70% in water.
9. Ficoll, 20% in water, with bromphenol blue 0.05%.
10. Low-melting temperature agarose.
11. Ethidium Bromide, 10 mg/mL in water (*see* **Note 9**).
12. 10X TBE: containing (per liter) 108 g Tris-base, 55 g boric acid, 9.3 g disodium EDTA.
13. T4 DNA ligase and its 10X buffer: 660 m*M* Tris-HCl, pH 7.5, 50 m*M* MgCl$_2$, 10 m*M* dithioerythritol (DTE), 10 m*M* ATP.

14. BSA, 1 mg/mL.
15. High-salt phenol: phenol buffered with 10 mM Tris, pH 8.0, 0.5 M NaCl (*see* **Note 8**).
16. 5 M NaCl.
17. $T_{10}E_{0.1}$: 10 mM Tris, pH 8.0, 0.1 mM EDTA.

2.2. PCR Amplification

1. Taq DNA polymerase, 5 U/µL, with its 10X buffer.
2. dATP, dCTP, dGTP, dTTP, sodium salts, 10 mM each (PCR grade).
3. [α-^{32}P]dATP or -dCTP, as trace label (1 µCi per 50-µL reaction; *see* **Note 10**).
4. $T_{10}E_{0.1}$.
5. 3 M Sodium acetate.
6. Chloropane (*see* **Subheading 2.1., item 6**; also, *see* **Note 8**).
7. 2-propanol.
8. Ethanol 100% and ethanol 70% in water.
9. PCR primer oligonucleotides:
   T1: 5'-Biotin-CCA GGA TCC AAC CGA TC-3' (17-mer + biotin; used for PCR generation of tester DNA, biotinylated, contains *Bam*HI site; *see* **Note 11**).
   D1: 5'-GGT CCA TCC AAC CGA TC-3' (17-mer; used for PCR generation of driver DNA, unbiotinylated, contains no *Bam*HI site).
   Both oligos stored concentrated, with a working stock diluted to 0.1 mM in $T_{10}E_{0.1}$.
10. DNA thermal cycler.

2.3. PERT Hybridization

1. 3 M Sodium acetate.
2. 2-propanol.
3. Ethanol, 70% in water.
4. 5 mM Tris-HCl, pH 8.0.
5. 5X phosphate-EDTA: 600 mM sodium phosphate, pH 6.8, 50 mM EDTA.
6. 5 M sodium thiocyanate in water.
7. Phenol, highest quality, stored under nitrogen, freshly melted at 55°C immediately before use (*see* **Note 8**).
8. DNA thermal cycler with "hot lid" attachment permitting cycling reactions without oil overlay.

2.4. SABRE Selection

1. S1 nuclease buffer: 30 mM sodium acetate, pH 4.6, 250 mM NaCl, 3 mM ZnSO4.
2. S1 nuclease, diluted in nuclease buffer to 12.5 U/µL (*see* **Note 12**).
3. Salmon sperm DNA, sonicated, 4–10 mg/mL in water
4. 1 M Tris-HCl, pH 8.0.
5. 0.5 M EDTA, pH 8.0.
6. Streptavidin-coated paramagnetic beads (Dynal Inc.), binding capacity 400 pmol biotinylated cDNA per milliliter.

7. "1M TENT" buffer: 10 mM Tris, pH 8.0, 1 mM EDTA, 0.1% Triton X-100, 1 M NaCl.
8. Magnetic separator, with at least two places for Eppendorf tubes.
9. Heparin, sodium salt Grade 1-A (Sigma catalog number H-3393), 50 mg/mL in water.
10. Wash solution: 0.1X SSC, 0.1% SDS (*see* **Note 13**).
11. Restriction enzyme *Bam*HI and its 10X reaction buffer.
12. 10 % Triton X-100, in water.
13. Bovine serum albumin (BSA) 10–20 mg/mL.
14. Materials for PCR, as in **Subheading 2.2.**

2.5. Analysis

1. Apparatus for preparation and running of acrylamide sequencing gels.
2. Acrylamide:bis-acrylamide, 38%:2% in water (*see* **Note 14**).
3. TEMED.
4. Ammonium persulfate (APS), 25% (w/v) in water.
5. 10X TBE : per liter: 108 g Tris-base, 55 g boric acid, 9.3 g disodium EDTA.
6. Formamide loading dyes (formamide 95%, EDTA 10 mM, Bromphenol blue 0.05%, xylene cyanol FF 0.05%).
7. Gel dryer.
8. X-ray film.
9. X-ray exposure cassettes.
10. Agarose (normal melting temperature).
11. Ethidium bromide, 10 mg/mL in water (*see* **Note 9**).

3. Methods
3.1. Generating Libraries

1. Digest 0.5–4 µg of double-stranded cDNA from tester and driver samples with 20 U of restriction enzyme *Sau*3AI or *Mbo*I, in 1X reaction buffer plus BSA 0.1 mg/mL, in a final volume of 20–50 µL for 2 h at 37°C.
2. Stop the reaction by adding SDS to 0.2%, EDTA to 20 mM, in a final volume of 200 µL.
3. Extract twice with 2 vol chloropane, and twice with CIA.
4. Adjust sodium concentration to 0.3 M with 3 M sodium acetate; mix well.
5. Add 2.5 vol 100% ethanol; mix well, and store at -20°C >1 h.
6. Spin at maximum speed in a microcentrifuge, 15 min, at 4°C.
7. Remove supernatant, and wash pellet with ethanol 70%.
8. Resuspend cDNA to 0.5 mg/mL in $T_{10}E_{0.1}$.
9. Set up ligation reaction using 0.4–1 µg of digested cDNA and a 50-fold molar excess of annealed linkers. Assuming an average size of 250 bp for the digested cDNA, 1 µg of digested cDNA is equal in molar quantity to 72 ng of the 32-base, partially double-stranded linker. Thus, a 3.6-fold mass excess of linkers to cDNA will equal a 50-fold molar excess. Therefore, a ligation using 0.4 µg of digested cDNA should contain 1.44 µg of annealed linkers. *Essential:* Include a control

reaction containing water in place of cDNA. This will serve as a ligation control to ensure that the reaction was successful.

Per reaction:

cDNA, 0.2–0.5 mg/mL	2 µL
Linker, 2 mg/mL	0.5–2.5 µL
10X Ligase buffer	2 µL
1 mg/mL BSA	2 µL
T4 DNA ligase, 5 U/µL	2 µL
Water	to 20 µL

10. Incubate overnight at 14°C.
11. Stop by adding EDTA to 20 mM, and 1/10 volume ficoll/bromphenol blue.
12. Prepare a 2% low-melt agarose / 1X TBE minigel (35 mL, usually 8–10 cm long) with 0.5 µg/mL ethidium bromide.
13. Load the ligation reaction, including the control (linkers alone) reaction. Also, load molecular-weight markers (approx 100 bp to >4 kbp) and the equivalent amount of *unligated* linkers (in 1X ligation buffer) next to the control linker reaction.
14. Run the gel under constant voltage at approx 40 V, until the bromphenol blue is approximately halfway to the bottom of the gel (4–5 cm).
15. Photograph the gel under low-intensity UV light. The ligated linkers should migrate at twice the apparent molecular weight of the unligated linkers (as linker dimers). If the ligated and unligated linkers comigrate, the ligation reaction has not occurred successfully.
16. With a clean razor or scalpel blade, cut out the region of the gel corresponding to 150 bp to approx 2 kbp, and transfer to clean, preweighed Eppendorf tubes (*see* **Note 15**). Be careful to cut cleanly across the top, taking the same size range for all reactions.
17. Weigh the filled tubes to estimate the mass (and thus volume) of the gel slices. If the mass is greater than 0.45 g, split into multiple tubes. Assuming 1 mg of gel is approximately equal to 1 mL of volume, adjust the volume of each tube to 0.45 mL with $T_{10}E_{0.1}$.
18. Add 50 µL of 5 M NaCl per tube.
19. Heat at 70°C for 10 min to melt the agarose.
20. Immediately add to each tube 0.5 mL high-salt phenol. Close tubes carefully.
21. Mix by gentle vortex mixing or rotation on a tube rotator for 20 min at room temperature.
22. Centrifuge for 10 min at 4°C at maximum speed in a microcentrifuge.
23. Remove supernatant to a clean tube, leaving behind the white agarose interphase.
24. Re-extract as in **steps 20** and **21** with 0.5 mL high-salt phenol, and centrifuge as in **step 22**.
25. Remove supernatant to a clean tube; add 1 mL 100% ethanol and mix well.
26. Store at -20°C for > 2 h.
27. Centrifuge for 15 min at 4°C at maximum speed in a microcentrifuge.
28. Remove supernatant, and wash pellet with ethanol 70%.
29. Resuspend pellet in 1 mL $T_{10}E_{0.1}$. If multiple tubes were used, pool all pellets in 1 mL final $T_{10}E_{0.1}$. This is the master stock for PCR amplification and should be stored in aliquots at –20 or –70°C.

3.2. PCR Amplification

A dependable, reproducible PCR reaction is essential for the success of SABRE. Detailed below are the conditions used in our laboratory for reproducible library amplification. However, optimal conditions will vary between laboratories; thus the conditions presented should be used as guidelines for the optimization of the reaction in each laboratory. Although the manufacturer's recommendations are a good starting point, one should keep in mind that they were probably developed for the optimal amplification of a single-sized band, rather than a complex mixture of species with different sizes and abundances. The PCR reaction should consistently reproduce the pattern of the cDNA digested with *Sau*3AI, in relation to average size and relative intensity of individual bands. To test this, a small amount of the *Sau*3AI-digested cDNA fragments can be radiolabeled at their 3' termini using the DNA polymerase Klenow fragment, and the labelled population compared with the products of PCR reactions on a denaturing polyacrylamide sequencing gel (*see* **Subheading 3.5.**; *see* **Note 16**).

Important parameters to be optimized include the following:

• Taq polymerase concentration: Too much Taq DNA polymerase leads to the generation of very large, artifactual species, whereas too little will not yield enough DNA.

• Primer concentration: If the PCR primer becomes limiting during the reaction, spurious priming may result; however, too much primer risks the formation of primer multimer products.

• Taq buffer composition: Although it is best to use the buffer recommended by the manufacturer, some suppliers recommend testing various Mg^{+2} concentrations, which might, in turn, be influenced by other parameters, such as concentration of dNTPs.

• Number of amplification cycles: Overamplification can lead to DNA degradation, primer multimers, and other artifacts, whereas too few cycles reduces the yield.

• Primer annealing temperature: This may change depending on the concentration of primer or other parameters and, thus, should be optimized. A thermal cycler that permits a gradient of annealing temperatures is best for this application.

1. Set up PCR reactions for amplification of libraries: 4X 50 µL reactions of "tester" library amplified with PCR primer T1; 4X 50 µL reactions of "driver" library amplified with PCR primer T1; and 10X 50 µL reactions of "driver" library amplified with PCR primer D1. Use standard laboratory precautions for working with radioactivity. Our laboratory's standard conditions per 50 µL reaction are the following:

Per reaction

cDNA PCR library	5 μL
10X PCR buffer (50 mM KCl)	5 μL
DMSO (*see* **Note 17**)	5 μL
25mM dNTPs	0.8 μL
0.1 mM PCR primer	1 μL
Taq DNA polymerase, 1 U/μL	2 μL
[α-^{32}P]dATP, 10 μCi/μl	0.1 μL
Water	to 50 μL

2. Run PCR amplification in a thermal cycler. Our standard conditions with a Stratagene (La Jolla, CA) Robocycler Gradient 96 machine are: (1 min. at 95°C, 2 min. at 52°C, 2 min. at 72°C) for 25–30 cycles.
3. At the end of the reaction, combine the multiple reactions in clean 1.5-mL Eppendorf tubes, and bring the final volume to 600 μL with $T_{10}E_{0.1}$.
4. Remove 6 μL (1/100) from each sample for determination of radioactivity; place in a scintillation vial or tube for counting afterwards.
5. Extract each sample (600 μL) twice with 0.6 mL chloropane, transferring the supernatant to a fresh tube each time.
6. Add 60 μL of 3 M sodium acetate; mix well.
7. Add 430 μL 2-propanol (0.65 vol). Mix very well by inversion and by vortex mixing (*see* **Note 18**).
8. Store at –20°C for > 1 hr.
9. Centrifuge in a microcentrifuge at maximum speed for 15 min at 4°C.
10. Carefully remove supernatant; dispose of as radioactive waste.
11. Wash pellet with 0.75 mL of 70% ethanol; let air-dry.
12. Resuspend each pellet in 200 μL $T_{10}E_{0.1}$.
13. Remove 2 μL (1/100) from each sample for determination of incorporated radioactivity. Place the 2 μL in a scintillation vial or tube.
14. Measure radioactivity in total (**Subheading 3.2., step 4**) and incorporated (**Subheading 3.2., step 3**) samples by liquid scintillation or Cerenkov counting.
15. Determine yield: (cpm incorporated)/(cpm total) (20 nmol dNTP/reaction)(number of reactions) = nmol dNTP incorporated in pooled sample. As 1 nmol dNTP = approx 1.32 μg, (nmol dNTP incorporated) × 1.32 μg/nmol dNTP = yield in μg.

3.3. Thermal Cycler PERT Hybridization

This method is based on that of Miller and Riblet *(4)*.

1. Set up two hybridization reactions per analysis: Tester DNA amplified with T1 + driver DNA amplified with D1; and driver DNA amplified with T1 + driver DNA amplified with D1. This second hybridization is the autohybridization control discussed above (*see* Introduction). Combine 10 μg of driver/D1 DNA with 0.33 μg tester/T1 DNA. Likewise, combine 10 μg of driver/D1 DNA with 0.33 μg driver/ T1 DNA. Add sodium acetate to 0.3 M, and 2.5 vol of 100% ethanol. Mix well, and store at –20°C for >30 min.

2. Centrifuge 15 min at maximum speed in a microcentrifuge at 4°C.
3. Remove supernatant and discard (check to be sure there are very few cpm in supernatant).
4. Wash with ethanol 70% and air-dry.
5. Resuspend each pellet in 21 µL of 5 m*M* Tris-HCl, pH 8.0 (*see* **Note 19**).
6. Denature by heating at 95°C for 3 min.
7. Transfer to ice for 30 s.
8. Centrifuge briefly (5 s) in microcentrifuge to bring condensation to the bottom of the tube.
9. Transfer to a thermal cycler reaction tube (0.2 mL).
10. Add 10 µL 5X phosphate-EDTA, 15 µL of 5 *M* sodium thiocyanate.
11. Melt a small amount of phenol crystals in an Eppendorf tube by heating in a water bath or heat block at 50°C.
12. Remove 4 µL of melted phenol and transfer immediately into sample. Quickly recap the tube and gently vortex to ensure that all phenol has entered the aqueous mixture. Check to make no phenol crystals have formed, and that no drops are adhering to the sides of the tube.
13. Transfer to a thermal cycler with "hot lid" attachment, and run PERT cycles (*see* **Note 20**):
    65°C, 10 min for 1 cycle
    (37°C, 15 min; 65°C, 2 min) for 85 cycles (approx 24 h).

3.4. SABRE Selection of Tester Homohybrids

1. At the end of the hybridization, transfer reactions to clean Eppendorf tubes, and centrifuge 1 min at room temperature to separate aqueous and phenol phases.
2. Remove aqueous solution from underneath the upper phenol phase, and transfer to a clean tube (*see* **Note 21**).
3. Add 25 µL of each reaction to 500 µL 1X S1 buffer, containing 100 µg/mL BSA, 25 µg of double-stranded salmon sperm DNA, and 25 U of S1 nuclease.
4. Digest at 37°C for 30 min.
5. Stop reaction by adding 50 µL of 1 *M* Tris, pH 8.0, 25 µL of 0.5 *M* EDTA.
6. Add 11 µL heparin 50 mg/mL.
7. Prepare streptavidin-coated magnetic beads: to remove azide from the beads, dilute 50 µL/reaction of beads in four to five volumes of 1 *M* TENT solution in an Eppendorf tube. Concentrate the beads using the magnetic attractor, and resuspend in the original volume of 1 *M* TENT solution.
8. Add 50 µL of the washed streptavidin beads to each reaction. Close the tube and seal with parafilm.
9. Place on a laboratory rocker or rotator, and allow to agitate gently for 1 h at room temperature.
10. Place the tubes in the magnetic attractor, and gently rock the attractor in order to capture all of the beads (approx 1 min; *see* **Note 22**).
11. Carefully remove the supernatant and dispose of as radioactive waste.

12. Wash the beads five times with 400 µL per wash of 0.1X SSC, 0.1% SDS, by removing the tube from the attractor, gently resuspending the magnetic bead pellet with the wash solution, transferring to a fresh tube, and replacing in the attractor (*see* **Note 23**).

13. Wash once in 800 µL of 1X *Bam*HI reaction buffer containing 0.1% Triton X-100 and 100 µg/mL BSA, at 37°C for 15 min.

14. Remove wash supernatant by attraction of beads as above; discard supernatant, and gently resuspend bead pellet in 100 µL 1X *Bam*HI reaction buffer with 0.1% Triton X-100, 100 µg/mL BSA.

15. Add 30 U *Bam*HI; incubate for 1 h at 37°C, with gentle agitation by hand every 5–10 min to prevent the beads from settling.

16. Place the reaction tubes in the magnetic attractor, and allow ample time for the beads to form a pellet.

17. Carefully remove the supernatant from the beads, being sure to take only the supernatant and to not touch the bead pellet.

18. Transfer the supernatant to a fresh tube; place this tube in the magnetic attractor and repeat the magnetic separation to ensure that the maximum amount of beads are removed from the supernatant.

19. Carefully remove the supernatant to a fresh tube.

20. Heat the recovered supernatants at 95°C for 5–10 min, to inactivate the *Bam*HI enzyme. These supernatants represent the selected tester homohybrid products from each hybridization.

21. Use 2.5 µL of the selected products for PCR amplification as in **Subheading 3.2., step 1.** For analysis of selected products, only one primer need be used to amplify each sample (T1 or D1). If another selection round is to be performed, driver and tester DNA can be directly amplified from the selected materials. The tester/T1 + driver/D1 hybridization (experimental) selected product and the driver/T1 + driver/D1 hybridization (control) selected product are used in the place of the original tester and driver DNA, respectively. Thus, for the next round, 4X 50 µL reactions of experimental/T1, 4X 50 µL reactions of control/T1, and 10X 50 µL reactions of control/D1 will be PCR amplified to generate the DNA for round 2 hybridizations, including the autohybridization control.

3.5. Analysis

1. Prepare a 5% acrylamide, 1X TBE sequencing gel (0.4-mm thickness), with wells of sufficient size to load 10 µL each. (Per 100 mL: 48 g urea, 12.5 mL acrylamide:bisacrylamide 38:2, 10 mL 10X TBE, water to 100 mL; *see* **Note 14**). Agitate the gel mixture while heating gently to dissolve the urea, then add 65 µL TEMED, 250 µL APS 25%, swirl to mix, and cast between gel plates. Add comb and allow the gel to polymerize (usually 1–2 h or more.).

2. Prepare a 1.6% agarose (normal, not low-gelling temperature), 1X TBE minigel containing 0.5 µg/mL ethidium bromide, with wells of sufficient size to load at least 7 µL each (*see* **Note 9**).

3. At the end of the PCR reactions, take 5 μL of each sample to a fresh tube, add 2 μL of ficoll/bromphenol blue, and load on the 1.6% agarose minigel, along with 0.5 μg of DNA molecular weight markers in the 150 bp to 4 kbp range.

4. Allow the gel to run at 50–100 V (depending upon gel system), until bromphenol blue is approximately halfway to the bottom of the gel.

5. Examine the gel under UV light. Intensity of ethidium bromide-stained PCR reaction products should be approximately half that of the size standard, or greater. If not, reactions may need to be repeated, with an increase in the number of PCR cycles.

6. If PCR reaction products show approximate expected yield and size distribution on the agarose minigel, take 3 μL of each PCR reaction, and add to 7 μL of formamide loading dyes in a fresh Eppendorf tube.

7. Heat to 75°C for 3 min; transfer to ice.

8. Load on a 5% sequencing gel which has been prerun to 45–50°C (approx 45 min to 1 h), according to the manufacturer's suggestions.

9. Run gel according to manufacturer's suggestions (usually up to 55 W for large gels), until xylene cyanol dye reaches the bottom of the gel.

10. Dismantle the gel, separate the glass plates, and transfer the gel to filter paper (*see* **Note 24**).

11. Dry gel on gel dryer, and expose to X-ray film. Length of exposure, with or without intensifying screen, will depend on the specific activity of the radioactivity used, as well as the investigator's patience. Normally, with the above standard conditions, 3 μL of PCR products can be clearly seen after exposure to Fuji X-ray film for 5 h with an intensifying screen at –80°C, or overnight without screen.

12. Comparison of the experimental and control reaction products will indicate whether any species are clearly enriched after the selection (see example in **Fig 3**). If so, these species can be isolated directly from the gel, as has been described previously *(5)*. If, however, the species common to both the experimental and control populations still represent virtually all of the experimental species, subsequent rounds of hybridization should be performed (*see* **Note 25**). Finally, to confirm that the enrichment seen after SABRE selection reflects true differences in mRNA accumulation, the total RNA used to make the SABRE libraries can be analysed by RNase protection (*see* example in **Fig. 4**)

4. Notes

1. Other selection procedures utilizing competitive hybridization include RDA *(6)* and EDS *(7)*.

2. Whereas increasing the ratio of driver DNA to tester DNA in the hybridization increases enrichment efficiency, the enrichment of species with different amplitudes is not equally affected by different ratios. The theoretical enrichment of two species, one overexpressed by fivefold in the tester and the other by 1000-fold, can be calculated following selection with either a fivefold or a 1000-fold excess of driver DNA to tester DNA. As seen below, the 1000-fold

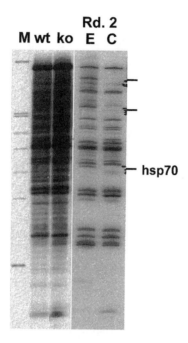

Fig. 3. Enrichment of cDNA species by SABRE. Liver cDNA SABRE libraries were prepared as described in the text from wild type mice ("wt") or mice homozygous for a disrupted allele of the transcription factor TEF ("ko"; P. Fonjallaz, J. Zakany, D. Duboule, and U. Schibler, unpublished). Comparison of the starting libraries demonstrates no detectable differences between the two. The libraries were used for SABRE selection as described in the text, with the wild-type mouse library used as the source of tester DNA and the mutant mouse library used as the source of driver DNA for the experimental (E) hybridizations. For control (C) hybridizations, the mutant mouse library was used to generate both tester and driver DNA. Following two rounds of SABRE selection (Rd.2), selected products from the experimental and control hybridizations were displayed on a polyacrylamide sequence gel. Bars at right indicate three species enriched in the experimental but not control hybridizations, which were isolated directly from the gel and analyzed. Sequence analysis of the species labelled "hsp70" indicated it displayed 99% sequence identity to a 327 base pair *Sau*3AI fragment of the mouse hsp70 cDNA.

overexpressed species greatly benefits from the higher driver-to-tester ratio. In contrast, enrichment of the fivefold overexpressed species is increased by less than twofold. This slight increase may not even be realized, because of increased contamination from the use of such a vast excess of driver DNA in the hybridization.

1 2 3 4

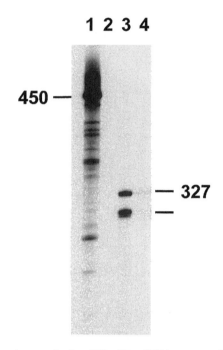

450 —

— 327

Fig. 4. Rnase protection analysis of Hsp70 mRNA accumulation in wild-type and knockout mice. To confirm that the enrichment of the hsp70 cDNA *Sau*3AI fragment by SABRE selection reflected differences in accumulation of its mRNA between wild-ype and mutant mice, the same total liver RNA from wild-type and mutant mice used to make the SABRE libraries was analyzed by RNase protection analysis. A 450-base, uniformly labeled RNA probe antisense to the hsp70 cDNA *Sau*3AI fragment was generated (lane 1), and hybridized with either yeast RNA (lane 2), or 10 µg of total liver RNA isolated from wild-type (lane 3) or homozygous knockout mice (lane 4). Following RNase digestion, analysis on a 6% sequencing gel demonstrates two protected probe bands, one of 327 bases representing full-length protection of the hsp70 mRNA ("327"), the other a smaller, partially protected band (lower bar at right). This smaller band may result from a sequence mutation introduced in the cloned cDNA fragment during PCR amplification. The results indicate that, consistent with the SABRE selection results, hsp70 mRNA accumulation was greater in liver RNA wild type mice than in the knockout mice.

Overexpressed species	Enrichment following SABRE selection round with D:T ratio of:	
	5:1	1000:1
5X Tester:driver	threefold	fivefold
1000X Tester:driver	5.9-fold	500-fold

From the enrichment ratio (**Subheading 1.1.**), it can be calculated that when the driver to tester ratio R tends towards infinity, the theoretical limit of enrichment is equal to N ($\lim[R \longrightarrow \infty]$ $(1 + R)/(1 + R/N) = N$). Thus, no matter how high the ratio of driver DNA to tester DNA, the fold-enrichment will not exceed the amplitude of the difference in abundance of the tester-overexpressed species between the tester and driver populations. Empirically, we have found that in our experimental systems, best results are generally obtained with a driver-to-tester ratio of 30:1.

3. The enrichment factor can be calculated in the following way Within two populations of cDNA, T and D, are two species: y, whose abundance (A) in both T [$A(yT)$] and D [$A(yD)$] is equal to Y:

$$A(yT) = Y; A(yD) = Y$$

and x, whose abundance in T [$A(xT)$], X, is greater than that in D [$A(xD)$] by the factor N:

$$A(xT) = X; A(xD) = X/N$$

If T and D are mixed, with D in excess by a factor of R:

$$At_{total} = AT + (R{\times}AD)$$

then, the total abundance of x and y in the combined populations will be:

$$Ay_{total} = A(yT) + [R{\times}A(yD)] = Y + RY$$

$$Ax_{total} = A(xT) + [R{\times}A(xD)] = X + RX/N = X + [R/N]X$$

Hybridization of the two populations will lead to a proportion of tester homohybrids (T^2) equal to

$$T^2 = [AT/(A_{total})]\ ([AT/(A_{total})]) = (AT)^2 / [AT + (R{\times}AD)]^2$$

(This is equal to the probability of randomly selecting two tester molecules from the mix of all tester and driver molecules.)

Multiplying this proportion by the total pool of molecules of y or x (Ay_{total} or Ax_{total}) will give the abundance of y or x tester homohybrids (yT^2 or xT^2):

$$yT^2 = Y^2/(Y + RY)^2 \times (Y + RY) = Y^2/(Y + RY) = Y/(1 + R)$$

$$xT^2 = X^2/(X + [R/N]X)^2 \times (X + [R/N]X) = X^2/(X + [R/N]X) = X/(1 + R/N)$$

Thus the ratio of x tester homohybrids to y tester homohybrids is:

$$[X/(1 + R/N)]\ /\ [Y/(1 + R)] = (X/Y) \times (1 + R)/(1 + R/N)$$

As the ratio of x to y in the starting tester population was X/Y, the ratio of x to y in the tester homohybrids has changed after one round of selection by the factor:

$$(1 + R)/(1 + R/N)$$

4. The combination of enzymes *Sau*3AI (or isoschizomers *Mbo*I and *Nde*II) for library preparation and *Bam*HI for SABRE selection differs from the original

SABRE protocol, which used *Sau*3AI and *Eco*RI, respectively *(2)*. This is because the site recognized by *Bam*HI (GGATCC) is a subset of the sites recognized by *Sau*3AI (GATC). Thus, the use of *Bam*HI in the selection procedure ensures that no species exist which contain *Bam*HI sites within the cDNA insert, which would be lost during the selection procedure. Any such "internal" *Bam*HI sites would have been already digested by *Sau*3AI to generate the library. This is not the case if another enzyme such as *Eco*RI is used. Indeed, when *Bam*HI digestion was used to reanalyze libraries which had been analyzed previously using *Eco*RI digestion, the first new fragment identified was found to contain an *Eco*RI site (D. Lavery, *unpublished observations*).

5. All reagents must be PCR grade and dedicated for SABRE selection where possible.

6. While we describe the analysis of cDNA populations, genomic DNA can also be used for SABRE selection in order to detect restriction polymorphisms or deletions/insertions in the genome. This was the original objective of the RDA protocol *(6)*. The recommended cDNA quantity is for RNA samples which permit large-scale isolation of poly(A)+ RNA and cDNA synthesis. However, in our laboratory, samples of total RNA as small as 100 ng have been used to generate cDNA for SABRE libraries. In this case, proportionately lower amounts of linkers are used in the ligation reaction. Our laboratory routinely follows the published protocol of Gubler *(8)*, though commercial cDNA synthesis kits may work as well.

7. To anneal linkers, combine an equimolar mixture of the two oligos (for instance, 10 nmoles of each) in 10 m*M* Tris, pH 8.0, 0.1 m*M* EDTA. Heat the mixture at 65°C for 10 min, add NaCl to 0.3 *M*, and allow the mixture to cool slowly to room temperature. The mixture is then placed on ice, and precipitated with 2.5 vol of 100% ethanol. After precipitation (> 1 h at -20°C), centrifuge, and wash with ethanol 70%. The resulting pellet is resuspended at 2 mg/mL in 50 m*M* NaCl, 10 m*M* Tris, pH 8.0, 0.1 m*M* EDTA. The annealed oligos will generate a 5'-GATC overhang at one end (for ligation to *Sau*3AI fragments) and a 3' overhang at the other (to prevent linker multimerization).

8. Caution! Solutions containing phenol are caustic and poisonous.

9. Ethidium bromide is a carcinogen and should be handled and disposed of with care.

10. When using radioactive materials, follow institutional safety guidelines for handling and disposal.

11. The degree of biotinylation should be tested for each batch of biotinylated oligos. This is essential because the reactants for biotinylation are highly labile, and if a supplier does not have a high enough turnover of these reagents, a large fraction of the oligos may in fact be nonbiotinylated. Biotinylation may be checked by electrophoresis through a high percentage polyacrylamide gel with a size standard (the biotin group causes the oligo to migrate as though it were 2–3 bases longer), or by functional testing by binding to streptavidin-coated magnetic beads. To test this, the oligos can be incubated with streptavidin beads in a standard binding reaction (without any detergent, which strongly absorbs light of wavelength around 260 nm), and the amount of oligos remaining unbound can be mea-

sured by absorbance at 260 nm. Alternatively, a sample can be PCR amplified using the biotinylated oligo and [α-^{32}P]dCTP, and the purified products can be bound to the streptavidin-coated beads. By measuring the bound and free cpm, the degree of biotinylation of the oligos can be estimated. If necessary, the biotinylated oligos can be purified from the nonbiotinylated oligos by gel electrophoresis and elution. However, provided the biotinylation is efficient (>95%), standard purification by the supplier (for example, reverse-phase chromatography) is sufficient. If the biotinylation is not efficient, the supplier should be pressured to resynthesize the oligo correctly.

12. Many protocols, including the first published SABRE protocol *(2)*, have used mung bean nuclease rather than S1 nuclease. However, as mung bean nuclease is sensitive to the salts and phenol present in the PERT hybridization solution, the hybridization reaction must be purified by organic extraction and ethanol precipitation before digestion. In contrast, S1 nuclease functions well in the presence of these contaminants, provided bovine serum albumin and carrier DNA are present in the reaction.

13. 20X SSC = 3 *M* NaCl, 0.3 *M* sodium citrate, pH 7.0.

14. Acylamide and bis-acrylamide are neurotoxins. Take adequate precautions when handling and disposing of solutions containing unpolymerised acrylamide.

15. As a control, the same area of the gel lane for the control, linker ligation reaction can also be extracted in parallel to the extraction of the other libraries. This can then serve as a control for PCR amplification, to detect species specific to amplification of linkers alone. While this protocol uses hot phenol extraction for DNA isolation, other procedures, such as absorption to glass beads, glass filters, or other polycationic matrices, are also compatible.

16. When comparing end-labelled cDNA fragments with uniformly labeled PCR products, two points should be kept in mind. First, the size of all of the species will be greater in the PCR samples by the length of the linkers at each end (26 bases). Second, the apparent intensity of the PCR amplified species will increase with size, because of the greater number of radiolabeled bases incorporated. Among the end-labeled fragments, in contrast, each strand will receive only one radiolabelled phosphate and will thus be labeled in an equimolar fashion.

17. The presence of DMSO improves amplification of sequences high in G/C content. Concentrations as high as 20% of reaction volume can be used. However, DMSO also appears to increase polymerase activity; thus, reaction conditions should be carefully monitored to prevent artifacts of overamplification.
Assembly of reactions should be performed using a master mix of components common to all reactions, to avoid pipetting errors and differences. The master mix should be assembled for N + 2 reactions when N are required. The master mix should be kept on ice, with the Taq polymerase added last, to minimize creation of primer multimer artifacts.

18. Occasionally, some groups have reported problems with precipitation of PCR reaction products using these conditions. This may be the result of measuring error, or to the use of an old supply of 2-propanol. If PCR reactions are yielding

microgram quantities of DNA per tube (as they should), a pellet should always be clearly visible. Likewise, a significant proportion (10–20% or more) of input radioactivity should precipitate with the pellet. When first starting these experiments, it is strongly recommended that all supernatants be saved until it is certain that the precipitation has succeeded. Should problems arise, new reagents should be used, or the 2-propanol proportion should be increased to 0.75-volume, or even to 1 vol. Care should be taken to completely mix the propanol and aqueous mixture, and to remove the supernatant without dislodging the nucleic acid pellet. As a pellet from propanol often does not adhere well to the tube wall, the supernatant may be removed using a Pasteur pipet with a fine tip drawn out in a Bunsen burner. This will give greater control than a plastic yellow tip in the removal of the supernatant.

19. This resuspension should be performed carefully. If the DNA pellet is not resuspended completely, random reassociation of complementary strands between the driver and tester populations will not occur.

20. The "hot lid" attachment is required because paraffin oil, which is normally used to prevent condensation, would absorb the phenol of the PERT hybridization mix, and no acceleration would take place. If no "hot lid" is available, the samples must be briefly centrifuged every few hours to collect the condensation within the cap.

21. Some phenol may carry over with the reaction samples. If preferred, the phenol can first be removed from the hybridization reaction by transferring the reaction to a fresh tube and extracting with two volumes of chloroform:iso-amyl alcohol (be sure the tube is chloroform-resistant). The 25-µL samples for S1 digestion can then be taken from the aqueous phase (in this case, the aqueous phase is the upper phase).

22. When gently rocking the attractor, care should be taken to allow the liquid in the tube to wash the inside of the cap. This will allow the recovery of beads which have a tendency to adhere there.

23. Introduction of bubbles should be avoided during washes (during pipeting, transferring between tubes, etc.), as these can dislodge the pellet of magnetic beads when the supernatant is being removed.

24. The gel should not be fixed with acetic acid/ethanol or methanol before drying. This will lead to acid depurination and degradation of DNA, making PCR amplification much more difficult. Gel fixation is not necessary when working with [^{32}P], since urea will not quench its signal. Saran wrap can be left on the gel during exposure, to prevent sticking of the gel to the film.

25. Once selected species are discernible in the experimental population, a sample of this population (as well as the control population) should be digested with *Sau*3AI and analyzed by sequencing gel electrophoresis alongside the undigested population. The enriched species should be visible in both the digested and undigested population, except that the digested species should migrate faster due to the removal of the PCR liners. This test will exclude any species which are enriched due to artifacts such as multiple inserts or multiple PCR linkers selected during

the procedure, which will not show the expected size after *Sau*3AI digestion. Those containing an insert of the expected size can be further analyzed by isolation from the undigested sample gel lane and PCR amplification *(5)*.

References

1. Morimoto, R. I., Kline, M. P., Bimston, D. N., and Cotto, J. J. (1997) The heat-shock response: regulation and function of heat-shock proteins and molecular chaperones. *Essays Biochem.* **32,** 17–29.
2. Lavery, D.J., Lopez-Molina, L., Fleury-Olela, F., and Schibler, U. (1997) Selective amplification via biotin- and restriction-mediated enrichment (SABRE), a novel selective amplification procedure for the detection of differentially expressed mRNAs. *Proc. Natl. Acad. Sci. USA* **94,** 6831–6836.
3. Kohne, D. E., Levison, S. A., and Byers, M. J. (1977) Room temperature method for increasing the rate of DNA reassociation by many thousandfold: the phenol emulsion reassociation technique. *Biochemistry* **16,** 5329–5341.
4. Miller, R. D. and Riblet, R. (1995) Improved phenol emulsion DNA reassociation technique (PERT) using thermal cycling. *Nucleic Acids Res.* **23,** 2339–2340.
5. Wan, J. S. and Erlander, M. G. (1997) Cloning differentially expressed genes by using differential display and subtractive hybridization. *Methods Mol. Biol.* **85,** 45–68.
6. Lisitsyn, N., Lisitsyn, N., and Wigler, M. (1993) Cloning the differences between two complex genomes. *Science* **259,** 946–951.
7. Zeng J., Gorski R. A., and Hamer, D. (1994) Differential cDNA cloning by enzymatic degrading subtraction (EDS). *Nucleic Acids Res.* **22,** 4381–4385.
8. Gubler, U. (1988) A one tube reaction for the synthesis of blunt-ended double-stranded cDNA. *Nucleic Acids Res.* **16,** 2726.

22

UVB-Regulated Gene Expression in Human Keratinocytes

Analysis by Differential Display

Harry Frank Abts, Thomas Welss, Kai Breuhahn, and Thomas Ruzicka

1. Introduction

The ozone layer of the earth effectively absorbs high-energy solar ultraviolet (UV) radiation below 290 nm. Consequently UVC (100–280 nm) radiation is completely blocked by the upper atmosphere while UVB (from 290–320) and UVA (320–400 nm) radiation reaches the earth's surface *(1,2)*. Of this terrestrial UV light, the midrange ultraviolet radiation (UVB) is the most relevant with respect to physical injury to human skin and causes severe damage, including cutaneous inflammation, immunosuppression, as well as the induction and promotion of cancer (for review, *see* **ref. 3**). In searching for the physiological targets of UVB radiation, it is important to note that 99% of this radiation is absorbed within the outermost 0.03 mm of the epidermis *(4)*. Therefore, keratinocytes probably constitute a major cellular target. Early events in the UV response of mammalian cells are the activation of transcription factors such as AP-1 *(5–7)* and nuclear factor-κB (NF-κB) *(6)* (*see also*, Chapter 16, this volume) as well as the initiation of signal-transduction events mediated by tyrosine kinases *(8,9)* (*see also*, Chapter 7, this volume). Although there is an increasing amount of information about cellular components that lie both upstream and downstream of these proteins within UV-activated signal-transduction pathways, investigations of the response of mammalian cells to UV exposure have mainly been restricted to a small number of known target genes.

In order to identify and clone UV-inducible genes without any *a priori* knowledge of their sequence or function, differential hybridization and subtractive hybridization can be used (for review, *see* **refs. 10** and *11*). The only

From: *Methods in Molecular Biology, vol. 99: Stress Response: Methods and Protocols*
Edited by: S. M. Keyse © Humana Press Inc., Totowa, NJ

requirement for these techniques to be of use, is that the induction of a particular gene is reflected by the relative abundance of its corresponding mRNA. For the systematic identification of UV-regulated genes in keratinocytes, subtractive cDNA hybridization in combination with differential cDNA screening has been used in the past *(12,13)*. Although the use of these methods led to the isolation of novel UV-induced genes these techniques have intrinsic limitations which restrict their application. First, large quantities of poly(A)$^+$ RNA are required, although this requirement can be overcome by incorporating polymerase chain reaction (PCR) amplification of the cDNA into the protocol *(14,15)*. Second, only two cell populations can be compared in parallel. Finally, the very nature of the subtraction and screening approach restricts each analysis to either induced or repressed genes. To cover both modes of regulation, two independent subtractions and screenings must be performed.

Differential display PCR (DD-PCR) and the related technique, termed random arbitrary primed PCR (RAP-PCR), were first described by Liang and Pardee *(16)* and Welsh et al. *(17)* respectively. Both provide a potentially powerful tool in the analysis of UVB-modulated gene expression *(18)*. Like subtractive hybridization, DD-PCR allows the isolation of unknown genes on the basis of the cellular abundance of their corresponding transcripts. Since its first application for the analysis of genes which are differentially expressed in tumors as opposed to normal tissue *(19)*, differential display has been used to analyze differential gene expression in a wide variety of biological systems (for review, *see* **refs.** *20* and *21*). Many modifications to the original protocol has been described but basically the technique consists of three steps (**Fig. 1**). After isolation of total RNA from the cells that are to be compared, single-

---

Fig. 1. Differential Display PCR Analysis: Schematic Representation of the Method Using Single-Base anchor Primers. (**I**) Reverse transcription is performed on total RNA. The poly(A)$^+$ RNA fraction is reverse transcribed in three separate reactions using three different single-base anchor primers. (**II**) The single-stranded cDNA is then used in a PCR reaction using the original anchor primer in combination with an arbitrary primer (13 mer). For labeling of the PCR product, a radioactive nucleotide is included. The specificity of the arbitrary primer is determined by a heptameric sequence to which a restriction site is added. The use of the TG-anchor primer is shown as an example. According to the priming specificity of the arbitrary primer, cDNA fragments for a subset of mRNA species with a characteristic sizes are generated. (**III**) These RNA fingerprints are generated for all the cell populations analyzed for differential gene expression and separated on a sequencing gel side by side. After autoradiography, the differentially expressed bands can be identified. Position of the bands corresponding to RNA specifically expressed in A (A1, A2) or B (B1, B2) are indicated by arrows.

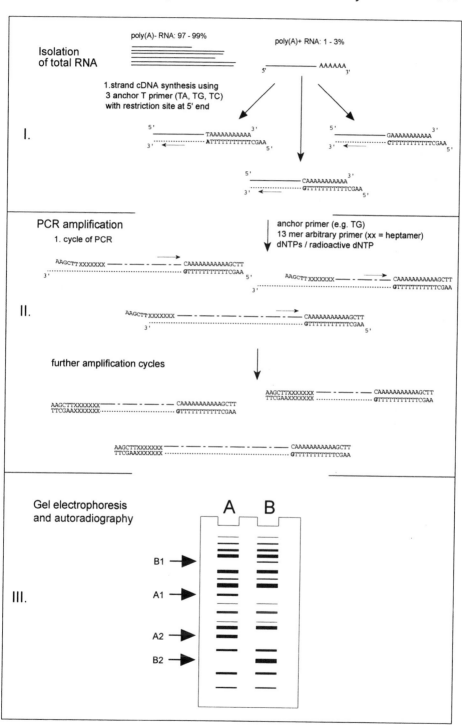

stranded cDNA is prepared by reverse transcription using an anchor primer. The use of different anchor primers allows a reduction of the complexity of the analyzed cDNA population. Whereas the early protocols used 2 base-anchored primers and thus needed 12 separate reverse-transcription reactions, recent modifications have included the use of single base-anchored primers (22), thus reducing the number of reactions from 12 down to 3. The single-stranded cDNA molecules are then PCR amplified using the anchor primer used for the reverse transcription and a second, arbitrary primer. By using different arbitrary primers, in principle each mRNA species in a cell can be displayed as a distinct band with a characteristic size. During PCR, a radioactive nucleotide is incorporated. The PCR products from the cells that are to be analyzed are separated on a sequencing gel side by side and visualized by autoradiography. By simply comparing the band pattern in the adjacent lanes, corresponding RNAs with differential expression can be identified.

One of the major advantages of differential display is that differentially expressed genes in both populations are detected. This means that during analysis of UVB-modulated gene expression both UV-induced and UV-repressed genes can be identified. Moreover, differential display is not limited to the comparison of just two different cell populations. This gives the opportunity to compare not only unirradiated cells with irradiated cells at one time point but to perform a kinetic analysis of between 50 and 100 different genes per primer combination. This is an important advantage as there are not only pronounced differences in the mRNA levels in an all or nothing fashion following UV irradiation but also time-dependent increases and decreases in the levels of particular transcripts. In addition, the parallel analysis of more than two related RNA populations helps to discriminate bands that show random variation between lanes, thus reducing the risk of isolating false positives.

2. Materials

2.1. Cell Lines and Cell Culture

1. The spontaneously transformed human epidermal cell line HaCaT (23).
2. Medium: Dulbecco's modified Eagle's medium (DMEM) with sodium pyruvate, (Gibco-BRL, Eggenstein, Germany) containing 100 µg/mL penicillin/streptomycin, 2 mM L-glutamine and supplemented with 10% fetal calf serum (FCS).
3. Phosphate-buffered saline (PBS): 120 mM NaCl, 2.7 mM KCl, phosphate buffer 10 mM, pH 7.4.

2.2. UVB-Irradiation

UVB irradiation equipment: Bank of four FS20 sunlamp bulbs (Westinghouse Electric. Corp., Pittsburgh, PA).

2.3. RNA Preparation

1. RNaseZap (Ambion, Austin, TX).
2. Trizol reagent (Gibco-BRL).
3. Chloroform.
4. Isopropanol.
5. Diethylpyrocarbonate (DEPC). Use under a fume hood and wear gloves (suspected carcinogen!).
6. DEPC-treated ddH$_2$O: 0.1% DEPC is prepared and left for at least 1 h before autoclaving.

2.4. Reverse Transcription

1. M-MLV reverse transcriptase (Superscript II, Gibco-BRL).
2. 5X First-strand buffer: 240 mM Tris-HCl, pH 8.3, 375 mM KCl, 15 mM MgCl$_2$ (Gibco-BRL, supplied with Superscript II).
3. RNAse inhibitor (40 U/µL, Boehringer Mannheim).
4. 100 mM dithiothreitol (DTT) (Gibco-BRL, supplied with Superscript II).
5. dNTP mix (400 µM of each nucleotide).
6. Anchor-Primer: one-base-anchored 16 mer oligo-dT primer *(22)*:
   ```
   TA:  5'-AAGCTTTTTTTTTTTA-3'
   TC:  5'-AAGCTTTTTTTTTTTC-3'
   TG:  5'-AAGCTTTTTTTTTTTG-3'
   ```
   Prepare 10 µM stock solution in TE (10 mM Tris-HCl, 0.1 mM EDTA). Store at –20°C.

2.5. Differential Display PCR

1. Arbitrary 13 mer primer (RNAimage Kit 4, GenHunter, Nashville, TN):
   ```
   AP25:  5'-AAGCTTTCCTGGA-3'
   AP26:  5'-AAGCTTGCCATGG-3'
   ```
   Prepare 10 µM stock solution in TE and store at –20°C.
2. 10X PCR-buffer: 100 mM Tris-HCl, pH 8.3, 500 mM KCl, 15 mM MgCl.
3. 0.1% Gelatin.
4. dNTP mix (25 µM of each nucleotide).
5. Taq polymerase.
6. Thermocycler (Trio-Block; Biometra, Germany).

2.6. Polyacrylamide Gel Electrophoresis

1. Sigmacote (Sigma, Deisenhofen, Germany) or BlueSlick (Boehringer-Ingelheim Bioproducts, Heidelberg, Germany).
2. Binding coat: Add the following to 12.5 mL 100% EtOH, 250 µL 10% acetic acid, and 60 µL silan, prepare fresh.
3. Acrylamide/bis solution 19:1 40%.
4. 5X TBE loading buffer: 0.1% xylene cyanol, 0.1% Bromphenol blue, 10 mM EDTA, pH 8.5, 95% deionized formamide.
5. TrackerTape (Amersham, Braunschweig, Germany).
6. Conventional DNA sequencer.

2.7. Reamplification of cDNA Fragments

1. 3 *M* Sodium acetate (NaOAc), pH 5.2.
2. Glycogen (Boehringer Mannheim).
3. Jetsorb (GENOMED, Bad Oeynhausen, Germany) or QIAquick spin columns (Qiagen, Hilden, Germany).

2.8. Cloning

TA Cloning Kit (Invitrogen, San Diego, CA).

2.9. Northern Blot Analysis

1. Hybond N$^+$ (Amersham, Braunschweig, Germany).
2. Multi Tissue Northern (MTN) blots (Clontech, Palo Alto, CA).
3. 20X SSC: 3 *M* NaCl, 0.3 *M* sodium citrate, pH 7.0.
4. Random prime labeling Kit (Megaprime, Amersham, Pharmacia Biotech, Freiburg, Germany).
5. [α-^{32}P]dCTP (specific activity 3000 Ci/mmol).
6. S-400 spin column (Amersham Pharmacia Biotech, Freiburg, Germany).
7. Dig Easy Hyb (Boehringer Mannheim).

2.10. Sequence Analysis

1. ABI PRISM DyeDeoxy Terminator Kit (Perkin-Elmer, Weiterstadt, Germany).
2. Sequencing primers:
   M13 universal: 5'-TGTAAAACGACGGCCAGT-3'
   M13 reverse: 5'-CAGGAAACAGCTATGACC-3'

3. Methods
3.1. Cell Culture

1. Culture the HaCaT cells in 100×20-mm tissue culture dishes with DMEM containing 10% FCS at 37°C, with 5% CO_2 in a humidified atmosphere (*see* **Note 1**).
2. At approximately 70% confluence, UVB irradiate the cells (*see* **Subheading 3.2.**) and then culture them for the desired time in DMEM before harvesting for total RNA preparation.

3.2. UVB Irradiation

Replace the tissue culture medium with phosphate-buffered saline (PBS) and expose the cells without the plastic dish lid to UVB (100 J/m^2) using a bank of four FS20 sunlamp bulbs (Westinghouse Electric. Corp.), which emit primarily in the UVB range *(24)* (*see* **Note 2**).

3.3. RNA Preparation

1. Wash the HaCaT cells grown in a 100-mm tissue culture dish twice with PBS (4°C) and lyse them by adding 1.6 mL Trizol Reagent (*see* **Note 3**).

2. Collect the lysate with a cell scraper and transfer it to a 2-mL Eppendorf tube. At this point, the lysates may be stored frozen at –80°C for several months.
3. Add 320 µL chloroform to the lysate at room temperature (RT) and vortex vigorously in order to shear genomic DNA. Incubate at RT for 3 min.
4. Centrifuge (15,000g) for 15 min at 4°C and transfer the supernatant to a fresh tube. In order to minimize DNA contamination, the aqueous phase of the first extraction is re-extracted with Trizol Reagent.
5. Add 400 µL Trizol Reagent and 80 µL chloroform to the supernatant from **step 4**; vortex and centrifuge as in **step 4**.
6. Precipitate the RNA by adding 800 µL isopropanol. Incubate 10 min at RT. Centrifuge (15,000g) for 15 min at 4°C. Air-dry the pellet and resuspend in 50 µL DEPC–ddH$_2$O.
7. Quantitate RNA in a spectrophotometer by reading the optical density (OD) at 260 nm and 280 nm *(25)*. Check integrity and quantification by running 1 µg of each sample on a 1% agarose gel containing formaldehyde.
8. Store RNA at –80°C.

3.4. Reverse Transcription

One microgram of total RNA from each cell population to be analysed is divided into three separate reactions (333ng per reaction) and reverse transcribed in a volume of 20 µL using each of the three one-base anchored 16mer oligo-dT-primers (TA, TC, TG) with Superscript II reverse transcriptase.

1. To the total RNA (333 ng in 9 µL DEPC–ddH$_2$O) add 2 µL anchor primer (2 µ*M*) and denature the RNA by heating to 70°C for 10 min. Put on ice for 5 min and recover any condensate by brief centrifugation.
2. Add on ice 8 µL from a master mix containing the following components for each reaction:
   2 µL DTT (0.1 *M*)
   4 µL 5x reaction buffer
   1 µL dNTP (400 µ*M* of each nucleotide)
   1 µL RNAse inhibitor (40 U/µL)
3. Prewarm to 37°C for 2 min before adding 1 µL (200 U) Superscript II reverse transcriptase.
4. Incubate for 40 min at 37°C and 20 min at 40°C.
5. Inactivate the enzyme by heating to 70°C for 5 min.
6. Store the reaction mix at –20°C.

3.5. Differential Display PCR

For the display PCR reactions, upstream 13mer arbitrary primers are used in combination with the same anchored oligo-dT-primer used for reverse transcription.

1. Use 2 µL of reverse transcription mix for each PCR. From a master mix add 16 µL containing for each reaction:

2.0 μL oligo-dT-primer (2 μ*M*)
2.0 μL arbitrary primer (2 μ*M*)
1.8 μL 10X PCR buffer
2.0 μL gelatin (0.1%)
1.6 μL dNTP mix (25 μ*M*)
0.4 μL [α-^{32}P]dCTP (3000 Ci/mmol)
6.2 μL ddH$_2$O

3. Cover reaction mix with 30 μL mineral oil.
4. Perform a "hot-start" by adding 2 U Taq polymerase in 2 μL 1X PCR buffer in the first PCR cycle after denaturation for 1 min at 95°C, during a pause at 42°C. Annealing is then continued at 41°C for 1 min and elongation for 2 min at 72°C.
5. Continue the PCR for 39 cycles with 50 s at 94°C, 1 min at 42°C (+ temperature increment of 0.1°C/cycle), and 2 min at 72°C, and then extend at 72°C for 5 min (*see* **Note 4**).

3.6. Polyacrylamide Gel Electrophoresis

1. To fix the gel to one of the glass plates treat one plate with binding coat and the other with Sigmacote or BlueSlick.
2. Prepare a 6% denaturing polyacrylamide sequencing gel (*see* **Note 5**).
3. Flush out urea and gel debris from the wells and prerun at 40 W until the gel temperature is 50°C.
4. Mix 4–6 μL of the PCR reaction mix with 0.25 volume of 5X loading buffer, incubate at 75°C for 2 min, and keep on ice until loading.
5. Flush urea out of the wells again and load equal amount of samples.
6. Electrophorese for approx 4 h at 40–45 W constant power until the xylene cyanol marker reaches the bottom of the gel (*see* **Note 5**).
7. Let the gel cool to room temperature and separate the glass plates. Cover the gel adhering to the glass plate with Saran Wrap.
8. Use TrackerTape to unequivocally determine the position of the X-ray film.
9. Expose the gel to X-ray film (Kodak XAR or Amersham Hyperfilm) without screen at –70°C for 8–24 h (*see* example in **Fig. 2**).

3.7. Reamplification of cDNA Fragments

1. Identify DD-PCR products on the exposed X-ray film and mark side borders by two small dots (*see* example in **Fig. 3**).
2. Align the film with the gel by the help of the TrackerTape labels and fix it with adhesive tape.
3. Transfer your marks to the gel by punching through the film with a needle.
4. Cut the marked portion of the gel with a clean scalpel and transfer to 1.5-mL reaction tube .
5. Add 100 μL ddH$_2$O and elute the DNA from the gel by heating the tube to 80–90°C for 20 min.
6. Pellet the gel by centrifugation (15,000*g*) for 2 min in microcentrifuge and transfer the supernatant to a new tube.

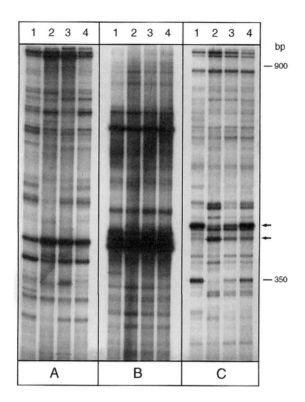

Fig. 2. Differential Display PCR Analysis with HaCaT Cells. Differential display with unirradiated cells harvested at the beginning of the experiment (lane 1), mock-irradiated cells cultivated in parallel to the irradiated cells for 24 h (lane 4), UVB-irradiated cells (100 J/m^2) 8 h (lane 2) and 24 h (lane 3) postirradiation. Differential display PCR has been performed on the same cDNAs using different primer combinations: (A) TA/AP25; (B) TA/AP26; (C) TC/AP25. Arrows indicate positions of differential cDNAs further analyzed as shown in **Figs. 2** and **4**. Modified from **ref. 18** with permission (copyright 1997, American Society for Photobiology).

7. Precipitate the DNA by adding 10 μL of 3 M NaOAc, 2.5 μL glycogen (20 mg/mL) and 450 μL of cold 100% EtOH. Incubate over night at –20°C or 2 h at –80°C. Pellet the DNA by centrifugation (15,000g) for 30 min at 4°C. Carefully remove the supernatant and rinse the pellet with 300 μL cold 75% EtOH. Air-dry the pellet and redissolve in 10 μL TE (10 mM Tris-HCl, 0.1 mM EDTA). Use 4 μL for reamplification.
8. Reamplify the cDNA in a 40 μL PCR reaction using the same primer combination (0.2 μM) used to generate the fingerprint but with 20 μM of each dNTP by using a touch-down cycling protocol (annealing temperature was decreased from 50°C in the first cycle to 45°C during the following five cycles and then kept constant for 34 additional cycles) and hot start (*see* **Note 6**).

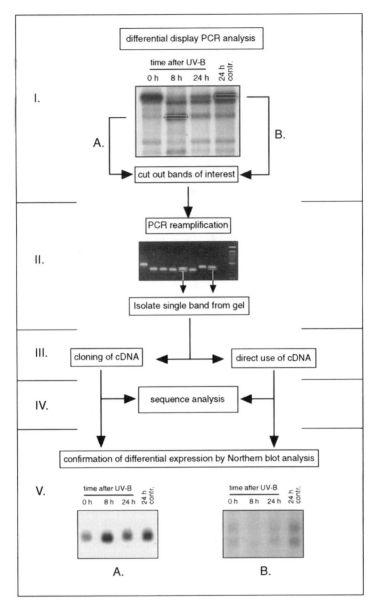

Fig. 3. Example for the use of Differential Display PCR for the Identification and Cloning of UVB modulated Genes. The human keratinocyte cell line HaCaT has been irradiated with 100 J/m^2 and harvested 8 h and 24 h after irradiation. Unirradiated cells at the beginning and after 24 h in the culture served as control. (**I**) Differential display analysis has been performed as described in **Fig. 1.** A UVB-induced (A) and a UVB repressed (B) band has been isolated from the gel. (**II**) The cDNAs eluted from the gel fragments have been reamplified and purified by agarose gel electrophoresis. (**III**) After TA cloning, the cDNAs are (**IV**) sequenced. Alternatively, direct sequencing is possible. (**V**) Confirmation of the differential expression of the isolated cDNAs is done by Northern blot analysis with total RNA using the cloned cDNA as probe.

3.8. Cloning

1. Separate reamplified cDNA fragments by electrophoresis of 35 μL PCR reaction mix through a 1.5% agarose gel stained with ethidium bromide (*see* example in **Fig. 2**).
2. Cut out a single band of the expected size and purify the DNA using Jetsorb or QIAquick spin columns.
3. Clone the cDNA fragment into pCR II via the T/A cloning procedure using the TA-cloning kit. Prior to plasmid preparation, individual colonies are checked for inserts of the expected size by PCR analysis using the same primer combinations and cycling conditions as for reamplification (*see* **Note 7**).
4. Transfer bacteria from three independent single colonies to a master plate in an ordered array using sterile pipet tips. Put the rest of bacteria adhering to the pipet tip in 50 μL ddH$_2$O. Liberate the plasmid DNA by incubation of the bacteria for 15 min at 95°C in a water bath or a thermocycler with a heated lid. Use 5 μL for PCR amplification of the insert in a 20-μL reaction mix with the same concentration of all components and the same PCR conditions as described in **Subheading 3.7., step 8**.
5. Separate PCR reaction on a 1.5% agarose gel stained with ethidium bromide. Identify recombinant plasmids on the basis of the presence of a PCR product of the expected size.

3.9. Northern Blot Analysis

1. Size separate 10 μg total RNA from HaCaT cells in a 1% denaturing formaldehyde gel and transfer to Hybond N$^+$ nylon membrane by capillary blot *(25)*.
2. Multiple tissue Northern blots (MTN) containing 2 μg poly(A$^+$) RNA from pancreas, kidney, skeletal muscle, liver, lung, placenta, brain, and heart can be purchased from Clontech (Palo Alto, CA) (*see* **Note 8**).
3. Isolate the cDNA insert from the recombinant plasmid by PCR reamplification or restriction enzyme digestion (*Eco*RI) and purify by agarose gel electrophoresis.
4. Label cDNA insert with [α-^{32}P]dCTP by random priming using the Megaprime labeling kit.
5. Remove unincorporated nucleotides by chromatography over a S-400 spin column.
6. Hybridize the membrane in Dig-Easy Hyb solution with the labeled probe (10^6 cpm/mL) overnight at 42°C. Nonspecific bound probe is removed by several washes with increasing stringency using initially 2X SSC, 0.1% SDS at RT followed by 0.1X SSC and 0.1% SDS at 42–60°C depending on the probe. Background radioactivity is checked with a hand held β-counter.
7. Expose the washed membrane to X-ray film at –70°C with intensifying screen overnight to 72 h (*see* examples in **Figs. 3–6**).

3.10. Sequence Analysis

1. Purify recombinant plasmids via Qiagen columns (QIAGEN, Hilden, Germany) .
2. Cycle sequence 500 ng plasmid DNA with the ABI PRISM DyeDeoxy Terminator Kit (Perkin-Elmer, Weiterstadt, Germany) using M13 forward and reverse primers on an automated sequencer (ABI 373).

3. Homology search against the latest EMBL Nucleotide Sequence Database with the FASTA program *(26,27)* can be performed via the world wide web (URL: http://www2.ebi.ac.uk/fasta3).

3.11. Examples

The usefulness of the differential display approach to identify UV-regulated genes has been evaluated by analyzing the modulation of gene expression after UVB-irradiation (100 J/m^2) in the human epidermal cell line HaCaT. Differential displays were generated from UV-B irradiated cells 8 h and 24 h after irradiation and compared with those from unirradiated cells that were harvested at the beginning of the experiment or mock irradiated cells cultivated in parallel to the irradiated cells for 24 h. **Figure 2** shows an example of a differential display using three different primer combinations. The subsequent processing and analysis of one UV-B induced gene (A) and one UVB-repressed gene (B) is shown as an example (**Fig. 3**).

After isolation of the differentially expressed band the cDNA can be reamplified using the same primer combination as for the generation of the display. Isolating the major reamplification product from the gel helps to avoid contamination with undesired cDNAs generated during the reamplification. Since this differential cDNA could be also contaminated by non differential cDNA of the same size, cloning of the reamplified material may be useful to isolate a single cDNA species. Alternatively, the reamplified cDNA may be used directly for further analysis. The identity of the unknown cDNA clones is determined by sequence analysis of the cloned cDNA or by direct sequencing of the reamplified cDNA. The UV-induced sequence in **Fig. 3** A was identified as the human acidic ribosomal phosphoprotein P0 *(28)*. The ribosomal phosphoprotein P0 has been found to be induced after treatment of cells with a variety of DNA damaging agents and is thought to be related to DNA repair *(29)*. The UV-repressed sequence in **Fig. 3** B is a new sequence with partial sequence homology to serine protease inhibitors (serpins) *(30)*.

As a result of the stochastic nature of the PCR amplification used in differential display, it is necessary to confirm the differential expression of the isolated sequences by a second method. Preferably, this is done by Northern blot analysis. In the example given in **Fig. 3** the induction of band A and the repression of band B could be confirmed. The complete Northern blot analysis of an UVB-repressed gene (HUR 7) and an UVB-induced gene (HUI 13) is shown in **Fig. 4A**. The ability of the differential display to detect both induction as well as repression of genes leads to the finding that approximately one-third of the UVB-regulated sequences identified so far are repressed. **Figure 4B** shows another example of this UVB-mediated repression in an independent Northern blot analysis. The clone HUR F95 was isolated from a different differential

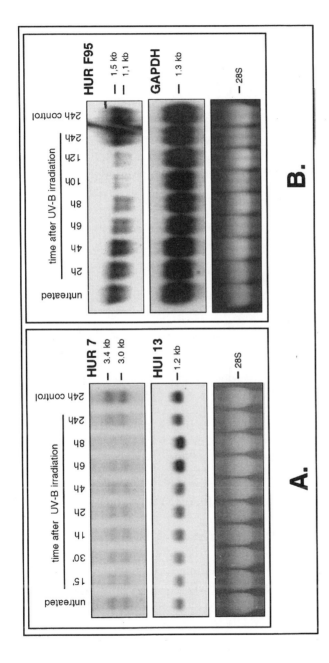

Fig. 4. Northern blot confirmation of differential expression of isolated UVB modulated sequences. Ten micrograms total RNA from UVB-irradiated HaCaT cells (100 J/m²) cultivated for different periods of time after irradiation and from unirradiated controls where hybridized to cDNA clone HUR 7 and HUI 13 (A) or to HUR F95 (B). Control hybridization was done with a GAPDH-specific probe. Ethidium bromide (EtBr) staining of 28S RNA was used as loading and integrity control. [Modified from ref. 18 with permission (copyright 1997, American Society for Photobiology).]

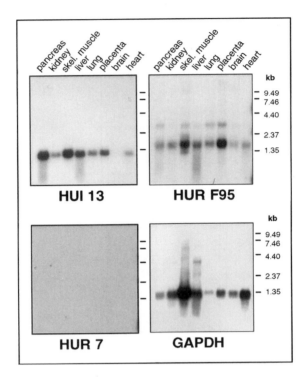

Fig. 5. Comparison of denaturing and nondenaturing gel system. Differential display (primer combination TC/AP25) with unirradiated primary human keratinocytes (lane 1) and HaCaT cells harvested at the beginning of the experiment (lane 2), mock irradiated cells cultivated in parallel to the irradiated cells for 24 h (lane 5) and UVB-irradiated cells (100 J/m^2) 8 h (lane 3) and 24 h (lane 4) postirradiation. Aliquots of the same reaction have been separated on a non-denaturing (**A**) and a denaturing (**B**) gel. Positions of identical bands are indicated (*see* **Note 5**).

display analysis (primer combination TA/AP18) and shows an even more pronounced transient UVB-repression.

In summary, these examples demonstrate that the differential display PCR is a useful approach for identifying and cloning genes which are differentially regulated during the UVB response of human keratinocytes. In particular, the ability to identify genes that are negatively regulated by UVB in addition to UV-induced gene sequences will lead to the identification of new targets and mediators of UVB-irradiation effects in the skin.

4. Notes

1. Take into account that the HaCaT cells harbor mutations in the gene for p53 (*31,32*). Because p53 is a major regulator of the response to UV-induced damage

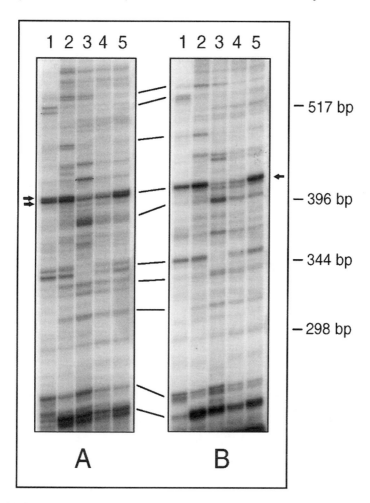

Fig. 6. Tissue distribution of UVB-regulated sequences. Expression of UV-induced (HUI) and UV-repressed (HUR) cDNA clones in different human tissues. A multi tissue Northern blot containing 2 μg poly(A⁺) RNA from the indicated human tissues was sequentially hybridized to the clones depicted below each autoradiograph. Hybridization with a GAPDH specific probe served as a control.

in human cells *(33,34)*, HaCaT cells might not be representative of all aspects of the UV response in human keratinocytes. Nevertheless, the described method also works with normal human keratinocytes (**Fig. 5**), thus permitting analysis of the UVB response in primary cells.

2. The UVB radiation source must be calibrated in order to determine the time and distance from the bulbs needed to achieve the desired UVB dose. To monitor the UVB output an IL1700 research radiometer and SEE 240 UVB photodetector

(International Light, Newburyport, MA) can be used. The FS20 sunlamp bulbs used as UVB device emit mainly in the UVB range with a peak emission at 313 nm. This characteristic made this lamp a commonly used UVB source in photobiology. Nevertheless, one should keep in mind that these lamps emit a continuous spectrum from 250–400 nm.

3. Bear in mind that the hands of the investigator are a rich source of RNases. Wear gloves at all times and change these often. Use fresh plastic tips and tubes. The bench, automatic pipets, and plastic tubes can be easily decontaminated of RNases using RNaseZAP. Use only DEPC-treated H_2O for RNA work. The procedure used for the preparation of total RNA, is based on the acid guanidinium thiocyanate–phenol–chloroform protocol described by Chomczynski and Sacchi *(35)*. Because of the repeated extraction of the RNA contamination with genomic DNA is rarely seen. If necessary, the RNA can be treated with RNase free DNase I (10 U/µL, Boehringer). To do this, 10 µg RNA is incubated in a total volume of 100 µL containing 10 mM $MgCl_2$, 10 U DNase I, 1mM DTT, and 16 U RNase inhibitor at 37°C for 30 min. Phenol–chloroform extraction of the RNA is then done using 1 mL Trizol Reagent and 200 µL chloroform as described in **Subheading 3.3.**

4. The sequences of the primers used in this protocol are essentially those provided in the RNAimage Kit 1–4 (GeneHunter, Brookline, MA). A total of 80 arbitrary primers are provided by GeneHunter. The quality of the generated differential display patterns (background) varies between the different primers. In general, primers with two or more G or C at their 3' end are assumed to generate a higher background. The incremental increase in annealing temperature used in the PCR gives higher stringency and is used to promote the generation of distinct bands with low background. For enhanced reproducibility, elongation of the primers at their 5' ends by adding restriction sites and the use of higher annealing temperature has been proposed *(36)*. In our system, elongation of the GeneHunter primers TA and AP25 resulting in a 24 mer anchor primer and a 22 mer arbitrary primer was not beneficial because they produced an increased background compared to the original primers. To determine the size of the bands in the differential display, a labeled DNA size standard can be used.

5. Good results are also obtained using a denaturing 5% gel prepared using LongRanger™ (AT Biochem, PA) acrylamide with modified crosslinkers. This is available as a 50% stock solution which can be stored at room temperature. LongRanger gels provide enhanced resolution, even spacing of large and small DNA fragments, and allow shorter prerun and run times and lower running temperatures (40–50°C). The binding of the gel to the glass plate is especially helpful when using conventional combs. If shark-tooth combs are intended to be used, binding of the gel is not necessary and the gels may be transferred to 3M paper and dried. However, copying the position of differential bands from the film to the gel is sometimes easier when backlight illumination can be used, which, of course, is possible only with a gel attached to a glass plate. Casting the gel on a transparent polyester support film (e.g., GEL-FIX, Boehringer-Ingelheim

Bioproducts, Heidelberg, Germany) works very well, but we notice that the bands become "fuzzy" in comparison to gels fixed on the glass plate. Some protocols favor the use of native gels for differential display *(37)*. As shown in **Fig. 5**, the general band pattern in denaturing and nondenaturing gels is very similar. Nevertheless, the denaturing gels yielded more distinct and sharper bands. The doublets obtained in the nondenaturing gel most likely represent double-stranded cDNA and incompletely annealed single-stranded cDNA of the same gene. This has been proven for one UVB-repressed gene that appeared in the denaturing gel as a single band and in the nondenaturing gel as two closely migrating bands (depicted by arrows in **Fig. 5**). All three bands were excised from the gels, reamplified, cloned and sequenced as described. For all three individual bands the same sequence could be confirmed representing the HaCaT UVB-repressed clone 7 *(18)*.

6. The touch-down PCR protocol usually resulted in one major PCR product. Reamplification of excised cDNAs in the size range of 150–910 bp yielded a PCR product of the expected size. Nevertheless, cDNAs larger than 600 bp sometimes generated additional bands. Because these were reproducible even when the template fragment was gel purified before reamplification, the shorter fragments may result from internal priming.

7. Although the use of the original differential display primers for checking the recombinant plasmids permits insert-specific reamplification, it is also possible to use the M13 universal and reverse sequencing primer for this purpose (different cloning vectors may require different primers). This allows the use of a large master-mix for all clones to be tested regardless of the original primer combination used to generate the cloned fragment. Because an approx 259 bp PCR fragment (PCR II vs B) is also generated from plasmids carrying no insert, this gives a further control for the presence of plasmid DNA. Instead of the touch-down protocol, annealing at 55°C for 1 min is performed.

8. Tissue distribution of the isolated cDNAs can be easily determined by Northern blot analysis using Multi Tissue Northern (MTN) blots. The example in **Fig. 6** shows two ubiquitously expressed clones (HUI 13, HUR F95) and a clone whose expression turned out to be restricted to keratinocytes (HUR 7). Northern blot analysis using total RNA may be unable to detect certain transcripts because of the low abundancy of the corresponding mRNA *(38,39)*. Nevertheless, determining the transcript size of a new gene can often be accomplished by using the higher sensitivity afforded by the poly(A$^+$) RNA MTN blots. Confirmation of differential expression can then be attempted using semiquantitative RT-PCR on the original total RNA samples.

References

1. Rosen, C. F., Gajic, D., and Drucker, D. J. (1990) Ultraviolet radiation induction of ornithine decarboxylase in rat keratinocytes. *Cancer Res.* **50,** 2631–2635.
2. Shea, C. R. and Parrish, J. A. (1991) Nonionizing radiation and the skin, in *Physiology, Biochemistry and Molecular Biology of the Skin.* Oxford University Press, New York, pp. 910–927.

3. Ullrich, S. E. (1995) Cutaneous biologic responses to ultraviolet radiation. *Curr. Opin. Dermatol.* **2**, 225–230.
4. Buzzell, R. A. (1993) Effects of solar radiation on the skin. *Otolaryngol. Clin. North Am.* **26**, 1–11.
5. Radler-Pohl, A., Sachsenmaier, C., Gebel, S., Auer, H. P., Bruder, J. T., Rapp, U., et al. (1993) UV-induced activation of AP-1 involves obligatory extranuclear steps including Raf-1 kinase. *EMBO J.* **12**, 1005–1012.
6. Fisher, G. J., Datta, S. C., Talwar, H. S., Wang, Z. Q., Varani, J., Kang, S., and Voorhees, J. J. (1996) Molecular basis of sun-induced premature skin ageing and retinoid antagonism. *Nature* **379**, 335–339.
7. Devary, Y., Gottlieb, R. A., Lau, L. F., and Karin, M. (1991) Rapid and preferential activation of the c-jun gene during the mammalian UV response. *Mol. Cell Biol.* **11**, 2804–2811.
8. Devary, Y., Gottlieb, R. A., Smeal, T., and Karin, M. (1992) The mammalian ultraviolet response is triggered by activation of Src tyrosine kinases. *Cell* **71**, 1081–1091.
9. Coffer, P. J., Burgering, B. M., Peppelenbosch, M. P., Bos, J. L., and Kruijer, W. (1995) UV activation of receptor tyrosine kinase activity. *Oncogene* **11**, 561–569.
10. Cochran, B. H., Zumstein, P., Zullo, J., Rollins, B., Mercola, M., and Stiles, C. D. (1987) Differential colony hybridization: molecular cloning from a zero data base. *Methods Enzymol.* **147**, 64–85.
11. Schweinfest, C. W., Nelson, P. S., Graber, M. W., Demopoulos, R. I., and Papas, T. S. (1995) Subtraction hybridization cDNA libraries. *Methods Mol. Biol.* **37**, 13–30.
12. Rosen, C. F., Poon, R., and Drucker, D. J. (1995) UVB radiation-activated genes induced by transcriptional and posttranscriptional mechanisms in rat keratinocytes. *Am. J. Physiol.* **268**, 846–855.
13. Kartasova, T., Cornelissen, B. J., Belt, P., and van de Putte, P. (1987) Effects of UV, 4-NQO and TPA on gene expression in cultured human epidermal keratinocytes. *Nucleic Acids Res.* **15**, 5945–5962.
14. Belyavsky, A., Vinogradova, T., and Rajewsky, K. (1989) PCR-based cDNA library construction: general cDNA libraries at the level of a few cells. *Nucleic Acids Res.* **17**, 2919–2932.
15. Christoph, T., Rickert, R., and Rajewsky, K. (1994) M17: a novel gene expressed in germinal centers. *Int. Immunol.* **6**, 1203–1211.
16. Liang, P. and Pardee, A. B. (1992) Differential display of eukaryotic messenger RNA by means of the polymerase chain reaction. *Science* **257**, 967–971.
17. Welsh, J., Chada, K., Dalal, S. S., Cheng, R., Ralph, D., and McClelland, M. (1992) Arbitrarily primed PCR fingerprinting of RNA. *Nucleic Acids Res.* **20**, 4965–4970.
18. Abts, H. F., Breuhahn, K., Michel, G., Köhrer, K., Esser, P., and Ruzicka, T. (1997) Analysis of UVB modulated gene expression in human keratinocytes by mRNA differential display PCR (DD-PCR). *Photochem. Photobiol.* **66**, 363–367.
19. Liang, P., Averboukh, L., Keyomarsi, K., Sager, R., and Pardee, A. B. (1992) Differential display and cloning of messenger RNAs from human breast cancer versus mammary epithelial cells. *Cancer Res.* **52**, 6966–6968.

20. Liang, P. and Pardee, A. B. (1995) Recent advances in differential display. *Curr. Opin. Immunol.* **7,** 274–280.
21. Liang, P., Bauer, D., Averboukh, L., Warthoe, P., Rohrwild, M., Muller, H., Strauss, M., and Pardee, A. B. (1995) Analysis of altered gene expression by differential display. *Methods Enzymol.* **254,** 304–321.
22. Liang, P., Zhu, W., Zhang, X., Guo, Z., O'Connell, R. P., Averboukh, L., Wang, F., and Pardee, A. B. (1994) Differential display using one-base anchored oligo-dT primers. *Nucleic Acids Res.* **22,** 5763–5764.
23. Boukamp, P., Petrussevska, R. T., Breitkreutz, D., Hornung, J., Markham, A., and Fusenig, N. E. (1988) Normal keratinization in a spontaneously immortalized aneuploid human keratinocyte cell line. *J. Cell Biol.* **106,** 761–771.
24. Krutmann, J., Bohnert, E., and Jung, E. G. (1994) Evidence that DNA damage is a mediate in ultraviolet B radiation-induced inhibition of human gene expression: ultraviolet B radiation effects on intercellular adhesion molecule-1 (ICAM-1) expression. *J. Invest. Dermatol.* **102,** 428–432.
25. Sambrook, J., Fritsch, E. F., and Maniatis, T. (1989) *Molecular Cloning: A Laboratory Manual.* Cold Spring Harbor Laboratory, Cold Spring Harbor, NY.
26. Pearson, W. R. (1990) Rapid and sensitive sequence comparison with FASTP and FASTA. *Methods Enzymol.* **183,** 63–98.
27. Pearson, W. R. and Lipman, D. J. (1988) Improved tools for biological sequence comparison. *Proc. Natl. Acad. Sci. USA* **85,** 2444–2448.
28. Rich, B. E. and Steitz, J. A. (1987) Human acidic ribosomal phosphoproteins P0, P1, and P2: analysis of cDNA clones, in vitro synthesis, and assembly. *Mol. Cell Biol.* **7,** 4065–4074.
29. Grabowski, D. T., Pieper, R. O., Futscher, B. W., Deutsch, W. A., Erickson, L. C., and Kelley, M. R. (1992) Expression of ribosomal phosphoprotein PO is induced by antitumor agents and increased in Mer- human tumor cell lines. *Carcinogenesis* **13,** 259–263.
30. Potempa, J., Korzus, E., and Travis, J. (1994) The serpin superfamily of proteinase inhibitors: structure, function, and regulation. *J. Biol. Chem.* **269,** 15,957–15,960.
31. Magal, S. S., Jackman, A., Pei, X. F., Schlegel, R., and Sherman, L. (1998) Induction of apoptosis in human keratinocytes containing mutated p53 alleles and its inhibition by both the E6 and E7 oncoproteins. *Int. J. Cancer* **75,** 96–104.
32. Lehman, T. A., Modali, R., Boukamp, P., Stanek, J., Bennett, W. P., Welsh, J. A., et al. (1993) p53 mutations in human immortalized epithelial cell lines. *Carcinogenesis* **14,** 833–839.
33. Smith, M. L. and Fornace, A. J., Jr. (1997) p53-mediated protective responses to UV irradiation. *Proc. Natl. Acad. Sci. USA* **94,** 12,255–12,257.
34. Sanchez, Y. and Elledge, S. J. (1995) Stopped for repairs. *Bioessays* **17,** 545–548.
35. Chomczynski, P. and Sacchi, N. (1987) Single-step method of RNA isolation by acid guanidinium thiocyanate-phenol-chloroform extraction. *Anal. Biochem.* **162,** 156–159.
36. Zhao, S., Ooi, S. L., and Pardee, A. B. (1995) New primer strategy improves precision of differential display. *Biotechniques* **18,** 842–846, 848, 850.

37. Bauer, D., Muller, H., Reich, J., Riedel, H., Ahrenkiel, V., Warthoe, P., and Strauss, M. (1993) Identification of differentially expressed mRNA species by an improved display technique (DDRT-PCR). *Nucleic Acids Res.* **21,** 4272–4280.
38. Blanchard, R. K. and Cousins, R. J. (1996) Differential display of intestinal mRNAs regulated by dietary zinc. *Proc. Natl. Acad. Sci. USA* **93,** 6863–6868.
39. Blok, L. J., Kumar, M. V., and Tindall, D. J. (1995) Isolation of cDNAs that are differentially expressed between androgen- dependent and androgen-independent prostate carcinoma cells using differential display PCR. *Prostate* **26,** 213–224.

IV

Analysis of Stress Protein Function

23

Heme Oxygenase Activity

Current Methods and Applications

Stefan W. Ryter, Egil Kvam, and Rex M. Tyrrell

1. Introduction

1.1. The Heme Oxygenase Enzymes

The heme oxygenase enzymes (HO-1 and HO-2) oxidize heme to biliver-din-IXα (BVIXα), releasing carbon monoxide (CO) and iron (**Fig. 1**). HO enzymes control the rate of heme degradation and, consequently, also control the redistribution of the heme iron *(1)*. The CO generated from the HO reaction affects signal transduction pathways in neuronal and vascular systems *(2)*.

Heme oxygenase-1 (HO-1), the inducible HO isozyme, identified as the 32-kDa mammalian stress protein, participates in cellular defense mechanisms. The induction of the HO-1 gene occurs ubiquitously in mammalian cells as an indicator of oxidative stress *(3,4)*. The extracellular stimuli that induce the HO-1 response includes ultraviolet-A radiation (UVA: 320–380 nm) and hydrogen peroxide *(3–6)*, photosensitizers *(7)*, hypoxia *(8)*, nitric oxide *(9)*, metalloporphyrins *(10)*, heavy metals and thiol-reactive substances *(5,11,12)*, and others *(13,14)*. The mechanism by which oxidants, including UVA radiation, trigger the transcription of the HO-1 gene involves a decrease in cellular reducing equivalents *(15)*; illustrated by potentiation with glutathione depleting agents *(12,16)* or attenuation by antioxidants and iron chelators *(6)*. The activation of HO-1 by UVA radiation or photosensitizing agents involves singlet molecular oxygen ($^1O_2$), demonstrated by enhancement in deuterium oxide and attenuation by semispecific $^1O_2$ reactive agents *(7,17)*.

A constitutive isozyme heme oxygenase-2, (HO-2) catalyzes a reaction identical to that of HO-1, with different reaction kinetics *(18)*. HO-2 contributes to basal HO activity and participates in constitutive cellular defense mechanisms *(19)*. HO-1 and HO-2 originate from two divergent genes *(20,21)*. HO-2 occurs

From: *Methods in Molecular Biology, vol. 99: Stress Response: Methods and Protocols*
Edited by: S. M. Keyse © Humana Press Inc., Totowa, NJ

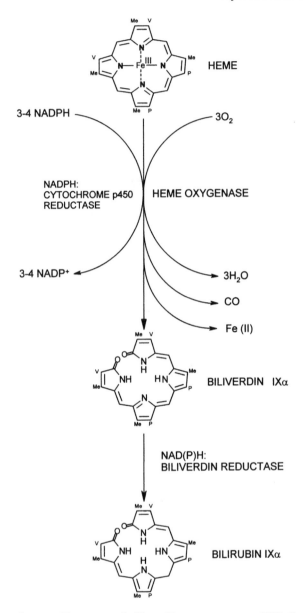

Fig. 1. The pathway of heme metabolism. Heme oxygenase (HO-1), first described as a microsomal mixed-function oxygenase (E.C. 1:14:99:3, heme-hydrogen donor:oxygen oxidoreductase), catalyzes the rate-determining step in heme metabolism. Both heme oxygenase isozymes (HO-1 and HO-2) oxidize heme (ferriprotoporphyrin IX) to the bile pigment biliverdin-IXα, in a reaction requiring 3 mol of molecular oxygen. The accessory enzyme, NADPH:cytochrome p-450 reductase, reduces the ferric heme iron as a prerequisite for each cycle of oxygen binding and

abundantly in brain, liver, and testes *(13,21,22)* and appears extensively in vascular endothelium and neural ganglia *(23)*. Unlike HO-1, the HO-2 gene does not respond to xenobiotic induction *(13)*.

1.2. Heme Oxygenase: Possible Mechanisms of Functional Significance

Heme oxygenase activity generates three reactive products and therefore would be expected to have multiple functional consequences (**Fig. 2**). The substrate of HO activity, heme, facilitates vital cellular functions in hemoprotein form *(13)* but may catalyze iron-dependent oxidation reactions in a nonprotein form. Although HO activity may prevent the heme-iron chelate from directly catalyzing toxic cellular processes such as membrane lipid peroxidation *(24,25)*, HO also releases catalytic iron directly into the cytosol.

The in vitro and serum antioxidant properties of the major HO reaction product biliverdin and its metabolite bilirubin led Stocker et al. to suggest an antioxidant role for HO enzymes *(26)*. Unlike cellular antioxidants such as glutathione and α-tocopherol; bilirubin accumulation is associated with pathological conditions and is eliminated by biliary excretion following hepatic glucuronide conjugation *(13)*.

Heme oxygenase may function to regulate intracellular iron distribution, by initiating the transfer of catalytic heme iron to the ferritin-iron pool *(27,28)*. The heme oxygenase mediated iron release indirectly leads to an acquired cellular resistance to oxidants in the long term. The proposed mechanism involves the iron-mediated stimulation of ferritin synthesis *(29)*, which, in turn, sequesters the excess iron in an inert Fe(III) state. Evidence for the relationship between heme oxygenase and ferritin in cytoprotection appears in several recent studies *(30,31)*. On the other hand, the released heme iron may be transiently available to catalyze deleterious free-radical reactions prior to its sequestration *(31a)*.

In transgenic mouse models, HO-1 homozygous knockout mice HO-1 (–/–) are anemic yet accumulate iron in the liver and kidneys *(32)*. Embryonic HO-1(–/–) cells display oxidant hypersensitivity *(32)*. HO-2 (–/–) mice display sensitivity to hyperoxia despite a normal HO-I response and accumulate iron in their lungs *(33)*. In HO-1 (–/–) or HO-2 (–/–) mice, heme is degraded by compensatory mechanisms that are uncoupled from normal iron reutilization path-

---

Fig. 1. *(continued)* oxygen activation. The cleavage of the heme ring frees the coordinated iron, as well as the α-methene bridge carbon as carbon monoxide. The principle HO reaction product, biliverdin-IXα, is further metabolized by divalent reduction to form bilirubin-IXα, by NAD(P)H:biliverdin reductase. Heme side chains are designated as M, methyl; V, vinyl; and P, propionate.

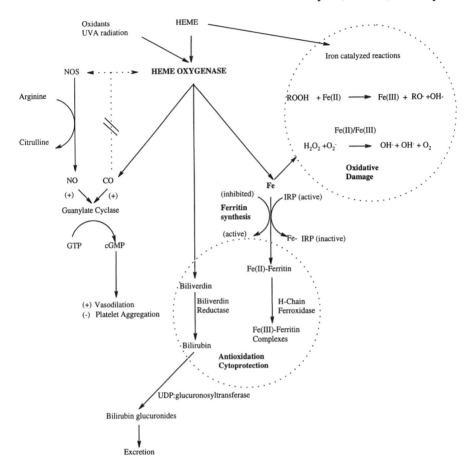

Fig. 2. The functional consequences of HO activity. These include heme degradation, bile pigment generation, anti-oxidation, pro-oxidation, ferritin synthesis, and the
stimulation of cGMP-dependent signal-transduction pathways. The abbreviations used
are as follows: cGMP, cyclic 3':5' guanosine monophosphate; CO, carbon monoxide;
Fe(III), ferric iron; Fe(II), ferrous iron; GTP, guanosine triphosphate, H_2O_2, hydrogen
peroxide; IRP, iron regulatory protein; NADPH, nicotinamide adenine dinucleotide
phosphate (reduced form); NO, nitric oxide; NOS, nitric oxide synthase; O_2^-, superoxide anion; $\cdot$OH, hydroxyl radical; RO$\cdot$, alkoxyl radical; ROOH, organic hydroperox-
ide; UDP, uridine diphosphate, UVA, ultraviolet (320–380 nm) radiation.

ways (32-33). However, targeted gene deletions often force the development
of compensating mechanisms not otherwise important in wild-type organisms.

   Heme oxygenase activity produces carbon monoxide (CO), which was considered a waste metabolite until potential roles for this compound in neuronal
signal transduction were discovered (34–37). Endothelial-derived carbon mon-

oxide (CO) and nitric oxide (NO) stimulate vasorelaxation *(23,38,39)*. NO and CO coordinate the heme iron of soluble guanylate cyclase (sGC) (although NO has a higher affinity than CO), inducing sGC activity and cyclic 3':5' guanosine monophosphate (cGMP) formation *(39,40)*. In transgenic models, HO-2 (–/–) mice are consistent with a role for HO-2 derived CO in (1) intestinal nonadrenergic-noncholinergic (NANC) relaxation *(36)* and (2) male sexual function *(41)*, but (3) not in hippocampal long-term potentiation (LTP) *(42)*.

1.3. Analyzing HO Expression by Classical Molecular Biology Techniques

1.3.1. Heme Oxygenase: Messenger RNA Detection

Techniques involving messenger ribonucleic acid (mRNA) detection are commonly used to monitor the in vitro HO-1 response. Such approaches provide an endpoint for gene regulation studies and are useful for screening potential HO-1 inducing agents; however, they will not measure how much mRNA, if any, is translated into functional enzyme protein. A Northern blot analysis of HO-1 mRNA steady-state levels has been previously described *(43)*, using cultured human fibroblasts subjected to oxidative stress as a model, and further exemplified in refs. *(4,6,16,17)*. The technique requires isolation of total mRNA, using the method of Chomczynski *(44)*, and a nucleic acid hybridization probe. A cDNA probe corresponding to the human heme oxygenases-1 mRNA (clone 2/10) is available *(5)*, as well as other cDNAs corresponding to human, rat, mouse, porcine, and chicken HO-1 mRNA *(45–49)*.

In the nuclear run-off technique, which is rigorous proof of transcriptional activation, the rate of mRNA transcription is estimated by the metabolic labeling of total mRNA followed by selection of the species of interest using a cold-specific filter-bound hybridization probe. Nuclear run-off experiments have demonstrated the transcriptional activation of HO-1 by many agents, including heme and heavy metals, UVA radiation, and sodium arsenite *(50,51)*.

The reverse transcriptase–polymerase chain reaction (RT-PCR) may be applied to estimate HO-1 activation, using noninducible HO-2 as an internal standard *(52)*. RT-PCR obviates the need for a cDNA clone, relying instead on synthetic oligonucleotide primers. The technique is only semiquantitative and prone to artifacts introduced by nonlinearity of the PCR process with increasing reaction cycle.

1.3.2. Heme Oxygenase: Protein Separation Techniques

Many studies of HO-1 and HO-2 have utilized protein separation and analysis techniques. Because HO-1 requires membrane integration and accessory enzymes (NADPH:cytochrome P-450 reductase), analysis of protein synthesis or levels does not truly estimate functional enzymatic activity.

Protein electrophoresis techniques such as one-dimensional sodium dodecyl sulfate polyacrylamide gel electrophoresis (SDS-PAGE) *(3,6,11,12,53,54)* and two-dimensional gel electrophoresis (2-D PAGE) *(8,53)* have been used to study HO-1 (30- to 34-kDa protein) induction. These techniques resolve proteins by approximate molecular weight. Thus, Western immunoblotting analysis, which immunochemically identifies the detected species, has become a popular method for studying the HO-1 induction phenomenon (e.g., **refs. *12,53***). The technique requires a specific antibody [rabbit anti-rat HO-1 or HO-II *(18,55)*, polyclonal rabbit anti-chicken HO-1 *(56)*, and rabbit anti-bovine HO-1 *(57)* antisera have all been described; anti-rat and anti-human (HO-1 or HO-2) rabbit polyclonal antisera as well as anti-rat and anti-human HO-1 mouse monoclonal antisera are commercially available (Stressgen, Victoria BC, Canada)].

1.4. HO Activity Analysis

Enzymatic activity assays are important tools in studying the functional significance of HO. This chapter provides detailed protocols for performing HO activity assays, based on the detection of bile pigments (biliverdin or bilirubin) by high-performance liquid chromatography (**Subheading 3.3.1.**), visible spectrophotometry (**Subheading 3.3.3.**), or radiochemical methods (**Subheading 3.3.2.**). HO reactions typically include a crude extract (15,000–18,000g supernatant fraction) from cultured cells, or require further purification to obtain microsomes (105,000g pellet). Indirect, coupled HO assays that rely on the detection of bilirubin-IXα require complete conversion of biliverdin by supplementation with an excess of NAD(P)H:biliverdin reductase (BVR). The BVR may be prepared from animal tissues as previously described, in partially purified form (30,000–105,000g supernatant) *(58)*.

Bilirubin-IXα (BR) can be measured by visible spectrophotometry. One variant of the HO assay measures the rate of BR formation directly in reaction mixtures (± NADPH) at 37°, by visible difference spectroscopy (λ_{max} 464–468 nm) *(1,59)* (**Subheading 3.3.3.2.**). Alternatively the reaction mixture may be extracted in chloroform *(60)* such that BR is calculated from the difference in optical density (OD) at 464 nm and 530 nm (OD_{464}-OD_{530} nm) of the extract (**Subheading 3.3.3.1.**).

If radiolabeled (^{14}C) heme is used, then the (^{14}C) bilirubin can be isolated by recrystallization from chloroform extracts *(61)* or separated by thin-layer chromatography, and then quantified by radiochemical methods (**Subheading 3.3.2.**) *(62)*.

High-performance liquid chromatography (HPLC) may be conveniently used to detect bilirubin formation *(63–68)*. We have modified the HPLC methods to allow the simultaneous detection of both biliverdin and bilirubin, obviating the need for BVR supplementation (**Subheading 3.3.1.**) *(67,68)*. Alternatively, CO detection by gas chromatography can provide an estimate of HO activity *(69)*.

The principle drawback of HO activity assays is that they measure HO activity under the condition of excess substrate, rather than in vivo activity under conditions of endogenous substrate availability. Furthermore, regardless of the sensitivity of the assay method, HO activity assays cannot distinguish between HO-1 and HO-2. To achieve that distinction, HO activity assays require an immunoprecipitation step to selectively eliminate one of the species.

2. Materials
2.1. Preparation of Cell-Free Extracts for HO Assays

1. Phosphate-buffered saline (PBS), pH 7.4: 8.0 g NaCl, 0.2 g KCl, 0.2 g KH_2PO_4, 1.13 g/L Na_2HPO_4 .
2. $1M$ EDTA, pH 8.0 (Sigma, St. Louis, MO).
3. 1 M Tris-HCl, pH 7.4 (Sigma).
4. PBS–EDTA, 1 mM, pH 8.0. To 500 mL PBS, add 0.5 mL of 1 M EDTA, pH 8.0. Store at 4°C.
5. Solution A: 0.25 M sucrose, 20 mM Tris-HCl, pH 7.4 (85.6 g/L sucrose, 20 mL/L of 1 M Tris-HCl, pH 7.4) Autoclave and store at 4°C.
6. Protease inhibitor cocktail (1000 X stock solutions); 4-(2-aminoethyl) benzolsulfonylfluoride hydrochloride (AEBSF, Pefabloc SC) (50 mg/mL in distilled water); leupeptin (4 mg/mL in distilled water), and pepstatin (4 mg/mL in 100% methanol)(Boehringer Mannheim, Indianapolis, IN). Aliquot and store at –20°C. Add to buffer A immediately before use at 1X final concentration (*see* **Note 1**).
7. Tabletop sonicator (Branson sonifier B-12 sonicator, Branson Sonic Power, Danbury, CT).
8. Ultracentrifuge apparatus (e.g., Centrikon T20-60 ultracentrifuge with TFT65.13 rotor.
9. 13.5 mL polyallomar ultracentrifuge tubes, Kontron Instruments, Zurich, Switzerland).
10. Protein assay reagent kit (Coomassie Brilliant Blue #G-250, Bio-Rad, Hercules, CA).
11. Bovine serum albumin (BSA) (Sigma).

2.2. Heme Oxygenase Activity

1. 40 mM Glucose 6-phosphate (Sigma). Dissolve in solution A and store at –20°C.
2. 20 mM Nicotinamide adenine dinucleotide phosphate, reduced form (β-NADPH)(Sigma). Dissolve in solution A.
3. Glucose-6-phosphate dehydrogenase, Type XV from Baker's yeast (Sigma). Dilute to 1 U/μL in solution A. Aliquot and store at –20°C).
4. 10 mM Hemin. Dilute bovine hemin (Sigma) in 100% dimethyl sulfoxide (DMSO). Store at –20°C. Immediately before use, gradually dilute the stock to 2.5 mM with distilled water (*see* **Note 2**).

5. 10 m*M* SnProtoporphyrin-IX (Sn-PPIX) (Porphyrin Products, Logan, UT). Dissolve in 100% DMSO. Reagent is light sensitive so prepare in the dark, aliquot, and store in foil-wrapped tubes at –20°C. Immediately before use, gradually dilute the stock solution to 2.5 m*M* with distilled water (*see* **Note 2**).

2.3. Detection of HO Reaction Products

2.3.1. Detection of Biliverdin and Bilirubin by HPLC

1. High-performance liquid chromatography (HPLC) system [e.g., Kontron Instruments (Zurich, Switzerland)], equipped with a data acquisition system (e.g., Kontron MT 450 MS-DOS), two solvent reservoirs, and an autosampler (e.g., Kontron HPLC 360).
2. Visible detector (e.g., Kontron UVIKON 720 LC microdetector).
3. Waters Novapak reverse phase C-18 steel cartridge HPLC column (3.9 mm × 15 cm, cat. no. WAT036975, or phenyl reverse-phase column, cat. no. WAT036970) (Waters Chromatography, Milford, MA) (*see* **Note 3**).
4. Compressed helium tank.
5. PCR microtubes (e.g., 0.5 mL, Treff, Degersheim, Switzerland) (*see* **Note 4**).
6. 200 µ*M* Mesoporphyrin (Sigma). Dissolve in 100% DMSO. Prepare in the dark, aliquot, and store in foil-wrapped tubes at –20°C.
7. Stop solution; ethanol:dimethylsulphoxide 95:5 (v/v), 0.4 µ*M* mesoporphyrin. Add 20 µL of the 200 µ*M* mesoporphyrin stock solution to a 10-mL aliquot of ethanol: dimethylsulphoxide 95:5 (v/v) immediately before use.
8. 10X Biliverdin-IX standard (Sigma) Prepare 200 µ*M* biliverdin in 100% DMSO. Dilute to 20 µ*M* in 100% DMSO. Store at –20°C.
9. 10X Bilirubin-IXα standard: prepare 200 µ*M* bilirubin-IXα in 100% DMSO (Sigma). Dilute to 20 µ*M* in 100% DMSO (*see* **Note 5**). Prepare freshly before use. Protect from light.
10. 1 *M* Ammonium acetate, pH 5.2: 77g/L ammonium acetate, 26 mL/L glacial acetic acid. Adjust to 1 L with Millipore filtered water, adjust pH to 5.2 with glacial acetic acid. Filter and store at 4°C.
11. Methanol (HPLC grade) (Fisher Scientific, Pittsburgh, PA).
12. HPLC buffer A: 100 m*M* ammonium acetate, pH 5.2, 60% methanol (v/v). Combine 100 mL of 1*M* ammonium acetate, pH 5.2, and 600 mL methanol. Adjust to 1 L with Millipore filtered water and Degas for 2 min with compressed helium (*see* **Note 6**).
13. HPLC buffer B: 100% methanol, HPLC grade. Degas 2 min with compressed helium.

2.3.2. Detection of Bilirubin
by Radiochemical Thin-Layer Chromatography (TLC)

1. [^{14}C] Hemin (54 Ci/mmol) (Leeds Radioporphyrins, UK). Dissolve to 10 m*M* in 100% DMSO. Dilute to 100X stock (2.5 m*M*) with distilled water (*see* **Note 2**). Store at –70°C.
2. TLC tank (Fisher Scientific).

3. TLC silica gel sheet (Eastman Kodak, Rochester, NY).
4. Glacial acetic acid (Fisher Scientific).
5. Chloroform, spectrophotometric grade (Fisher Scientific).
6. 20:1 Chloroform: acetic acid (v/v).
7. X-ray film (X-OMAT, Eastman Kodak).
8. Scintillation counting apparatus.
9. Scintillation fluid.
10. En^{3}Hance spray (New England Nuclear)(*see* **Note 7**).

2.3.3. Detection of Bilirubin by Visible Spectrophotometry

2.3.3.1. VISIBLE SPECTROSCOPY OF CHLOROFORM EXTRACTS

1. Visible dual-beam spectrophotometer (e.g., Model U-2000, Hitachi, San Jose, CA).
2. Cuvets (*see* **Note 8**).
3. Chloroform (Fisher Scientific).

2.3.3.2. VISIBLE DIFFERENCE SPECTROSCOPY

1. Visible dual-beam spectrophotometer (e.g., Model U-2000, Hitachi).
2. Constant-temperature chamber (Thermocirculator C005-0405, Perkin Elmer, Oak Brook, IL).

2.4. Preparation of Partially Purified Biliverdin Reductase

1. Liver (rat or bovine).
2. Isotonic saline (0.9% NaCl).
3. Tissue homogenizer (e.g., Janke-Kunkel Ultraturrax, Tekmar Co., Cinncinnati, OH).
4. Solution A: 0.25 M sucrose (Sigma), 20 mM Tris-HCl, pH 7.4 (85.6 g/L sucrose, 20 mL/L of 1 M Tris-HCl, pH 7.4). Autoclave and store at 4°C.
5. Ammonium sulfate (Sigma).
6. Dialysis bags (10,000–15,000 Dalton mol wt cutoff [MCO], SpectraPor).
7. 0.1 M KPO$_4$, pH 7.4
8. Biliverdin (Sigma). Prepare 10 mM stock solution in 100% DMSO. Dilute to 2.5 mM with distilled water. Store at –20°C.

3. Methods
3.1. Preparation of Cell-Free Extracts for HO Assays

The following protocol generates crude extracts (**steps 1–6**) or high speed (microsomal) extracts (**steps 1–12**) from cultured mammalian cells, (*see* **Note 9**) suitable for use with HO activity assays. Perform all steps at 4°C using pre-cooled solutions and equipment.

1. Aspirate culture media from 150-mm plates and rinse the monolayers 2X with ice-cold PBS. To each dish, add 4 mL ice cold PBS-EDTA, 1 mM, pH 8.0, containing 50 µg/mL protease inhibitor (AEBSF). Detach cells on ice with a rubber cell-scraper.

2. Recover the suspension from each dish and combine into 50-mL tubes. Rinse the plates with 2 mL of PBS and pool.
3. Pellet the cells at 150g at 4°C.
4. Resuspend the cells in 800 μL solution A containing 1X protease inhibitor cocktail (*see* **Note 1**).
5. Transfer the suspension to 1.5- mL microcentrifuge tubes and sonicate on ice 2X 15 s at 20 W using a tabletop sonicator (*see* **Note 10**).
6. Centrifuge the homogenates at 4°C at 15,000g for 20 min. Repeat the centrifugation step until the supernatant is clear (*see* **Note 11**). If crude extracts are desired, proceed to **step 12**.
7. Transfer the supernatant to ultracentrifuge tubes. Balance the tubes with solution A containing freshly added protease inhibitors (*see* **Note 1**).
8. Centrifuge at 105,000g for 1 h at 4°C, in a fixed-angle rotor.
9. Drain the microsomal pellets at 4°C. Resuspend the pellet in 200 μL buffer A (*see* **Note 1**) by vigorous manual pipetting. Transfer the suspension to microcentrifuge tubes. If necessary, use another 100–200 μL buffer A to recover residual particles and pool (*see* **Note 12**).
10. Sonicate the crude suspensions on ice for 10 s at low power (20 W) (*see* **Note 10**).
11. Centrifuge the suspensions at 4°C for 5 min at 15,000g (*see* **Note 13**).
12. Decant the supernatant to a fresh microcentrifuge tube and store on ice until use (*see* **Note 14**).
13. Calculate the protein concentration of the supernatant using a commercial protein assay reagent kit according to the manufacturer's instructions. Standardize with bovine serum albumin (*see* **Notes 9** and **15**).

3.2. Heme Oxygenase Activity

The following typical reaction mixture works well for the HPLC detection protocol (**Subheading 3.3.1.**). Variations are described in the following sections (**Subheadings 3.3.2.** and **3.3.3.**). The final reaction concentrations are 5 mg/mL fresh microsomal protein, 1 mM β-NADPH, 2 mM glucose 6-phosphate, 1 U glucose 6-phosphate dehydrogenase, 25 μM hemin, 0.25 M sucrose, 20 mM Tris, pH 7.4, in a final volume of 100 μL. If desired, include 0.5–2 mg/mL partially purified biliverdin reductase (*see* **Note 16**) prepared as described in **Subheading 3.4.**

1. Assemble the following reagents in order in eppendorf microcentrifuge tubes on ice.
2. Add 500 μg fresh microsomal protein suspension for each sample tube in <88 μL volume. Adjust volume to 88 μL with buffer A (0.25 M sucrose, 20 mM Tris-HCl, pH 7.4, omit protease inhibitors)(*see* **Notes 15** and **16**).
3. Add 5 μL of 40 mM glucose-6-phosphate solution.
4. Add 5 μL of 20 mM nicotinamide adenine dinucleotide phosphate reduced form (β-NADPH).
5. Add 1 μL glucose-6-phosphate dehydrogenase (1 U/μL).
6. Add 1 μL bovine hemin (2.5 mM stock in 25% DMSO).

7. Initiate the reactions by pipetting the reaction mix up and down several times. Vortex 5 s and place tubes in a circulating water bath at 37°C in the dark.
8. Prepare negative reaction controls (*see* **Note 17**).
9. Incubate reactions for a chosen time interval, typically 5–60 min (*see* **Note 18**).
10. Stop the reaction under dark ambient conditions according to the specific method for each detection strategy (*see* **Subheadings 3.3.1.–3.3.3.**).

3.3. Detection of HO Reaction Products

3.3.1. Detection of Biliverdin and Bilirubin by HPLC

In this protocol, heme oxygenase enzymatic activity is calculated from the rate of formation of bilirubin equivalents [Biliverdin-IXα (BV) + Bilirubin-IXα (BR)] at 37°C. BV and BR are separated by HPLC and detected at 405 nm using visible absorbance spectroscopy (*67,68*). As little as 10 pmol of each authentic bile pigment standard can be detected in alcohol:DMSO extracts of protein solution. Complete conversion of BV to BR is unnecessary, as both pigments are quantifiable.

1. In advance, prepare the HPLC apparatus as follows: Program a linear gradient (0–14 min) starting from 100% HPLC buffer A (100 mM ammonium acetate, pH 5.2, 60% methanol) and ending at 100% buffer B (100% methanol), followed by reversion to 100% buffer A at 14 min until the termination at 19 min. The total flow rate is 1.5 mL/min. Preset the detector for continuous absorbance monitoring at 405 nm (*see* **Note 19**). Prime the HPLC with 100% HPLC buffer A for 2 min prior to the first injection. Rinse the columns after each series with 100% methanol for 60 min (*see* **Note 20**).
2. Terminate the heme oxygenase reactions (from **Subheading 3.2.**) by adding 1 volume (100 µL) stop solution (95:5 ethanol:DMSO, 0.4 µM mesoporphyrin) to each tube at 20°C. Vortex 5 s.
3. Centrifuge at 15,000g at 20°C (*see* **Note 21**).
4. Recover supernatants and transfer to PCR microtubes (*see* **Note 4**). Store the ethanol:DMSO extracts at room temperature in the dark and inject into the HPLC system as soon as possible (*see* **Note 22**). A standard chromatograph is shown in **Fig. 3**.
5. Record the integrated peak areas of biliverdin-IXα, bilirubin-IXα, and mesoporphyrin from each sample chromatograph (*see* **Note 23**).
6. If there are no peaks, *see* **Note 24**. If additional degradation peaks occur, *see* **Note 25**.
7. From the linear standard curves (*see* **steps 12–19**), calculate the bile pigment concentrations in an injected sample in µM (pmol/µL).
8. Convert concentration values to total picomoles for each pigment assuming an original stopped reaction volume of 200 µL.
9. Calculate bilirubin equivalents (BReq) [(pmol BR + pmol BV) = picomoles of substrate oxidized].

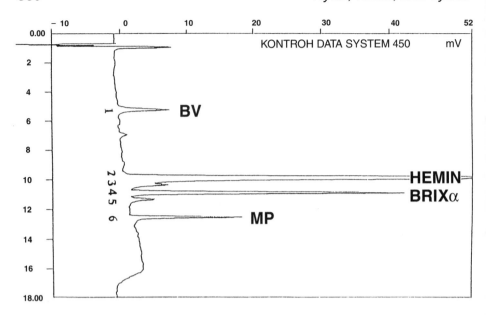

Fig. 3. HPLC chromatograph showing the resolution of tetrapyrrole standards in an alcohol extract of sonicated microsomal protein solution. Final concentrations were 2 μM biliverdin (BV), 2 μM Bilirubin-IXα (BRIXα), 0.35 μM BRIIIα, 0.27 μM BRXIIIα, 12.5 μM hemin, and 0.4 μM mesoporphyrin (MP), corresponding to injected amounts of 240 pmol BV, 240 pmol BRIXα, and 48 pmol MP. The tetrapyrroles were eluted with the following typical retention times, (1) biliverdin (BV), 5.29 min; (2) hemin 9.8 min; (3) bilirubin structural isomer (IIIα or XIIIα), 10.39 min; (4) bilirubin-IXα (BRIXα), 10.9 min; (5) bilirubin structural isomer (IIIα or XIIIα), 11.38 min; and (6) mesoporphyrin (MP) 12.54 min. Optical density was measured at 405 nm. [Reprinted from Ryter, S.W., Kvam, E., Richman, L., Hartmann, F., and Tyrrell. R. M. (1998) A chromatographic assay for heme oxygenase activity in cultured human cells: application to artificial heme oxygenase overexpression. *Free Radic. Biol. Med.* **24,** 959–971, copyright with the permission of Elsevier Science.]

10. Normalize the BReq value against the relative mesoporphyrin recovery for each sample. Multiply the BReq (MP*/MP$_S$); where MP* is the average mesoporphyrin recovery for the entire run and MP$_S$ is the sample mesoporphyrin recovery (*see* **Note 26**).

11. Express HO activity as BReq (pmol BV + BR)/mg protein/h, (*see* **Note 27**) using a reaction time in the linear range of the assay (*see* **Note 18**).

12. (***Steps 12–19** concern the construction of standard curves*) Under dark conditions, prepare a 10X dilution series for biliverdin and bilirubin (2–20 μM) by diluting the standard stock solutions with 100% DMSO.

13. To Eppendorf microtubes at room temperature, add 500 μg microsomal protein (in <100 μL) and adjust to 100 μL with solution A.

14. Heat the protein to 37°C for 10 min in a circulating water bath.
15. Add 100 µL stop solution and vortex 10 s.
16. Centrifuge at 15,000g for 5 min at room temperature.
17. Recover 135 µL of the supernatant. Before each injection, in the dark, add 15 µL BV or BR standard from 10X concentrated stock solutions to final concentrations ranging from 0.2 to 2 µM. Inject immediately into the HPLC system (**Fig. 3**).
18. To determine recovery, add the standard to the protein before alcohol extraction (*see* **Note 28**).
19. Construct standard curves by plotting the detector response (integrated peak area) (*see* **Note 23**) as a function of biliverdin or bilirubin concentration (*see* **Note 29**).

3.3.2. Detection of Bilirubin by Radiochemical TLC (62)

1. Prepare crude cell extracts (**Subheading 3.1., steps 1–5** and **12**) (*see* **Note 30**).
2. Assemble reaction components as stated in **Subheading 3.2.**, however, scale down the final reaction volume to 10 µL. Use 25–100 µg crude extract in 8 µL solution A. Initiate the reaction with [^{14}C] hemin (25 µM final), instead of cold heme.
3. Incubate reactions for a chosen time interval, typically 5–60 min (*see* **Note 18**).
4. Stop the reaction on ice, with the addition of a molar excess of cold heme and cold bilirubin (1 µL each of 1 mM stock).
5. Spot 1–2 µL onto silica gel TLC sheet. Dry with a portable hairdryer. Repeat.
6. Develop in a TLC tank containing 20:1 (v/v) chloroform-acetic acid, under a ventilated solvent hood, until the solvent front nearly approaches (1–2 cm) the top of the sheet.
7. Let the sheet dry under the fume hood.
8. Spray the TLC sheet with one coat of En^{3}Hance spray under a fume hood (*see* **Note 7**).
9. Autoradiograph in a film cassette against Kodak X-Omat film at –70°C (1–7 d).
10. Line up the X-ray film with the original plate and cut out section of the plate corresponding to bilirubin spots (*see* **Note 31**).
11. Recover the excised sections to scintillation vials and add scintillation fluid.
12. Count [^{14}C] cpm (dpm) in the sample in a scintillation counter. Convert to µCi.
13. Calculate picomoles of product formed (corrected for original reaction volume) from the known specific activity of the heme.
14. Express HO activity as pmol bilirubin/mg protein/h.

3.3.3. Detection of Bilirubin by Visible Spectrophotometry

3.3.3.1. VARIANT 1 (CHLOROFORM EXTRACTION)

1. Assemble the heme oxygenase reaction mixtures to the following final concentrations:
2. 1 mg/mL crude extract (**Subheading 3.1., steps 1–6**) or microsomal extract (**steps 1–12**) (*see* **Note 32**); 0.5-2 mg/mL partially purified biliverdin reductase (**Subheading 3.4.**),1 mM β-NADPH, 2 mM glucose 6-phosphate, 1 U glucose 6-phosphate dehydrogenase, 25 µM hemin, 0.25 M sucrose, 20 mM Tris-HCl, pH 7.4. Scale up the final reaction volume up to 500 µL. Incubate reactions for a chosen time interval, typically 5–60 min at 37°C in a circulating water bath in the dark (*see* **Note 18**).

3. Terminate the reactions by the addition of 1 volume chloroform. Vortex 30 s.
4. Centrifuge at room temperature 10 min at 15,000g.
5. Recover the (lower) chloroform phase.
6. Read OD_{464} and OD_{530} of the chloroform phase (or scan 400–530 nm)
7. Calculate OD_{464}–OD_{530}.
8. Calculate bilirubin concentration, assuming an extinction coefficient of 40 mM^{-1}/ cm in chloroform (*see* **Note 33**).
9. Express HO activity as pmol bilirubin/mg protein/h.

3.3.3.2. VARIANT 2 (VISIBLE DIFFERENCE SPECTROSCOPY)

1. Set the spectrophotometer for continuous absorbance monitoring (464 nm) at 37°C. Prewarm the cuvette chamber to 37°C.
2. Assemble two reaction mixtures inside two spectrophotometer cuvettes in the following order, (omitting NADPH): 1 mg/mL protein (crude extract, **Subheading 3.1., steps 1–6**) or microsomal extract (**Subheading 3.1., steps 1–12**) (*see* **Note 32**); 0.5–2 mg/mL partially purified biliverdin reductase (**Subheading 3.4.**), 2 mM glucose 6-phosphate, 1 U glucose 6-phosphate dehydrogenase, 25 µM hemin, 0.25 M sucrose, 20 mM Tris, pH 7.4., in a final volume of 950 µL.
3. Preincubate the two cuvettes for 5 min at 37°C, in a constant-temperature cuvet chamber.
4. Initiate the reactions by adding 50 µL of 20 mM β-NADPH solution in the sample cuvet; add 50 µL solution A to the reference cuvet.
5. Monitor OD_{464} on continuous (time-rate) mode.
6. Calculate the rate of bilirubin formation in the linear phase of the reaction. In this case, the extinction coefficient must be determined empirically (*see* **Note 34**).
7. Express HO activity as pmol bilirubin/mg protein/h.

3.4. Preparation of Biliverdin Reductase

3.4.1. Partial Purification

This protocol describes the partial purification of bilirubin reductase and stops at the third step of the original procedure described by Tenhunen et al. *(58)*. Further purification of BVR as described is not strictly necessary for the purpose of HO reactions.

1. Obtain a rat liver or a bovine liver (or kidney) from the local slaughterhouse.
2. Perfuse the liver through the hepatic portal vein with cold 0.9% NaCl. Perform all steps at 4°C.
3. Cut the tissue into small pieces with surgical scissors and place in a tube.
4. Homogenize the tissue chunks with an electric tissue homogenizer.
5. Centrifuge at 150g for 5 min. Save the supernatant.
6. Centrifuge at 18,000g for 10 min and save the supernatant.
7. Centrifuge at 30,000g for 30 min and save the supernatant.
8. Add ammonium sulfate to the 30,000g supernatant to attain 40% saturation. Let stand on ice 30 min.

9. Centrifuge at 4°C for 10 min at 10,000g.
10. Recover supernatant to a new tube.
11. Gradually add ammonium sulfate to increase the saturation to 60%.
12. Centrifuge at 4°C for 10 min at 10,000g and discard supernatant
13. Resuspend pellet in <1 mL of 0.01 M KPO$_4$, pH 7.4.
14. Dialyze in distilled water for 24 h at 4°C.
15. Centrifuge dialysate for 10 min at 10,000g. Discard the pellet.
16. Calculate the protein concentration of the supernatant using a commercial protein assay reagent kit according to manufacturer's instructions. Standardize with bovine serum albumin.
17. Adjust concentration to 10 mg/mL with 0.1 M KPO$_4$, pH 7.4 (solution A, 0.25 M sucrose, 20 mM Tris-HCl, pH 7.4, may also be used).
18. Aliquot and store frozen at –20°C (*see* **Note 35**).

3.4.2. Biliverdin Reductase Activity

The biliverdin reductase activity of a protein extract can be determined by a spectrophotometric assay *(58)* (*see* **Note 36**). The final reaction concentrations are 0.5 mg/mL BVR extract, 25 µM biliverdin, 100 µM NADPH in 0.1 M KPO$_4$, pH 7.4.

1. Prewarm the spectrophotometer cuvet chamber to 37°C. In two cuvettes assemble the reactions by adding (0.5 mg) BVR extract in 0.1 M KPO$_4$, pH 7.4,
2. Add 10 µL biliverdin stock solution (2.5 mM in 25% DMSO) (*see* **Note 37**) and adjust volume to 995 µL in 100 mM KPO$_4$, pH 7.4.
3. Preincubate the reaction mixtures for 5 min at 37°C.
4. Initiate the reaction (sample cuvet) with 5 µL of a 20 mM β-NADPH solution, or substitute buffer in the reference cuvet. Incubate at 37°C.
5. Determine the rate of absorbance gain at 464 nm (bilirubin formation).
6. Alternatively, scan the 350- to 550-nm range at fixed time intervals (10–20 min) following the initiation.
7. Calculate bilirubin formation from standard curves (*see* **Note 38**) and express BVR activity as pmol BR/mg protein/h.
8. Alternatively, measure the rate of NADPH consumption (Δ OD$_{360}$ nm) at 37°C.

4. Notes

1. A broad spectrum of protease inhibitors (e.g., leupeptin, pepstatin, and AEBSF) increases the yield of active enzyme protein. AEBSF (or phenylmethylsulfonyl fluoride) alone are not sufficient to prevent degradation during extraction. Add the protease inhibitors immediately before each use to an aliquot of solution A as needed.
2. Hemin (or SnPPIX) dissolves completely at 10 mM in 100% DMSO and will not precipitate from DMSO during gradual dilution with water.
3. The WAT036975 column used to develop the assay generated the standard chromatograph displayed in **Fig. 3**. The WAT036970 column further improves peak separation.

4. If an autosampler is used, make certain that plastic tube lids are compatible with the instrument. Reinforced (boiling) tubes may cause autosampler malfunction.

5. Weigh bilirubin mixed isomers (Sigma) according to percent weight composition of the IXα isomer. For HPLC applications, the appearance of BRIIIα or BRXIIIα structural isomers in standard chromatograms is of no consequence.

6. Ammonium phosphate should not be used as the mobile phase because it precipitates in the HPLC system at low pH.

7. This material is volatile and toxic by inhalation or skin contact. Use only in a well-ventilated fume hood. Coated surfaces may emit vapors for months. The material may also damage film cassettes and intensifying screens. Handle with extreme care, or omit from the protocol.

8. Make sure that the cuvets and tubes resist chloroform. Chloroform melts polystyrene.

9. Tissue culture parameters must be determined experimentally for each cell type. Use approx 5×10^7 monolayer cells from two to four 150-mm plates grown to near confluence. This approximation yields 1-2 mg microsomal protein from human HeLA cells. If crude extract is chosen for the assay, then one to two 100-mm near-confluent plates are sufficient to yield 1–2 mg protein.

10. Place the sonicator head inside the cell suspension while retaining the tube completely immersed in ice. Make certain that the sonicator does not heat the sample.

11. The protein extraction procedure may be stopped at this point, and the crude supernatant used in the enzyme assay instead of microsomal protein. The crude extract contains HO activity and may contain biliverdin reductase activity depending on the cell type. This substitution may be acceptable for radiochemical (**Subheading 3.3.2.**) and spectrophotometric-based assays (**Subheading 3.3.3.**), but may increase background in the latter. The use of crude extract is discouraged for the HPLC method (**Subheading 3.3.1.**) because it reduces the column lifetime.

12. The resulting pellets should be yellowish and translucent. The viscosity makes it difficult to resuspend the pellet. Hold the tube immersed in ice and repeatedly pipet the suspension (20–50 times) until it can be transferred to clean tubes.

13. The final suspensions should be homogenous, with a slight pellet. If the supernatant is not homogenous, or if the pellet is too large, repeat **steps 10** and **11**.

14. Finish the entire assay on the day of protein extraction. Freezing microsomal or crude extracts at –20°C may cause activity loss.

15. If the protein preparations are too dilute, increase starting material or reduce protein resuspension volume.

16. Because both biliverdin and bilirubin are detected by the HPLC method, this assay does not depend on the complete conversion of biliverdin to bilirubin. Furthermore, some human cell types contain endogenous BVR. Therefore, the addition of exogenous BVR has been omitted from the basic HPLC activity protocol. BVR may be included as follows: Add both microsomal (250–500 μg) and BVR extract 50–200 μg (*see* **Subheading 3.4.**) in 1 volume of < 88 μL. Adjust to 88 μL with buffer A.

17. Prepare negative controls by (1) assembling reaction mixtures in stated order with the additional inclusion of 1 μL Sn-protoporphyrin-IX solution (2.5mM), to a final concentration of 25 μM; (2) substituting the NADPH solution in the reaction assembly with 5 μL buffer A, or (3) using boiled protein extract.

18. Determine the linear range for measured activity. A 20-min reaction time is recommended from previous studies with various human cell types. Reaction time may have to be increased up to 1 h to detect activity from control (uninduced) cells that express low basal activity. Most mammalian cell types should contain basal HO activity detectable by the methods described under **Subheading 3.3.1.** and **3.3.2.**

19. The 405-nm setting is submaximal for detection of biliverdin or bilirubin, however it allows simultaneous detection of both pigments.

20. Reversing the column direction when peak spreading occurs can extend column lifetime. Excess injected hemin may be retained in the column and elute during subsequent methanol rinsing. The binding of excess heme inside the column may also affect column lifetime. The bile pigments, however, do not elute in subsequent column rinses when injected at < 2 μM.

21. The resulting colored pellet retains approx 80–85% of the heme. The supernatant is clear with a faint yellow color. Ensure that samples are free of particulates.

22. Freezing the alcohol-extracted HO reaction mixtures (final step) at –70°C will result in bile pigment and internal standard (mesoporphyrin) degradation. Furthermore, there is a significant dark decay of BR-IXα when the samples are stored in plastic tubes. Thus, long serial repetitions of equivalent samples in an autosampler result in gradual signal reduction. We recommend limiting autosampler runs to less than 20 samples. The rate of dark decay can be determined experimentally from the repetition of bilirubin standards in alcohol extracted protein (**Subheading 3.3.1., steps 12–19**).

23. If an integration program is not available, then detector response in peak height (arbitrary units) may also be used.

24. In the absence of HO activity (i.e., using boiled extract or NADPH blanks), only the heme and the mesoporphyrin peaks will be visible. If there is no activity, review the protein-extraction steps and safeguard against degradation. Make certain that the reaction took place at 37°C in a circulating water bath. Re-evaluate the quality of the reagents.

25. If the bilirubin is degraded, additional small peaks may occur in the chromatogram preceding the heme peak. Ensure the reactions were kept in the dark after their initiation. Additional unidentified peaks may also occur if the mesoporphyrin stock solution is frozen and thawed more than once.

26. Calculate the mesoporphyrin recovery ratio directly from arbitrary detector units.

27. If BVR occurs in excess, then the HO activity equation reduces to pmol BR/mg protein/ h.

28. Prepare biliverdin and bilirubin standards in protein solution for best results. For this step, protein extracts that have been stored frozen at –20°C may be used.

Bilirubin standard is unstable when prepared in HPLC buffer A alone. The assay protocol assumes that 100% of the BV and BR will be recovered from HO reactions by extraction in 95:5 v/v ETOH:DMSO, which is the case for HO reactions containing 500 μg protein and less than 2 μ*M* BV or BR. Recovery efficiency may be assessed by dissolving 20 μL 10X standard in 80 μL protein solution at 37°C at (0.2–2 μ*M*)(assuming a final volume of 200 μL). Add 100 μL 95:5 v/v EtOH:DMSO. Centrifuge 5 min at 15,000*g*. Inject supernatant immediately into HPLC. Compare with result obtained in **steps 12–19**.

29. The detector response should be linear with BV or BR concentration up to 10 μ*M*.
30. Crude extract may be prepared by a rapid freeze-thaw lysis protocol *(62)*. One to two confluent monolayers in 100-mm dishes provides excess starting material for crude extraction.
31. The reported *RF* = 0.56, for bilirubin in chloroform acetic acid (20:1) *(62)*.
32. The protocol works with 1 mg/mL crude extract. Addition of higher concentration (> 1 mg/mL) of crude extract tends to increase background. Reactions using up to 5 mg/mL microsomal protein have been reported.
33. More correctly, construct a standard curve using crude extract as the diluent. Extract the standard series with chloroform. Plot $OD_{464} - OD_{530}$ against starting bilirubin concentration. This will generate a standard curve corrected for extraction efficiency.
34. Extinction coefficients vary depending on the protein composition (tissue source) of the reaction. Construct standard curves using mock reaction mixtures (–NADPH) spiked with bilirubin standard.
35. Activity loss occurs with storage at –20°C. Do not freeze-thaw more than once.
36. The HPLC-based HO assay may be conveniently converted into a BVR assay. Assemble BVR reactions as described in (**Subheading 3.4.2., steps 1–4**) assuming a 100-μL final volume. Substitute the BVR reaction for the HO reaction in **Subheading 3.3.1., steps 1–10**). Express BVR activity as pmol BR/mg protein/h.
37. This results in a final reaction composition of 0.25% DMSO. The BVR will be partially inhibited by DMSO (at 2% v/v) in the reaction, therefore this parameter should not be increased.
38. Construct bilirubin standard curves using the BVR protein extract as the diluent (omit NADPH).

Acknowledgments

Many studies described herein were supported by core grants to R. M. Tyrrell from the International Association for Cancer Research (UK) and the Department of Health (UK), contract no. 121/6378, with additional support from the League Against Cancer of Central Switzerland, the Neuchateloise League Against Cancer, and the Swiss National Science Foundation. E. K. received a postdoctoral fellowship from the European Molecular Biology Organization (EMBO). We would also like to thank the colleagues who have contributed to the studies described herein, including Sharmila Basu-Modak,

Françoise Hartmann, Stephen Keyse, Patrick Lüscher, Alexander Noel, Charareh Pourzand, Olivier Reelfs, Larry Richman, Marco Soriani, and Glen Vile.

References

1. Tenhunen, R., Marver, H. S., and Schmid, R. (1969) Microsomal heme oxygenase, characterization of the enzyme. *J. Biol. Chem.* **244,** 6388–6394.
2. Maines, M. D. (1997) The heme oxygenase system: a regulator of second messenger gasses. *Annu Rev. Pharmacol. Toxicol.* **37,** 517–554.
3. Keyse, S. M. and Tyrrell, R. M. (1987) Both near ultraviolet radiation and the oxidizing agent hydrogen peroxide induce a 32-kDa stress protein in normal human skin fibroblasts. *J. Biol. Chem.* **262,** 14,821–14,825.
4. Applegate, L. A., Luscher, P., and Tyrrell, R. M. (1991) Induction of heme oxygenase: a general response to oxidant stress in cultured mammalian cells. *Cancer Res.* **51,** 974–978.
5. Keyse, S. M. and Tyrrell, R. M. (1989) Heme oxygenase is the major 32-kDa stress protein induced in human skin fibroblasts by UVA radiation, hydrogen peroxide, and sodium arsenite. *Proc. Natl. Acad. Sci. USA* **86,** 99–103.
6. Keyse, S. M. and Tyrrell, R. M. (1989) Induction of the heme oxygenase gene in human skin fibroblasts by hydrogen peroxide and UVA (365nm) radiation: evidence for the involvement of hydroxyl radical. *Carcinogenesis* **11,** 787–791.
7. Ryter, S. W. and Tyrrell, R. M. (1998) Singlet molecular oxygen: a possible effector of eukaryotic gene expression. *Free Radic. Biol. Med.* **24,** 1520–1534.
8. Murphy, B. J., Laderoute, K. R., Short, S. M., and Sutherland, R. M. (1991) The identification of heme oxygenase as a major hypoxic stress protein in Chinese hamster ovary cells. *Br. J. Cancer* **64,** 69–73.
9. Motterlini, R., Foresti, R., Intaglietta, M., and Winslow, R. M. (1996) NO-mediated activation of heme oxygenase: endogenous cytoprotection against oxidative stress to endothelium. *Am. J. Physiol.* **270(1 Pt 2),** H107–H114.
10. Kappas, A. and Drummond, G. S. (1984) Control of heme and cytochrome P-450 metabolism by inorganic metals, organometals, and synthetic metalloporphyrins. *Env. Health Perspect.* **57,** 301–306.
11. Caltabiano, M. M., Koestler, T. P., Poste, G., and Grieg, R. G. (1986) Induction of 32- and 34-kDa stress proteins by sodium arsenite, heavy metals and thiol-reactive agents. *J. Biol. Chem.* **261,** 13,381–13,386.
12. Taketani, S., Sato, H., Yoshinaga, T., Tokunaga, R., Ishii, T., and Bannai, S. (1990) Induction in mouse peritoneal macrophages of a 34 kDa stress protein and heme oxygenase by sulfhydryl reactive agents. *J. Biochem. (Tokyo)* **108,** 28–32.
13. Maines, M. D. (1992) *Heme Oxygenase: Clinical Applications and Functions.* CRC, Boca Raton, FL.
14. Ryter, S. W. and Tyrrell, R. M. (1997) The role of heme oxygenase-1 in the mammalian stress response: molecular aspects of regulation and function, in *Oxidative Stress and Signal Transduction,* (Forman, H. J. and Cadenas, E., eds.),Chapman and Hall, New York, pp. 343–386.

15. Tyrrell, R. M. (1997) Approaches to define pathways of redox regulation of a eukaryotic gene: the heme oxygenase 1 example. *Methods* **11**, 313–318.
16. Lautier, D., Luscher, P., and Tyrrell, R. M. (1992) Endogenous glutathione levels modulate both constitutive and UVA radiation/hydrogen peroxide inducible expression of the human heme oxygenase gene. *Carcinogenesis* **13**, 227–232.
17. Basu-Modak, S. and Tyrrell, R. M. (1993) Singlet oxygen: a primary effector in the ultraviolet A/near-visible light induction of the human heme oxygenase gene. *Cancer Res.* **53**, 4505–4510.
18. Maines, M. D., Trakshel, G. M., and Kutty, R. K. (1986) Characterization of two constitutive forms of rat liver microsomal heme oxygenase. Only one molecular species of the enzyme is inducible. *J. Biol. Chem.* **261**, 411–419.
19. Applegate, L. A., Noel, A., Vile, G., Frenk, E., and Tyrrell, R. M. (1995) Two genes contribute to different extents to the heme oxygenase enzyme activity measured in cultured human skin fibroblasts and keratinocytes: implications for protection against oxidant stress. *Photochem. Photobiol.* **61**, 285–291.
20. Trakshel, G. M., Kutty, R. K., and Maines, M. D. (1986) Purification and characterization of the major constitutive form of testicular heme oxygenase. The noninducible isoform. *J. Biol. Chem.* **261**, 11,131–11,137.
21. Cruse, I. and Maines, M. D. (1988) Evidence suggesting that the two forms of heme oxygenase are products of different genes. *J. Biol. Chem.* **263**, 3348–3353.
22. Trakshel, G. M., Kutty, R. K., and Maines, M. D. (1988) Resolution of rat brain heme oxygenase activity: absence of a detectable amount of the inducible form (HO-1). *Arch. Biochem. Biophys.* **260**, 732–739.
23. Zakhary, R., Gaine, S., Dinerman, J., Ruat, M., Flavahan, N., and Snyder, S. (1996) Heme oxygenase 2: Endothelial and neuronal localization and role in endothelium dependent relaxation. *Proc. Natl. Acad. Sci. USA* **93**, 795–798.
24. Vercellotti, G. M., Balla, G., Balla, J., Nath, K., Eaton, J. W., and Jacob, H. S. (1994) *Artif. Cells Blood Substit. Immobil. Biotechnol.* **22**, 207–213.
25. Stocker, R. (1990) Induction of haem oxygenase as a defense against oxidative stress. *Free Rad. Res. Comm.* **9**, 101–112.
26. Stocker, R., Yamamoto, Y., McDonagh, A. F., Glazer, A. N., and Ames, B. N. (1987) Bilirubin is an antioxidant of possible physiological importance. *Science* **235**, 1043–1045.
27. Vile, G. F., Basu-Modak, S., Waltner, C., and Tyrrell, R. M. (1994) Heme oxygenase 1 mediates an adaptive response to oxidative stress in human skin fibroblasts. *Proc. Natl. Acad. Sci. USA* **91**, 2607–2610.
28. Eisenstein, R. S. and Munro H., (1990) Translational regulation of ferritin synthesis by iron. *Enzyme* **44**, 42–58.
29. Eisenstein, R. S., Garcia-Mayol, D., Pettingel, W., and Munro, H. (1991) Regulation of ferritin and heme oxygenase synthesis in rat fibroblasts by different forms of iron. *Proc. Natl. Acad. Sci. USA* **88**, 688–692.
30. Vile, G. F. and Tyrrell, R. M. (1993) Oxidative stress resulting from ultraviolet A irradiation of human skin fibroblasts leads to a heme oxygenase-dependent increase in ferritin. *J. Biol. Chem.* **268**, 14,678–14,681.

31. Lin, F. and Girotti, A. W. (1998) Hemin-enhanced resistance of human leukemia cells to oxidative killing: antisense determination of ferritin involvement. *Arch. Biochem. Biophys.* **352,** 51–58.

31a. Ryter, S. and Tyrrell, R. M. (2000) *Free Radic. Biol. Med.* **28,** in press.

32. Poss, K. D. and Tonegawa, S. (1997) (I) Heme oxygenase 1 is required for mammalian iron reutilization. (II) Reduced stress defense in heme oxygenase 1-deficient cells. *Proc. Natl. Acad. Sci. U S A* **94,** 10,919–10,930.

33. Dennery, P. A., Spitz D. R., Yang, G., Tatarov, A., Lee, C. S., Shegog, M. L., and Poss, K. D. (1998) Oxygen toxicity and iron accumulation in the lungs of mice lacking heme oxygenase-2. *J. Clin. Invest.* **101,** 1001–1011.

34. Verma, A., Hirsch, D. J., Glatt, C. E., Ronnett, G. V., and Snyder, S. H. (1993) Carbon monoxide: a putative neural messenger. *Science* **259,** 381–384.

35. Dawson, T. and Snyder, S. H. (1994) Gasses as biological messengers: nitric oxide and carbon monoxide in the brain. *J. Neurosci.* **14,** 5147–5159.

36. Zakhary, R., Poss, K. D., Jaffery, S. R., Ferris, C. D., Tonegawa, S., and Snyder, S. H. (1997) Targeted gene deletion of heme oxygenase 2 reveals neural role for carbon monoxide. *Proc. Natl. Acad. Sci. USA* **94,** 14,848–14,853.

37. Stevens, C. F. and Wang Y. (1993) Reversal of long-term potentiation by inhibitors of haem oxygenase. *Nature* **364,** 147–149.

38. Johnson, R. A., Lavesa, M., Askari, B., Abraham, N., and Nasjletti, A. (1995) A heme oxygenase product, presumably carbon monoxide, mediates a vasodepressor function in rats. *Hypertension* **25,** 166–169.

39. Furchgott, R. F. and Jothianandan, D. (1991) Endothelium dependent and independent vasodilation involving cyclic GMP: Relaxation induced by nitric oxide, carbon monoxide, and light. *Blood vessels* **28,** 52–61.

40. Stone, J. and Marletta, M. (1994) Soluble guanylate cyclase from bovine lung: activation with nitric oxide and carbon monoxide and spectral characterization of the ferrous and ferric states. *Biochemistry* **33,** 5636–5640.

41. Burnett, A. L., Johns, D. G., Kriegsfeld, L. J., Klein, S. L., Calvin, D. C., Demas, G. E., Schramm, L. P., Tonegawa, S., Nelson, R. J., Snyder, S. H., and Poss, K. D. (1998) Ejaculatory abnormalities in mice with targeted disruption of the gene for heme oxygenase-2. *Nat. Med.* **4,** 84–87.

42. Poss, K. D., Thomas, M. J., Ebralidze, A. K., Odell, T. J., and Tonegawa, S. (1995) Hippocampal long term potentiation is normal in heme oxygenase-2 mutant mice. *Neuron* **15,** 867–873.

43. Tyrrell, R. M. and Basu-Modak, S. (1994) Transient enhancement of heme oxygenase-1 mRNA accumulation: a marker of oxidative stress to eukaryotic cells. *Methods. Enzymol.* **234,** 224–235.

44. Chomczynski, P. and Sacchi, N. (1987) Single-step method of RNA isolation by acid guanidinium thiocyanate-phenol-chloroform extraction. *Anal. Biochem.* **162,** 156–159.

45. Yoshida, T., Biro, P., Cohen, T., Müller, R. M., and Shibahara, S. (1988) Human heme oxygenase cDNA and induction of its mRNA by hemin. *Eur. J. Biochem.* **171,** 457–461.

46. Shibahara, S., Müller, R. M., Taguchi, H., and Yoshida, T. (1985) Cloning and expression of cDNA for rat heme oxygenase. *Proc. Natl. Acad. Sci. USA* **82,** 7865–7869.

47. Kageyama, H., Hiwasa, T., Tokunaga, K., and Sakiyama, S. (1988) Isolation and characterization of a complementary DNA clone for a M_R 32,000 protein which is induced with tumor promoters in BALB/c 3T3 cells. *Cancer Res.* **48,** 4795–4798.

48. Suzuki, T., Sato, M., Ishikawa, K., and Yoshida, T. (1992) Nucleotide sequence of cDNA for porcine heme oxygenase and its expression in *Escherichia coli. Biochem. Int.* **28,** 887–893.

49. Evans, C. O., Healey, J. F., Greene, Y., and Bonkovsky, H. L. (1991) Cloning, sequencing and expression of cDNA for chick liver haem oxygenase. *Biochem. J.* **273,** 659–666.

50. Alam, J., Shibahara, S., and Smith, A. (1989) Transcriptional activation of the heme oxygenase gene by heme and cadmium in mouse hepatoma cells. *J. Biol. Chem.* **264,** 6371–6375.

51. Keyse, S. M., Applegate, L. A., Tromvoukis, Y., and Tyrrell, R. M. (1990) Oxidant stress leads to transcriptional activation of the human heme oxygenase gene in cultured skin fibroblasts. *Mol. Cell. Biol.* **10,** 4967–4969.

52. Kutty, R. K., Kutty, G, Nagineni, C. N., Hooks, J. J., Chader, G. J., and Wiggert, B. (1994) RT-PCR assay for heme oxygenase-1 and heme oxygenase-2: a sensitive method to estimate cellular oxidative damage. *Ann. NY Acad. Sci.* **738,** 427–430.

53. Saunders, E. L., Maines, M. D., Meredith, M. J., and Freeman, M. J. (1991) Enhancement of heme oxygenase-1 synthesis by glutathione depletion in Chinese hamster ovary cells. *Arch. Biochem. Biophys.* **288,** 368–373.

54. Hiwasa, T., and Sakiyama, S. (1986) Increase in the synthesis of a M_R 32,000 protein in BALB/c 3T3 cells after treatment with tumor promoters, chemical carcinogens, metal salts and heat shock. *Cancer Res.* **46,** 2474–2481.

55. Ishizawa, S., Yoshida, T., and Kikuchi, G. (1983) Induction of heme oxygenase in rat liver. Increase of the specific mRNA by treatment with various chemicals and immunological identity of the enzymes in various tissues as well as the induced enzymes. *J. Biol. Chem.* **258,** 4220–4225.

56. Greene, Y. J., Healey, J. F., and Bonkovsky, H. L. (1991) Immunochemical studies of haem oxygenase. Preparation and characterization of antibodies to chick liver haem oxygenase and their use in detecting and quantifying amounts of haem oxygenase protein. *Biochem. J.* **279 (Pt 3),** 849–854.

57. Schacter, B., Cripps, V., Troxler, R. F., and Offner, G. D. (1990) Structural studies on bovine spleen heme oxygenase. *Arch. Biochem. Biophys.* **282,** 404–412.

58. Tenhunen, R., Ross, M. E., Marver, H. S., and Schmid, R. (1970) Reduced nicotinamide-adenine dinucleotide phosphate dependent biliverdin reductase: Partial purification and characterization. *Biochemistry* **9,** 298–303.

59. Schacter, B. (1978) Assay for microsomal heme oxygenase in liver and spleen. *Methods Enzymol.* **52,** 367–372.

60. Kutty, R. K. and Maines, M. D. (1982) Oxidation of heme C derivatives by purified heme oxygenase. Evidence for the presence of one molecular species of heme oxygenase in the rat liver. *J. Biol. Chem.* **257,** 9944–9952.
61. Tenhunen, R. (1972) Method for microassay of microsomal heme oxygenase activity. *Anal. Biochem.* **45,** 600–607.
62. Sierra, E. and Nutter, L. (1992) A microassay for heme oxygenase activity using thin layer chromatography. *Anal. Biochem.* **200,** 27–30.
63. Lincoln, B., Aw, T. Y., and Bonkovsky, H. (1989) Heme catabolism in cultured hepatocyte: Evidence that heme oxygenase is the predominant pathway and that a proportion of the synthesized heme is converted rapidly to biliverdin. *Biochim. Biophys. Acta* **992,** 49–58.
64. Lee, T. C. and Ho, I. C. (1994) Expression of heme oxygenase in arsenic-resistant human lung adenocarcinoma cells. *Cancer Res.* **54,** 1660–1664.
65. Lincoln, B. C., Mayer, A. and Bonkovsky, H. (1988) Microassay of heme oxygenase by high performance liquid chromatography. Application to assay of needle biopsies of human liver. *Anal. Biochem.* **170,** 485–490.
66. Bonkovsky, H. L., Wood, S. G., Howell, S. K., Sinclair, P. R., Lincoln, B., Healy, J. F., and Sinclair, J. F., (1986) High performance liquid chromatography separation and quantitation of tetrapyrroles from biological materials. *Anal. Biochem.* **155,** 56–64.
67. Ryter, S. W., Kvam. E., and Tyrrell, R. M. (1999) Determination of heme oxygenase activity by high performance liquid chromatography. *Methods Enzymol.* **300,** 322–336.
68. Ryter, S. W., Kvam, E., Richman, L., Hartmann, F., and Tyrrell, R. M. (1998) A chromatographic assay for heme oxygenase activity in cultured human cells: application to artificial heme oxygenase overexpression. *Free Radic. Biol. Med.* **24,** 959–971.
69. Vremen, H. and Stevenson, D. (1988) Heme oxygenase activity as measured by carbon monoxide production. *Anal. Biochem.* **168,** 31–38.

24

Analysis of Molecular Chaperone Activities Using In Vitro and In Vivo Approaches

Brian C. Freeman, Annamieke Michels, Jaewhan Song, Harm H. Kampinga, and Richard I. Morimoto

1. Introduction

Molecular chaperones function in a range of protein homeostatic events, including cotranslational protein folding, assembly and disassembly of protein complexes, and protein transport across membranes. Many molecular chaperones are also known as heat-shock proteins, which refers to their regulation by stress conditions as diverse as infection with viral and bacterial agents, exposure to transition heavy metals, heat shock, amino acid analogs, drugs, toxic chemicals, and pathophysiologic and disease states including oxidative stress, fever, inflammation, infection, myocardial stress and ischemia, neurodegenerative diseases, aging, and cancer. The heat-shock response through the elevated expression of heat-shock proteins (Hsp's) protects cells and tissues against the deleterious effects of stress. Pre-exposure to mild, nontoxic stresses such as lower heat-shock temperatures, and reduced levels of metals, arsenite, ethanol, or oxidants confers a transient resistance (thermotolerance) to a subsequent, otherwise lethal, exposure to stress. A common feature of most, if not all, stresses against which Hsp's have protective capacity are effects on protein folding and protein aggregation.

Molecular chaperones represent a large class of proteins including Hsps 110, 104, 90, 70, 60, 40 (hdj-1), and 27, which are functionally related based on common properties of influencing the conformation and activities of protein substrates *(1)* (*see also* Chapter 25, this volume). Generally, molecular chaperones discriminate among distinct folded states of a substrate protein and interact with only a subset of these states (e.g., non-native intermediates) to influence folding pathway kinetics. Whereas some chaperones such as Hsp70

From: *Methods in Molecular Biology, vol. 99: Stress Response: Methods and Protocols*
Edited by: S. M. Keyse © Humana Press Inc., Totowa, NJ

and Hsp60 (groEL) facilitate refolding to the native state, other chaperones, including α-crystallins, small heat-shock proteins, immunophilins, cdc37, and Hsp90 prevent aggregation but do not fully restore unfolded proteins to their final native state *(2–8)*.

The in vivo properties of specific Hsp's have been examined in cells engineered to overexpress Hsp's either transiently or stably *(9–11)*. The activities of Hsp's can be assessed using several assays, including resistance to stress-induced cell death, effects on the fraction of soluble protein following heat shock, and the use of ectopically expressed proteins such as the *Photinus pyralis* firefly luciferase as an enzymatic reporter of in vivo protein damage. The luciferase reporter assay is particularly effective; the advantages include the following: (1) the enzyme emits light upon addition of specific substrates and has a very low noise-to-signal ratio; (2) the enzyme activity is easily measured by a luminometer; (3) the enzyme activity is thermolabile *(12–14)*; (4) enzyme activity is protected by high levels of molecular chaperones *(12–14)*; [e.g., cells transiently or stably overexpressing Hsp70 protect luciferase against thermodenaturation *(11,15)*; (5) enzyme activity and solubility can be assessed in parallel using biochemical fractionation and western blot analysis *(12–14)*; (6) luciferase can be targeted either to the cytoplasm or nucleus to measure the effects of heat shock in either compartment *(14)*.

The methods described in this chapter will describe protocols for examining the in vitro and in vivo activities of the molecular chaperones with particular emphasis on Hsp70 and its cochaperone hsp40 (hdj-1). Hsp70 was among the original proteins classified as a molecular chaperone; furthermore, it has the capacity to refold a wide array of proteins from a non-native, enzymatically inactive state to their native state independent of the size of the substrate protein.

2. Materials and Reagents

2.1. In Vitro Assays

1. Isopropyl β-ᴅ-thiogalactopyranoside (IPTG) (Sigma).
2. DEAE fast flow sepharose (Pharmacia-LKB).
3. ResourceQ resin (Pharmacia-LKB).
4. ResourceS resin (Pharmacia-LKB).
5. Superdex-200 size exclusion column (Pharmacia-LKB).
6. Polyethyleneimine cellulose-thin layer chromatography (PEI-TLC) plates (Aldrich).

2.2. Cell Culture

In this chapter, all procedures will be described for the hamster fibroblast (O23) cell line. If other cell lines are used, it may be necessary to adapt the protocols accordingly. For cell culturing, the following general reagents are required:

1. Complete growth medium (cell line dependent): DMEM/10% FCS: Dulbecco's modified Eagles's medum (DMEM) supplemented with 10% (v/v) fetal calf serum (FCS) (Life Technologies, Inc). Store at 4°C and warm to 37°C before use.
2. Sterile H_2O.
3. 60-mm-diameter tissue culture dishes.
4. 100-mm-diameter tissue culture dishes.
5. Trypsin (cell-line dependent): stock 2.5% (Gibco-BRL [Gaithersburg, MD], Life Technologies Inc.). Dilute the stock 10-fold in sterile phosphate-buffered saline (PBS) to 0.25%. Store at 4°C.
6. PBS: dissolve 0.204 g KH_2PO_4, 1.157 g $Na_2HPO_4.2H_2O$, 8.01 g NaCl, 0.201 g KCl in 800 mL H_2O, make volume up to 1 L and adjust pH to 7.4. Sterilize by autoclaving.
7. T25 tissue culture flasks.
8. Sterile tubes 4.5 mL (12.4 × 75 mm; Greiner [Solingen, Germany] 120180) and 12.4 ml (18.0 × 95 mm; Greiner 191180).

2.3. Specific Equipment Required

1. Fast-performance liquid chromatography (FPLC) system (Pharmacia-LKB).
2. Chromatography chamber for thin-layer chromatography.
3. PhosphorImager (Molecular Dynamics).
4. IsoData 20/20 γ-counter.
5. Precision water baths for which the temperature can be regulated to ± 0.1°C (we use Julabo Labor Technik, Germany). For long-term experiments such as heat-shock and recovery studies, the use of controlled CO_2 flow is recommended.
6. Flow cytometer (Becton Dickinson [Rutherford, NJ] FACS STAR).
7. Luminometer (Berthold Lumat 9501).

2.4. Expression of Heat-Shock Proteins in Bacterial or Sf9 Cells

1. Expression of recombinant wild-type human Hsp70 is best accomplished with the pMS-hsp70 plasmid *(7)* rather than pET-hsp70 for several reasons. The expression level is higher with the pMS plasmid with the additional advantage of a wider host range (the pET construct is limited to strains that express T7 polymerase). The preferred bacterial strain is BB1553 *(16)*, which has a disrupted *dnaK* gene and thus avoids copurification of DnaK. Purification of human Hsc70, however, can be done from a wild-type DnaK background as Hsc70 can be easily separated from DnaK on a weak anion exchange column (e.g., DEAE sepharose).
2. For recombinant expression of Hdj-1 (Hsp40, DnaJ), we use BL21/DE3 cells transformed with the pET-hdj-1 construct *(7)*.
3. Human Hsp90-β is expressed in Sf9 cells using a baculovirus stock containing the cDNA for human hsp90-β (a gift from Dr. Gerald Litwack, Thomas Jefferson University, Philadelphia, PA).
4. Luria-Bertani media: dissolve 10 g bacto-tryptone, 5 g bacto-yeast extract and 10 g NaCl in dH_2O (total volume 1 L), then autoclave.

5. Terrific broth: dissolve 12 g bacto-tryptone, 24 g bacto-yeast extract and 5 mL glycerin in dH_2O (total volume 1 L) then autoclave.

6. $TEN_{0.1}$: 20 mM Tris pH 6.9 (room temperature), 0.1 mM EDTA, and 100 mM NaCl; the 0.1 in the $TEN_{0.1}$ refers to the molarity of the NaCl.

7. $TMgN_{0.1}$: 20 mM Tris, pH 6.9, 5 mM $MgCl_2$, and 100 mM NaCl.

8. $HEN_{0.05}$: 20 mM HEPES, 0.1 mM EDTA, and 50 mM NaCl.

2.5. ATPase Assay

1. Buffer C: 20 mM HEPES, pH 7.2, 25 mM KCl, 2 mM MgAc, 10 mM NH_4SO_2, and 0.1 mM EDTA.

2. [γ-^{32}P]ATP (Amersham, Arlington Heights, IL).

2.6. Protein Substrate Interactions

1. Buffer B: 20 mM HEPES, pH 7.2, 5 mM $MgCl_2$, and 100 mM NaCl.

2. Na[^{125}I] (NEN, DuPont, Boston, MA).

3. IodoBeads (Pierce, Rockford, IL).

2.7. Analysis of the Protein Substrate

1. β-Galactosidase (Sigma).

2. Firefly luciferase (Sigma)

3. Denaturation buffer: 25 mM HEPES, pH 7.5, 50 mM KCl, 5 mM $MgCl_2$, 5 mM β-mercaptoethanol, and 6 M guanidine-HCl.

4. Refolding buffer: 25 mM HEPES, pH 7.5, 50 mM KCl, 5 mM $MgCl_2$, 10 mM dithiothreitol (DTT), and 1 mM ATP. The DTT and ATP are added just prior to use of the buffer.

5. Na–borate buffer: 0.1 M sodium borate and 0.1 M sodium acetate.

6. SDS sample buffer (5X): 0.312 M Tris-HCl, pH 6.8, 50% (w/v) glycerol, 25% (v/v) β-mercaptoethanol, 10% (w/v) sodium dodecyl sulfate (SDS), 0.05% (w/v) bromophenol blue. Dilute to 1X with dH2O where required.

2.8. Expression of Heat-Shock Proteins
and the Reporter Enzyme Luciferase in Mammalian Cells

1. For expression of cytoplasmic luciferase, we use plasmid pRSVLL/V (17), and for nuclear luciferase we use plasmid pRSVnlsLL/V (15). For expression of the inducible Hsp70 (18) or Hsp40/Hdj1 (19), we use plasmid pCMV70 and pCMV40 respectively (11). These plasmids are derived from the eukaryotic expression vector pCMV5 (a gift of Dr. M. Stinsky, University of Iowa, Iowa City, IA). The cDNAs of the respective heat-shock proteins are cloned into the multiple cloning site, which lies downstream of a CMV promoter. For the production of stable cell lines with inducible chaperones, we use plasmid pUHD15-1 and pUHC13-3 (gift from Dr. H. Bujard, University of Heidelberg, Germany). Plasmid pUHD15-1 encodes the tetracycline responsive transactivator (tTA) under control of the human CMV promoter. pUHC13-3 encodes firefly luciferase under control of a tTA-dependent promoter. Chaperones are cloned

downstream from the tTA-regulated promoter in vector pTBC-1 (gift from Dr. H. Hauser, Gesellschaft für Biotechnologische Forschung, Braunschweig, Germany). Plasmids phGR272 (Bujard) and pSV2neo (Clontech, Palo Alto, CA) encode genes that confer hygromycin and geneticin/neomycin resistance respectively.

2. HBS 2X: Dissolve 5 g HEPES, 8 g NaCl, 0.37 g KCl, 0.125 g $Na_2HPO_4 \cdot 2H_2O$, and 1.1 g D-glucose (monohydrate) in H_2O and make up to 500 mL. Adjust pH to 6.95 with 1 M NaOH at 25°C. Filter sterilize. Store at –20°C in plastic sterile (culture) flasks (*see* **Note 1**).

3. 286 mM $CaCl_2$: Stock 2 M: 5.88 g $CaCl_2 \cdot 2$ H_2O to 20 mL H_2O. Filter-sterilize and store at –20°C. Dilute stock seven fold in H_2O immediately before use.

4. Culture tubes 100 × 15 mm (Nunc, Wiesbaden-Biebrich, Germany, cat. no. 145470).

5. Selective media: Geneticine (G418; Life Technologies Inc.) or hygromycin (Life Technologies Inc.) stock at 50 mg/mL. The optimal working solution depends on the cell line. For O23 cells: 1 mg/mL in DMEM/10% FCS.

6. Tetracycline (Sigma) stock 10 mg/mL. For inhibition of expression from the tTA responsive promoter, a final concentration of 3 µg/mL is required.

7. Sterile gloves.

8. Sterile glass cloning rings, approximate size: inner diameter 4 mm, outer diameter 6 mm, height 5 mm.

2.9. In Vivo Assays for Chaperone Activity

2.9.1. Protein Insolubility Assay

1. TTN buffer: 0.1 % Triton X-100, 5 mM Tris-HCl, and 10 mM NaCl, pH 8.0: dissolve 0.788 g Tris-HCl and 0.58 g NaCl in H_2O and make up to 1 L, add 1 mL Triton X-100 and adjust pH to 8.0. Store at 4°C.

2. TNMP buffer: 10 mM Tris-base, 10 mM NaCl, 5 mM $MgCl_2$, and 0.1 mM PMSF (phenylmethylsulfonylfluoride), pH 7.4. Dissolve 1.211 g Tris-base, 1.0165g $MgCl_2$, and 0.5844g NaCl in H_2O, make up to 1 L and adjust pH to 7.4. Just before use, add the PMSF from a 10 mM stock solution in dimethylsulfoxide (DMSO) (174.2 mg PMSF in 100 mL DMSO, store at room temperature) to a final concentration of 0.1 mM. Store at 4°C.

3. Propidium iodide (PI) stock solution: 70 µg/mL propidium iodide in TNMP + 0.1% sodium citrate: dissolve 35 mg propidium iodide in 500 mL TNMP and add 0.5 g Na_3–citrate. Store at 4°C, PI is light-sensitive and vessels should be stored in aluminum foil.

4. FITC stock solution: 30 µg fluorescein iso-thiocyanate (FITC) in 1 mL TNMP. Store at –20°C. FITC is light sensitive and should be stored in aluminum foil.

2.9.2. Luciferase as a Reporter Enzyme

1. 20 mM MOPS in DMEM/10% FCS: Stock 1 M MOPS: 20.93 g 4-morpholinepropanesulfonic acid (MOPS) made up to 100 mL in DMEM/10% FCS, adjust pH 7.0, filter-sterilize, and store at –20°C. Dilute stock 50-fold immediately before use in DMEM/10% FCS.

2. Cycloheximide stock 5 mg/mL in water (*see* **Note 2**), filter-sterilize, and store in 500-µL aliquots. Dilute immediately before use in DMEM/10% FCS to a 2X concentration of 40 µg/mL (final concentration is cell line dependent).

3. Luciferase lysis buffer BLUC. Buffer A: 25 mM H_3PO_4/Tris, pH 7.8. Add 1.7 mL H_3PO_4 (85%) to 600 mL H_2O, adjust to pH 7.8 with buffer B and make up to 1 L with H_2O. To avoid precipitation, filter-sterilize, and store in plastic (tissue culture) flasks. Buffer B: 1 M Tris-base, pH 8.0. Dissolve 121.1 g Tris-base in 800 mL H_2O, adjust to pH 8.0 with HCl, make up to 1 L with H_2O. Buffer C: 1 M $MgCl_2$. Dissolve 20.33 g $MgCl_2$.6 H_2O in 100 mL H_2O. Buffer D: 20% Triton X-100 (v/v) in H_2O, Buffer E: 0.5 M EDTA. Dissolve 186.1 g of disodium ethylenediaminetetraacetate in 800 mL of H_2O, adjust pH to 8.0 with NaOH, and make up to 1 L with H_2O; the EDTA will not go into solution until the pH is adjusted to approx 8.0 *(20)*. Mix 210 mL buffer A, 2.2 mL buffer C, 10.5 mL buffer D, 31.8 mL glycerol, and 702 µL buffer E. Add 0.5% (v/v) 2-mercaptoethanol directly before use. BLUC is stored at 4°C and the other components at room temperature.

4. Luciferase reaction buffer BRLUC. Buffer F: 100 mM ATP. Dissolve 300 mg ATP in 4.8 mL buffer A, store in 500-µL aliquots. Buffer G: Dissolve 50 mg D-luciferin (Sigma L-9504) in 168.6 mL H_2O + 10 mL buffer A while shielding it from light; store in 12.5-mL aliquots (*see* **Note 3**). To prepare BRLUC, mix 27.0 mL BLUC (without β-mercaptoethanol), 500 µL buffer F and 12.5 mL buffer G. Store buffers F and G and BRLUC at –20°C. Luciferin is light sensitive; wrap tubes containing BRLUC and buffer G in aluminum foil.

5. Stacking gel buffer (4X): Dissolve 6.06 g Tris-base in H_2O, add 4 mL 10% sodium dodecyl sulfate (SDS), and make up to 100 mL with H_2O, adjust pH to 6.8 with HCl.

6. Laemmli loading buffer (2X): mix 25 mL stacking gel buffer, 25 mL 10% (w/v) SDS, 20 mL glycerol, and 5 mL 1% (w/v) Bromophenol blue. Add 10 mL of 2-mercaptoethanol immediately before use and make up to 100 mL with H_2O.

2.10. Clonogenic Cell Survival

1. Alcohol-resistant marker.
2. 70% Alcohol in spray flacon.
3. 0.5 % (w/v) Crystal violet in water (tap water and not demineralized water should be used). Dissolve and filter. The solution can be reused after refiltering.

3. Methods
3.1. Protein Purification
3.1.1. Extract Preparation

1. Grow bacteria transformed with an appropriate expression vector in Terrific Broth supplemented with glucose (10 mM) and ampicillin (100 µg/mL) to an OD_{595} = 0.9 (this takes approx 4–6 h) then add IPTG to a final concentration of 1 mM (*see* **Notes 4–6**).

2. Incubate for an additional 4 h at 30°C and clarify the culture by centrifugation at 2000g for 5 min in a preweighed bottle.
3. Pool the complete culture in a single pre-weighed bottle, wash the cells in 250 mL: of TEN$_{0.1}$, and clarify by centrifugation at 2000g for 5 min. The supernatant is then removed and the bottle is reweighed.
4. For each gram of wet cells add 3 mL of TEN$_{0.1}$ to reconstitute the cell pellet (*see* **Note 7**).
5. Mix the sample gently for 30 min at 4°C (e.g., on a nutator).
6. Add the protease inhibitors leupeptin and pepstatin A to 1 µg/mL and flash-freeze the solution (*see* **Note 8**).
7. Thaw the sample by incubation at 37°C. This process of freezing and thawing is then repeated three more times, giving a total of four cycles (*see* **Notes 9** and **10**).
8. Sonicate the sample until the extract becomes nonviscous (approx 5 min) (*see* **Note 11**).
9. Following sonication, clarify the cell extract by centrifugation at 30,000g in the SA-600 rotor for 1 h (*see* **Notes 12** and **13**).

3.1.2. Purification of Hsp70

In describing the various methodologies for studying Hsp70 or Hsc70, we will indicate Hsp70; however, the protocols can be used for either protein. Any exceptions to this will be indicated in the text.

1. The extract is resolved over the weak anion-exchange resin DEAE (Pharmacia-LKB) (*see* **Note 14**).
2. Wash the column with 2 column volumes of TEN$_{0.1}$ and then elute the protein using a 100 to 300 mM NaCl gradient in TEN buffer over five column volumes (*see* **Notes 15** and **16**).
3. The fractions containing Hsp70 are pooled and recirculated over a 20 mL ATP agarose column for a total of three passes at 1 mL/min.
4. Following recirculation, wash the ATP agarose column with 100 mL of TEN$_{0.5}$, equilibrated with 50 mL of TMgN$_{0.1}$ and elute with 10 mL of TMgN$_{0.1}$ supplemented with 10% glycerol and 50 mM ATP that is flushed through with TMgN$_{0.1}$ (*see* **Note 17**).
5. The pooled DEAE fractions are then recirculated again and the elution process is repeated.
6. Pool the two ATP elutions and concentrate to 1 mL utilizing a Centriprep-30 (Amicon). The volume is then brought up to 15 mL with TEN$_{0.1}$ and reconcentrated to 1 ml.
7. Adjust the volume to 5 mL with TEN$_{0.1}$, place the sample in dialysis tubing (mol wt cut off 10,000–12,000 Da), and dialyze against 1.5 L of TEN$_{0.1}$ for 4 h.
8. Apply the sample to ResourceQ resin (Pharmacia-LKB) at 1 mL/min (*see* **Note 18**).
9. Elute the protein using a NaCl gradient of 100 to 400 mM over five column volumes (*see* **Note 15**).
10. Pool the appropriate fractions and concentrate to 5 mL utilizing a Centriprep-30 (Amicon). Transfer sample to dialysis tubing (mol wt cut off 10,000–12,000 Da)

and dialyze against 1.5 L of $TEN_{0.1}$ for 60 h changing the buffer every 20 h (*see* **Notes 19** and **20**).

11. Finally, reduce the sample volume using a Centriprep-30 until the protein concentration is approx 50 mg/mL.

3.1.3. Purification of Hdj-1

Growth of the transformants and preparation of the crude extract are accomplished as described in **Subheading 3.1.1.**

1. Resolve the crude extract using a DEAE sepharose column equilibrated with $HEN_{0.5}$ and eluted with a 50 to 500 mM NaCl gradient.
2. Pool the appropriate fractions, load onto a ResourceS column (Pharmacia-LKB) and elute with a 50–400 mM NaCl gradient.
3. Pool the appropriate fractions and add ammonium sulfate to a final concentration of 40% (w/v).
4. Clarify the solution by centrifugation at 30,000g for 30 min in a SA-600 rotor.
5. Reconstitute the pellet in $TEN_{0.1}$, resolve over a Sephadex-50 desalting column (to remove the excess ammonium sulfate), pool the appropriate fractions and dialyze against 1.5 L of $TEN_{0.1}$ for 4 h at 4°C.
6. Load the sample onto a ResourceQ column (Pharmacia-LKB) and elute with a 50–400 mM NaCl gradient.
7. Pool the appropriate fractions and reduce the sample volume using a Centriprep-10 until the protein concentration is approx 50 mg/mL.

3.1.4. Purification of Hsp90

1. Five days post infection of a log phase growing culture of Sf9 cells prepare a crude cell extract by dounce homogenization and sonication.
2. Clarify the extract by centrifugation at 10,000g for 1 h.
3. Resolve the crude extract over a DEAE sepharose column and elute the protein using a 50–500 mM NaCl gradient.
4. Pool the Hsp90 containing fractions, load onto a Heparin Sepharose column (Pharmacia-LKB) and elute with a 100–600 mM NaCl gradient.
5. Pool the Hsp90 containing fractions and dilute the sample with $TEN_{0.1}$ until the final NaCl concentration is approx 100 mM.
6. Load the equilibrated sample onto a ResourceQ column and elute the protein with a 100–500 mM NaCl gradient.
7. Pool the Hsp90 containing fractions and reduce the sample volume using a Centriprep-30 until the protein concentration is approx 50 mg/mL.

3.2. ATPase Assay

ATP hydrolysis by Hsp70 is determined by measuring the release of $[^{32}P]P_i$ from $[\gamma\text{-}^{32}P]$ATP according to the protocol of Sadis and Hightower *(20)*. In brief, the following protocol is used.

1. 2.5 µg of Hsp70 is diluted into 50 µL of buffer C supplemented with 100 µM ATP (10 mCi [γ-^{32}P]ATP) and preheated to 37°C.
2. At time 0 (prior to addition of Hsp70), 5, 10, 15, and 20 min of incubation at 37°C remove a 2-µL aliquot, spot it onto a PEI-TLC plate and air dry.
3. Resolve the spotted samples using 0.5 M lithium chloride and 0.5 M formic acid in a standard chromatography chamber.
4. Air-dry the TLC plate.
5. Visualize and quantitate the data by PhosphorImager analysis (Molecular Dynamics).
6. Calculate the hydrolysis rates as an average of [γ-^{32}P]ATP hydrolysis rate at each time point (5, 10, 15, and 20 min) from three separate experiments for each sample after the background hydrolysis has been subtracted (background rate being amount of free phosphate in a sample without Hsp70, *see* **Note 21**).

3.3. Protein Substrate Interactions

Interactions between Hsp70 and a nonnative substrate can be examined using a variety of protein substrates. We present three experimental approaches using the model unfolded protein substrate reduced carboxymethylated α-lactalbumin (RCMLA) or the native counterpart α-lactalbumin which can form complexes detected by native gel electrophoresis (*see* **Fig. 1A**) or in GST-pull down affinity chromatography assays (*see* **Fig. 1B**) *(22)*. The third assay involves separation of the chaperone-substrate complex from unbound substrate using gel filtration (*see* **Note 22**).

3.3.1. Gel Filtration Assay

1. Radio-iodinate reduced carboxymethylated α-lactalbumin (RCMLA, Sigma) using carrier free Na[^{125}I] (NEN, DuPont) and IodoBeads according to the manufacturers protocol (Pierce).
2. Combine Hsp70 (14 µM) and iodinated RCMLA at a 5:1 molar ratio in buffer B and incubate at 37°C for 30 min.
3. Resolve the sample by size exclusion chromatography [we use a Superdex-200 gel filtration column (Pharmacia-LKB)] with the column equilibrated in buffer B (*see* **Note 23**).
4. Quantify the [^{125}I] in each fraction using an IsoData 20/20 γ-counter.

3.3.2. Native Gel Electrophoresis Assay

1. Combine Hsp70 (14 µM) and iodinated RCMLA at a 5:1 molar ratio in buffer B and incubate at 37°C for 30 min.
2. Resolve through a 6% acrylamide gel at 150 V prepared in 0.5X TBE at 4°C (*see* **Note 24**).
3. Dry the gel and then visualize and quantify the signals by PhosphorImager analysis (Molecular Dynamics).

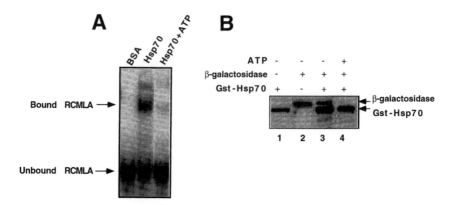

Fig. 1. Analysis of chaperone:substrate complex **(A)** Formation of a stable complex between Hsp70 and RCMLA was detected by gel-shift assay. ^{125}I-RCMLA
(2.8 μM) was incubated with Hsp70 (14 μM) in the presence (lane 2) or absence
(lane 3) of ATP (1 mM final) and analyzed by native gel electrophoresis followed by
PhosphoImager analysis. **(B)** Formation of a stable Gst-Hsp70: β-galactosidase
complex as detected by Gst pull-down assay. Chemically denatured β-galactosidase
(124 nM final) was incubated with Gst-Hsp70 (3.2 μM) alone in the absence (lane 3)
or presence (lane 4) of ATP (1 mM final). Protein complexes were pulled down by
glutathione-Sepharose beads and analyzed by Western blot analysis using anti-
Hsp70 and anti-β-galactosidase antibodies.

3.3.3. Gst-Pull Down Assay

1. Prepare a stock solution of β-galactosidase at 10 mg/mL in 1 M glycylglycine
   (pH 7.4).
2. Denature the β-galactosidase by fivefold dilution into denaturation buffer and
   incubation at 30°C for 30 min.
3. Initiate complex formation by 125-fold dilution of denatured β-galactosidase into
   refolding buffer with or without ATP at 4°C, supplemented with 3.2 μM
   Gst-Hsp70 , and incubate at 37°C for 30 min.
4. Add Glutathione without Sepharose 4B (Pharmacia-LKB) to a final concentration of 10% (v/v) and incubate for an additional 30 min at 4°C to form Gst-Hsp70
   complexes.
5. Following centrifugation at 10,000g for 10 s, wash the pelleted material
   corrresponding to Gst-Hsp70 complexes with 100 volumes of refolding buffer
   with or without ATP, and elute the bound material with 3 volumes of the same
   buffer containing 100 mM glutathione.
6. Elute the associated proteins from the glutathione-Sepharose and resolve by
   SDS-PAGE on a 12% gel. Perform Western blotting using anti-Hsp70 monoclonal antibody (5A5) or anti-β-gal antibody using standard techniques *(23)*.

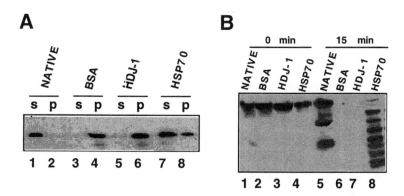

Fig. 2. Analysis of the solubility and conformational change of the substrate in the presence of the chaperone Hsp70. **(A)** The solubility of the β-galactosidase intermediate is established by fractionation of the substrate solution into soluble or insoluble fraction. Chemically denatured β-galactosidase (3.4 nM final) was incubated with indicated proteins (3.2 µM of BSA, Hdj-1, or Hsp70). Soluble (S) or aggregated (P) proteins were separated by centrifugation and analyzed by Western blot analysis using anti-β-galactosidase antibodies. **(B)** The conformation of the unfolded substrate is assessed using proteolysis to probe protein structure. Chemically denatured β-galactosidase (3.4 nM final) was incubated with the indicated proteins (3.2 µM of BSA, Hdj-1, or Hsp70). Following removal and quenching of one-half of the reaction in loading buffer (0 min), chymotrypsin was added to allow partial proteolysis (15 min). Samples were analyzed by Western blotting with anti-β-galactosidase antibodies.

3.4. Analysis of Biochemical Properties of the Nonnative Substrate

Three straightforward assays are presented to examine the properties of unfolded proteins incubated in the presence of chaperones. These are maintenance of the protein substrate in a soluble state, (*see* **Fig. 2A**), analysis of the partially folded soluble protein by native gel electrophoresis, and following digestion with proteases (*see* **Fig. 2B**). These methods rely on the propensity of guanidinium hydrochloride denatured β-galactosidase to form large aggregates upon dilution into aqueous buffers and the ability of certain chaperones to suppress this aggregation *(8,22)*.

3.4.1. Solubility Assay

1. Prepare β-galactosidase stock solution as in **Subheading 3.3.3., step 1**.
2. Denature the β-galactosidase by 20-fold dilution into 1 M glycylglycine (pH 7.4) followed by 10-fold dilution into denaturation buffer and incubation at 30°C for 30 min.

3. Initiate the reaction by 125-fold dilution of the denatured β-galactosidase into refolding buffer at 37°C supplemented with the desired chaperone or BSA (typically at 3.2 μ*M*).

4. Following a 30–120 min incubation at 37°C, separate the soluble and insoluble β-galactosidase species by centrifugation at 10,000*g* for 5 min.

5. Resolve the supernatant and pellet fractions on a 10% SDS-PAGE and visualize the β-galactosidase by Western-blot analysis using standard techniques *(23)*.

3.4.2. Native Gel Electrophoresis Assay

1. The β-galactosidase is denatured and diluted into refolding buffer as described in **Subheading 3.4.1., steps 1–3**.

2. Following a 1- to 2-h incubation at 37°C, mix the reactions with glycerol (10% final concentration) and Bromophenol dye (0.01% final concentration).

3. Resolve the reactions on a 4% acrylamide gel in Na-borate buffer at 1 mm/min until the bromophenol dye front reaches the edge of the gel *(see* **Notes 25** and **26**).

4. Following electrophoresis, electroblot the samples onto nitrocellulose membrane and detect the β-galactosidase by Western-blot analysis using standard techniques *(23)*.

3.4.3. Limited Proteolysis Assay

1. The β-galactosidase is denatured and diluted into refolding buffer as described in **Subheading 3.4.1., steps 1–3**.

2. Following a 2-h incubation at 37°C, add 1.0 μg of chymotrypsin to each reaction.

3. Remove aliquots (15 μL) prior to, and following, a 15 min incubation with the chymotrypsin, mix with 5X SDS sample buffer *(see* **Subheading 2.7., step 6**), and resolve immediately by SDS-PAGE using a 10% gel *(see* **Note 27**).

4. Transfer proteins to nitrocellulose and detect the β-galactosidase by Western-blot analysis using standard techniques *(23)* *(see* **Note 28**).

3.5. Luciferase Refolding Assay

The protocol for refolding of guanidine-HCl denatured firefly luciferase is a modified version of Buchberger et al. *(24)* as described in Freeman et al. *(25)*.

1. Prepare a stock solution of firefly luciferase at 4 mg/mL in 1 *M* glycylglycine (pH 7.4).

2. Denature the luciferase by twofold dilution into 1 *M* glycylglycine (pH 7.4) followed by 6.4-fold dilution into unfolding buffer and incubation for 30 min at 37°C.

3. Mix the denatured luciferase (1 μL) with 124 μL (final concentration 40 n*M* luciferase) of refolding buffer supplemented with Hdj-1 (1.6 μ*M*) and Hsp70 (0.8 μ*M*) *(see* **Note 29**).

4. Remove 1 μL of the refolding reaction at time 0, 5, 15, 30, 60, and 120 min, mix with 50 μL of refolding buffer and measure luciferase activity using a luminometer.

5. Calculate the activities as a percentage of the activity of 1 μL of luciferase which has been diluted to the same extent in refolding buffer supplemented with BSA (3.2 μ*M*) in the final dilution.

3.6. β-*Galactosidase Refolding Assay*

The chaperone-dependent refolding of β-galactosidase as described by Freeman and Morimoto *(8)* uses ONPG as the chromogenic assay *(26)* and is represented as a time dependent reactivation of unfolded β-galactosidase in the presence of Hsp70, Hdj-1, and ATP (**Fig. 3A**). Incubation in the presence of each chaperone alone, BSA, or in the absence of ATP does not result in refolding.

1. The β-galactosidase stock solution is prepared and denatured as described in **Subheading 3.4.1., steps 1** and **2**.
2. Initiate the refolding reaction by dilution of denatured β-galactosidase 125-fold into refolding buffer at 4°C supplemented with 1.6 μ*M* Hsp70 and 3.2 μ*M* Hdj-1.
3. Incubate at the desired temperature and assay the recovery of β-galactosidase activity at the desired time points (*see* **Note 30**).
4. To assay β-galactosidase activity mix 10 μL of the refolding reaction with 10 μL of refolding buffer supplemented with ONPG (final concentration 0.8 mg/mL) and incubate at 37°C for 15 min. Stop the chromogenic reaction by the addition of 50 μL of 0.5 *M* sodium carbonate and determine the absorbance of each sample at λ = 412 nm using a spectrophotometer.
5. Calculate the refolding activity relative to the enzymatic activity of native β-galactosidase diluted 20-fold with 1 *M* glycylglycine (pH 7.4) and then 125-fold with refolding buffer supplemented with BSA (3.2 μ*M*) (*see* **Note 31**).

3.7. *Maintenance of* β-*Galactosidase in Folding Competent State Assay*

Previous sections of this chapter describe methods for monitoring protein aggregation (**Subheadings 3.4.1.–3.4.3.**). However, these assays do not establish whether the substrate is in a "functional" state, experimentally defined as "on-pathway" intermediates, which can be refolded to the native state. An assay for the ability of a chaperone to maintain a substrate protein in a refoldable state is to assay whether the substrate remains soluble and can be refolded to its native enzymatically active state upon addition of one or more chaperones and ATP (**Fig. 3B**). The basic protocol is similar to the β-galactosidase refolding method described above except that a test chaperone (such as Hsp90) is added first. Hsp90 can maintain an unfolded substrate in a folding competent state and the Hsp70 and Hdj-1 are then mixed with the denatured β-galactosidase and the level of "refoldable" substrate is measured.

1. The β-galactosidase is denatured and diluted into refolding buffer as described in **Subheading 3.6., steps 1** and **2**.
2. Mix the denatured β-galactosidase with refolding buffer supplemented with a single test chaperone (e.g., Hsp90).
3. Incubate the reaction for 2 h at 37°C.
4. Add Hsp70 (1.6 μ*M*) and Hdj-1 (3.2 μ*M*) and monitor the recovery of β-galactosidase as described in **Subheading 3.6., steps 5–8** (*see* **Note 32**).

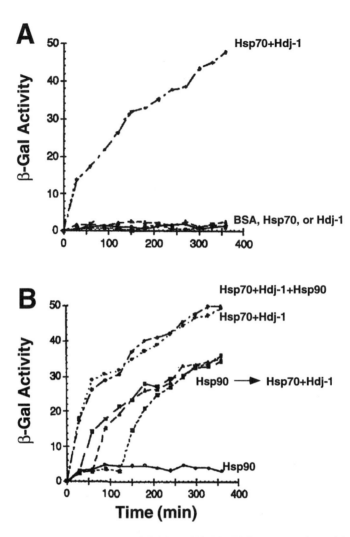

Fig. 3. Chaperones as holders and folders. (**A**) Hsp70 in cooperation with Hdj-1 can efficiently refold chemically (guanidine hydrochloride) denatured β-galactosidase. Unfolded β-galactosidase was diluted to a final concentration of 3.4 nM in refolding buffer supplemented with 1.6 μM Hsp70 and 3.2 μM Hdj-1. The level of refolding was determined relative to the activity of native β-galactosidase (3.4 nM) in refolding buffer supplemented with BSA (3.2 μM). β-galactosidase activity was measured using ONPG as the chromogenic substrate. Approximately 50% of the denatured substrate was refolded after 4 h in the presence of Hsp70, Hdj-1 and ATP. In the presence of BSA, the unfolded substrate aggregates. (**B**) Hsp90 maintains the denatured β-galactosidase in a refoldable state independent of nucleotide, which is subsequently refolded to the native state upon addition of the Hsp70, Hdj-1 and ATP. Denatured β-galactosidase was diluted into refolding buffer (3.4 nM final) supplemented with 1.6 μM Hsp90. After 30 min, 1- or 2-h incubation, 1.6 μM Hsp70, 3.2 μM Hdj-1, and ATP (1 mM final) were added to the reactions. The activity of β-galactosidase was monitored using ONPG as the chromogenic substrate.

3.8. Expression of Heat-Shock Proteins and the Reporter Enzyme Luciferase in Mammalian Cells

3.8.1. Transient Expression

This approach circumvents the requirement to establish stable cell lines. Firefly luciferase is cotransfected with putative chaperones and used as the reporter enzyme. Transient transfections do not require selection but each experiment requires a new transfection round and only a small proportion of the cell population will express the protein(s) of interest.

The following method is an adaptation of the method described by Kriegler *(27)* (*see* **Note 33**). Generally, transfection efficiencies of 1–5% are obtained for O23 hamster cells. Higher levels of protein per cell are generally achieved in these transient transfections than for stably transfected or inducible cell lines.

1. Culture O23 hamster fibroblasts (*see* **Note 34**) in complete growth medium and seed 1.5×10^5 cells per 60-mm culture dish.
2. Incubate for 24 h (*see* **Note 35**) before transfection with the appropriate plasmids (*see* **Note 36**).
3. Incubate HBS2X, H_2O and diluted $CaCl_2$ in a water bath at 25°C (*see* **Note 37**).
4. Add 10 μg plasmid (*see* **Note 38**) to 250 μL HBS2x in sterile 4.5-mL tubes with lid (Greiner) and incubate for 10 min at 25°C.
5. Add $CaCl_2$ over 1 min to the DNA/HBS solution while vortexing continuously (*see* **Note 39**).
6. Leave the DNA/calcium phosphate mix for 20 min at 25°C to allow formation of the DNA precipitate.
7. Wash the cells once for 2 min with HBS2x and aspirate all fluid from the cells.
8. Carefully add the DNA precipitate to the cells and leave the dishes for 20 min at room temperature (*see* **Note 40**).
9. Add 5 ml complete growth medium (*see* **Note 41**) to the cells and incubate overnight at 37°C in a CO_2 incubator (*see* **Note 42**).
10. After 24 h, remove the precipitate and process the cells for specific applications. For determination of luciferase enzyme activity about 2×10^4 cells in aliquots of 500 μL are placed in culture tubes.

3.8.2. Stable Transfections

When a whole cell population is to be studied, for example for cellular survival or measurements on the solubility of endogenous proteins, clonal cell lines capable of over-expressing individual Hsps are used. We prefer inducible systems such as the inducible-tetracycline promoter *(28)* in lieu of constitutive overexpression of chaperones which is likely to have deleterious effects on cellular metabolism. Moreover, the tetracycline-inducible system allows accurate titration of the protein of interest and the addition of tetracycline itself is not stressful for the cells.

 1. Follow **steps 1–10** as described in **Subheading 3.8.1.** However, to allow drug selection cells must be co-transfected with the appropriate plasmids in **step 2**, either 1 µg of a plasmid encoding the neomycin resistance gene (pSV2neo) or 8 µg of a plasmid encoding the hygromycin resistance gene (pHGR272). This protocol is optimized for O23 hamster fibroblasts. For the establishment of cell lines with tetracycline inducible chaperones specific selection and tests are required (*see* **Note 43**).
 2. Replace the DNA-precipitate by fresh complete growth medium 24 h after transfection.
 3. Transfer approx 3×10^6 cells into 100-mm tissue culture dishes and add selective medium containing geneticin or hygromycin (*see* **Note 44**).
 4. Check dishes and replace the selective medium regularly.
 5. Colonies of reasonable size are picked up at approx 7–10 d after transfection. Using sterile gloves, rinse the plate with PBS prewarmed at 37°C. Add 2 mL trypsin and cover the cells with it. Remove excess trypsin and place sterile glass cloning rings over the colonies of choice. After 3–4 min, add 50 µL medium to the ring, resuspend the cells and transfer them to a T25 flask containing the selective medium.
 6. When the cells in these flasks reach confluence, freeze an aliquot of the cells in liquid nitrogen, and analyze an aliquot for expression of the protein of interest (*see* **Note 45**).

3.9. In Vivo Assays for Chaperone Activity
3.9.1. Overall Protein Insolubility

The method described here is based on the observations by the Roti Roti laboratory (Washington University, St. Louis, MO) who showed a reduced solubility of (nuclear) proteins after a heat treatment of cells as determined by an increase in the protein-to-DNA ratio of nuclei after cell fractionation. The nucleus contains both water/detergent soluble proteins and insoluble ("structural") proteins. Many aqueous/detergent soluble proteins leak out of the nuclei when isolated from non-stressed cells but remain bound to (aggregate with) the structural, insoluble fraction when isolated from heat stressed cells. There are many possible variations to this assay that can depend on the type of cell fractionation one prefers and the type of cell line that is used (*see* **Note 46**). Here, we will describe a detergent based fractionation scheme for O23 cells in which a nuclear (pellet) fraction is isolated without contamination with other major cell organelles and cytoplasmic proteins and with a minimum of cytoskeletal remnants. For measurement of the protein content of these nuclei, a flow cytometry based method will be presented which is a slight modification of the protocol described by Blair et al. *(29)*. Using this and comparable methods, in vivo chaperone activities have been demonstrated for Hsp27 and Hsp70 *(30,31)*.

 1. Culture cells (about 5×10^6 cells) in 25-cm^2 flasks.
 2. Heat shocked cells (*see* **Note 47**). **Steps 3–14** must be carried out either on ice or at 4°C.

3. Wash cells twice with PBS.
4. Scrape cells in 2.5 mL TTN buffer (see **Note 48**) and collect in tube.
5. Rinse the flask with another 2.5 mL TTN and add to tube. If desirable, one can check using light microscopy whether the pellets contain clean nuclei (*see* **Note 49**).
6. Centrifuge for 5 min at 550*g*.
7. Pour off supernatant and dry inside the neck of the tube with tissue.
8. Resuspend the pellet in 5 mL TTN buffer.
9. Centrifuge for 5 min at 550*g*.
10. Wash the pellet twice in 5 mL TNMP (centrifuge for 5 min at 550*g* after each wash).
11. Dissolve the pellet in 0.4 mL TNMP.
12. Add 0.1 ml FITC solution and 0.5 mL PI solution.
13. Allow staining for a minimum of 1 h (preferably overnight).
14. Centrifuge for 5 min at 550*g* and resuspend the pellet in 0.5 mL TNMP.
15. Analyze 10000 nuclei per treatment on a Flow cytometer. Set the windows on the basis of Forward (FCS) and Sideward Scatter (SSC) to ensure that single particles are counted. Analyze the protein (FITC signal)-to-DNA (PI signal) ratio for each sample. The ratio of the FITC:PI signal of nuclei from treated cells can then be compared to that of untreated cells (*see* **Notes 50** and **51**).

3.9.2. Luciferase as a Reporter Enzyme

3.9.2.1. LUCIFERASE ENZYME INACTIVATION AND REACTIVATION IN VIVO

The use of foreign reporter proteins to assay chaperone activity in vivo was introduced by the Bensaude Laboratory (Ecole Normale Superieur, Paris, France). The rate and extent of in vivo luciferase enzyme inactivation and reactivation depends on several parameters, such as the cell line, the temperature used for heat shock, the intracellular localization of the luciferase, and the concentration of luciferase per cell (*see* **Note 52**). These conditions should be reoptimized for each new experimental condition. For example, the extent of chaperone protection against thermal stress often varies depending on the specific heat shock temperature or conditions of recovery (*see* **Note 53**).

For luciferase inactivation and reactivation experiments, cells are always transiently transfected with the luciferase encoding plasmids, cultured in tubes and heat shocked 24–48 h after seeding. For chaperone measurements, one can either cotransfect chaperone encoding plasmids or transfect luciferase encoding plasmids into clonal cell lines constitutively expressing the chaperones (*see* **Note 54**).

For luciferase inactivation, the following protocol is used *(11,14)*.

1. Incubate tubes containing cells expressing luciferase in a water bath at 37°C for 5 min.
2. Take eight tubes (*see* **Note 55**) as controls and place them on ice.
3. Close tubes tightly to avoid loss of CO_2-buffering capacity.
4. Transfer the tubes to a second water bath at the appropriate heat shock temperature.
5. Remove tubes at various times and place on ice.

For luciferase reactivation experiments, the following protocol is used *(11,14)*:

1. Replace the medium in tubes containing cells expressing luciferase by medium containing 20 mM MOPS (*see* **Note 56**) 24 h before the experiment.
2. Add 500 µL medium containing cycloheximide 2X (*see* **Note 2**) to the 500 µL medium in the tubes 30 min prior to heat shock (this prevents *de novo* synthesis of luciferase).
3. Take the tubes out of the incubator and place them in a water bath at 37°C for 5 min.
4. Take eight tubes as control for 100% luciferase activity and place them on ice.
5. Transfer the tubes to a water bath (preferably with CO_2 flow) at the desired heat shock temperature (*see* **Note 57**) for an appropriate time.
6. Immediately after heat shock, transfer the tubes to a water bath at 37°C to allow rapid cooling to the culture/recovery temperature. Take samples at $t=0$ by immediately placing the tubes on ice.
7. Place the tubes in a CO_2 incubator at 37°C and take samples at different times for assay of luciferase activity (*see* **Note 58**).

Dectection of Luciferase activity is as follows:

1. Aspirate the medium and wash the cells by gentle addition of 500 µL ice-cold PBS.
2. Aspirate the PBS and lyse the cells by addition of 500 µL ice-cold BLUC. Lysates can be analyzed immediately or can be stored at –20°C without affecting enzyme activity for at least 3 wk before assay.
3. Measure luciferase activity using a luminometer. Use 150 µL of lysate. After injection of 100 µL of substrates (BRLUC) luciferase activity is determined immediately by recording relative light units (RLU) emitted during the first 10 s (*see* **Note 59**).

3.9.2.2. LUCIFERASE INSOLUBILIZATION IN VIVO

Loss of luciferase enzyme activity is paralleled by a loss of luciferase solubility *(12)*, which is indicative of aggregation. Rather than measuring overall protein aggregation, this assay can also be employed in transient transfection assays in which the reporter is coexpressed with the chaperone of interest. Here, we describe a cell-fractionation protocol that can be used for determination of luciferase solubility.

1. Culture O23 cells expressing Hsp's and/or luciferase in 35 mm dishes.
2. Seal dishes tightly with parafilm and place them in a water bath at 37°C for 5 min.
3. Transfer the dishes to a water bath at the appropriate heat-shock temperature (*see* **Note 60**).
4. Stop heat shock by placing the dishes immediately on ice.
5. Wash briefly with ice-cold PBS.
6. Lyse cells by scraping them in 300 µL ice-cold BLUC.

7. Fractionate cell lysates into supernatants and pellets by centrifugation at 12,000g for 15 min at 4°C.
8. Add 300 μl 2X Lemmli SDS-PAGE loading buffer to the supernatants.
9. Analyze the supernatants by SDS-PAGE on a 12.5 % gel and visualize luciferase by Western-blot analysis using standard techniques *(22)* (*see* **Note 61**).

3.10. Clonogenic Cell Survival

For proliferating cells, one of the most sensitive endpoints for cytotoxicity is the loss of the capacity for sustained proliferation (reproductive cell death). Such a "dead" cell may still be biochemically active for quite some time and may even undergo one or two mitoses, but it has lost its reproductive integrity. We employ the definition that a cell which divides more than five times to form a macroscopically visible colonies (consisting of more than 50 cells) is considered clonogenically vital. For reasons that are still poorly understood, even within a nontreated growing cell population not all cells will be capable of forming colonies. Cellular survival (surviving fraction, SF) after a given treatment is therefore calculated, taking into account this plating efficiency, using the following formula:

$$SF = \frac{(\text{Colonies counted/Cells seeded})_{\text{treated}}}{(\text{Colonies counted/Cells seeded})_{\text{untreated}}}$$

The following protocol describes a variant of the assay used for anchorage-dependent cells (*see* **Note 62**).

1. Fill 60-mm premarked dishes (*see* **Note 63**) with 5 mL culture medium and place in a CO_2 incubator at 37°C until use, this will keep the medium warm and well buffered.
2. Trypsinize cells (*see* **Note 64**) and dilute to 1×10^6 cells/mL in warm (37°C) medium. Fill tubes (with caps) with 1 mL cells.
3. Heat cells at appropriate temperature and time interval (*see* **Note 65**).
4. Dilute cells (*see* **Note 66**) and plate 0.1 mL (diluted) cells into the prepared dishes in triplicate (*see* **Note 67**).
5. Place dishes in the CO_2 incubator and leave completely undisturbed (*see* **Note 68**) for a period of between 8 and 14 d to allow colonies of 50 or more cells to form (*see* **Note 69**).
6. At the end of this incubation period carefully remove all of the growth medium.
7. Fix the cells by spraying with 70% alcohol and allow to air-dry.
8. Stain colonies by overlaying dishes with a film of 0.5 % (w/v) crystal violet in water. Remove this carefully by flushing with water. Take particular care that colonies are not washed away. Allow dishes to air-dry.
9. Count all colonies containing 50 cells or more. We routinely use a binocular microscope (5–10× magnification).

4. Notes

1. The pH of HBS2x is crucial for transfection efficiency. It is recommended that several batches be prepared and tested to select those that yield the highest transfection efficiency. After freeze–thawing, two layers are visible that should be mixed before use.

2. The concentration of cycloheximide necessary to completely abolish protein translation must be established for every cell line. Cycloheximide is an irritant, so gloves should be worn when handling this reagent.

3. Luciferin does not dissolve in H_2O but will dissolve upon addition of buffer A, turning the solution yellow instead of white.

4. To increase the expression level, the transformants are restreaked twice and grown at 30°C. The second streaking should cover the entire surface of the plate to produce a bacterial lawn (one plate is streaked for every 2 L of media).

5. A similar protocol is followed for expression of Hsc70, except the transformants are grown at 22°C.

6. Prior to addition of the IPTG, a 1-mL sample is removed, clarified, the cell pellet reconstituted in 300 μL of 1X SDS sample buffer, and boiled for 5 min (uninduced sample).

7. Half the total volume of $TEN_{0.1}$ is used to reconstitute the cell pellet, the remaining fraction is supplemented with lysozyme to 100 μg/mL. Once the cell pellet is completely reconstituted in $TEN_{0.1}$ the solution is transferred to a fresh 50-mL conical flask and the bottle used for clarification is rinsed with the $TEN_{0.1}$ supplemented with lysozyme. The $TEN_{0.1}$ containing lysozyme is then transferred to the 50-mL flask and the contents are gently mixed.

8. Flash-freezing is accomplished with either a dry ice/methanol bath or liquid nitrogen.

9. The sample is incubated at 37°C until a fraction (approx 20%) of the solution remains frozen at which time the remaining fraction of the solution is thawed by gentle inversion of the tube at room temperature.

10. If you need to stop the preparation for any reason it is best to leave the cells frozen prior to the fourth thaw.

11. A 2-μL sample is removed and mixed with 58 μL of 1X SDS sample buffer (induced sample).

12. A 4-μL sample is removed and mixed with 116 μL of 1X SDS sample buffer (crude extract sample).

13. The efficiency of the induction is checked by analysis of 15 μL of the uninduced, induced, and crude extract samples by SDS-PAGE.

14. We use a 200-mL DEAE column to ensure complete retention of the sample.

15. The fractions containing Hsp70 are identified by SDS-PAGE.

16. Typically Hsp70 elutes between 150 and 225 mM NaCl.

17. The sample is collected once 5 mL of the $TMgN_{0.1}$ + glycerol + ATP has been loaded onto the column and collection is continued until a total of 35 mL has been obtained.

18. We use a 35-mL column of ResourceQ resin.
19. Following dialysis, the integrity of the Hsp70 preparation is checked by SDS-PAGE; if quality does not meet expectations (>98% homogeneity), the sample is reapplied and eluted from a ResourceQ column.
20. If further purification is required (e.g., the 45-kDa ATPase domain is present), the sample is resolved over a size exclusion column (a Superdex-200 column is preferable). Prior to column chromatography, the sample should be heated at 37°C for 15 min to allow fragments of Hsp70 to disassociate from full-length protein.
21. In addition to measuring the basal ATPase activity of Hsp70, one can assess the effects of protein substrates or DnaJ homologs. We find that a 10–20 M excess of substrate (reduced carboxy methylated α-lactalbumin; RCMLA) is required for efficient substrate induced hydrolysis and a 2–5 M excess of the DnaJ homolog Hdj-1 is required. As a control for the substrate induced ATPase activity, the effect of native α-lactalbumin should be evaluated.
22. As an important technical note, the efficiency of detection of complexes between Hsp70 and a given substrate is enhanced if the reaction is incubated at temperatures above 22°C prior to resolution through a matrix maintained at 4°C.
23. Based on the apparent K_d (9.5 μM) between wild-type Hsp70 and RCMLA, a molar ratio of 5:1 Hsp70:RCMLA is necessary to achieve approx 50% RCMLA binding.
24. The samples are electrophoresed until the Bromophenol Blue dye elutes from the bottom of the gel.
25. Resolution and detection of non-native species of β-galactosidase is dependent on the type of gel buffer (0.1 M sodium borate and 0.1 M sodium acetate), temperature (4°C), and speed (bromophenol dye resolved at 1 mm/min) utilized for electrophoresis. Deviation from these conditions results in inadequate or no resolution/detection of non-native species of β-galactosidase.
26. An inability to resolve/detect non-native β-galactosidase usually occurs because of retention of the β-galactosidase at the top of the gel.
27. The reactions are mixed within the Hamilton syringe needle used for sample loading and are immediately resolved by SDS-PAGE.
28. The proteolytic pattern is critically dependent on the amount of chymotrypsin (1.25 μg) used along with the concentrations of BSA or chaperones (3.2 μM).
29. The concentration of DTT in the refolding buffer is critical. A minimum of 10 mM DTT is necessary for maximum recovery.
30. Efficient refolding is obtained between 22°C and 41°C and the refolding reaction is complete between 2 and 3 h.
31. The β-galactosidase refolding assay has been optimized for pH, Mg^{2+}, K^+, DTT, and chaperone concentrations. Efficient refolding of β-galactosidase is obtained between pH 6.4 and 8.2, $[K^+] > 25$ mM, $[Mg^{2+}] > 1$ mM, and $[DTT] > 10$ mM. The given concentrations of Mg^{2+}, K^+, and DTT represent minimum concentrations required; the maximum concentrations have not been established. Concentrations of Hsp70 and Hdj-1 that are efficient for protein refolding are

codependent. The optimal concentration range for Hsp70 is 0.8–6.4 μM and 0.4–12.8 μM for Hdj-1. At the lower concentration range for Hsp70 (0.8–3.2 μM), Hdj-1 should be present at a twofold molar excess, however, at the upper range (6.4 μM or above) the concentration of Hdj-1 can be lowered. If the level of Hdj-1 is not lowered at the higher concentration of Hsp70, then a reduction in protein refolding yield can be expected, as Hdj-1 induces oligomerization of Hsp70.

32. Prior to addition of Hsp70 and Hdj-1, the activity of β-galactosidase is measured to check for the level of spontaneous or test chaperone-mediated protein refolding.

33. Other transfection methods are available that may enhance transfection efficiencies (e.g., lipofectamine [Life Technologies, Inc.]). When using alternative transfection protocols, the levels of transfected proteins per cell might be different and other quantitative results for chaperone effects might be obtained. It is important that the transfected cells do not express aberrant levels of chaperones or reporter enzymes as this may lead to aggregation, disturbed intracellular localization, uncontrolled cell division, and, ultimately, cell death. This can be visualized by immunofluorescence detection of the transfected protein.

34. Use exponentially growing monolayer cells. We use O23 hamster fibroblasts that are CHO-like. Rapidly growing cells take up DNA more readily, resulting in high transfection efficiencies. For cells growing in suspension specific transfection protocols are required.

35. Cells should be in log-phase and well attached but not confluent.

36. DNA is isolated and purified using Qiagen Q500 columns.

37. Keep the temperature at exactly 25°C, as this is optimized for formation of the precipitate.

38. Mix the appropriate vectors and equalize the plasmid quantity to 10 μg with an empty vector such as pSP64 (Promega). For luciferase experiments, we transfect 2 μg of luciferase plasmid and 1 μg of heat-shock protein-encoding plasmid. Large quantities of the plasmid-encoding nuclear luciferase might lead to a saturated nuclear transport system and a proportion of the luciferase will remain in the cytoplasm. Confirm luciferase localization by indirect immunofluorescence. Large quantities of heat-shock protein encoding plasmids might lead to excessive expression and disturbance of cellular metabolism.

39. This process partially determines the size of the precipitate and the cellular take-up. Therefore, for reproducibility between transfections, maintain the speed of the vortex constant and add the calcium slowly.

40. To allow optimal contact between DNA and the cells, do not stir the culture dishes but leave them in the laminar flow cabinet.

41. Avoid medium that contains high phosphate levels (e.g., RPMI), as this results in uncontrolled formation of precipitates.

42. This procedure may be toxic. To avoid this, the incubation can be shortened to a minimum of 4 h.

43. To establish a cell-line stable expressing the tetracycline responsive activator (tTA), use geneticin resistance and culture in presence of 3 μg/mL tetracycline to suppress promoter activity. Select the cell line with the lowest leakiness and high-

est inducibility of the promoter, using transiently transfected luciferase downstream of a tTA responsive promoter. Transfect the selected cell line stable with the chaperone of interest under control of the tTA responsive promoter. Use hygromycin and geneticin as selective agent and culture in the presence of 3 µg/mL tetracycline. Test several cell lines for their expression by withdrawal of tetracycline. Withdrawal of tetracycline will induce full expression; intermediate tetracycline concentrations will result in intermediate levels of expression *(28)*.

44. For O23 cells geneticin (G418) and hygromycin, concentrations of 1 mg/mL are sufficient to kill all nonresistant cells. The optimal concentration of these antiboiotics should be established for each new cell line.

45. Although all selected cells are geneticin resistant, they may not always express the protein of interest. Variation of expression levels between selected colonies is always observed and several lines should be tested and compared functionally. Sometimes multiple selection rounds are necessary to establish a clonal cell line.

46. To isolate microscopically pure nuclei, the concentration or type of detergent is often critical and may vary from cell type to cell type. Also, salt requirements may differ drastically. For some adherent cell lines (e.g., CHO cells, rat fibroblasts), we have used repeated (two to three times) washes in an alternative nuclear isolation buffer (50 mM NaCl, 10 mM EDTA, 0.5% Nonidet NP-40, 50 mM Tris-HCl, adjusted to pH 7.4) instead of the TTN buffer described here. For nonadherent suspension cells (like HeLa cells or Ehrlich ascites cells) we have routinely used a 1% TX-100 detergent solution (0.08 M NaCl, 0.01 M EDTA, 1% Triton X-100, adjusted to pH 7.2) to isolate pure nuclei *(32)*.

47. To obtain significant insoluble nuclear protein, a range of 42–46°C for 15–180 min can be used for most cells, including O23 cells.

48. TTN buffer can be left on the cells for a few minutes only. Prolonging the incubation may lead to collapse/lysis of nuclei.

49. The quality of nuclear preparations can be checked by phase-contrast microscopy. Clean nuclei have a regular shape and should not have cytoplasmic tags at the poles. Such tags could cause an increase in the protein-to-DNA ratio because of collapse of proteins on nuclei rather than intranuclear protein aggregation. To double check, one can use immunofluorescence analysis of the isolated nuclei stained with PI and FITC (as for the flow cytometric analysis) in which cytoplasmic tags will show up as a green fluorescence signal (protein only) around an orange (PI/FITC) signal (DNA/protein) of the nuclei.

50. Check for normal DNA based on the PI signal to ensure that alterations in cell cycle distribution of the cell population has not occurred. The protein content (FITC signal) of nuclei is different for cells in different phases of the cell cycle. Treatment-induced alterations in the cell-cycle distribution will therefore have an effect on the outcome of the experiments; this can be circumvented by comparing cells in a specific cell-cycle phase [e.g., a mid-S phase analysis has proven to be quite useful *(31)*]. In such cases, one has to analyze more than 10,000 nuclei in total to yield sufficient numbers of mid-S phase nuclei.

51. For details of flow cytometry, *see* **ref. 33**.

52. Luciferase inactivation follows an exponential curve corresponding to first-order kinetics with the second phase of the inactivation curve departing from exponential decay *(14)*. Nuclear luciferase inactivates more rapidly and recovers slowly relative to cytoplasmic luciferase *(14)*. To obtain similar luciferase inactivation kinetics, Ltk mouse fibroblasts and O23 hamster fibroblasts should be heat shocked at 43°C and 42°C, respectively. Under these conditions, luciferase will denature with log-linear kinetics during the first 8 min. *(11,14)*. When expressed in cells growing at lower temperatures than 37°C, luciferase will denature at lower heat-shock temperatures *(12)*. Extremely high cellular concentrations of luciferase will exhibit different rates of heat inactivation *(14)*.

53. Cooperation of Hsp40 with Hsp70 in the refolding of cytoplasmic luciferase in O23 hamster cells was only observed during recovery from a heat shock of 30 min at 45°C but not when milder heat shocks were given *(11)*.

54. For protein (re)folding, the chaperone/substrate ratio is important. With the different transfection protocols, different ratios may be obtained and this may explain variations in the magnitude of chaperone effects observed.

55. The data are expressed relative to luciferase activity in untreated samples; therefore, it is essential to have an accurate 100% value. To accomplish this, we take eight samples at $t=0$ before heat shock for inactivation and reactivation experiments. We take four tubes at every other time point.

56. Reactivation experiments take relatively more time outside the CO_2 incubator, which could interfere with results. The pH of the tissue culture medium should be constantly monitored and stabilized by addition of 20 m*M* MOPS to the medium on the day before the experiment is initiated and by performing the heat shock in a water bath with CO_2 flow, followed by recovery in a CO_2 incubator. The latter also allows performance of the experiments using uncapped tubes.

57. For reactivation studies, the level of initial damage, from which recovery of enzyme activity is studied, is critical. If damage is too high, no recovery will be observed, whereas insufficient damage will lead to very rapid recovery kinetics, which are difficult to follow and more variable, making it likely that subtle differences in reaction kinetics may go undetected. For O23 cells, a heat shock of 30 min at 42–43°C enables luciferase recovery to be studied over a relatively broad range. Chaperone effects are usually measured after higher heat shocks to allow better discrimination of the effect. For O23 cells, we heat shock for 30 min at 44°C or 45°C to detect cooperation of Hsp40 and Hsp70 in the reactivation of nuclear or cytoplasmic luciferase, respectively.

58. During reactivation experiments, protein synthesis is arrested and luciferase activity decays even at 37°C. For O23 cells, luciferase activity decays 10–15% during the first 60 min, with 50% of the luciferase activity no longer detected after 4 h. If possible, avoid experiments at time-points exceeding 60 min, where decay is high and luciferase decay might prevail over reactivation. Check the stability of luciferase at 37°C in the presence of cycloheximide and measure the reactivation kinetics before choosing one time-point at which to measure recovery.

59. During the luciferase enzyme reaction, a flash of light is emitted that is most intense during the first 10 s. Use this interval for optimal detection.

60. Using this protocol, we found that nuclear luciferase expressed in O23 cells insolubilizes completely during a heat shock of 20 min at 43°C, whereas cytoplasmic luciferase is only partially insolubilized *(11)*.

61. Antiluciferase antibody can be obtained from Promega.

62. For anchorage-independent (suspension) cells, a soft agar equivalent of the technique is available *(34)*.

63. Dishes should be labeled with an alcohol-resistant marker.

64. It is extremely important to obtain a single cell suspension, as populations of undispersed cells invalidate the calculations.

65. For O23 cells, 30-min heat-shock treatments at 42, 43, 44, or 45°C will typically result in about 10, 50, 90, and 99% cell killing, respectively.

66. Estimate the fraction of surviving cells for each temperature and time-point and make dilutions so that approx 100 colonies will form on the dishes. It is recommended to make several dilutions, as the surviving fraction is dependent on cell line, heat-shock temperature, and other variables such as the level of chaperone overexpression.

67. Cells that are not capable of generating colonies (clonogenically dead) may remain biochemically active and, in fact, may produce factors in the medium that stimulate growth and cell division of neighboring cells (feeder effect). Different dilutions will contain different concentrations of dead cells and these may affect colony formation. To avoid confounding effects (different dilutions will contain a different number of dead cells), we advise adding 1×10^6 feeder cells. Feeder cells have been exposed to doses above 50 Gy of X-radiation or heat treatments for >3 h at 45°C. To assure that they are indeed clonogenically dead, always plate 1×10^6 feeder cells as a control.

68. Stirring of the dishes might detach dividing cells originating from only one surviving cell. This will result in the formation of satellite colonies that may be erroneously counted as independent colonies.

69. The time required for colony formation depends on the cell doubling time (Td). For Td = 12–18 h, it takes about 8–10 d; for Td > 24 h, it takes more than 2 wk. Always check the size of the colonies. Stop the experiment at a time when longer incubations will only result in existing colonies becoming larger rather than the formation of any new colonies containing 50 cells or more.

References

1. Morimoto, R. I., Tissieres, A., and Georgopoulos, C. (1994), *The Biology of Heat Shock Proteins and Molecular Chaperones*, 2nd ed. CSH, Cold Spring Harbor Laboratory Press, Cold Spring Harbor, New York.

2. Skowyra, D., Georgopoulos, C., and Zylicz, M. (1990) The *E. coli* dnaK gene product, the hsp70 homolog, can reactivate heat-inactivated RNA polymerase in an ATP hydrolysis-dependent manner. *Cell* **62,** 939–944.

3. Hwang, D. S., Crooke, E., and Kornberg, A. (1990) Aggregated dnaA protein is dissociated and activated for DNA replication by phospholipase or dnaK protein. *J. Biol. Chem.* **265,** 19,244–19,248.

4. Ziemienowicz, A., Skowyra, D., Zeilstra-Ryalls, J., Fayet, O., Georgopoulos, C., and Zylicz, M. (1993) Both the Escherichia coli chaperone systems, GroEL/GroES and DnaK/DnaJ/GrpE, can reactivate heat-treated RNA polymerase. Different mechanisms for the same activity. *J. Biol. Chem.* **268,** 25,425–25,431.

5. Hartl, F.-U. (1996) Molecular chaperones in cellular protein folding. *Nature* **381,** 571–580.

6. Johnson, J. L. and Craig, E. A. (1997) Protein folding in vivo: Unraveling complex pathways. *Cell* **90,** 201–204.

7. Freeman, B. C. and Morimoto, R. I. (1996) The human cytosolic molecular chaperones Hsp90, Hsp70 (Hsc70), and Hdj-1 have distinct roles in recognition of a non-native protein and protein refolding. *EMBO J.* **15,** 2969–2979.

8. Freeman, B. C., Toft, D. O., and Morimoto, R. I. (1996) Molecular chaperone machines: Chaperone activities of the cyclophilin Cyp-40 and the steroid aporeceptor associated protein, p23. *Science* **274,** 1718–1720.

9. Landry, J., Chrétien, P., Lambert, H., Hickey, E., and Weber, L. A. (1989) Heat shock resistance conferred by expression of the human HSP27 gene in rodent cells. *J. Cell Biol.* **109,** 7–15.

10. Li, G. C., Li, L. L., Liu, Y.-K., Mak, J. Y., Chen. L., and Lee, W. M. F. (1991) Thermal response of rat fibroblasts stably transfected with the human 70kDa heat shock protein-encoding gene. *Proc. Natl. Acad. Sci. USA* **88,** 1681–1685.

11. Michels, A. A., Kanon, B., Konings, A.W.T., Ohtsuka, K., Bensaude, O., and Kampinga, H.H. (1997) Hsp70 and Hsp40 chaperone activities in the cytoplasm and the nucleus of mammalian cells. *J. Biol. Chem.* **272,** 33,283–33,289.

12. Nguyen, V. T. Morange, M., and Bensaude, O. (1989) Protein denaturation during heat shock and related stress. *Escherichia coli* β-galactosidase and Photinus pyralis luciferase inactivation in mouse cells. *J. Biol. Chem.* **264,** 10,487–10,492.

13. Pinto, M., Morange, M., and Bensaude, O. (1991) Denaturation of proteins during heat shock. In vivo recovery of solubility and activity of reporter enzymes. *J. Biol. Chem.* **266,** 13,941–13,946.

14. Michels, A. A., Nguyen, V.-T., Konings, A. W. T., Kampinga, H. H., and Bensaude, O. (1995) Thermostability of a nuclear-targeted luciferase expressed in mammalian cells. Destabilizing influence of the intranuclear environment. *Eur. J. Biochem.,* **234,** 382–389.

15. Nollen, E. A. A., Brunsting, J. F., Song, J., Kampinga, H. H., and Morimoto, R. I. (2000) Bag1 functions in vivo as a negative regulator of Hsp70 chaperone activity. *Mol. Cell Biol.* **20,** 1083–1088.

16. Bukau, B. and Walker, G. C. (1990) Mutations altering heat shock specific subunit of RNA polymerase suppress major cellular defects of *E. coli* mutants lacking the DnaK chaperone. *EMBO J.* **9,** 4027–4036.

17. De Wet, J. R, Wood, K. V., DeLuca, M., Helinski, D. R., and Subramani, S. (1987) Firefly luciferase gene: structure and expression in mammalian cells. *Mol. Cell. Biol.* **7,** 725–737.

18. Hunt, C., and Morimoto, R. I. (1985) Conserved features of eukaryotic Hsp70 genes revealed by comparison with the nucleotide sequence of human Hsp70. *Proc. Natl. Acad. Sci. USA* **82,** 6455–6459.

19. Ohtsuka, K. (1993) Cloning of a cDNA for heat shock protein Hsp40, a human homologue of bacterial DnaJ. *Biochem. Biophys. Res. Commun.* **197,** 235–240.

20. Sambrook, J. Fritsch, E. F., and Maniatis, T. (1989) in *Molecular Cloning, A Laboratory Manual* 2nd ed. Cold Spring Harbor Laboratory Press, Cold Spring Harbor, New York.

21. Sadis, S. and Hightower, L. E. (1992) Unfolded proteins stimulate molecular chaperone Hsc70 ATPase by accelerating ADP/ATP exchange. *Biochemistry* **31,** 9406–9412.

22. Bimston, D., Song, J., Winchester, D., Takayama, S., Reed, J. C., and Morimoto, R. I. (1998) BAG-1, a negative regulator of Hsp70 chaperone activity, uncouples nucleotide hydrolysis from substrate release. *EMBO J.* **17,** 6871–6878.

23. Harlow, E. and Lane, D. (1988) *Antibodies: A Laboratory Manual* Cold Spring Harbor Laboratory Press. Cold Spring Harbor, NY.

24. Buchberger, A., Schroder, H., Buttner, M., Valencia, A., and Bukau, B. (1994) A conserved loop in the ATPase domain of the DnaK chaperone is essential for stable binding of GrpE. *Struct. Biol.* **1,** 95–101.

25. Freeman, B. C., Myers, M. P., Schumacher, R., and Morimoto, R. I. (1995) Identification of a regulatory motif in Hsp70 that affects ATPase activity, substrate binding, and interaction with HDJ-1. *EMBO J.* **14,** 2281–2292.

26. Cohn, M. (1957) Contributions of studies on the β-galactosidase of *Escherichia coli* to our understanding of enzyme synthesis. *Bacteriol. Revs.* **21,** 140-152.

27. Kriegler, M. (1990) in *Gene Transfer and Expression. A Laboratory Manual* Stockton Press, New York, pp. 96–98.

28. Gossen, M. and Bujard, H. (1992) Tight control of gene expression in mammalian cells by tetracycline-responsive promoters. *Proc. Natl. Acad. Sci. USA* **89,** 5547–5551.

29. Blair, O. C., Winward, R. T., and Roti Roti, J. L. (1979) The effect of hyperthermia on the protein content of HeLa nuclei: a flow cytometric analysis. *Radiat. Res.* **78,** 474–484.

30. Kampinga, H. H., Brunsting, J. F., Stege, G. J. J., Konings, A. W. T., and Landry, J. (1994) Cells overexpressing hsp27 show accelerated recovery from heat-induced nuclear protein aggregation. *Biochem. Biophys. Res. Comm.* **204,** 1170–1177.

31. Stege, G. J. J., Li, L., Kampinga, H. H., Konings, A. W. T., and Li, G. C. (1994) Importance of the ATP-binding domain and nucleolar localization domain of hsp72 in the protection of nuclear proteins against heat-induced aggregation. *Exp. Cell Res.* **214,** 279–284.

32. Kampinga, H. H., Turkel-Uygur, N., Roti Roti, J. L., and Konings, A. W. T. (1989) The relationship of increased nuclear protein content induced by hyperthermia to killing of HeLa cells. *Radiat. Res.* **117,** 511–522.

33. Shapiro, H. M. (1988) In *Practical Flow Cytometry*. Alan R. Liss, New York.

34. Jorritsma, J. B. M. and Konings, A. W. T. (1983) Inhibition of repair of radiation-induced strand breaks by hyperthermia and its relationship to cell survival after hyperthermia alone. *Int. J. Radiat. Biol.* **43,** 506–516.

25

Analysis of Chaperone Properties of Small Hsp's

Monika Ehrnsperger, Matthias Gaestel, and Johannes Buchner

1. Introduction

Small Hsp's (sHsp's) are an ubiquitous but diverse class of proteins that differ from other Hsp families in that only certain short-sequence motifs, the so-called α-crystallin domains and some sequence in the N-terminal parts, are conserved. Characteristic features in common are their low molecular mass (15–42 kDa), their oligomeric structure ranging from 9 to 50 subunits and their chaperone function in vitro. Additional features attributed to small Hsps range from RNA storage in heat shock granules to inhibition of apoptosis, actin polymerization and contribution to the optical properties of the eye lens in the case of α-crystallin (reviewed in **refs.** *1* and *2*). At the moment, it is unclear how these seemingly different functions can be explained by a common mechanism. However, as most of the observed phenomena involve non-native protein, the repeatedly reported chaperone properties of sHsp's might be a key feature for further understanding of their function.

Recent evidence suggests that, at least under heat-shock conditions, sHsp's bind a maximum of one non-native polypeptides on the surface of their oligomeric complex *(3–6)*. The function of sHsp's in this context is to create a reservoir of non-native proteins for subsequent refolding under physiological conditions. Refolding may occur in cooperation with other, ATP-dependent chaperones *(5,6)* (*see also* Chapter 24, this volume). To analyze the chaperone function of the sHsps several experimental in vitro systems using different "model substrates" (e.g., citrate synthase, β-crystallin and insulin) have been established to study the mechanism of sHsp chaperone function.

Citrate synthase (CS) is a commercially available homodimeric (2×48.969 kDa) mitochondrial enzyme readily inactivated upon incubation at elevated temperatures with a midpoint of transition at 48°C *(7,8)*. Like many other ther-

From: *Methods in Molecular Biology, vol. 99: Stress Response: Methods and Protocols*
Edited by: S. M. Keyse © Humana Press Inc., Totowa, NJ

mally denaturing proteins CS shows a tendency to aggregate upon heating. This is based on nonspecific hydrophobic interaction resulting in the formation of high-molecular-weight particles which can be detected by monitoring the turbidity and light scattering of the protein solution.

Small Hsp's form stable complexes with aggregation-prone unfolding intermediates, thus preventing aggregation and light scattering of the protein solution. However, inactivation of CS is not influenced by the presence of Hsp25, indicating the formation of a stable complex with CS *(6)*. Binding of the substrate molecules oxaloacetic acid (OAA) and acetyl-CoA induces a conformational change in the structure of CS so that the midpoint of thermal transition is shifted to 66.5°C *(9)*. This stabilization can be exploited to observe the refolding of two defined dimeric unfolding intermediates after thermal inactivation in the presence of molecular chaperones *(6,10)*.

Another substrate protein used for analyzing the chaperone activity of sHsp's is insulin *(4)*. Upon reduction of the interchain disulfide bonds, the insulin B chain will readily aggregate and precipitate while the A chain stays in solution *(11)*. The precipitation of the insulin B chain can be followed by monitoring the apparent absorbance resulting from the increase in light scattering. As aggregation can be measured at different temperatures, the temperature dependence of chaperone activity can be tested. An increase of protective activity with temperature has been proposed for α-crystallin *(12)*.

2. Materials

2.1. Equipment

1. Centricon-30 microconcentrators (Amicon, Danvers, MA).
2. Stirrable quarz cuvet (3 mL).
3. Microcuvets (100 µL).
4. Microcentrifuge with temperature control.
5. Fluorescence spectrophotometer with stirrable and temperature-controlled cuvet holder.
6. UV-Vis spectrophotometer with temperature control unit.

2.2. Reagents

1. Acetyl coenzyme A (Acetyl-CoA) (Boehringer Mannheim, IN).
2. Citrate synthase (CS) (EC 4.1.3.7.): Ammonium sulfate suspension of CS from porcine heart mitochondria (Boehringer Mannheim).
3. Dithiothreitol (DTT) (ICN, Costa Mesa, CA).
4. 5,5'-Dithio-bis(2-nitrobenzoic acid) (DTNB) (Sigma, St. Louis, MO).
5. Insulin: Lyophilized; from bovine pancreas (Sigma).
6. Oxaloacetic acid (OAA) (Sigma).
7. TE-buffer: 50 mM Tris-HCl, 2 mM EDTA, pH 8.0.

8. sHsp's: Recombinant Hsp25 and Hsp27 can be obtained from StressGen (Victoria, Canada), bovine lens α-crystallin used in the described experiments (*see* **Figs. 1** and **2**) was a generous gift from Prof. Josef Horwitz (UCLA).
9. Hsp70, Hsc70, or DnaK can be obtained from StressGen (Victoria, Canada).

3. Methods
3.1. Preparation of CS Stock Solution

1. Dialyze an ammonium sulfate suspension of CS against at least 3000 volumes of TE buffer overnight at 4°C.
2. Concentrate to about 17 mg/mL using Centricon-30 microconcentrators
3. To remove precipitated protein centrifuge the sample in a microcentrifuge at 18,000g for 30 min at 4°C.
4. Determine exact protein concentration using the published extinction coefficient of 1.78 for a 1-mg/mL solution in a 1-cm cuvet at 280 nm *(13)*.
5. Prepare aliquots and store at –20°C.

3.2. Preparation of Insulin Stock Solution

1. Prepare a 5-mg/mL solution of insulin in 10 mM sodium phosphate buffer, 100 mM NaCl, pH 7.0.
2. Add 1 M HCl until a pH of <2.0 is reached. Quickly titrate back to pH 6.5 with 1 M NaOH. Make sure that the solution is still clear. Check titration with pH indicator strips.
3. To remove precipitated protein, centrifuge the sample in a microcentrifuge at 18,000g for 30 min at 20°C.
4. Determine exact protein concentration using the calculated extinction coefficient of 1.06 for a 1-mg/mL solution in a 1-cm cuvet at 280 *(14)*.
5. Store protein at room temperature during measurement. For long-term storage freeze the solution at –20°C. Insulin solutions can be frozen and thawed several times. After thawing, centrifuge and determine protein concentration again.

3.3. Thermal Aggregation of Citrate Synthase

1. For each set of experiments, supplement an aliquot of CS (thawed on ice) with TE buffer to a final concentration of 30 μM (monomer). Store sample on ice (*see* **Note 1**).
2. Set the fluorescence spectrophotometer to an excitation and emission wavelength of 500 nm with spectral bandwidths of 2 nm each (*see* **Note 2**). Record data points at every 0.2–0.5 s.
3. Preparation of molecular chaperones: The concentrations of the molecular chaperones used should not be less than 2 mg/mL to avoid significant changes in the incubation temperature upon addition. If possible the molecular chaperone should be dialyzed against the assay buffer (40 mM HEPES-KOH, pH 7.5) (*see* **Note 3**). When other storage buffers have to be used, controls for the influence of the buffer substances on CS are essential (*see* **Note 4**) .

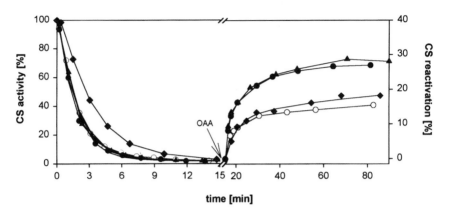

Fig. 1. Thermal inactivation and reactivation of CS. CS (75 nM) was inactivated at 43°C for 15 min in the absence (○) and presence of 0.3 μM Hsp25 (▲), 0.3 μM Hsp27 (●), and 0.3 μM authentic bovine lens α-crystallin (◆). CS activity was measured at the time-points as described in the text. To start reactivation, at time 15 min, OAA was diluted 1:100 into the samples (final concentration: 1 mM). The tubes were then transferred to the folding permissive temperature of 25°C and reactivation was monitored by measuring CS activity. The concentrations of Hsp25 and Hsp27 refer to 16-mers. α-crystallin concentration was calculated for a 25-mer.

4. The preincubated (43°C) assay buffer (*see* **Notes 5** and **6**) is pipetted into a thermostated (*see* **Note 7**), stirrable (*see* **Note 8**) quarz cuvet. When a constant baseline is reached, molecular chaperones (*see* **Note 9**) and/or other components are added (*see* **Notes 4**, **10**, and **11**). The signal is observed for about 5 min to constancy before, under stirring (see **Note 8**), CS is diluted 1:200 (final concentration: 0.15 μM; monomer) into the cuvet. Light scattering is then monitored for at least 20 min (*see* **Notes 12** and **13**).

3.4. Thermal Inactivation and Refolding of CS and Cooperation with Hsp70

3.4.1. Inactivation

1. CS and chaperones are prepared as described above with the exception that CS is diluted to a final concentration of 15 μM (monomer). Store CS on ice.
2. Samples (200–500 μL) (*see* **Note 14**) are prepared in 2-mL Eppendorf tubes equipped with a small stirring bar (7 × 2 mm). Assay buffer (40 mM HEPES-KOH, pH 7.5) with or without chaperone (*see* **Note 15**) is equilibrated at 25°C in the absence of CS.
3. CS is added under stirring by diluting the 15 μM stock solution 1:100 into the sample.
4. While still at 25°C, CS activity is measured as a reference for 100% activity.
5. To start the inactivation reaction, the tube is placed in a 43°C water bath. The temperature should be adjusted after 30 s when the first aliquot is taken to determine

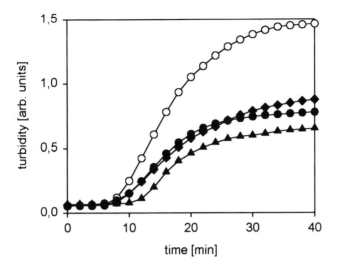

Fig. 2. Aggregation of insulin B chain at 30°C. Samples were prepared as described in the text. The reaction was started at time 0 min by the addition of DTT to a final concentration of 20 m*M*. Then, 45 µ*M* insulin were incubated in the absence (○) and presence of 0.1 µ*M* Hsp25 (▲), 0.1 µ*M* Hsp27 (●), and 0.23 µ*M* authentic bovine lens α-crystallin (◆). The scattering was recorded in a 1-cm cuvet (120 µL) in a spectrophotometer at 400 nm and 30°C. The concentrations of Hsp25 and Hsp27 refer to 16-mers. α-crystallin concentration was calculated for a 25-mer.

remaining activity. At least 10 more aliquots have to be analyzed during the time-course of inactivation to be able to follow the kinetics of the reaction (*see* **Fig. 1**).

6. Activity assay: The activity assay is performed according to Srere et al. (*7*). It is based on the first step of the Krebs cycle, where CS catalyzes the condensation of oxaloacetic acid and acetyl-CoA. The resulting coenzyme-A reduces Ellman's reagent (dithio-1.4.-nitrobenzoic acid; DTNB). This stoichiometric reaction can be monitored by the associated increase in absorption at 412 nm.

To determine the activity, 20-µL aliquots are taken from the inactivating samples and added to 980 µL of the reaction mixture:

930 µL TE buffer
10 µL of 10 m*M* oxaloacetic acid (in 50 m*M* Tris, pH not adjusted)
10 µL DTNB (in TE buffer)
30 µL acetyl-CoA (in TE buffer)

3.4.2. Refolding of CS with OAA and in Cooperation with Hsp70

1. To induce reactivation of defined CS folding intermediates (*6,10*), the ligand OAA is added to the inactivating sample (*see* **Note 16**). Under stirring, dilute OAA 1:100 into the sample (final concentration 1 m*M*) (*see* **Note 17**).
2. Transfer the tube immediately to 25°C to avoid further inactivation.

3. Monitor CS activity as described in **Subheading 3.4.1.** throughout the time-course of reactivation. Take more time-points at the beginning of reactivation to be able to fit the collected data (*see* **Note 18**) (*see* **Fig. 1**).
4. The stability of CS–sHsp complexes can be determined by performing a temperature shift to 25°C without adding OAA. If the enzyme stays stably bound to the Hsp no increase in activity is observed after the temperature shift. Reactivation by addition of OAA can then be triggered at time-points as late as hours after the temperature shift. A stable complex has been shown for CS-murine Hsp25 *(6)*.
5. To monitor reactivation with the aid of the ATP-dependent chaperone Hsp70, the experiment is performed similarly to OAA-induced reactivation. After 15 min incubation at 43°C Hsp70, $MgCl_2$, KCl, and ATP are added quickly before the sample is shifted to 25 °C and activity measurement is continued (*see* **Note 19**). Example for suitable concentrations: 0.3 μM sHsp
                 0.6 μM Hsp70
                 10 μM $MgCl_2$
                 10 μM KCl
                 5 μM ATP

3.5. Reduction-Induced Aggregation of Insulin

1. Chaperones are prepared as described in **Subheading 3.3.** (*see* **Note 3**).
2. The experiments are performed in 20 mM sodium phosphate buffer pH 7.0, 100 mM NaCl.
3. Samples can be prepared either directly in the cuvets or in reaction vessels.
4. Insulin is added to a final concentration of 45 μM (0.25 mg/mL) to the buffer in the presence or absence of sHsp and further additives (*see* **Note 20**).
5. The sample is preincubated at the chosen reaction temperature (e.g., 30°C) (*see* **Note 21**) for about 5 min.
6. The reaction is started by addition of DTT to a final concentration of 20 mM (stock solution: 0.5 M in assay buffer) (*see* **Notes 22** and **23**) (*see* **Fig. 2**).
7. Turbidity resulting from aggregation of the insulin B chain is monitored at 400 nm in a UV-Vis spectrophotometer equipped with a temperature control unit using microcuvets (100 μL) to keep protein consumption low (*see* **Note 24**).

4. Notes

1. Thawed aliquots of CS must not be refrozen.
2. To detect smaller aggregates of either insulin or CS and to increase the signal, a lower wavelength (e.g., 360 nm) may be chosen.
3. If the chaperone protein is stored in a buffer different from the assay buffer, controls with equal volumes lacking the chaperone must be included, as buffer compositions can alter the aggregation behavior of proteins *(15)*.
4. Unfolding CS is sensitive to a number of additives like Mg^{2+}, ATP glycerol, detergents, and so forth. Controls for the effect of such substances on CS folding are thus necessary.

5. Filtration and degassing of the test buffer prior to thermal aggregation assays reduces the background signal.
6. Buffer substances different to HEPES can be used. The influence of these buffers on aggregation behavior should be thoroughly tested.
7. Continuous temperature monitoring in a parallel cuvet is necessary, as small changes in incubation temperature might influence the aggregation behavior of CS.
8. Make sure that the sample is stirred vigorously and at the same rate throughout the experiment.
9. Suitable sHsp concentrations for inactivation/reactivation assays are in the range of 75 nM to 0.6 μM.
10. For all chaperone assays, both negative and positive control reactions should be performed. Suitable negative control proteins include bovine serum albumin (BSA), immunoglobulin G (IgG) and lysozyme. As positive controls use known chaperones such as GroEL.
11. For CS reactivation experiments in the presence of ATP, concentrations of nucleotide as well as salts should be kept as small as possible, as CS is affected by those additives. To make sure that the chosen concentrations are sufficient, add ATP again at a later time-point during reactivation. If no change in the kinetics of reactivation occurs, the original nucleotide concentration was adequate.
12. During aggregation processes both the number and the size of particles increases. Therefore, kinetics cannot be analyzed quantitatively.
13. When monitoring aggregation processes resulting from thermal unfolding in the presence of uncharacterized chaperones, always analyze the behavior of the chaperone alone at the given temperature. Chaperones might aggregate themselves at elevated temperatures, which renders them unsuitable for the this test system. Changing the storage and/or the assay buffer might circumvent such problems.
14. Samples should not exceed 500 μL to ensure rapid adaptation to the respective temperature.
15. If possible, the molecular chaperone should be dialyzed against the assay buffer (40 mM HEPES-KOH, pH 7.5). When other storage buffers have to be used, controls for the influence of used buffer substances on CS are essential.
16. Addition of OAA to thermally inactivating CS leads to the reactivation of a certain amount of folding intermediates. The amount of reactivatable intermediates is time dependent, with a maximum at about 6 min incubation at 43°C for CS without further additives. Because of aggregation, the amount of folding intermediates that are in equilibrium with the native state declines. So that after 10 min, reactivation can no longer be observed. In the presence of sHsp this situation might be significantly altered providing information about the mode of interaction with the thermally inactivated substrate.
17. Oxaloacetic acid used for reactivation reactions has a concentration of 100 mM. After dissolving OAA into 50 mM Tris, the pH has to be adjusted to pH 7.0.
18. The kinetics of spontaneous inactivation of CS at 43°C follows a single exponential curve corresponding to a first-order reaction (k_{app} at 9×10^{-3}/s). Different sHsp's might lead to different effects on the inactivation behavior of CS (e.g.,

Hsp25 does not alter the kinetics of CS inactivation because the chaperone forms a rather stable complex with CS intermediates). These folding intermediates of CS can be reactivated by the addition of OAA *(6)*. On the other hand, α-crystallin decelerates the loss of CS activity during incubation at elevated temperatures. This is probably due to a more transient interaction of the lens protein with CS, which, upon release from α-crystallin, allows refolding of unfolding intermediates during the inactivation reaction. A similar behavior was described for Hsp90 *(10)*. With this transient interaction of the chaperone, folding intermediates are not accumulated, so that reactivation in the presence of OAA does not exceed the spontaneous level.

19. For reactivation with the aid of Hsp70 the concentrations of Mg^{2+}, K^+, and ATP are kept low because these additives at higher concentrations are influencing CS folding.

20. The insulin assay is rather insensitive to additives. Nucleotides as well as many peptides do not change reaction kinetics. It is thus a good assay for monitoring ATP dependence or for competition experiments. Nevertheless, it is still necessary to check every used additive carefully for influence on spontaneous insulin aggregation.

21. The assay can be performed at different temperatures ranging from at least 20°C to 45°C. The kinetics of insulin aggregation are faster at higher temperatures. Effects can still be quantified by normalizing maximal signal to 100%. A convenient temperature for standard assays is 30°C. Here, light scattering will be observable about 8 min after DTT addition.

22. Prepare fresh DTT stock solutions daily. Store on ice.

23. Some sHsp's like Hsp25 and Hsp27 contain a single cysteine residue per subunit. These can form intersubunit disulfide bonds in the oligomer, which do not alter the functional properties of the protein *(16)*. Thus, reducing conditions in the insulin aggregation assay should thus not influence the chaperone properties of sHsp's.

24. Controls and samples including chaperones should be measured in parallel, as the maximum turbidity signal might vary slightly in different experiments.

References

1. Ehrnsperger, M., Buchner, J., and Gaestel, M. (1998) Structure and function of small heat shock proteins, in *Molecular Chaperones in the Life Cycle of Proteins* (Fink, A. L. and Goto Y., eds), Marcel Dekker, New York. pp. 533–575.
2. Gaestel, M., Buchner, J., and Vierling, E., (1997) The small heat shock proteins-an overview, in *Molecular Chaperones and Protein-Folding Catalysts* (Gething, M.-J. ed.), Oxford University Press, Oxford, pp. 269–272.
3. Carver, J. A., Guerreiro, N., Nicholls, K. A., and Truscott, R. J. W. (1995) On the interaction of α-crystallin with unfolded protein, *Biochim. Biophys. Acta* **1252,** 251–258.
4. Farahbakhsh, Z. T., Huang, Q. L., Ding, L. L., Altenbach, C., Steinhoff, H. J., Horwitz, J., and Hubbell, W. L. (1995) Interaction of α-crystallin with spin-labeled peptides. *Biochemistry* **34,** 509–516.

5. Lee, G. J., Roseman, A. M., Saibil, H. R., and Vierling, E. (1997) A small heat shock protein stably binds heat-denatured model substrates and can maintain a substrate in a folding-competent state. *EMBO J.* **16,** 659–671.

6. Ehrnsperger, M., Gräber, S., Gaestel, M., and Buchner, J. (1997) Binding of non-native protein to Hsp25 during heat shock creates a reservoir of folding intermediates for reactivation. *EMBO J.* **16,** 221–229.

7. Srere, P. A., Brazil, H., and Gonen, L. (1963) The citrate-condensing enzyme of pidgeon breast muscle and moth flight muscle. *Acta Chem. Scand.* **17,** 129–134.

8. Srere, P. A. (1966) Citrate-condensing enzyme–oxaloacetate binary complex. Studies on its physical and chemical properties. *J. Biol. Chem.* **241,** 2157–2165.

9. Zhi, W., Srere, P. A., and Evans, C. T. (1991) Conformational stability of pig citrate synthase and some active-site mutants. *Biochemistry* **30,** 9281-9286.

10. Jakob, U., Lilie, H., Meyer, I., and Buchner, J. (1995) Transient interaction of Hsp90 with early unfolding intermediates of citrate synthase. Implications for heat shock in vivo. *J. Biol. Chem.* **270,** 7288–7294.

11. Sanger, F. (1949) The terminal peptides of insulin. *J. Biol. Chem.* **45,** 563–574.

12. Raman, B., Ramakrishna, T., and Rao, C. M. (1995) Temperature dependent chaperone-like activity of α-crystallin. *FEBS Lett.* **365,** 133–136.

13. West, S. M., Kelly, S. M., and Price, N. C. (1990) The unfolding and attempted refolding of citrate synthase from pig heart. *Biochim. Biophys. Acta* **1037,** 332–336.

14. Gill, S. C., and von Hippel, P. H. (1989) Calculation of protein extinction coefficients from amino acid sequence data. *Anal. Biochem.* **182,** 319–326.

15. Rudolph, R., Böhm, G., Lilie, H., and Jaenicke, R. (1997) *Folding Proteins in Protein Function—A Practical Approach* 2nd ed., (Creighton, T. E., ed.), Oxford University Press, New York. pp. 57–100.

16. Zavialov, A., Benndorf, R., Ehrnsperger, M., Zavyalov, V., Dudich, I., Buchner, J., and Gaestel, M. (1998) The effect of the intersubunit disulfide bond on the structural and functional properties of the small heat shock protein Hsp25. *Int. J. Biol. Macromol.* **22,** 163–173.

26

Analysis of Small Hsp Phosphorylation

Rainer Benndorf, Katrin Engel, and Matthias Gaestel

1. Introduction

Small heat-shock proteins (sHsps) are involved in diverse biological phenomena and can act as molecular chaperones in vitro (reviewed in **ref. 1**, *see also* Chapter 25). At the posttranslational level, several modifications of sHsp's have been detected, including phosphorylation, deamidation, acylation, as well as mixed intermolecular disulfide formation, oxidation, and glycation (reviewed in **ref. 2**).

Covalent modification of sHsps by phosphorylation has been described for bovine αA-crystallin *(3)* and αB-crystallin *(4)*, for rat Hsp25 *(5)*, human Hsp27 *(6)*, and mouse Hsp25 *(7)*. Phosphorylation of the mammalian Hsp25 and Hsp27 is increased in response to a wide variety of extracellular stimuli (reviewed in **ref. 1**). The main enzymes responsible for stress-dependent phosphorylation of Hsp25 and Hsp27 are MAPKAP (mitogen-activated protein kinase-activated protein) kinase 2 *(8)* and 3, which are activated by p38 MAP kinase (reviewed in **ref. 9**) (*see also* Chapter 11, this volume). In all sHsps analyzed so far, phosphoserines are the exclusively phosphorylated amino acid residues. These serines in Hsp25 and Hsp27 are located in a MAPKAP kinase recognition motif HXRXXS(P), where H represents an amino acid carrying a hydrophobic side chain. Mouse Hsp25 is phosphorylated at two sites, S15 and S86 *(10)*, which are also present in rat Hsp25. Hence, Hsp25 can exist in four different isoforms (S15,86, S15-P, S86-P, S15,86-P) represented by three different bands or spots in isoelectric focusing (IEF), (non-, mono- and bis-phosphorylated). Human Hsp27 contains three phosphorylation sites, S15, S78, and S82 *(11)* leading to eight possible, differentially phosphorylated isoforms that migrate in four different bands or spots in IEF. So far, phosphorylation site-specific antibodies have been developed only for αB-crystallin (K. Kato, per-

From: *Methods in Molecular Biology, vol. 99: Stress Response: Methods and Protocols*
Edited by: S. M. Keyse © Humana Press Inc., Totowa, NJ

sonal communication). Recently, phosphorylation specific antibodies against S15-P and S78-P of Hsp 27 have been described *(12)*. Hence, to analyze the degree of phosphorylation of mouse/rat Hsp25 and human Hsp27, isoelectric focusing or two-dimentional (2D) electrophoresis combined with immunodetection by Western blotting must be applied. In addition, the activity of MAPKAP kinase 2 and 3 can be determined in the appropriate cell extracts. However, both methods will not necessarily provide the same results, because net sHsp phosphorylation reflects a balance between phosphorylation and dephosphorylation events over time whereas the kinase assay provides a snapshot of the sHsp phosphorylating activity at a distinct time-point.

2. Materials
2.1. Western Blot Experiments
2.1.1. Equipment

1. Blotting system, semidry (NovaBlot, Pharmacia, Uppsala, Sweden).
2. Electrode paper (Pharmacia).
3. Electrophoresis unit, Multiphor II (Pharmacia) with accessories, including sample applicator holder and sample applicator (*see* **Note 1**).
4. Dounce tissue grinder (Fischer, Pittsburgh, PA).
5. Isoelectric focusing cell (111 Mini IEF Cell, Bio-Rad, Hercules, CA) with gel casting tray (*see* **Note 1**).
6. Liquid nitrogen.
7. Mortar and pestle (small).
8. PVDF [poly(vinylidene fluoride)] membrane (Immobilon-P, Millipore, Bedford, MA).
9. Two-chamber gradient mixer (model 385, Bio-Rad).
10. Vertical electrophoresis system (Protean II xi Cell, Bio-Rad) with suitable glass plates, spacers, and casting stand (*see* **Note 1**).
11. Power supply (*see* **Note 1**).
12. X-ray films.
13. X-ray film developing apparatus.

2.1.2. Reagents

1. Acrylamide solution A (T=25%, C=3%): 24.25% (w/v) acrylamide; 0.75% (w/v) bis(N,N′-methylene-bis-acrylamide) in H_2O, store in refrigerator (*see* **Note 2**).
2. Acrylamide solution B (T=32%, C=2.5%): 31.2% (w/v) acrylamide; 0.8% (w/v) bis(N,N′-methylene-bis-acrylamide) in H_2O, store in refrigerator (*see* **Note 2**).
3. Acrylamide solution C (T=32%, C=5%): 30.4% (w/v) acrylamide; 1.6% (w/v) bis(N,N′-methylene-bis-acrylamide) in H_2O, store in refrigerator (*see* **Note 2**).
4. 1% (w/v) agarose in H_2O.
5. 10% (w/v) APS (ammonium persulfate) in H_2O (*see* **Note 3**).
6. 40% ampholytes: Bio-Lyte 3/10 and Bio-Lyte 5/7 (Bio-Rad).
7. Blocking solution: 5% (w/v) non-fat powdered milk in PBST.

8. DC Protein Assay kit, detergent compatible (Bio-Rad 500-0116).
9. Silicone oil (Pharmacia).
10. ECL-Western blotting detection reagents, includes peroxidase-linked secondary antibodies (Amersham RPN 2108).
11. Electrophoresis buffer: 25 mM Tris base, 192 mM glycine, 0.1% (w/v) SDS in H_2O.
12. Equilibration buffer: 2% (w/v) SDS, 6 M urea, 1 mM EDTA, 30% (v/v) glycerol, 0.01% (w/v) bromophenol blue, 60 mM DTT (dithiothreitol), 50 mM Tris-HCl, pH 6.8, in H_2O.
13. Extraction solution (prepare immediately before use): 6 M urea, 10 mM DTT, 2% (v/v) Triton X-100, 2% ampholytes 3/10, 0.01 vol. PMSF stock solution (1 mM final concentration), 0.1 volume protease/phosphatase inhibitor mixture, in H_2O.
14. 75% (v/v) glycerol in H_2O.
15. IEF standard proteins (Bio-Rad 161-0310).
16. Immobiline DryStrip pH 3-10L, 180 mm (Pharmacia) or Immobiline DryStrip pH 4-7, 180 mm (Pharmacia).
17. Isobutyl alcohol (2-methyl-1-propanol), water-saturated.
18. Molecular weight standard proteins, coloured (Amersham RPN 756).
19. Phosphate-buffered saline (PBS): Prepare i) 130 mM NaCl/10 mM Na_2HPO_4 and ii) 130 mM NaCl/10 mM NaH_2PO_4; mix to obtain 1X PBS, pH 7.2.
20. Phosphate-buffered saline (PBS)/Tween (PBST): 0.05% Tween-20 in PBS.
21. 100 mM phenylmethylsulfonyl fluoride (PMSF) in anhydrous isopropanol (2-propanol).
22. Protease/phosphatase inhibitor mix (10X): 250 mM NaF, 50 mM EDTA, 10 mM benzamidine, 50 mM $Na_4P_2O_7$, 5 mM Na_3VO_4, 10 µg/mL aprotinin, 10 µg/mL leupeptin, 50 µg/mL pepstatin, in H_2O.
23. Rehydration solution: 8 M urea, 0.5% (w/v) Triton X-100, 10 mM DTT, in H_2O.
24. 0.1 % (w/v) (FMN riboflavin-5'-phosphate) in H_2O.
25. Sephadex G-150 Superfine (Pharmacia).
26. 20% (w/v) SDS (sodium dodecyl sulfate) in H_2O.
27. TEMED (N,N,N´,N´-tetramethylethylenediamine) (*see* **Note 2**).
28. Transfer buffer: 39 mM glycine, 48 mM Tris base, 0.375% (w/v) SDS, 20% (v/v) methanol in H_2O.
29. 2 M Tris-HCl in H_2O, pH 8.8.
30. 1 M Tris-HCl in H_2O, pH 6.8.
31. Triton X-100.
32. Urea (ultrapure).
33. Rabbit anti-Hsp25 antibody (StressGen) diluted 1:4000 in blocking solution.

2.2. Kinase Assays

2.2.1. Equipment

1. Cell culture facilities.
2. Microcentrifuge and vortex.
3. Microtiter plate reader and appropriate plates.
4. Thermostatically controlled water bath set to 30°C.

5. Safety box for β-emitters including rack filled with water ice.
6. Heating block set to 96°C.
7. Vertical electrophoresis system (Mini-Protean II Cell, Bio-Rad) with suitable glass plates, spacers, casting stand and power supply (*see* **Note 1**).
8. Vacuum pump.
9. Tuberculin syringe with a 27-gauge needle.
10. Rotating wheel.
11. Filter paper and Saran wrap.
12. Gel dryer and vacuum pump.
13. PhosphoImager or X-ray films and film developing apparatus.

2.2.2. Reagents

1. Phosphate-buffered saline (PBS): 20 mM Na-phosphate, pH 7.4, 154 mM NaCl.
2. Trypsin/EDTA (Gibco-BRL).
3. Lysis buffer: 20 mM Tris-acetate, pH 7.0, 0.1 mM EDTA, 1 mM EGTA, 10 mM Na-β-glycerophosphate, 50 mM NaF, 5 mM Na-pyrophosphate, 0.27 M sucrose, 1% (v/v) Triton X-100, 1 mM benzamidine, 0.1% (v/v) β-mercaptoethanol, 0.2 mM PMSF (Phenylmethylsulfonyl fluoride), store in 1-mL aliquots at –20 °C (*see* **Note 4**).
4. Bio-Rad Protein Assay solution (Bio-Rad 500-0006).
5. Bovine plasma albumin (Bio-Rad).
6. Assay buffer (10X): 500 mM Na-β-glycerophosphate, pH 7.4, 1 mM EDTA, store in 1-mL aliquots at –20°C.
7. 10 mM ATP (diluted from 0.1 M ATP, pH 7.0, Boehringer Mannheim), store in 0.1-mL aliqots at –20°C.
8. 1 M Mg-acetate (store at 4°C).
9. [γ-^{33}P]ATP (NEN Du Pont NEG 302H, 10 mCi/mL).
10. Recombinant Hsp25 10 mg/mL (StressGen).
11. Anti-MAPKAPK-2-antiserum (*see* **Note 5**).
12. IP buffer: 20 mM Tris-HCl, pH 7.4, 154 mM NaCl, 50 mM NaF, 1% Triton X-100, 1% (w/v) bovine serum albumin (BSA), store at 4°C (*see* **Note 6**).
13. 50% Slurry of Protein A-Sepharose (Pharmacia) in PBS (*see* **Note 7**).
14. 4X SDS-sample buffer: 0.1 % (w/v) Bromophenol blue, 20% (v/v) glycerol, 12% (w/v) SDS, 20% (v/v) β-mercaptoethanol.
15. Molecular-weight standard proteins, colored (Amersham RPN 756).

3. Methods

3.1. Collection and Extraction of Protein Samples from Research Animals

1. After sacrificing the animal (*see* **Note 8**), the heart, skeletal muscle (hind limb), uterus, kidney (cortex), eye lens, spleen, lungs, and ovaries are harvested, washed in ice cold PBS and placed on a chilled support (e.g., culture dish). Tissues are then cut with a scalpel into pieces of about 25 mg weight and immediately snap-frozen in liquid nitrogen (*see* **Note 9**).

2. Extraction of proteins from the more rigid tissues of the heart, skeletal muscle, skin, and uterus is facilitated by grinding under liquid nitrogen. In this procedure, powder about 200 mg frozen tissue under liquid nitrogen using a precooled mortar and pestle. After evaporation of liquid nitrogen, transfer the remaining frozen tissue powder with a precooled spatula into a Dounce tissue grinder and homogenize with 300 µL of extraction solution at room temperature.

3. For disintegration of the more fragile tissues (eye lens, cortex of kidney, ovary, lung), place about 100 mg directly into the Dounce tissue grinder and homogenize with 300 µL extraction solution.

4. Centrifuge the homogenates (5 min at 18,000g) at room temperature, snap-freeze the clear supernatants in liquid nitrogen and store them at –140°C. For protein determination, dilute aliquots of each supernatant 10-fold with water and determine the protein content with the DC Protein Assay kit (tolerates all the added components) following the manufacturer's instructions.

3.2. One-Dimensional Analysis of Hsp25 Isoforms Using Isoelectric Focusing Gels

This is a quick and simple technique to separate Hsp25 isoforms on the basis of their different isoelectric points. Large numbers of samples may easily be handled. However, if samples contain more than the three major isoforms of Hsp25 due to further modifications, as has been reported to occur occasionally *(13,14)*, the results may be confusing and difficult to interpret.

1. Place two IEF glass plates (111 Mini IEF Cell) on the casting tray (*see* **Note 10**). Prepare 8 ml of the IEF gel solution (sufficient for two gels) by dissolving 2.64 g urea (5.5 M final concentration) in 0.3 mL of 40% ampholytes 3/10 and 0.1 mL of 40% ampholytes 5/7 (2% ampholytes final concentration), 0.53 mL of 75% glycerol (5% final concentration), 1.6 mL acrylamide solution A (5% final concentration), and 3.26 mL water.

2. Degas the solution under vacuum for 5 min and initiate polymerization by addition of 50 µL 0.1% FMN, 12 µL 10% APS, and 3 µL TEMED. Pipette the solution under the glass plates and allow polymerization to proceed for 1 h under light.

3. Remove the glass plates with the attached gels from the casting tray using a suitable wedge by lifting the glass plates carefully at one corner until the gels detach from the casting tray.

4. Place a suitable sample loading template (provided with the IEF cell) on the gel (leave approx 1.5 cm distance between the wells and one edge of the gel). Load the samples (0.5–5µL, approx 3–30 µg total protein) into the wells (*see* **Note 11**) and allow them to absorb (this may require 30 min). Load 3 µL IEF standard proteins (five of the nine proteins are naturally coloured) in one well to follow the separation and the transfer of the proteins onto the PVDF-membrane.

5. After absorption of the samples, remove the template carefully and place the gel (gel side down) on the electrodes of the IEF cell (*see* **Note 12**). Place the samples close to the anode.

6. Run the gel for 15 min at 100 V for 15 min at 200 V, and for 1–2 h at 450 V.
7. Carefully lift the glass plate to detach the gel from the electrodes.
8. To detach the gel from the glass plate, place a piece dry blotting paper (matching the size of the gel) against the gel and rub it lightly to achieve uniform contact. Use the paper to peel the gel from the glass plate and wet both paper and gel with transfer buffer. Place the gel on the PVDF-membrane for the protein transfer (*see* **Subheading 3.4.**).

3.3. Two-Dimensional Analysis of Hsp25 Isoforms Using Immobilized pH Gradients in the First Dimension and Vertical Linear Pore Gradient Gels in the Second Dimension

The two-dimensional analysis (2D) of Hsp25 isoforms is more labor intensive, but it provides additional information on their apparent molecular masses. This is especially helpful when more than the three differentially phosphorylated major Hsp25 isoforms are to be studied *(13,14)*. The physical basis for appearance of the additional isoforms is largely unknown, but other protein modifications and degradation might be involved. Two-dimensional electrophoresis combines the separation of proteins by their isoelectric points (first dimension) with the separation by their molecular mass (second dimension). In the following protocol, the use of commercially available preformed immobilized pH gradient (IPG) strips is described for the first dimension, which saves time and is highly reproducible (cf. **ref. *15***). For the second dimension, the use of standard vertical linear pore-gradient gels is described.

3.3.1. First Dimension Using IPG Strips

1. Rehydrate the IPG strips overnight in rehydration solution, rinse them briefly with water and keep them between two sheets of wet filter paper until used.
2. Place the IPG strip container on the cooling plate (coated with silicone oil) of the Multiphor II chamber and place the rehydrated IPG strips in the grooves. Place the electrodes in the contact grooves, position the sample applicator holder at the anodic side, and finally, place the sample applicator in the holder in such a way that the cups are gently pressed onto the IPG strips. Pour silicone oil in the container to immerse the IPG strips. The gel area inside the cups should remain free of oil.
3. Set the temperature of the cooling plate to 15°C and load the samples (10–100 µg total protein) into the cups (*see* **Notes 11** and **13**). To facilitate the penetration of the proteins into the gel, apply the samples with three volumes of a "gel slurry" (30 mg Sephadex G-150 Superfine in 1 mL extraction solution).
4. Allow the proteins to enter the gel for 1 h at 650 V, then separation is performed for approx 16 h (overnight) at 3500 V. Finally, increase the voltage to 5000 V for 1 h.
5. After the run, either immediately equilibrate the IPG strips in equilibration buffer and load them onto the slab gels, or freeze, and process them later.

3.3.2. Preparation of Vertical Linear Pore Gradient Gels

Glass plates (180×160 mm, with spacers of 0.75 mm) are used as part of the Bio-Rad vertical electrophoresis system Protean II xi.

1. Assemble the glass plates and spacers in the casting stand and position a standard two-chamber gradient mixer on a magnetic stirrer slightly above the upper edge of the glass plates. Connect the outlet of the gradient mixer with the assembled glass plates with suitable tubing.
2. Mix the separation gel (20 mL: 6–16 % acrylamide; 0–15% glycerol; 0.1 % SDS; 0.375 M Tris-HCl, pH 8.8; all final concentrations) from two gel solutions (I, II) using the gradient mixer. Ten milliliters of gel solution I is made from 1.875 mL acrylamide solution B, 0.05 mL 20% SDS, 1.875 mL 2 M Tris-HCl, pH 8.8, and 6.2 mL distilled water. Ten milliliters of gel solution II is made from 5.0 mL acrylamide solution C, 0.05 mL 20% SDS, 1.875 mL 2 M Tris-HCl, pH 8.8, 2.0 mL 75% glycerol, and 1.075 mL distilled water. Degass the mixtures and after polymerization has been initiated by adding 4 μL TEMED and 20 μL 10% APS, cast the gel using the gradient mixer (stir gel solution II). Add the polymerization mixture to a level 25-mm below the upper edge of the inner glass plate and gently overlay it with 1 mL isobutyl alcohol. Allow the mixture to polymerize for 2 h.
3. Remove the isobutyl alcohol and unpolymerized gel mixture from the gel surface. Prepare 5 mL of the stacking gel mixture (5.0% acrylamide; 0.1% SDS; 0.125 M Tris-HCl, pH 6.8) from 0.78 mL acrylamide solution B, 25 μL 20% SDS, 0.625 mL 1 M Tris-HCl, pH 6.8, and 3.57 mL distilled water. After degassing, initiate polymerization by adding 5 μL TEMED and 50 μL 10% APS and pour the mixture on the separation gel to a level 3-mm below the upper edges of the inner glass plate. Overlay the mixture with isobutyl alcohol and allow it to polymerize for 30 min. After removing the isobutyl alcohol from the gel surface the IEF strips can be mounted.

3.3.3. Loading of IEF Strips onto the Slab Gel

1. Cut the IEF strips from both sides to a length of 15 cm. For equilibration, place the IEF strips in a suitable test tube and incubate them for 10 min in 10 mL of equilibration buffer. Remove the excess buffer from the equilibrated IPG strips by blotting them onto filter paper and place the strips on the outer glass plate approx 2 cm above the stacking gel (strips adhere to the plates) in a way that they can be easily shifted down onto the stacking gel using two spatulas.
2. Make 1% agarose in a microwave oven and apply it on the top of the stacking gel (approx 6 mm height). Immediately thereafter shift the IEF strip downwards into the agarose until it reaches the top of the stacking gel. To run molecular weight standard proteins in parallel, insert a plastic strip (0.75 mm width, 5 mm wide) into the agarose close to one of the edges of the slab gel.
3. Leave the agarose 5 min to solidify, fill the upper (cathode) and lower (anode) buffer tank with running buffer, and remove the plastic strip. Load 4 μL of the

coloured molecular weight standard proteins into the well and start the run. Use 60 V until the bromophenol blue reaches the separation gel, then increase the voltage to 200 V (about 6 h running time) or to 80 V (for a run overnight). During the run the separation of the coloured standard proteins can be easily followed. After the run, disassemble the chamber and immediately use the gel for blotting.

3.4. Semi-Dry Blotting and Immunological Detection of Hsp25

1. Wet a piece of PVDF-membrane matching the size of the gel in methanol and place it in transfer buffer to equilibrate. Soak 5 pieces of blotting paper, each matching the size of the gel, in transfer buffer and place them on the anode (bottom) plate of the blotting system, then place the equilibrated membrane on top of the stack. Place the gel (either the paper-backed IEF gel or the SDS gradient pore gel) on top of the membrane and subsequently place 5 further sheets of blotting paper (soaked in transfer buffer) on top of the stack. Keep the stack wet using sufficient transfer buffer. During assembly of the stack care should be taken to avoid trapping air bubbles. Any air bubbles must be forced out by rolling the stack with a glass rod.
2. Place the cathode (top) plate of the blotting system onto the stack and connect the apparatus to a power supply. Set the current at 1.2 mA/cm^2 and perform the transfer for 40 min.
3. Disassemble the stack. Check if the coloured marker proteins were transferred onto the membrane.
4. Wash the membrane for 5 min in PBST and incubate it at room temperature for 2 h each with (1) blocking solution, (2) the anti-Hsp25-specific antibody (raised in rabbit) diluted 1:4000 in blocking solution (*see* **Note 14**), and after repeated washing in PBST (5 × 10 min), with (3) the secondary anti-rabbit antibody conjugated with horseradish peroxidase (ECL-Western blotting detection system). Alternatively, each of these incubations can be performed overnight at 4°C. Wash the membrane 5 × 10 min in PBST. Use a suitable shaker for all incubations and washing steps.
5. To develop the blot with the ECL-Western blotting detection system, pipet 0.3 mL of each reagent 1 and 2 onto a piece of thin plastic film (approx 30 × 30 cm), mix well and place the membrane (surface with bound proteins downwards) onto the plastic film assuring that the reagent mix is evenly distributed over the whole area of the membrane. Wrap the membrane into the film and immediately use it for exposing X-ray films for various time periods (1, 3, and 10 min).
6. Develop the X-ray film with a film developing apparatus. Typical results are shown in **Fig. 1**.

3.5. Assays for Small Hsp-Kinase Activity

3.5.1. Assay of Whole-Cell Extracts

1. After subjecting in vitro cultured cells to different stress stimuli, the cell dishes are immediately transferred on ice. Cells are collected in a centrifugation tube

A

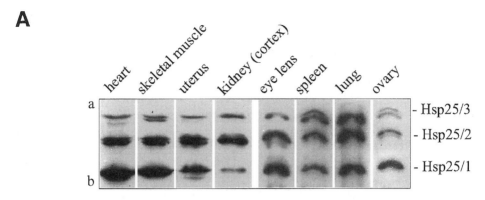

B

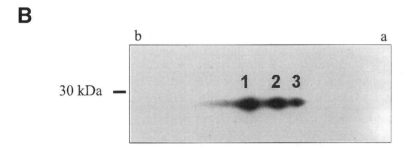

Fig. 1. (**A**) Blot of IEF gels loaded with protein extracts of different rat tissues. The acidic (a) and basic (b) sides of the gels are indicated. Hsp25 isoforms were detected with a polyclonal anti-Hsp25 antibody (primary antibody) and an anti-rabbit antibody conjugated to peroxidase (secondary antibody) using the ECL detection system. In heart, skeletal muscle, eye lens, and ovary, the nonphosphorylated isoform Hsp25/1 dominates. The uterus contains more Hsp25/1 and monophosphorylated Hsp25/2 (which have a similar abundance) than Hsp25/3, whereas in the spleen and lung all three isoforms occur with similiar abundance. Note that in some of the tissues (skeletal muscle, spleen, ovary), the bis-phosphorylated isoform Hsp25/3 is resolved into two bands. The physical basis for this is not known, but protein degradation during protein extraction and/or electrophoresis might be involved. In the cortex of the kidney, the predominant isoform is Hsp25/2, a finding which has been described previously (*16*). Because the degree of phosphorylation of Hsp25 is known to change within minutes in response to a variety of environmental factors, the distribution of Hsp25 isoforms may be different under other experimental settings. (**B**) Western blot of a 2D gel loaded with a rat heart protein extract. Hsp25 isoforms were detected as described in (A). Note the typical distribution of the three major isoforms of Hsp25: HSP25/1 (nonphosphorylated), HSP25/2 (monophosphorylated), and HSP25/3 (bis-phosphorylated) as has been described previously (*17*). The corresponding isoelectric points (pI) were estimated to be 6.2, 5.8, and 5.6, respectively (kindly supplied by R. R. Gilmont).

and washed twice with ice-cold PBS (*see* **Note 15**). Washed and pelleted cells can be frozen in liquid nitrogen and stored at –80°C.

2. Cell pellets are resuspended in lysis buffer (100 μL/10⁶ cells) by pipetting up and down and incubated on ice for 15 min with occasional vortexing for 10 s. The lysate is centrifuged at 4°C in a microcentrifuge at 15,000g for 10 min. The supernatant, i.e., the cell extract, is carefully removed and transferred to a fresh tube.

3. The protein concentration of the extract is determined with the Bio-Rad Protein Assay (*see* **Note 16**). Equal amounts of protein are used for assaying MAPKAPK-2/3 kinase activity.

4. Add 2.5 μL of cell extract (containing about 30 μg of protein) to 22.5 μL kinase assay mix (2.5 μL 10X assay buffer, 0.25 μL 10 mM ATP, 0.25 μL 1 M Mg-acetate, 0.5 μL Hsp25, 1 μCi [γ-³³P]ATP, 19 μL H₂O). The kinase reaction is performed at 30°C for 10 min (*see* **Note 17**). After addition of 8.3 μL SDS-sample buffer and heating at 96°C for 3 min, the sample is ready for SDS-PAGE (*see* **Note 18**).

5. Run a SDS-PAGE and dry the mini gel (*see* **Note 19**).

6. After the gel has been dried, it is used to expose an imager plate for approx 1 h. Quantification is performed using the software provided with the PhosphoImager (*see* **Note 20**). Alternatively, the gel can be exposed on a X-ray film over night (*see* **Fig. 2A**).

3.5.2. IP Kinase-Assay

1. For native immunoprecipitation of MAPKAPK-2, 50 μL of extract (*see* **Subheading 3.5.1.**) is diluted with 0.5 mL ice-cold IP buffer and 4 μL of anti-MAPKAPK-2-antiserum is added. This mixture is incubated on a rotating wheel at 4°C for at least 2 h (*see* **Note 21**).

2. Add 25 μL of Protein A Sepharose slurry and incubate the sample for another hour on a rotating wheel at 4°C.

3. Spin the sample in a microcentrifuge at 15,000g for 2 min at 4°C. Wash three times with 0.5 mL IP buffer without BSA (*see* **Note 22**).

4. Add 25 μL kinase assay mix (2.5 μL 10X assay buffer, 0.25 μL 10 mM ATP, 0.25 μL 1 M Mg-acetate, 0.5 μL Hsp25, 1 μCi [γ-³³P]ATP, 21.5 μL H₂O). Incubate the samples at 30°C in a water bath for 10 min (*see* **Note 17**). After addition of 8.3 μL SDS-sample buffer and heating at 96°C for 3 min, the sample is ready for SDS-PAGE (*see* **Note 18**).

5. Run SDS-PAGE and dry the mini gel (*see* **Note 19**).

6. After the gel has been dried, it is used to expose an imager plate for approx 1 h. Quantification is performed using the software provided with the PhosphoImager (*see* **Note 20** and **Fig. 2B**).

4. Notes

1. The safety of all electrophoresis equipment including power supplies, cables, plugs etc. must be checked before use.

A

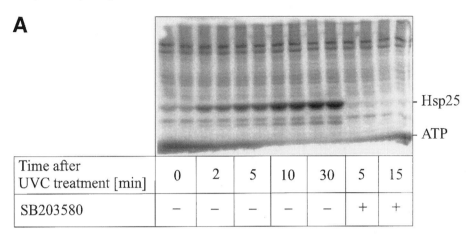

Time after UVC treatment [min]	0	2	5	10	30	5	15
SB203580	−	−	−	−	−	+	+

B

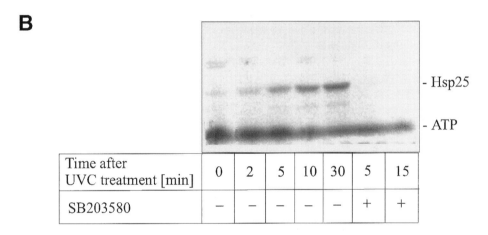

Time after UVC treatment [min]	0	2	5	10	30	5	15
SB203580	−	−	−	−	−	+	+

Fig. 2. sHsp kinase detection in HeLa cells treated with 300 J/M^2 UV light. sHsp kinase activity was determined in 2.5 µL of whole cell extracts (10^6 cells lysed in 100 µL buffer) before and different times after UV treatment (**A**) in the absence (−) and presence (+) of 20 µM of the p38 MAP kinase inhibitor SB203580. The assay was carried out twice for each extract. In parallel, 25 µL of the same cell extracts were used for the IP-kinase assay (**B**). A stimulation of the p38 MAP kinase-dependent sHsp kinase activity by UV as described in **ref. 18** can be detected.

2. Handle toxic and cancerous substances (e.g., unpolymerized acrylamide, organic solvents, TEMED) with care. Wear protective clothing and use a hood for volatile substances and fine powders.

3. Remains of acrylamide solutions are detoxified by polymerization with an excess of APS.

4. Benzamidine, PMSF and β-mercaptoethanol are added freshly to the lysis buffer from stock solutions of 1 M in water, 0.2 M in anhydrous isopropyl alcohol, and 14.2 M, respectively. Handle these toxic substances with care.

5. The antiserum was obtained after immunization of a rabbit with recombinant glutathione S-transferase-fused mouse MAPKAPK-2 expressed in *Escherichia coli (19)*. Crossreaction of the antiserum with human MAPKAP kinase 2 and with MAPKAP kinase 3 has been detected. Alternatively, anti-human MAPKAPK-2 antibody C-18 from Santa Cruz (Sc-6221) can be used with good results. This antibody shows cross-reactivity with mouse and rat MAPKAP kinase 2 but not with MAPKAP kinase 3.

6. IP buffer is prepared and stored without BSA. Prior to the experiment an aliquot of IP buffer with BSA is freshly made up by adding BSA.

7. The Protein A Sepharose slurry is obtained by swelling the substance in PBS at room temperature for 1 h. After settling of the swollen Sepharose the slurry is adjusted with PBS to 50% (v/v).

8. Care and use of research animals must be in accordance with federal, state and local laws and regulations. In the experiment described here a female Sprague-Dawley rat (200 g weight) was used.

9. Handling of liquid nitrogen requires protective gloves and glasses.

10. For casting the IEF gels, only thoroughly cleaned glass plates and casting trays should be used. This assures that the gels attach to the glass plates and detach easily from the tray. Since occasionally gels cannot be removed properly from the casting tray, pouring of an extra gel is advised.

11. The amount of Hsp25 in different tissues varies. Additionally, the efficiency of extraction of Hsp25 may be a source for variability. Therefore, run gels loaded with varying amounts of protein (e.g., 3, 10, and 30 µg total protein for IEF gels, 10, 30, and 100 µg total protein for 2D gels). The concentrations of primary and secondary antibodies may be varied in a wide range (range of dilution: 1:1,000–1:15,000) for optimal performance.

12. To facilitate both a good contact between the gel and the electrodes and the detachment of the gels from the electrodes after the run, it is important to keep both gels and electrodes wet. Applying a thin film of water to the area of the gels that is not covered by sample loading template prevents drying of the gels during sample loading. Soak the electrodes for 5 min in water before the gels are placed on them. After the gels are placed on the electrodes, pipet a few drops of water between the glass plate and the electrodes in a way that the water flows along the entire length of the electrodes.

13. When large amounts of proteins are to be loaded onto IEF strips, it is advisable to dilute the samples and load them in several steps.

14. Do not use antibodies containing sodium azide as a preservative when using a peroxidase-based detection system.

15. Cell suspensions can easily be collected by centrifugation at 100g (4°C, 5 min). Washing is achieved by resuspending the pellet in PBS and subsequent centrifu-

gation. The pellet is easy to resuspend by flicking the tube prior to addition of PBS. Adherent cells are treated with trypsin/EDTA for 5 min at 4°C, scraped in PBS and washed as described above.

16. Pipet 150 µL of 1:5 in water diluted Bio-Rad Protein Assay dye reagent in the wells of a 96-well microtiter plate. Add 2.5 µL of serial dilutions (1:20, 1:10, 1:5) of the extract in water. Blank is the appropriate dilution of lysis buffer in water. Bovine plasma albumin diluted in water in the range from 0.2 mg/mL to 0.8 mg/mL is used as standard. After 15 min incubation at room temperature absorption at 595 nm is measured in a microtiter plate reader.

17. Handling of radioactive materials has to be performed in accordance with federal, state and local laws and regulations. Working with radioactive materials requires protection of yourself by protective glasses, gloves, a labcoat, and acrylic work shields for β-emitters. Prepare a master kinase assay mix for all samples. Master mix and extracts have to be precooled and mixed well by pipeting on ice. Avoid air bubbles. Setting the rack from the safety box to the water bath starts the reaction of all samples simultaneously. Vice versa, the reaction is stopped by transfer of the rack back to the safety box on ice. Reaction times might be prolonged when quantification of phosphorylation is not necessary.

18. Tubes tend to open during boiling. Therefore the use of Eppendorf Safe-Lock tubes is recommended. Alternatively, punch a small hole in the lid of the tube.

19. SDS-PAGE: 7.5–22.5% acrylamide (stock solution: 30% acrylamide/0.8% bisacrylamide) gradient SDS-Gels (resolving gel: 85×50×1.5 mm) are prepared according to **ref. 20**. Use 5 mL of the coloured molecular weight standard proteins for checking the position of Hsp25 which runs with an apparent molecular weight of 25.000. Run the gel at 150 V for about 45 min until the dye front is located 5 mm above the bottom of the gel. Cut the dye front plus 2 mm above from the gel with a razor blade, because it contains most of the free [γ-^{33}P]ATP. Put the gel on filter paper, cover it with Saran Wrap (do not wrap both gel and filter paper), and dry the gel using a gel dryer connected to a vacuum pump at 80°C for 2 h.

20. To calculate the stoichiometry of [γ ^{33}P]-incorporation into Hsp25, 5 µL of diluted kinase assay mix (1:1000 and 1:100) are spotted onto the filter paper next to the dried gel. Air-dry these spots. Expose the imaging plate to the gel. This allows to estimate the ratio between photostimulated pixels (PSL) and pmol ATP.

21. Alternatively, the incubation may be performed overnight.

22. Supernatants are carefully removed by the use of a 27-gauge needle on a tuberculin syringe connected to a vacuum pump. Use of this needle ensures that no Protein A-Sepharose is removed.

Acknowledgments

The authors would like to thank Drs. R. R. Gilmont and R. F. Ransom for critical reading of the text.

References

1. Ehrnsperger, M., Buchner, J., and Gaestel, M. (1998) Structure and function of small heat shock proteins, in *Molecular Chaperones in the Life Cycle of Proteins* (Fink, A. L. and Goto, Y., eds.), Marcel Dekker, NY, pp. 533–575.
2. Groenen, P. J., Merck, K. B., deJong, W. W., and Bloemendal, H. (1994) Structure and modifications of the junior chaperone alpha-crystallin. From lens transparency to molecular pathology. *Eur. J. Biochem.* **225,** 1–19.
3. Voorter, C. E., de Haard Hoekman, W. A., Roersma, E. S., Meyer, H. E., Bloemendal, H., and deJong, W. W. (1989) The in vivo phosphorylation sites of bovine alpha B-crystallin. *FEBS Lett.* **259,** 50–52.
4. Chiesa, R. and Spector, A. (1989) The dephosphorylation of lens alpha-crystallin A chain. *Biochem. Biophys. Res. Commun.* **162,** 1494–1501.
5. Kim, Y. J., Shuman, J., Sette, M., and Przybyla, A. (1984) Nuclear localization and phosphorylation of three 25-kilodalton rat stress proteins. *Mol. Cell. Biol.* **4,** 468–474.
6. Arrigo, A. P. and Welch,W. J. (1987) Characterization and purification of the small 28,000-dalton mammalian heat shock protein. *J. Biol. Chem.* **262,** 15,359–15,369.
7. Benndorf, R., Kraft, R., Otto, A., Stahl, J., Bohm, H., and Bielka, H. (1988) Purification of the growth-related protein p25 of the Ehrlich ascites tumor and analysis of its isoforms. *Biochem. Int.* **17,** 225–234.
8. Stokoe, D., Engel, K., Campbell, D. G., Cohen, P., and Gaestel, M. (1992) Identification of MAPKAP kinase-2 as a major enzyme responsible for the phosphorylation of the small mammalian heat shock proteins. *FEBS Lett.* **313,** 307–313.
9. Kyriakis, J. M. and Avruch, J. (1996) Sounding the alarm: protein kinase cascades activated by stress and inflammation. *J. Biol. Chem.* **271,** 24,313–24,316.
10. Gaestel, M., Schroder, W., Benndorf, R., Lippmann, C., Buchner, K., Hucho, F., Erdmann, V.A., and Bielka, H. (1991) Identification of the phosphorylation sites of the murine small heat shock protein hsp25. *J. Biol. Chem.* **266,** 14,721–14,724.
11. Landry, J., Lambert, H., Zhou, M., Lavoie, J. N., Hickey, E., Weber, L. A., and Anderson, C. W. (1992) Human HSP27 is phosphorylated at serines 78 and 82 by heat shock and mitogen-activated kinases that recognize the same amino acid motif as S6 kinase II. *J. Biol. Chem.* **267,** 794–803.
12. Eyers, P. A., vanden IJssel, P., Quinlan, R. A., Goedert, M., and Cohen, P. (1999) Use of a drug-resistant mutant of stress-activated protein kinase 2a/p38 to validate the in vivo specificity of SB 203580. *FEBS Lett.* **451,** 191–196.
13. Scheler, C., Müller, E.-C., Stahl, J., Müller-Werdan, U., Salnikow, J., and Jungblut, P. (1997). Identification and characterization of heat shock protein 27 protein species in human myocardial two-dimensional electrophoresis patterns. *Electrophoresis* **18,** 2823–2831.
14. Benndorf, R., Hayess, K., Stahl, J., and Bielka, H. (1992) Cell-free phosphorylation of the murine small heat shock protein hsp25 by an endogenous kinase from Ehrlich ascites tumor cells. *Biochim. Biophys. Acta* **1136,** 203–207.

15. Westermeier, R. (1993) *Electrophoresis in Practice*. VCH Verlagsgesellschaft mbH Weinheim, Germany.
16. Müller, E., Neuhofer, W., Ohno, A., Rucker, S., Thurau, K., and Beck F.-X. (1996) Heat shock proteins Hsp25, HSP60, HSP72, HSP73 in isoosmotic cortex and hyperosmotic medulla of rat kidneys. *Eur. J. Physiol.* **431,** 608–617.
17. Lutsch, G., Vetter, R., Offhauss, U., Wieske, M., Gröne, H.-J., Klemenz, R., Schimke, I., Stahl, J., and Benndorf, R. (1997) Abundance and location of the small heat shock proteins Hsp25 and αB-crystallin in rat and human heart. *Circulation* **96,** 3466–3476.
18. Iordanov, M., Bender, K., Ade, T., Schmid, W., Sachsenmaier, C., Engel, K., Gaestel, M., Rahmsdorf, H. J., and Herrlich, P. (1997) CREB is activated by UVC through a p38/HOG-1-dependent protein kinase. *EMBO J.* **16,** 1009–1022.
19. Plath, K., Engel, K., Schwedersky, G., and Gaestel, M. (1994) Characterization of the proline-rich region of mouse MAPKAP kinase 2: influence on catalytic properties and binding to the c-abl SH3 domain *in vitro*. *Biochem. Biophys. Res. Commun.* **203,** 1188–1194.
20. Laemmli, U. K. (1970) Cleavage of structural proteins during the assembly of the head of bacteriophage T4. *Nature* **227,** 680–685.

27

Analysis of Multisite Phosphorylation of the p53 Tumor-Suppressor Protein by Tryptic Phosphopeptide Mapping

David W. Meek and Diane M. Milne

1. Introduction
1.1. The p53 Protein

The p53 protein (reviewed in **refs. *1–5***) is a latent transcription factor that is activated in response to a variety of cellular stresses, including DNA damage, mitotic spindle damage, heat shock, hypoxia, cytokines, metabolic changes, viral infection, and activated oncogenes, and is considered to be an *integration point* for these signals *(4,6–9)*. Activated p53 can induce growth arrest or apoptosis, events that prevent the survival of genetically damaged cells. p53 also plays a central role in promoting early senescence in response to unregulated mitogenic signaling *(10)*. The transactivation function of p53 is mediated through sequence-specific binding of the central domain of the protein to cis-acting elements within the promoters or introns of the responsive genes. At present, there are more than 20 genes known to be activated by p53, some of which are involved in growth arrest or apoptotic pathways. Similarly, there is a growing list of promoters, many of which are viral promoters or promoters of growth-stimulatory genes, which are repressed by p53. Consequently, the downstream effects of activating p53 are complex and there is, as yet, no known *single* pathway that mediates its function in full.

1.2. Multisite Phosphorylation of p53 and the Integration of Stress Signals

The p53 protein is subject to tight regulation through a number of routes including protein-protein association and rapid degradation. Intertwined with

From: *Methods in Molecular Biology, vol. 99: Stress Response: Methods and Protocols*
Edited by: S. M. Keyse © Humana Press Inc., Totowa, NJ

these mechanisms, p53 is also regulated by a variety of phosphorylation events (reviewed in **ref. 9**) that mediate several functions, including site-specific DNA binding, transactivation, trans-repression, apoptosis, association with the negative regulator MDM2, and targeting of the p53 protein for degradation *(9,11,12)*. The *N*-terminus of p53 is phosphorylated by a range of protein kinases, including the chromosome segregation-associated CK1δ and CK1ε (which are isoforms of casein kinase 1), Jun-*N*-terminal kinases (JNKs), mitogen-activated protein (MAP) kinase (murine p53 only), the DNA-activated protein kinase (DNA-PK), the products of the ataxia telangiectasia gene (ATM) and ATM- Rad3-related gene products (ATR), and the cyclin-dependent protein kinase-activating kinase (CAK). The C-terminus of p53 is also subject to multiple modifications, including phosphorylation by protein kinase CK2, protein kinase C, and the G2/M active cyclin-dependent kinases cyclinB/CDC2 and cyclinA/CDK2. The C-terminus is additionally modified through acetylation by p300 and P/CAF. There are also phosphorylation events within the central DNA binding domain of p53 *(13)*, but very little is known about the nature or physiology of these modifications.

Understanding how each modification of the p53 protein contributes to the regulation of its function, and the signals and pathways to which these modifications respond, are key objectives in p53 research. The integration of these complex and interactive phosphorylation events and their relationship to the responsiveness of p53 to a wide range of environmental stimuli are of particular interest.

1.3. Phosphopeptide Mapping Studies of p53

As a result of the multisite nature of the phosphorylation of p53, many laboratories have used the technique of phosphopeptide mapping to dissect the roles of individual and multiple modifications of this protein. In this chapter, we describe in detail the tryptic phosphopeptide mapping procedure that we routinely use to assess the modification status of the p53 protein isolated from cells grown in culture. This procedure is based on the method published by Hunter's laboratory *(14)* and many of the concepts and advantages are discussed therein. The principle of the method is that phosphoproteins which have been radiolabeled (with $[^{32}P]$) either in vitro or in cultured cells can be isolated by immunoprecipitation and phosphopeptides generated through the action of proteases such as trypsin. The phosphopeptides have unique migration properties in electrophoresis and chromatography based on their size and net charge at any given pH, and their solubility in different solvents as determined by their constituent amino acids, respectively. Consequently, if placed at a common origin, these peptides can be separated in two dimensions on cellulose thin-layer plates, by electrophoresis followed at right angles by chromatography, and a distinctive and reproducible separation of the phosphopeptides can be achieved. The positions of these phosphopeptides can be monitored by autoradiography.

From our own point of view the procedure described in **Subheading 3.** has several advantages. First, because p53 is regulated by *multisite* phosphorylation, the procedure gives a comparative measure of the phosphorylation status at each of these different sites (because each of the modifications is potentially able to give rise to phosphopeptides with different migration properties). Second, selective modification of any individual site(s) can easily be discerned against a background in which modification at other sites remains unaltered (for examples, *see* **refs. *15*** and ***16***). Third, the phosphopeptides can be isolated from the thin-layer plate for further analysis. Finally, the method is very sensitive and although very low levels of radioactive material are often recovered (e.g., from radiolabeled cells with low levels of p53) very clear phosphopeptide maps can be obtained after autoradiography. The one possible disadvantage of the method is that radiolabeling of cells may cause induction/activation of p53 (*see also* Chapter 14). Consequently, shorter labelling times are advised in order to minimize the appearance of artifactual modification. However, in spite of this caveat we, and others, have used the technique extensively and successfully to detect specific changes in p53 modification in response to signaling events (*see* **refs. *13*, *15*,** and ***16***).

2. Materials

2.1. Reagents and Buffers

The following reagents are required to carry out cell labeling, immunoisolation, and tryptic phosphopeptide analysis of p53.

1. Cell lines plus appropriate media for normal growth and maintenance.
2. Phosphate-free minimal essential medium Eagle from Sigma (St. Louis, MO) (code M-3786)
3. Fetal bovine serum (FBS) that has been dialyzed extensively against Tris-buffered saline (TBS).
4. Tris-buffered saline (TBS): 25 mM Tris, pH 8.0, 140 mM NaCl, and 3 mM KCl.
5. [^{32}P]-Orthophosphate (HCl-free), specific activity 400–800 mCi/mL.
6. NP-40 lysis buffer: 10 mM sodium phosphate, pH 7.0, 0.15 M NaCl, 1% (v/v) Igepal CA-630 (Sigma), 2 mM EDTA, 50 mM NaF, 1 mM benzamidine.
7. RIPA buffer: 10 mM sodium phosphate, pH 7.0, 0.15 M NaCl, 1% (v/v) Igepal CA-630, 1% (w/v) sodium deoxycholate, 0.1% (w/v) sodium dodecyl sulfate (SDS), 2 mM EDTA, 50 mM NaF, 1 mM benzamidine.
8. Antibodies; we routinely use either monoclonal antibodies as culture media from hybridoma cells lines, or polyclonal antibodies as serum.

9. Protein A-Sepharose 4B suspension. This comprises 0.22 g dry Protein A-sepharose 4B beads (Sigma) and 3 mL sepharose 4B bead suspension (Sigma), resuspended in PBS to a final volume of 10 mL. Add sodium azide at 0.05% (w/v) as a preservative and store at 4°C.

10. SDS sample buffer: 0.063 M Tris-HCl, pH 6.8, 2% (w/v) SDS, 25% (v/v) 2-mercaptoethanol, 0.02% (w/v) Bromophenol blue, and 10% (v/v) glycerol.

11. Acrylamide solution: 30% (w/v) acrylamide, 0.8% (w/v) bis-acrylamide.

12. Western transfer buffer: 3 g/L Tris base, 14.4 g/L glycine, 1 g/L SDS and 20% (v/v) methanol.

13. 4 mg/mL Ammonium bicarbonate. Make up immediately before use.

14. TPCK-treated trypsin (Worthington, Freehold, NJ). This is made up as a 1 mg/mL solution in 1 mM HCl and stored in aliquots at –80°C.

15. Performic acid (freshly prepared). Mix 450 μL of formic acid with 50 μL of 30% (v/v) hydrogen peroxide. Allow to stand on ice for 1 h prior to use.

16. pH 1.9 electrophoresis "buffer" for two-dimensional (2D maps).

17. Phosphopeptide chromatography "buffer": 37.5% (v/v) *n*-butanol, 25% pyridine (v/v), and 7.5% (v/v) acetic acid. We routinely use 200 mL in a chromatography tank and add 0.5 mL of 0.5 M EDTA, pH 8.0

2.2. Equipment

1. Liquid scintillation counter
2. Tissue culture hood
3. Regular and dedicated CO_2 incubators.
4. Dedicated radioactivity suite.
5. Specialized shielding including a storage block. This shielding and its location for setting up the labeling of the cells is shown in **Fig. 1**. We use several 1-cm-thick upright perspex shields (dimensions 35 × 54 cm) to protect the user during the labeling procedure. In addition, a 18 × 12 × 6 cm perspex block containing holes suitable for carrying or storing up to 24 samples in Eppendorf tubes is required. This block is encased in a perspex box with a lid. Other shielding includes a large perspex box with the lid containing a rack for transportation and storage of 50-mL screw-cap tubes and a small cylindrical container with lid designed to hold a single 50-mL screw-cap tube. The latter is used for collecting the radioactive medium and washes generated when labeling the cells and preparing them for lysis. This container has a plastic screw in the side which, when tightened, holds the screw-cap tube firmly in place. A lidded perspex box of dimensions 24 × 24 × 52-cm that can accommodate radioactive plastic pipets is also useful. Finally, a perspex box, (dimensions 40 × 32 × 10 cm, with a hinged lid) is used to hold cell-culture dishes containing the radioactive medium during incubation in the CO_2 incubator. The box has a suitable catch on the side containing the hinges such that it is possible to keep the lid open and use the lid as an additional shield when working with the cell-culture dishes. There are also handles on the sides for easy and safe transportation.

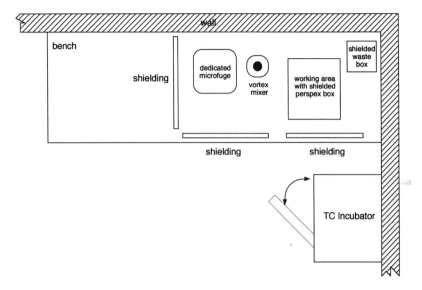

Fig. 1. Laboratory layout for radiolabeling of cells in culture. The labeling is performed in a dedicated radioactive facility in which the equipment is located in one corner of the room. A working area, with radioactive waste disposal box, vortex mixer, and microfuge are located behind the appropriate shielding. A dedicated cell culture incubator is located close to this work station.

6. Rotational mixer to which 1.5-mL Eppendorf tubes can be attached, and shielded space in a cold room for use of this apparatus.
7. Dedicated cupboard space with shielded boxes to allow medium to decay prior to disposal.
8. Pipetman.
9. Gilson pipets or equivalent.
10. Microcentrifuge (dedicated for use with radioactivity).
11. Heating block.
12. Sodium dodecyl sulfate-polyacrylamide gel electrophoresis (SDS-PAGE) apparatus.
13. Western transfer apparatus. We routinely use a semidry system.
14. Lightbox.
15. 37°C waterbath.
16. Speedivac or similar lyophilization equipment.
17. Flat-bed electrophoresis system. Any 20 × 20 cm flat bed unit can be used. We currently use a Maxi Horizontal Agarose Gel Unit model H2020 (Scotlab, UK).
18. Chromatography tank dimensions 27.5 × 27.5 × 7.5 cm. Model Z12,619-5 (Sigma).
19. Autoradiography cassettes with enhancing screens.
20. PhosphorImager (optional).

21. 10-cm Tissue culture dishes.
22. Disposable 10-mL, 5-mL, and transfer pipets.
23. 1.5-mL Screw-cap and snap-cap microfuge tubes.
24. 50-mL Screw-cap tubes.
25. Cell scrapers.
26. Disposable 10-mL syringes plus 23-gauge needles.
27. Transfer membranes (Immobilon-P).
28. Cellulose thin layer plates size 20×20 cm.
29. X-ray film.

3. Methods

3.1. Radiolabeling of Cells and Preparation of Extracts

Using the procedure described in this chapter, the phosphorylation status of p53 can be assessed in most cell lines. We recommend that researchers who are using the technique for the first time should carry out the procedure using SV3T3 cells, as these are a very good source of highly phosphorylated p53 (*see* **Note 1**).

1. Seed the cells in normal growth medium in 10-cm dishes at 1×10^6 cells per dish on the day prior to labeling. We normally use Dulbecco's modified Eagle's medium (DMEM) supplemented with 10% FBS, 2 mM L-glutamine, 50 IU/mL penicillin, and 50 µg/mL streptomycin; routinely we use 10 mL per 10-cm dish. Cells are grown in a humidified incubator at 37°C in the presence of 5% CO_2.
2. On the day of labeling, wash the cells once in phosphate-free DMEM supplemented with 5% dialyzed FBS. Then, preincubate the cells in 10 mL of this medium for 1 h in the incubator at 37°C.
3. Remove the medium from the cells and replace this with 5 mL of fresh phosphate-free medium. Take the dishes to the shielded area, place them in a shielded box constructed of perspex (plexiglass) and add 5 mCi of [^{32}P]-orthophosphate to each plate. Close the lid of the shielded box and transfer it to a dedicated CO_2 incubator. Place the box in the incubator such that the hinges of the lid are toward the incubator door. In that way, it is possible to raise the lid slightly to allow equilibration of the CO_2 without risk of exposure; (the lid can be kept open by placing a pipet tip horizontally between the lid and the box). Incubate the cells in the presence of the radioactive phosphate for up to 3 h; (*see* **Note 2**).
4. To harvest the cells, remove the shielded box from the incubator and place it behind the shielded area. Be very careful, as the box will be slippery after being in the humidified atmosphere. Take out the plates and place them on a tray packed with ice, again in the shielded area.
5. Remove the medium as quickly as possible and place it in disposable 50-mL screw-cap plastic tubes (*see* **Note 3**). Rinse the plates three times with ice-cold Tris-buffered saline (TBS) and drain them (prior to lysis) by tilting for 1 min; remove any residual TBS.

6. For lysis, add 0.5 mL of ice-cold "NP40 buffer" per plate (*see* **Note 4**). Tilt the plate back and forward, side to side to make sure that the lysis buffer covers all of the surface. Scrape the material off the surface of the plate using a cell scraper and transfer the extract to a 1.5-mL screw-cap tube.

7. Vortex the tube vigorously, then place it on ice for 10 min.

8. Place the samples in a dedicated microcentrifuge (*preferably cooled or in a cold room*) and centrifuge at 12,000*g* for 10 min.

9. Carefully remove the supernatant fraction with a pipet, discard the pellet, and transfer the supernatant to fresh 1.5-mL screw-cap tubes.

3.2. Immunoprecipitation of p53 and SDS-PAGE

1. To each of the radioactive extracts add excess of the antibody of choice. We routinely use about 5 µg of antibody per immunoprecipitation (*see* **Note 5**). Incubate the extract/antibody mixture on ice for 1 h.

2. Add 50 µL of a suspension of protein A-sepharose 4B beads and rock the samples for at least 1 h in a shielded area at 4°C.

3. Place the samples in the dedicated microcentrifuge and spin at full speed for 10–15 s. Remove and discard the supernatant fractions (*see* **Note 6**).

4. Add 1 mL of ice-cold RIPA buffer to each of the beads and vortex vigorously. Return the tubes to the microfuge and repeat this procedure three times, or until no further radioactive material is released from the beads.

5. After the final wash, resuspend the beads in 50 µL of SDS sample buffer, heat to 100°C for 2–3 min and pellet the beads in the microcentrifuge at full speed for 10 s. At this point the samples can be stored at –20°C until required.

6. Resolve the radiolabeled proteins by standard SDS-PAGE. We routinely run 10% polyacrylamide gels (acrylamide:bis ratio of 30:0.8) for the analysis and isolation of radiolabeled p53. Always include a lane with suitable prestained markers: (for this purpose we use Rainbow markers from Amersham). The electrophoresis equipment we routinely use is an ATTO Mini PAGE system.

7. The proteins in the gel should then be transferred to a suitable membrane. We routinely use Immobilon-P Transfer Membranes (Millipore) but good results can also be obtained by transferring the proteins to nitrocellulose (*see* **Note 7**). For transfer, we use the semidry blotting system (model IMM-1) manufactured by the W.E.P. Company (Seattle, WA). The membranes are prepared according to the manufacturer's instructions and proteins are transferred in 3 g/L Tris base, 14.4 g/L glycine, and 10% (w/v) SDS and 20% (v/v) methanol at 150 mA constant current for 45 min.

8. Following transfer, wrap the membrane in a layer of Saran Wrap (it is important that the membrane should not be allowed to dry out). Mount the membrane on a suitable surface and attach phosphorescent labels (Radtape, Diversified Biotech) to the periphery. This is very important as it will permit the membrane to be lined up accurately with the autoradiograph after exposure and facilitate removal of the radioactive protein from the membrane for further analysis.

9. Expose the membrane to X-ray film. We normally do this at –80°C in the presence of an intensifying screen (*see* **Note 8**).

3.3. Elution of Radioactive Protein from Membranes

1. To recover the radioactive protein from the membrane, place the autoradiograph on an illuminated lightbox. Place the transfer membrane containing the radioactive p53 on top of the autoradiograph and line up the phosphorescent labels with the appropriate exposed areas on the autoradiograph.
2. Mark on the membrane the outline of the radiolabeled p53 according to the signal on the autoradiograph.
3. Use sharp scissors or a scalpel to excise the piece of membrane containing the radioactive p53.

3.3.1. Trypsinization of the Radiolabeled Protein on the Membrane

1. Cut the excised membrane piece into small fragments and place these in the bottom of a 1.5-mL Eppendorf tube.
2. Use a liquid scintillation counter to measure the amount of radioactive material (i.e., the cpm) bound to the membrane (i.e., by Cerenkov radiation, *see* **Note 9**). Block the membrane fragment for 30 min at 37°C in 1 mL of 5% (w/v) polyvinylpyrrolidone in 0.1 M acetic acid. Wash the fragment four times in 1 mL of water followed by twice in 1 mL of 50 mM ammonium bicarbonate (Ambic).
3. Add 200 µL of Ambic plus 10 µg bovine serum albumin 10 µL of a 1 mg/mL solution of TPCK-treated trypsin to the tube, ensuring that the membrane fragments are completely immersed. Incubate overnight at 37°C.
4. The next morning add a further 10 µL of trypsin and incubate for a further 2 h.
5. Pellet the immobilon fragments by centrfugation at 12,000g for 1 min and transfer the supernatant to a fresh tube.
6. Wash the membrane fragments twice with 100 µL of Ambic to remove the remaining phosphopeptides. Measure the amount of radioactivity released from the membrane (in the Ambic/trypsin solution) and the amount remaining bound to the filter. The recovery of radioactive material at this stage should be approx 65%.
7. Add 300 µL of water to the solution containing the radioactive peptides. Substitute the cap on the eppendorf tube with a similar cap from another tube which has three or four holes in it (made by puncturing with a needle).
8. Freeze the sample on dry ice and place in a freeze drying apparatus until dry. Check that no radioactivity has been lost during the drying procedure by counting the sample in the scintillation counter (*see* **Note 10**). If there is a salt pellet remaining after the drying procedure, add another 300 µL of water and dry again.
9. To oxidise the peptides, remove the tube from the apparatus and add 50 µL of ice-cold, freshly made performic acid (*see* **Note 11**). Incubate on ice for 1 h.
10. Add 300 mL of water, freeze on dry ice then dry in the freeze-dryer. Check the amount of radioactive material in the sample again at this stage.

3.4. Two-Dimensional Tryptic Phosphopeptide Analysis

1. Resuspend the radioactive peptides in 5 µL of pH 1.9 buffer.
2. Pellet any remaining insoluble material at 12,000g in the microfuge for 5 min.

3. Carefully remove the supernatant fraction and check the recovery of radioactive material.

4. Spot this onto the origin on a cellulose thin layer plate, marked as described in **Fig. 2**; (always handle the plate wearing disposable gloves). Loading of the phosphopeptides should be carried out very carefully. We use a Gilson P2 pipet with appropriate tips. The phosphopeptide solution should be loaded incrementally using 0.2- to 0.4-µL aliquots. Each aliquot loaded on the plate should be dried with a hairdryer (*at the cold setting*; *see* **Note 12**) prior to loading the next aliquot (*see* **Note 13**).

5. Once the phosphopeptides are loaded, 2 µL of marker dye should be loaded at the appropriate position at the top of the plate as shown in **Fig. 2**.

3.4.1. Electrophoresis

1. In order to wet the plate, soak a blotter, prepared as described in **Fig. 3**, in pH 1.9 buffer. Mop of the excess liquid with a tissue but without drying out the blotter too much.

2. Place the thin layer plate on a light box (*see* **Note 14**) then place the blotter over the thin layer plate and press down quickly and evenly to ensure even wetting. In particular make sure that the buffer migrating towards the origin does so evenly from all sides.

3. Remove the blotter and place on the thin layer electrophoesis system as shown in **Fig. 4**. Prepare wicks from single sheets of 3MM paper; the dimensions of these should be 20×5 cm. Place one wick in each of the buffer reservoirs (making sure that they are completely wetted) and make even contact with each side of the thin layer plate adjacent to the reservoirs. A weight supplied with the electrophoresis system should be used to hold the wicks in place.

4. Cover the apparatus and carry out electrophoresis at 500 V for 1 h and 15 min (this is best done in a cold room). It is very important that the plate does not dry out during electrophoresis.

5. After electrophoresis, remove the thin layer plate and place it upright in a fume hood to allow the buffer to evaporate. The plate can be left in the fumehood overnight at this stage.

3.4.2. Chromatography

1. To prepare the chromatography tank, make up 200 mL of the chromatography "buffer" as described in the Materials section. Place this in the bottom of the tank and cover with a lid (*see* **Note 15**).

2. Prior to carrying out the chromatography, 2 µL of the marker dye mix should be spotted on the appropriate place on the thin layer plate (as shown in **Fig. 2**).

3. Open the chromatography tank and place the plate(s) in the tank such that the electrophoretically-separated peptides are at the bottom edge. The plate should rest against the wall of the tank. (It is possible to get two plates into the tank at any one time.)

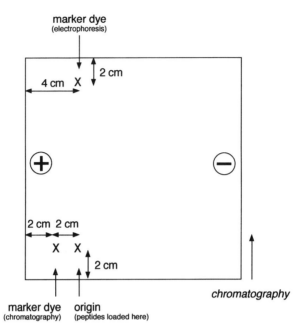

Fig. 2. Thin-layer plate preparation. Cellulose thin-layer plates are prepared for electrophoresis as shown in the diagram. The positions at which the sample and marker dyes are loaded on the plates are marked with a soft pencil as shown. It is often a good idea to mark the polarity of the plate and the direction of chromatography.

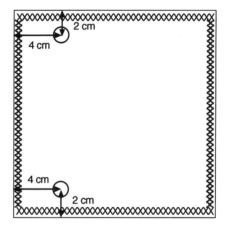

Fig. 3. Preparation of a blotter for the TLC Plates. We routinely use two sheets of Whatmann 3MM paper sewn together at the edges (using a sewing machine) as this strengthens the blotter for continued reuse. Holes that match the positions at which the sample and the first dimension marker dye are cut in the blotter as marked using a cork borer.

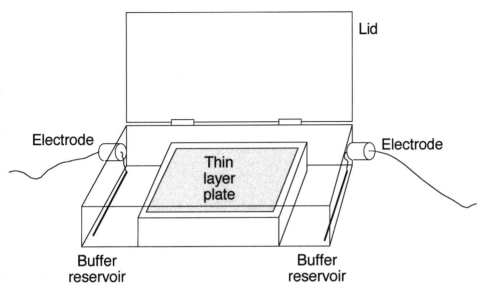

Fig. 4. The equipment used for separation in the first dimension (electrophoresis). Details of the choice of equipment are given in the text.

4. Place the lid on the tank, making sure that it is sealed, and allow the solvent to rise until it is about 1–2 cm from the top.
5. After chromatography, remove the plate and allow it to dry in the fume hood (*see* **Note 16**).
6. Attach appropriate phosphorescent labels to the plate and place it in an autoradiography (X-ray) cassette.
7. In the darkroom, place a piece of X-ray film over the plate and an enhancing screen over the film prior to closing the cassette.
8. Expose the cassette at –80°C for the desired length of time (*see* **Note 17**).

3.5. Interpretation of Data

A typical phosphopeptide map of p53 from SV3T3 cells is shown in **Fig. 5**. Under the separation conditions described above, peptides with a net negative charge will migrate towards the left hand side (anode), whereas peptides with a net positive charge migrate to the right hand side (cathode). At pH 1.9 (the pH at which the electrophoresis is carried out) lysine, arginine and histidine side chains will be protonated and therefore carry a positive charge. Glutamic acid and aspartic acids side chains will also have picked up protons and will be neutral. The N- and C-termini of any peptide will be positively charged and neutral respectively. Methionone sulphone and cysteic acid, the products of methionine and cysteine oxidation, will be neutral and negatively charged,

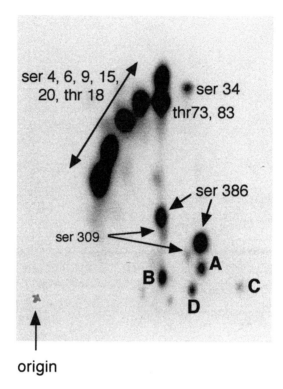

ser 4, 6, 9, 15, 20, thr 18

ser 34

thr73, 83

ser 386

ser 309

B

A

C

D

origin

Fig. 5. A typical phosphopeptide map of murine p53 from SV3T3 cells. The phosphopeptides shown are identified according to the phosphorylated residues that they contain. The origin at which the phosphopeptides were loaded is indicated. For a full explanation of the identity of these phosphopeptides, *see* **ref. 8**.

respectively. Phosphate groups on proteins will each carry a single negative charge. For full details of how pH influences the charge on residues and how the mobility of any peptide can be predicted based on its constituent amino acids, the reader is referred elsewhere *(14)*.

In **Fig. 5**, the *N*-terminal phosphopeptides of murine p53, which have a low charge/mass ratio, migrate readily with the solvent front. This includes three peptides: amino acids 1–24 containing serines 4, 6, 15, 20, and threonine 18, amino acids 25–59 containing serine 34, and amino acids 60–95 containing threonines 73 and 83. It should be noted that the addition of subsequent phosphate groups to a phosphopeptide shifts the migration of the phosphopeptide further towards the anode during electrophoresis and (being more hydrophilic) reduces the mobility with the solvent front during chromatography. Thus for the *N*-terminal peptide (amino acids 1–24) there are at least four spots lying on a diagonal in the map in **Fig. 5**. Moving from top right to bottom left over this diagonal the peptides contain one, two, three and four phosphates respectively. However, the technique gives no information about the residue(s) in this peptide at which the modification occurs.

The *C*-terminal phosphopeptides (at serines 309 and 386) lie in a basic domain of the protein and this is reflected in their mobilities. Similarly the peptides labeled A–D, which map to within the central DNA binding region of p53, are also located in a relatively basic region *(13)*. Trypsin digestion at *tandem* basic residues in a protein can give rise to more than one peptide containing the same phosphorylated residue(s). For example, tryptic digestion of p53 gives rise to two phosphopeptides each containing phosphoserine at position 386, one of which contains additionally a lysine residue at its N-terminus. This phosphopeptide migrates further toward the cathode during electrophoresis and has a reduced mobility during chromatography relative to the phosphopeptide lacking the additional lysine (*see* **Fig. 5**).

The picture obtained from the phosphopeptide analysis will also depend on the length of the exposure. Thus, longer exposure will reveal minor phosphopeptides which may not be visible in shorter exposures. The intensity of the individual phosphopeptides will give a relative (but not absolute) measure of the extent of phosphorylation, i.e. in comparison with other phosphopeptides on the same map. However, caution must be taken in interpreting these exposures, as a weak signal may reflect low turnover of phosphate at a particular site rather than low stoichiometry (for example, *see* **ref. *17***).

3.6. Applications and Advantages of the Method

Tryptic phosphopeptide analysis of p53 provides several advantages for the study of the regulation of this protein. First, the separation of individual phosphopeptides permits the analysis of multiple independent events and, provided that the site(s) of phosphorylation in any particular phosphopeptide is known, valuable information concerning the status of that site can be obtained independently of modifications at other residues. Second, the technique provides sensitivity in detection and separation and can be used to analyse the phosphorylation status of p53 even from cell lines in which the protein is rapidly turned over. Third, we also routinely examine p53, which has been phosphorylated in vitro using this technique. By using p53 proteins in which a potential phosphorylation site has been substituted using site-directed mutagenesis, we can then confirm the identity of that site by the absence of the appropriate phosphopeptide following phosphorylation of the protein in vitro or in vivo. Finally, when screening fractionated cell extracts, which may contain many activities that modify p53, the technique provides the ability to determine both the location and numbers of sites in the protein which are modified by activities in any particular fraction.

The are also disadvantages and limitations of the technique. First, it requires the use of the radioactive isotope [^{32}P]. As mentioned previously, this can be a potential disadvantage during cell labeling because the radiation itself can induce the p53 protein. Other techniques, such as the use of phosphospecific antibodies, can be used to assess the status of phosphorylation in western analyses without the need to use radiolabel (see Chapter 14). However, the nonradioactive approach requires a range of antibodies each of which is able to specifically recognize a particular phosphorylation site and is therefore more suited for the study of modification at individual sites rather than "global" modification of p53. Moreover, with short labeling times (e.g., 0.5–3 h), it is possible to obtain very clear phosphopeptide maps that show stimulus-related phosphorylation changes (for examples see **refs.** *12*, *15*, and *16*). In addition, one must use radioactivity to initially identify a phosphorylation site. Once the identity is confirmed, information about the biological role of the phosphorylation site can be obtained following the development of phosphospecific antibodies. The other issue associated with radiolabeling is the need to have the appropriate equipment and shielding for labeling with tens of millicuries.

4. Notes

1. Using the procedures described in this chapter, we have analysed the phosphorylation status of p53 in a range of different cell lines. We have obtained exceptionally good phosphopeptide maps of p53 from SV3T3 cells and we would recommend that researchers who are using the technique for the first time should first gain experience of the procedures using these cells. There are two principle reasons for the advantages provided by SV3T3 cells. First, because they are transformed by SV40, they express very high levels of p53 making detection of the protein and resulting phosphopeptides much more sensitive and providing ample radioactive material for analysis. Second, p53 phosphorylation is stimulated by T antigen and clear maps can be obtained which show most (if not all) of the phosphopeptides normally seen within the p53 protein.

2. The maximum period of incubation with [^{32}P] we use is 3 h. Using short labeling times minimises the induction of p53 by radiation-induced stress/damage.

3. We keep a shielded box with a rack for tubes of this size for this purpose. We also store the radioactive waste for at least 3 mo in such a box, behind lead bricks, in a locked shielded cupboard. This allows the radioactivity to decay sufficiently to permit safe disposal.

4. Lysis buffers other than the NP40 buffer can be used. We have also used RIPA buffer (which is essentially NP-40 buffer containing 0.1% [w/v] SDS and 1% [w/v] sodium deoxycholate). However, because RIPA effectively dissolves the nuclear structure and releases the DNA, the pellets obtained after centrifugation are not tightly packed and can be easily disturbed when removing the supernatant frac-

tion. This can give unacceptably high backgrounds when the immunoprecipitated p53 is analyzed by SDS-PAGE.

5. For immunoprecipitation of p53, it is possible to use any of a number of different anti-p53 antibodies. For example monoclonal antibodies such as PAb421, PAb242 (both of which are available commercially), and polyclonal antisera such as CM5, which was generated against murine p53 (a gift from David Lane, University of Dundee). Generally we use about 5 μg of antibody in each immunoprecipitation reaction.

6. It is important to remove all of the extract or buffer from the Protein A-sepharose beads following each microcentrifugation. We normally aspirate the supernatant fractions using a 10-mL syringe and a 23-gauge needle. The small bore of the needle minimizes loss of the beads from the pellet. However, extra care should be taken to prevent beads entering and blocking the needle.

7. Until recently, we used to dry down the polyacrylamide gel, expose to X-ray film and then extract the radiolabeled proteins from the gel. However, the procedure that we have now adopted and that is described **Subheading 3.2., step 7** involves transferring the proteins to a membrane prior to exposure to film. There are two major advantages to this approach. The first is that the background on the exposure is invariably lower. The second is that we do not have to extract the proteins from the gel (which would normally incur significant losses): The trypsinisation process, in addition to providing the peptide fragments for analysis, also releases them from the membrane. The only disadvantage to this procedure is that a small fraction of the protein usually fails to transfer to the membrane.

8. It is important to obtain an autoradiograph for subsequent isolation of the radioactive protein. Once the autoradiograph is available, the membrane may be exposed to a PhosphorImager screen if either quantitation or a digital image is required.

9. There is no need to add any scintillant as a measure of the Cerenkov radiation is ideal for determining recoveries at the different stages of preparation.

10. For lyophilization, if the sample has not been properly frozen prior to freeze-drying, it is possible to lose some material through the punctured lid owing to bubbling (release of dissolved gas) that occurs as the air pressure drops.

11. The oxidation procedure completely converts cyteine and methionine to the stable derivatives cysteic acid and methionine sufone, respectively. Without this procedure, artifactual and partial modification of these amino acids during the isolation and separation steps can lead to spurious results.

12. When drying the samples onto the thin-layer chromatography (TLC) plate only use the COLD setting on the hair dryer. If warm or hot, this will irreversibly bind the peptides to the plate. It is easier to observe the spots drying if the plate is placed on a lightbox during this procedure.

13. The repeated procedure of loading and drying very small volumes ensures that the diameter of the spot is kept to an absolute minimum and this in turn gives far tighter spots after the 2D separation on the final map.

14. Placing the thin-layer plate on a lightbox makes it easier to see whether all of the surface has been wetted by the blotter.
15. For chromatography, we normally seal the lid with vacuum grease to allow the solvent to equilibrate efficiently. It is also advizable to prepare the tank the day before it is required.
16. It is best to let the plates dry overnight.
17. The length of exposure is empirical. However, as a rule of thumb, a plate onto which 1000 Cerenkov cpm had been loaded would be given an initial exposure of about 2 d.

Acknowledgments

We wish to express our thanks to colleagues past and present in the laboratory, and, in particular, Uwe Knippschild, for developments, improvements and general feedback on the use of this procedure.

References

1. Agarwal, M. L. (1998) The p53 network. *J. Biol. Chem.*, **273,** 1–4.
2. Gottlieb, T. M. and Oren, M. (1996) p53 in growth control and neoplasia. *Biochem. Biophys. Acta*, **1287,** 77–102.
3. Ko, L. J. and Prives, C. (1996) p53: puzzle and paradigm. *Genes Dev.*, **10,** 1054–1072.
4. Levine, A. J. (1997) p53, the cellular gatekeeper for growth and division. *Cell*, **88,** 323–331.
5. Soussi, T. and May, P. (1996) Structural aspects of the p53 protein in relation to gene evolution: a second look. *J. Mol. Biol.*, **260**, 623–637.
6. Hall, P. A., Meek, D., and Lane, D. P. (1996) p53 - Integrating the complexity. *J. Path.*, **180,** 1–5.
7. Jacks, T. and Weinberg, R. A. (1996) Cell-cycle control and its watchman. *Nature*, **381,** 643–644.
8. Meek, D. W. (1997) Post-translational modification of p53 and the integration of stress signals. *Pathologie Biologie*, **45,** 804–814.
9. Meek, D. W. (1998) Multisite phosphorylation and the integration of stress signals at p53. *Cellular Signalling*, **10,** 159–166.
10. Lin, A. W., Barradas, M., Stone, J. C., van Aelst, L., Serrano, M., and Lowe, S. W. (1998) Premature senescence involving p53 and p16 is activated in response to constitutive MEK/MAPk mitogenic signaling. *Genes Dev.*, **12,** 3008–3019.
11. Shieh, S.-Y. , Ikeda, M., Taya, Y., and Prives, C. (1997) DNA damage-induced phosphorylation of p53 alleviates inhibition by MDM2. *Cell*, **91,** 325–334.
12. Siliciano, J. D. Canman, C. E., Taya, Y., Sakaguchi, K., Appella, E., and Kastan, M. B. (1997) DNA damage induces phosphorylation of the amino terminus of p53. *Genes Dev.* **11**, 3471–3481.
13. Milne, D. M. McKendrick, L., Jardine, L. J., Deacon, E., Lord, J. M., and Meek, D. W. (1996) Murine p53 is phosphorylated within the PAb421 epitope by protein kinase C in vitro, but not in vivo, even after stimulation with the phorbol ester o-tetradecanoylphorbol 13-acetate. *Oncogene*, **13,** 205–211.

14. Boyle, W. J., van der Geer, P., and Hunter, T. (1991) Phosphopeptide mapping and phosphoamino acid analysis by two-dimensional separation on thin-layer cellulose plates. *Methods Enzymol.*, **201,** 110–149.

15. Milne, D. M., Campbell, D. G., Caudwell, F. B., and Meek, D. W. (1994) Phosphorylation of the tumour suppressor protein p53 by mitogen activated protein (MAP) kinases. *J. Biol. Chem.*, **269,** 9253–9260.

16. Milne, D. M., Campbell, L., Campbell, D. G., and Meek, D. W. (1995) p53 is phosphorylated *in vitro* and *in vivo* by an ultra-violet radiation-induced protein kinase characteristic of the c-Jun kinase, JNK-1. *J. Biol. Chem.*, **270,** 5511–5518.

17. McKendrick, L., Milne, D. M., and Meek, D. W. (1999) Protein kinase CK2-dependent regulation of p53 function: evidence that the phosphorylation status of the serine 386 (CK2) site of p53 is constitutive and stable. *Mol. Cell. Biochem.* **191,** 187–199.

28

Development of Physiological Models to Study Stress Protein Responses

Ted R. Hupp

1. Introduction

Multicellular animals are exposed routinely to oxidizing chemicals and radiation from the environment, as well as endogenous metabolic by-products that can damage DNA and proteins over the life-span of the cell. Such damage may contribute to tissue injury, promote aging, and is implicated in many chronic degenerative human diseases, including cancer. The interplay of environmental agents with factors that control mammalian cell integrity have been most widely studied by employing tumor cell lines as a convenient source of homogeneous and rapidly growing cells. Although it could be argued that the use of tumor cell lines precludes the formation of an accurate understanding of the mechanisms regulating the normal cellular damage response, the use of such cultured systems has facilitated the discovery of a host of regulatory enzymes and repair factors.

However, recent analysis of widely studied stress proteins has shown that there is a pronounced cell-type specificity in their activation in vivo and this serves to underline the fact that very little is actually known about the in vivo function and regulation of the many well-characterized factors controlling damage responses. This review will focus on recently developed model systems that may be used to study the in vivo regulation of two of the most widely characterized stress-activated pathways. These are the recruitment of the tumor suppressor protein p53 and the heat-shock (HSPs) protein response (*See also* Chapters 17 and 27, this volume). New developments to be discussed include: the in vivo regulation of the tumor-suppressor protein p53, the role of HSPs in regulating normal stress responses in vivo, and dysregulation of the HSP and p53 pathways in models of cancer progression.

From: *Methods in Molecular Biology, vol. 99: Stress Response: Methods and Protocols*
Edited by: S. M. Keyse © Humana Press Inc., Totowa, NJ

2. The Tumor-Suppressor Protein p53 is a Key Regulator of Diverse Stress-Response Pathways In Vivo

2.1. Cell-Specific Activation of the Transcription Factor p53 After Whole-Body Exposure to Ionizing Radiation: The Guardian of the Genome Paradigm

The use of ionizing radiation as a chemotherapeutic agent has been recognized for almost a century and continues to be widely used for the treatment and palliation of many human cancers. Ionizing radiation can also be mutagenic or lethal to individual cells; thus, a critical balance must be achieved when using radiation as a form of anticancer treatment to ensure tumor cell death with minimal side-effects to normal tissue and organ function. Initial examination of tissues from animals exposed to whole-body radiation indicated that most organs did not exhibit any gross changes in morphology, although an unusual form of cell death involving "nuclear fragmentation" occurred rapidly at only a few sites; the spleen, thymus, bone marrow, and intestinal epithelium. This phenomenon of cellular and nuclear disintegration, now defined as apoptosis *(1)*, had long been known to be an outcome of whole-body irradiation in these mammalian cell types and has been most carefully studied in spleen, thymus, and intestinal epithelium *(2,3)*.

In contrast to extensive morphological studies, biochemical pathways that govern the differential survival or repair of normal cells exposed to ionizing radiation in vivo are only beginning to be defined. The discovery that ionising radiation-induced growth arrest in tissue culture systems is dependent upon the tumor suppressor protein p53 *(4)*, prompted further examination of the response of the p53 pathway to ionizing radiation injury in vivo. The biochemical activity most tightly linked to the tumor suppressor function of p53 is its sequence-specific-transactivation activity *(5)*, whereby p53 protein induces the expression of gene products that play a direct role in the cellular response to damage. Initial experiments had demonstrated that p53 protein can activate an ionising radiation-dependent apoptotic pathway in tissues known to be the acutely sensitive to radiation in vivo namely the thymus *(6,7)*, the bone marrow *(8)*, the intestinal epithelium *(9,10)*, and the spleen *(11)*. However, later studies demonstrated that the transcription factor function of p53 is also activated by ionising radiation in tissues that suffer no acute apoptotic responses in vivo, including the kidney, brain, lung, and the salivary gland duct epithelium *(12,13)*. Thus, despite the clear activation of the transactivation function of p53 after whole-body exposure to ionising radiation in many tissue types, the mechanism whereby p53 mediates cell death or the reason why its function is activated in nonapoptotic sites is not yet clear. However, the p53-dependent induction of O-6-alkyltransferase activity in liver relatively late after radiation exposure *(14)*, suggests a role for p53 in DNA repair and tissue maintenance in nonapoptotic sites after radiation injury.

Thus, p53 protein appears to be a key regulatory transcription factor used by an organism to protect from the proliferation of cells that have sustained potentially harmful mutations from ionising radiation. The ionizing radiation response in vivo will clearly involve other essential players; for example, a p53-independent apoptotic program is activated by ionizing radiation in intestinal epithelium *(15)* revealing the existence of a second pathway whose identification may have interesting consequences for radiation therapy. The cell specificity of stress-protein responses in vivo was further highlighted by studies comparing the induction of p53 protein in the gut of mice exposed to either ionizing radiation or 5-Fluorouracil *(16)*. Although ionizing radiation induces maximal apoptosis in the small intestine at a position in the crypt containing stem cells, 5-Fluorouracil dependent apoptosis occurred in rapidly proliferating transit cells. In contrast in the colon, 5-Fluorouracil promoted apoptosis at the base of the crypt whereas stem cells at the base of the crypt were insensitive to ionizing radiation.

In addition to p53, three relevant factors have been recently identified as playing a role in the radiation response in vivo. These are superoxide dismutase (SOD), catalase, and poly (ADP-ribose) polymerase (PARP). Ionizing radiation induces direct damage to DNA but also indirect DNA damage, lipid peroxidation, and protein oxidation from the generation of highly reactive oxygen intermediates. Cells respond to the production of reactive oxygen species by inducing antioxidant enzymes that play a role in scavenging these reactive oxygen intermediates thereby minimizing chemical damage to the cell. It is therefore not surprising that radiation-resistant strains of mice rapidly induce both SOD and catalase activities in the liver following irradiation, whereas radiation sensitive mouse strains show no such increases in the activities of these enzymes *(17)*. In addition to enzymes that affect the rate of reactive oxygen damage to cells there are factors which control the repair of direct DNA damage. One of the most notable of these is PARP (poly-ADP ribose polymerase). Transgenic animals that lack PARP are viable but show enhanced death after whole body exposure to ionizing radiation resulting from acute radiation toxicity to the small intestine *(18)*. A similar radiation sensitivity in the small intestine is observed in animals with severe-combined immunodeficiency (SCID) *(19)*. This may reflect the disruption of a more global damage response pathway which uses the DNA-dependent protein kinase (DNA-PK), p53, and PARP as modulators of the rates of DNA repair and apoptosis in the intestinal epithelium in vivo. Ongoing detailed studies of the factors which control the in vivo response of tissues to ionizing radiation will have important implications for developing diagnostic assays for defining radiation-sensitive patients and in developing more rational approaches in the use of ionizing radiation as a form of anticancer treatment.

2.2. Activation of the p53 Pathway in Regenerating Liver

The "Guardian of the Genome" metaphor that developed from research into the function of p53 (20) was an important milestone in p53 research as it focused attention on the inducible nature of the p53 pathway. Although irradiation-dependent DNA damage was the first signal shown to activate p53, it is clear that many non-genotoxic agents or "stresses" can also lead to p53 activation. Among these is a physiologically relevant pathway that plays a role in the control of hepatocyte proliferation. Tissue regeneration after death of hepatic cells is a basic biological response of the liver to injury and damage. Many cytokines and growth factors including EGF, TGF-β, IL-6, TNF-α, and norepinephrine, play an important role in the re-growth of injured liver (21). After biological stages of hepatic growth and restructuring, the regeneration process is complete within 72 h and falls under the control of growth inhibitory pathways.

Although the hepatic regeneration response does not involve p53, a damage-induced growth arrest pathway during regeneration recruits p53 for maximal effect (22). Using primary hepatocytes ex-vivo, a significantly greater proportion of the population of cells were cycling after receiving a proliferative stimulus in p53-deficient strains, compared to the wild-type cells. This indicates that p53-null cells are more likely to proliferate than normal hepatocytes under the control of non-genotoxic growth promoting pathways. In addition, TGF-β at concentrations which induce a growth arrest in normal cells, failed to abrogate proliferation in p53-null cells, providing a compelling link between the TGF-β pathway and p53-growth arrest responses. These p53-dependent proliferative control mechanisms were also shown to function in vivo, as animals treated with carbon tetrachloride or with a nongenotoxic mitogen promoted a greater degree of proliferation or hyperplasia in liver which lack p53 protein. In addition, ionizing radiation-dependent G1 and G2 checkpoints require p53 in the regenerating liver, providing a further role for p53 in inducible proliferation control in this organ. The most striking data to emerge from the use of hepatocytes as a model to study p53 regulation is the identification of a TGF-β signaling network that promotes p53 function. These studies have identified a model system likely to be fruitful in defining novel links between stress-activated kinase/phosphatase pathways, the mdm2-dependent p53 degradation pathways and kinase pathways which are thought to play a role in the regulation of p53.

2.3. Activation of the p53 Pathway in Injured Neuronal Tissue

Historically, studies aimed at understanding the physiological role of stress proteins in vivo (i.e., heat shock proteins [HSPs]; *see* **Subheading 3.**) have centered substantially on their expression in brain, as this organ is a common site of ischemic injury affecting organ function. For example, transient ischemic brain injury or intracerebral hemorrhage in animal models, can result in prolonged and elevated expression of heme oxygenase-1 (Hsp32) protein *(23,24)*, (*see also* Chapters 18 and 23, this volume) and overproduction of SOD in the brain reduces stress-induced expression of HSP70 *(25,26)*, suggesting a direct role for oxygen radicals in mediating HSP expression in injured neuronal tissue. Thus, neuronal tissue has provided a good in vivo model linking ischemic injury to expression of "classical" stress proteins. Independent studies showed that the induction of p53 protein correlates with apoptosis in a range of cells within the brain, including oligodendrocytes *(27,28)*. Most strikingly, although embryonic neurons from p53 null mice undergo programmed cell death in the absence of neurotrophins, stroke-induced neuronal damage is attenuated *(29)*, suggesting that p53 plays a direct role in ischemic damage and possibly also seizures *(30–32)*. As such, animal models have been developed that center on the regulation of neuronal apoptosis by exogenous agents and its relationship to the p53 pathway. One apoptotic model involves intrastriatial injection of the *N*-methyl-D-aspartate (NMDA) receptor agonist, quinolinic acid, which induces DNA fragmentation in both the striatum and neocortex, as well as mediating the induction of p53 protein, BAX, and GADD45 *(33)*. An intact adrenal gland appears to be required for p53-mediated neuronal cell death *(31)*, providing evidence for hormonal regulation of the p53 pathway in vivo.

Interestingly, the neurotoxicity elicited by quinolinic acid resembles many neurochemical and pathological characteristics of Huntington's disease *(34)*, and recent studies have recorded DNA fragmentation (i.e., apoptotic signatures) in tissue from patients with Huntington's disease *(35)*. Together, these data suggest that ischemia and chemically-induced neuronal damage is p53-dependent and defines a model system that is likely to provide new links between the ischemic/HSP responses and apoptotic pathways.

3. Regulation of the HSP Response In Vivo

3.1. Regulation and Function of the HSP Pathway

One of the most evolutionarily conserved and widely recruited cellular defense pathways comprises the stress protein or heat shock protein family members. These polypeptides, termed molecular chaperones, are classified by their apparent molecular weights and generally include proteins of 25, 40, 60,

70, 90, and 110 kDa. The precise role of these proteins in response to cellular stress is centered around their functions as molecular chaperones (*36, see also* Chapters 24 and 25), in which they can refold misfolded proteins or promote degradation of irreversibly damaged proteins, thus aiding cellular repair and survival following stress. Hsp25/27 and Hsp70 proteins are induced following exposure to many types of cellular damaging agents that can induce either genotoxic or nongenotoxic damage and their overproduction has been shown to confer protection from damage-induced death (*37,38*), providing evidence for an important role for many HSP family members in protecting cells from environmental damage. The biological consequences of stress protein induction in normal or tumor cells can involve, not only repair of damaged polypeptides and cellular survival after injury, but acquisition of thermotolerance or protection of cells form normally lethal levels of damage (*39*). The mechanism for the antiapoptotic activity of certain chaperones is not precisely clear. However, recent studies have shown that the GrpE-like protein BAG-1, which can stimulate the ATPase activity of Hsc70 (*40–42*), can counteract p53-dependent growth arrest pathways (*43*). The latter studies demonstrate the potent affect that chaperone components play in promoting cell survival after damage.

Organ models in mammals have also indicated a therapeutic effect of activation of the heat shock pathway on tissue integrity and repair. Whole body hyperthermia (*44,45*) or in vivo transfection with the Hsp70 gene (*46*) can protect against myocardial damage in response to ischemia and reperfusion. Heat shock by perfusion in kidney can protect the tissue from damage induced by warm ischemic injury (*47,48*). Strikingly, whole- body hyperthermia 20 h prior to a normally lethal dose of radiation can prevent chromatin fragmentation as well as thymocyte and bone marrow cell death (*49*), indicating that hyperthermia can be a potent radioprotector against p53-dependent apoptotic pathways in vivo.

The induction of heat shock proteins is controlled by the heat-shock transcription factor family of proteins (HSF-1 through HSF-4) (*50, see also* Chapter 17). The heat inducible HSF is regulated at multiple points and may act in conjunction with other HSF homologs. It is thought that vertebrate HSF1 is stored in an inactive monomeric form in undamaged cells and the negative regulation of HSF1 in undamaged cells is thought to occur via an intrinsic autoregulatory domain that when disrupted by heat shock converts the HSF into an active trimer. Further, chaperones themselves are thought to contribute to a negative regulation of the HSF family after heat shock (*51,52*) and phosphorylation plays a negative regulatory role in suppressing HSF function as a transcription factor (*53*). Thus, HSF1 appears to be controlled at multiple steps that ensures its activity is tightly controlled. In addition, the dependence of thermotolerance on HSF3 activity in a chicken B lymphocyte cells, suggests an interaction between HSF1 and HSF3 and highlights the complexity of the stress protein response in cultured cells *(54–56)*. Whether these conclusions hold true in vivo remains to be determined.

3.2. Activation of the Heat-Shock Pathway In Vivo

Although most studies on the stress-protein responses in mammals have been centered on in vitro studies using tissue culture cell lines, the effects of in vivo hyperthermia, inflammation, or damage on heat-shock protein induction are beginning to be unravelled using animals subjected to reversible hyperthermic treatment. Internal body temperature in rodents is regulated by the metabolic activity of brown adipose tissue and relatively small elevations in the external temperature over the internal body temperature can have profound effects on hyperthermic induction *(57)*. This internal rheostat is controlled, in part, by cyclic-AMP-dependent protein kinase pathways, as PKA-regulatory site knockout mice have elevated metabolic rates and elevated internal body temperatures *(58)*. Using this in vivo hyperthermic regime, one of the most striking results to emerge is the remarkable discordance in the expression of the HSP genes after heat shock in vivo *(59)*. The cell-specific expression of homologues of the transcription factor HSF *(56,60,61)* may account, in part, for such tissue-specific responses. Similarly, activation of the heat shock activated protein kinases, including SAPK/JNK, are not induced in many tissues during in vivo heat shock and display activation that is uncoupled from temperature changes *(62)*. Although SAPK/JNK activities are not induced by in vivo heat shock in spleen or lung, cells derived from these same organs, and cultured in vitro do induce SAPK/JNK after heat stress. These data further establish that heat shock induction of the SAPK/JNK pathway in vivo lacks the similar control that operates on cultured cells, suggesting the existence of physiological factors that regulate the tissue response to damage.

The HSP pathway can be activated not only by exogenous stresses, but normal physiological changes can affect the HSP pathway in vivo in a tissue-specific manner. Hsp70 protein levels increase during exercise in some tissues and Hsp70 protein levels are related to the content of slow oxidative or type I muscle fibers *(63–65)*. Phosphorylation of Hsp70 protein in skeletal muscle may play a role in early signal transduction events after exercise *(66)*, suggesting a link between protein kinase/phosphatase function, HSF activation, and the chaperone pathway in this tissue. Together these data indicate that environmentally or physiologically-generated signals can activate the HSP pathway in vivo, but this is cell-specific, suggesting the existence of cell-type specific factors that control the organ responses to stress.

4. Regulation of the HSP and p53 Damage Response Pathways in Clinical Models of Cancer Progression

4.1. Molecular Chaperones as Effective Anticancer Drug Targets: p53-HSP Interactions in Tumor Cells

There has been some evidence recently acquired for a direct interaction between the p53 and HSP pathways in normal cells. For example, p53 protein is involved in suppressing HSP70 protein levels in normal fibroblasts by many orders of magnitude *(67)*, presumably through direct inhibition of hsp70 gene transcription *(68)*. In primary murine fibroblasts the classic heat-shock-induced growth arrest pathway has an absolute requirement for p53 function *(69)*. It is not yet known whether p53 protein also functions in vivo as a heat shock protein. Nevertheless, these recent data have indicated for the first time that there may be some cooperation in normal cells between the p53 and HSP pathways to coordinately assist in repair of the growth arrested and heat-injured cell. What has been more evident is that the HSP and p53 pathways show direct interactions in tumour cells and that the two pathways can be coordinately altered at specific stages in the multi-step process of carcinogenesis (described below).

The first cellular protein shown to bind to p53 included a member of the HSP70 family of proteins *(70)*, whose associations with p53 have since been extended to include the molecular chaperones HSP40 *(71)* and HSP90 *(72,73)*. These three heat shock proteins form a holoenzyme complex that can coordinately refold denatured proteins or protect from unfolding and may sequester p53 protein in the cytoplasm in tissue culture cells and in vivo. These data provide a specific mechanism for p53 inactivation in tumors. HSPs have also been shown to prevent drug or radiation-dependent apoptosis in cells *(74–77)* and to block p53 function *(78,79)*, highlighting the role these proteins may play in tumour cell survival or anti-apoptotic survival pathways. The relevance of the interaction of p53 with heat-shock proteins in tumour cells have been unclear, but recent evidence described below suggests that one component of the antiapoptotic function of HSPs may be related to the control of the conformation and inactivation of p53. As a result, drugs that disrupt HSP-p53 interactions in tumours may hold promising therapeutic potential.

A striking breakthrough in dissecting molecular pathways that regulate mutant p53 protein conformation and stability in tumour cells came from independent studies examining the mechanism of function of the benzoquinone ansamycin class of anti-tumour compounds, which include Geldanamycin *(80)*. The anti-proliferative activity of Geldanamycin is attributed to depletion of oncogenic proteins including erbB2 receptor kinases and RAF-1 *(81)*. The mechanism whereby these kinases are inactivated by Geldanamycin appears to depend in part on the binding to and inhibition of the molecular chaperones HSP90 *(82)*, which is involved in assembly of the activated kinases.

Given that one of the major HSPs bound to mutant p53 in tumors appears to be a member of the HSP90 family of chaperones and that HSP90 is the major target of Geldanamycin, studies were initiated to examine whether mutant p53 folding in vitro and stability in vivo is regulated by molecular chaperones. A pronounced reduction in mutant p53 protein levels upon treatment of tumour cells with Geldanamycin *(83,84)* mediated by the proteosome *(85)* and the reassembly of the mutant and unfolded p53 into the wild-type conformation *(86)* demonstrates that HSP90 may play a significant role in modulating mutant p53 conformation and stability. Geldanamycin has been previously reported to be effective in halting tumor cell proliferation in animal models *(87–89)*, so it is evident that, at least within the context of models of tumor cell proliferation, the HSPs play a role in tumor cell survival in vivo and that they can be effective anticancer drug targets. Recent reports also show in models of colorectal cancer development that cytoplasmic sequestration of p53 and colocalization with HSP70 occurs before p53 mutation and adenocarcinoma development, suggesting HSPs play a direct role in p53 inactivation during the process of tumorigenesis.

4.1.2. Dysregulation of the Heat-Shock Protein and p53 Pathways at an Early Stage in a Clinical Model of Carcinogenesis

Despite the fact that p53 protein is mutated in most cancers, the multistep processes which give rise to such tumors have only been delineated in a small number. Two major clinical models that have given mechanistic insight into the multistep nature of human neoplasia include colorectal and oesophageal cancer *(90,91)*, primarily because patients can present complications to clinicians relatively early in the neoplastic sequence which allows clinical material to be biopsied for study. The development of neoplasia in both of these tissue types involve mutations in tumor suppressor genes such as APC and p53, but there are striking variations in the stage of the neoplastic sequence at which such mutations occur. In colorectal cancer, mutations in p53 occur very late in the sequence to adenocarcinoma (**Fig. 1A**), while in the oesophagus p53 mutations can occur earlier during metaplasia (**Fig. 1B**). Strikingly, mutations in p53 can be detected in diploid cells *(92)* and in metaplastic epithelium of the oeosphagus *(93)* prior to dysplastic lesions. The earlier mutation pattern of p53 in squamous epithelium suggests a unique requirement for inactivation of the p53-dependent damage-induced cell-cycle checkpoint pathway and presumably involves the type of environmental damage imposed upon cells of the oesophagus and oral mucosa by chemical oxidants. The types of naturally

A

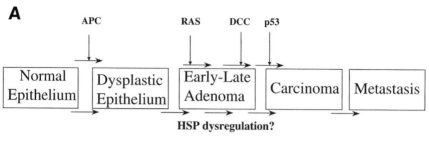

B

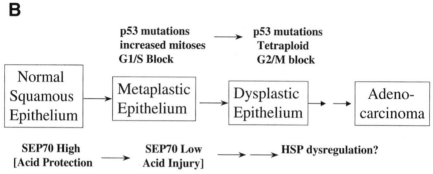

Fig. 1. (**A**) Colorectal cancer progression. The progression of colorectal tumourigenesis is driven by sequential mutation in oncogenes (K-RAS) and tumor suppressor genes (APC, DCC, and p53). At each stage in the progression sequence certain endogenous or environmental selection pressures promote mutation in growth regulatory genes and clonal expansion of the tumor. In the case of p53, hypoxic conditions in adenomas may be the type of stress-response that recruits the p53 pathway and cancers that expand under such conditions will have mutated and inactivated p53. It is not clear if the tumor suppressor function of p53 is required at earlier stages, prior to development of adenomas, as no evidence has been found using animal models for an increase in the rate of adenoma formation in APC ±, p53 –/– animals. Additional changes such as the inactivation of genes which control the rate of mutations (MSH2, MLH1, PMS, [not shown]) and other epigenetic events may play a role in accelerating the rate of tumor formation. Based on cultured tumour cell line studies, aberrant overexpression of HSPs will give selective survival advantage in growing tumors by blocking apoptotic pathways. Studies of the temporal activation of these HSP or related antiapoptotic pathways in vivo will be a fruitful area of future research. (**B**) Oesophageal cancer progression. When compared with colorectal cancer, the progression of oesophageal adenocarcinoma is less-well-defined genetically, but striking differences have emerged. Most notable is that mutation of the p53 gene occurs much earlier in the progression sequence, even prior to dysplasia. These data suggest that unique environmental pressures are being placed on oesophageal cells that normally involve activation of p53's tumour suppressor function to halt cancer development. Candidate damaging agents which presumably activate p53 under these conditions include chemical oxidants and thermal injury from food, beverage, and smoke, as well as acid from

occurring environmental agents that can activate p53 protein include; unmodified or "activated" genotoxins, radiation, and stresses such as lowered oxygen concentration, acid, and heat. Thus, to develop a physiological system to study p53 regulation, attempts must be made to blend these two observations; i.e., to identify sites that are damaged by agents relevant to a human model of carcinogenesis and to identify a tissue amenable to integrating biochemistry and clinical studies. The human oesophageal epithelium provides such a model.

This tissue is exposed routinely to chemicals from the environment, as well as refluxed acid and bile from the stomach *(94,95)*. Cellular damage or trauma induced by refluxed acid/bile, carcinogens, alcohol and thermal irritation to the oesophageal epithelium appears to be an initiating event in the evolution of ulceration, Barrett's oesophagus, p53 mutation, and dysplasia *(96–98)*. Recent clinical work has indicated that very hot beverage intake is one of the key factors that predispose to oesophageal cancer development *(99)*, suggesting that thermal injury and its rate of repair will affect cancer progression in this tissue. To begin to use the oesphageal epithelium as a physiologically relevant model systems to study regulation of p53 protein in mammalian tissue and how inactivation of the p53 pathway contributes to neoplasia, ex-vivo organ culture methods using human epithelial biopsies have been devised to study the biochemical response of human oesophageal squamous cells to damage with the long term goal of studying activation of p53 by acid, heat, and genotoxic injury.

The conclusions of this work include that this tissue has the unusual property of down-regulating the normal heat-shock protein HSP70 after stress *(100)*, and upregulating a stress protein named squamous epithelium p70 (SEP70), which is a functional homologue of the glucose regulated family of stress-proteins. This abundant stress protein has since been micropurified from normal human oesophageal biopsies, and using monoclonal antibodies generated to the pure protein, it has been shown that SEP70 is downregulated below the level of detection in metaplastic epithelium from the oesophagus (our unpub-

---

reflux. The types of genes activated by p53 in response to such stress may be redox regulatory genes such as glutathione-S-transferase or thioredoxins. Cancer progression will therefore involve mutation in p53 at an early stage in order to allow for the inactivation of this key stress-responsive regulatory pathway. Where and when in this progression sequence HSPs function in protection from the unusual type of damage imposed upon this tissue and whether HSPs provide antiapoptotic roles in cancer development in vivo remain an attractive area of research for the future. However, initial studies have shown that one unusual feature of oeosphageal squamous epithelium is that the major heat-shock protein is a glucose regulated stress protein that is induced by heat shock and not the "classical" HSP70 *(100)*.

lished observations). These latter data suggest that downregulation of stress-protein pathways can occur very early in a carcinogenesis sequence and further highlights the need to develop physiological models to develop a molecular understanding of the role of stress proteins in human disease progression or resistance. Further research into this system could involve understanding the whether reduction of stress protein levels in metaplastic epithelium further predisposes this tissue to acid and thermal injury and inactivation of p53 protein, both of which play a direct role in oesophageal disease progression.

5. Conclusions

Analysis of the pathways controlling the cellular response to damage in mammals has been possible, in part, as a result of the development of cultured tumour cell lines in vitro. The physiological significance of many of these stress-activated pathways remains unclear, but recent research has started to bridge the basic science with animal models of injury, to place the molecular data in a biological context and to identify novel factors implicated in vivo in assisting the regulation of stress-activated pathways. Using the p53 and HSP pathways as models, key observations reviewed here include that p53 and/or HSP activation in response to damage in vivo is cell-specific, suggesting the existence of tissue-specific regulatory factors that activate or suppress these pathways. The further development of animal and clinical models will be essential in defining the role of stress-protein pathways in normal and disease processes and will no doubt lead to the discovery of unexpected links between known and novel regulatory factors that contribute to the regulation of tissue repair.

References

1. Kerr, J. F. R., Wyllie, A. H., and Currie, A. R. (1971) Apoptosis: a basic biological phenomenon with wide ranging implications in tissue kinetics. *British J. Cancer* **26,** 239–257.
2. Quastler, H. (1956) The nature of intestinal radiation death. *Radiation Res.* **4,** 303–320.
3. Petrakis, N. L. (1957) Quantitative histological analysis of the early effects of whole–body irradiation on the mouse thymus. *Radiation Res.* **5,** 569–572.
4. Kastan, M. B., Onyekwere, O., Sidransky, D., Vogelstein, B., and Craig, R. W. (1991) Participation of p53 protein in the cellular response to DNA damage. *Cancer Res.* **51,** 6304–6311.
5. Pietenpol, J. A., Tokino, T., Thiagalingam, S., el–Deiry, W. S., Kinzler, K. W., and Vogelstein, B. (1994) Sequence–specific transcriptional activation is essential for growth suppression by p53. *Proc. Natl. Acad. Sci. USA* **91,** 1998–2002.
6. Clarke, A. R., Purdie, C. A., Harrison, D. J., Morris, R. G., Bird, C. C., Hooper, M. L., and Wyllie, A. H. (1993) Thymocyte apoptosis induced by p53–dependent and independent pathways. *Nature* **362,** 849–852.

7. Lowe, S. W., Schmitt, E. M., Smith, S. W., Osborne, B. A., and Jacks, T. (1993) p53 is required for radiation–induced apoptosis in mouse thymocytes. *Nature* **362,** 847–849.

8. Lee, J. M. and Bernstein, A. (1993) p53 mutations increase resistance to ionizing radiation. *Proc. Natl. Acad. Sci. USA* **90,** 5742–5746.

9. Clarke, A. R., Gledhill, S., Hooper, M. L., Bird, C. C., and Wyllie, A. H. (1994) p53 dependence of early apoptotic and proliferative responses within the mouse intestinal epithelium following gamma–irradiation. *Oncogene* **9,** 1767–1773.

10. Merritt, A. J., Potten, C. S., Kemp, C. J., Hickman, J. A., Balmain, A., Lane, D. P., and Hall, P. A. (1994) The role of p53 in spontaneous and radiation–induced apoptosis in the gastrointestinal tract of normal and p53–deficient mice. *Cancer Res.* **54,** 614–617.

11. Midgley, C. A., Owens, B., Briscoe, C. V., Thomas, D. B., Lane, D. P., and Hall, P. A. (1995) Coupling between gamma irradiation, p53 induction and the apoptotic response depends upon cell type in vivo. *J. Cell Sci.* **108,** 1843–1848.

12. Macleod, K. F., Sherry, N., Hannon, G., Beach, D., Tokino, T., Kinzler, K., Vogelstein, B., and Jacks, T. (1995) p53–dependent and independent expression of p21 during cell growth, differentiation, and DNA damage. *Genes Dev.* **9,** 935–944.

13. MacCallum, D. E., Hupp, T. R., Midgley, C. A., Stuart, D., Campbell, S. J., Harper, A., Walsh, F. S., Wright, E. G., Balmain, A., Lane, D. P., and Hall, P. A. (1996) The p53 response to ionising radiation in adult and developing murine tissues. *Oncogene* **13,** 2575–2587.

14. Rafferty, J. A., Clarke, A. R., Sellappan, D., Koref, M. S., Frayling, I. M., and Margison, G. P. (1996) Induction of murine O6–alkylguanine–DNA–alkyltransferase in response to ionising radiation is p53 gene dose dependent. *Oncogene* **12,** 693–697.

15. Merritt, A. J., Allen, T. D., Potten, C. S., and Hickman, J. A. (1997) Apoptosis in small intestinal epithelial from p53–null mice: evidence for a delayed, p53–independent G2/M–associated cell death after gamma– irra-diation. *Oncogene* **14,** 2759–2766.

16. Pritchard, D. M., Watson, A. J., Potten, C. S., Jackman, A. L., and Hickman, J. A. (1997) Inhibition by uridine but not thymidine of p53–dependent intestinal apoptosis initiated by 5–fluorouracil: evidence for the involvement of RNA perturbation. *Proc. Natl. Acad. Sci. USA* **94,** 1795–1799.

17. Hardmeier, R., Hoeger, H., Fang–Kircher, S., Khoschsorur, A., and Lubec, G. (1997) Transcription and activity of antioxidant enzymes after ionizing irradia-tion in radiation–resistant and radiation–sensitive mice. *Proc. Natl. Acad. Sci. USA* **94,** 7572–7576.

18. de Murcia, J. M., Niedergang, C., Trucco, C., Ricoul, M., Dutrillaux, B., Mark, M., et al. (1997) Requirement of poly(ADP–ribose) polymerase in recovery from DNA damage in mice and in cells. *Proc. Natl. Acad. Sci. USA* **94,** 7303–7307.

19. Biedermann, K. A., Sun, J. R., Giaccia, A. J., Tosto, L. M., and Brown, J. M. (1991) scid mutation in mice confers hypersensitivity to ionizing radiation and a

deficiency in DNA double–strand break repair. *Proc. Natl. Acad. Sci. USA* **88,** 1394–1397.

20. Lane, D P. (1992) p53, guardian of the genome. *Nature* **358,** 15–16.
21. Michalopoulos, G. K. and DeFrances, M. C. (1997) Liver regeneration. *Science* **276,** 60–66.
22. Bellamy, C. O., Clarke, A. R., Wyllie, A. H., and Harrison, D. J. (1997) p53 Deficiency in liver reduces local control of survival and proliferation, but does not affect apoptosis after DNA damage. *FASEB J.* **11,** 591–599.
23. Raju, V. S. and Maines, M. D. (1996) Renal ischemia/reperfusion up–regulates heme oxygenase–1 (HSP32) expression and increases cGMP in rat heart. *J. Pharmacol. Exp. Ther.* **277,** 1814–1822.
24. Nimura, T., Weinstein, P. R., Massa, S. M., Panter, S., and Sharp, F. R. (1996) Heme oxygenase–1 (HO–1) protein induction in rat brain following focal ischemia. *Brain Res. Mol. Brain Res.* **37,** 201–208.
25. Mikawa, S., Sharp, F. R., Kamii, H., Kinouchi, H., Epstein, C. J., and Chan, P. K. (1995) Expression of c–fos and hsp70 mRNA after traumatic brain injury in transgenic mice overexpressing CuZn–superoxide dismutase. *Brain Res. Mol. Brain Res.* **33,** 288–294.
26. Kamii, H., Kinouchi, H., Sharp, F. R., Koistinaho, J., Epstein, C. J., and Chan, P. H. (1994) Prolonged expression of hsp70 mRNA following transient focal cerebral ischemia in transgenic mice overexpressing CuZn–superoxide dismutase. *J. Cereb. Blood Flow Metab.* **14,** 478–486.
27. Eizenberg, O., Faber–Elman, A., Gottlieb, E., Oren, M., Rotter, V., and Schwartz, M. (1996) p53 plays a regulatory role in differentiation and apoptosis of central nervous system–associated cells. *Mol. Cell Biol.* **16,** 5178–5185.
28. Eizenberg, O., Faber–Elman, A., Gottlieb, E., Oren, M., Rotter, V., and Schwartz, M. (1995) Direct involvement of p53 in programmed cell death of oligodendrocytes. *EMBO J.* **14,** 1136–1144.
29. Hughes, P. E., Alexi, T., and Schreiber, S. S. (1997) A role for the tumour suppressor gene p53 in regulating neuronal apoptosis. *Neuroreport* **8,** 5–12.
30. Sakhi, S., Sun, N., Wing, L. L., Mehta, P., and Schreiber, S. S. (1996) Nuclear accumulation of p53 protein following kainic acid–induced seizures. *Neuroreport* **7,** 493–496.
31. Sakhi, S., Gilmore, W., Tran, N. D., and Schreiber, S. S. (1996) p53–deficient mice are protected against adrenalectomy–induced apoptosis. *Neuroreport* **8,** 233–235.
32. Sakhi, S., Bruce, A., Sun, N., Tocco, G., Baudry, M., and Schreiber, S. S. (1997) Induction of tumor suppressor p53 and DNA fragmentation in organotypic hippocampal cultures following excitotoxin treatment. *Exp. Neurol.* **145,** 81–88.
33. Hughes, P. E., Alexi, T., Yoshida, T., Schreiber, S. S., and Knusel, B. (1996) Excitotoxic lesion of rat brain with quinolinic acid induces expression of p53 messenger RNA and protein and p53–inducible genes Bax and Gadd– 45 in brain areas showing DNA fragmentation. *Neuroscience* **74,** 1143–1160.
34. Anderson, K. D., Panayotatos, N., Corcoran, T. L., Lindsay, R. M., and Wiegand, S. J. (1996) Ciliary neurotrophic factor protects striatal output neurons in an animal model of Huntington disease. *Proc. Natl. Acad. Sci. USA* **93,** 7346–7351.

35. Dragunow, M., Faull, R. L., Lawlor, P., Beilharz, E. J., Singleton, K., Walker, E. B., and Mee, E. (1995) In situ evidence for DNA fragmentation in Huntington's disease striatum and Alzheimer's disease temporal lobes. *Neuroreport* **6,** 1053–1057.
36. Hartl, F. U. (1996) Molecular chaperones in cellular protein folding. *Nature* **381,** 571–579.
37. Li, G. C., Mivechi, N. F., and Weitzel, G. (1995) Heat shock proteins, thermotolerance, and their relevance to clinical hyperthermia. *Int. J. Hyperthermia* **11,** 459–468.
38. Li, G. C., Li, L. G., Liu, Y. K., Mak, J. Y., Chen, L. L., and Lee, W. M. (1991) Thermal response of rat fibroblasts stably transfected with the human 70–kDa heat shock protein–encoding gene. *Proc. Natl. Acad. Sci. USA* **88,** 1681–1685.
39. Minowada, G. and Welch, W. J. (1995) Clinical implications of the stress response. *J. Clin. Invest.* **95,** 3–12.
40. Hohfeld, J. and Jentsch, S. (1997) GrpE–like regulation of the hsc70 chaperone by the anti–apoptotic protein BAG–1. *EMBO J.* **16,** 6209–6216. (*see also* erratum in *EMBO J.* 1998 **17**: 847).
41. Takayama, S., Bimston, D. N., Matsuzawa, S., Freeman, B. C., Aime–Sempe, C., Xie, Z., Morimoto, R. I., and Reed, J. C. (1997) BAG–1 modulates the chaperone activity of Hsp70/Hsc70. *EMBO J.* **16,** 4887–4896.
42. Bimston, D., Song, J., Winchester, D., Takayama, S., Reed, J. C., and Morimoto, R. I. (1998) BAG–1, a negative regulator of Hsp70 chaperone activity, uncouples nucleotide hydrolysis from substrate release. *EMBO J.* **17,** 6871–6878.
43. Matsuzawa, S., Takayama, S., Froesch, B. A., Zapata, J. M., and Reed, J. C. (1998) p53–inducible human homologue of Drosophila seven in absentia (Siah) inhibits cell growth: suppression by BAG–1. *EMBO J.* **17,** 2736–2747.
44. Donnelly, T. J., Sievers, R. E., Vissern, F. L., Welch, W. J., and Wolfe, C. L. (1992) Heat shock protein induction in rat hearts. A role for improved myocardial salvage after ischemia and reperfusion? *Circulation* **85,** 769–778.
45. Currie, R. W., Tanguay, R. M., and Kingma, J. G., Jr. (1993) Heat–shock response and limitation of tissue necrosis during occlusion/reperfusion in rabbit hearts. *Circulation* **87,** 963–971.
46. Suzuki, K., Sawa, Y., Kaneda, Y., Ichikawa, H., Shirakura, R., and Matsuda, H. (1997) In vivo gene transfection with heat shock protein 70 enhances myocardial tolerance to ischemia–reperfusion injury in rat. *J. Clin. Invest.* **99,** 1645–1650.
47. Chaston, G., Perdizet, G. A., Anderson, C., Pleau, C., Berman, M., and Schweiser, R. (1990) Heat shock protects rat kidneys against warm ischaemic injury. *Curr. Surg.* **47,** 420–422.
48. Perdizet, G. A., Heffron, T. G., Buckingham, F. C., Salciunas, P. J., Gaber, A. O., Stuart, F. P., and Thistlewaite, J. R. (1989) Stress conditioning: a novel approach to organ preservation. *Curr. Surg.* **46,** 23–25.
49. Shen, R. N., Crabtree, W. N., Wu, B., Young, P., Sandison, G. A., Hornback, N. B., and Shidnia, H. (1992) A reliable method for quantitating chromatin fragments by flow cytometry to predict the effect of total body irradiation and hyperthermia on mice. *Int. J. Radiat. Oncol. Biol. Phys.* **24,** 139–143.

50. Morimoto, R. I. (1998) Regulation of the heat shock transcriptional response: cross talk between a family of heat shock factors, molecular chaperones, and negative regulators. *Genes Dev.* **12,** 3788–3796.
51. Shi, Y., Mosser, D. D., and Morimoto, R. I. (1998) Molecular chaperones as HSF1–specific transcriptional repressors. *Genes Dev.* **12,** 654–666.
52. Satyal, S. H., Chen, D., Fox, S. G., Kramer, J. M., and Morimoto, R. I. (1998) Negative regulation of the heat shock transcriptional response by HSBP1. *Genes Dev.* **12,** 1962–1974.
53. Kline, M. P. and Morimoto, R. I. (1997) Repression of the heat shock factor 1 transcriptional activation domain is modulated by constitutive phosphorylation. *Mol. Cell. Biol.* **17,** 2107–2115.
54. Tanabe, M., Kawazoe, Y., Takeda, S., Morimoto, R. I., Nagata, K., and Nakai, A. (1998) Disruption of the HSF3 gene results in the severe reduction of heat shock gene expression and loss of thermotolerance. *EMBO J.* **17,** 1750–1758.
55. Nakai, A., Kawazoe, Y., Tanabe, M., Nagata, K., and Morimoto, R. I. (1995) The DNA–binding properties of two heat shock factors, HSF1 and HSF3, are induced in the avian erythroblast cell line HD6. *Mol. Cell. Biol.* **15,** 5268–5278.
56. Kanei–Ishii, C., Tanikawa, J., Nakai, A., Morimoto, R. I., and Ishii, S. (1997) Activation of heat shock transcription factor 3 by c–Myb in the absence of cellular stress. *Science* **277,** 246–248.
57. Blake, M. J., Fargnoli, J., Gershon, D., and Holbrook, N. J. (1991) Concomitant decline in heat–induced hyperthermia and HSP70 mRNA expression in aged rats. *Am. J. Physiol.* **260,** R663–667.
58. Cummings, D. E., Brandon, E. P., Planas, J. V., Motamed, K., Idzerda, R. L., and McKnight, G. S. (1996) Genetically lean mice result from targeted disruption of the RII beta subunit of protein kinase A. *Nature* **382,** 622–626.
59. Blake, M. J., Gershon, D., Fargnoli, J., and Holbrook, N. J. (1990) Discordant expression of heat shock protein mRNAs in tissues of heat– stressed rats. *J. Biol. Chem.* **265,** 15275–15279.
60. Nakai, A., Tanabe, M., Kawazoe, Y., Inazawa, J., Morimoto, R. I., and Nagata, K. (1997) HSF4, a new member of the human heat shock factor family which lacks properties of a transcriptional activator. *Mol. Cell. Biol.* **17,** 469–481.
61. Fiorenza, M. T., Farkas, T., Dissing, M., Kolding, D., and Zimarino, V. (1995) Complex expression of murine heat shock transcription factors. *Nucleic Acids Res.* **23,** 467–474.
62. Hu, Y., Metzler, B., and Xu, Q. (1997) Discordant activation of stress–activated protein kinases or c–Jun NH2– terminal protein kinases in tissues of heat–stressed mice. *J. Biol. Chem.* **272,** 9113–9119.
63. Locke, M., Noble, E. G., and Atkinson, B. G. (1990) Exercising mammals synthesize stress proteins. *Am. J. Physiol.* **258,** C723–729.
64. Locke, M., Atkinson, B. G., Tanguay, R. M., and Noble, E. G. (1994) Shifts in type I fiber proportion in rat hindlimb muscle are accompanied by changes in HSP72 content. *Am. J. Physiol.* **266,** C1240–1246.
65. Locke, M. and Noble, E. G. (1995) Stress proteins: the exercise response. *Can. J. Appl. Physiol.* **20,** 155–167.

66. Hernando, R. and Manso, R. (1997) Muscle fibre stress in response to exercise: synthesis, accumulation and isoform transitions of 70–kDa heat–shock proteins. *Eur. J. Biochem.* **243,** 460–467.

67. He, C., Merrick, B. A., Patterson, R. M., and Selkirk, J.K. (1995) Altered protein synthesis in p53 null and hemizygous transgenic mouse embryonic fibroblasts. *Appl. Theor. Electrophor.* **5,** 15–24.

68. Agoff, S. N., Hou, J., Linzer, D. I., and Wu, B. (1993) Regulation of the human hsp70 promoter by p53. *Science* **259,** 84–87.

69. Nitta, M., Okamura, H., Aizawa, S., and Yamaizumi, M. (1997) Heat shock induces transient p53–dependent cell cycle arrest at G1/S. *Oncogene* **15,** 561–568.

70. Pinhasi–Kimhi, O., Michalovitz, D., Ben–Zeev, A., and Oren, M. (1986) Specific interaction between the p53 cellular tumour antigen and major heat shock proteins. *Nature* **320,** 182–184.

71. Sugito, K., Yamane, M., Hattori, H., Hayashi, Y., Tohnai, I., Ueda, M., Tsuchida, N., and Ohtsuka, K. (1995) Interaction between hsp70 and hsp40, eukaryotic homologues of DnaK and DnaJ, in human cells expressing mutant–type p53. *FEBS Lett.* **358,** 161–164.

72. Sepehrnia, B., Paz, I. B., Dasgupta, G., and Momand, J. (1996) Heat shock protein 84 forms a complex with mutant p53 protein predominantly within a cytoplasmic compartment of the cell. *J. Biol. Chem.* **271,** 15,084–15,090.

73. Dasgupta, G. and Momand, J. (1997) Geldanamycin prevents nuclear translocation of mutant p53. *Exp. Cell Res.* **237,** 29–37.

74. Gordon, S. A., Hoffman, R. A., Simmons, R. L., and Ford, H. R. (1997) Induction of heat shock protein 70 protects thymocytes against radiation–induced apoptosis. *Arch. Surg.* **132,** 1277–1282.

75. Mehlen, P., Schulze–Osthoff, K., and Arrigo, A. P. (1996) Small stress proteins as novel regulators of apoptosis. Heat shock protein 27 blocks Fas/APO–1–and staurosporine–induced cell death. *J. Biol.Chem.* **271,** 16,510–16,514.

76. Polla, B. S., Kantengwa, S., Francois, D., Salvioli, S., Franceschi, C., Marsac, C., and Cossarizza, A. (1996) Mitochondria are selective targets for the protective effects of heat shock against oxidative injury. *Proc. Natl. Acad. Sci. USA* **93,** 6458–6463.

77. Li, W. X., Chen, C. H., Ling, C. C., and Li, G. C. (1996) Apoptosis in heat–induced cell killing: the protective role of hsp–70 and the sensitization effect of the c–myc gene. *Radiat. Res.* **145,** 324–330.

78. Kaul, S. C., Duncan, E. L., Englezou, A., Takano, S., Reddel, R. R., Mitsui, Y., and Wadhwa, R. (1998) Malignant transformation of NIH3T3 cells by overexpression of mot–2 protein. *Oncogene* **17,** 907–911.

79. Wadhwa, R., Takano, S., Robert, M., Yoshida, A., Nomura, H., Reddel, R. R., Mitsui, Y., and Kaul, S. C. (1998) Inactivation of tumor suppressor p53 by mot–2, a hsp70 family member. *J. Biol. Chem.* **273,** 29,586–29,591.

80. Schulte, T. W. and Neckers, L. M. (1998) The benzoquinone ansamycin 17–allylamino–17–demethoxygeldanamycin binds to HSP90 and shares important biologic activities with geldanamycin. *Cancer Chemother. Pharmacol.* **42,** 273–279.

81. An, W. G., Schnur, R. C., Neckers, L., and Blagosklonny, M. V. (1997) Depletion of p185erbB2, Raf-1 and mutant p53 proteins by geldanamycin derivatives correlates with antiproliferative activity. *Cancer Chemother. Pharmacol.* **40**, 60–64.

82. Stebbins, C. E., Russo, A. A., Schneider, C., Rosen, N., Hartl, F. U., and Pavletich, N. P. (1997) Crystal structure of an Hsp90–geldanamycin complex: targeting of a protein chaperone by an antitumor agent. *Cell* **89**, 239–250.

83. Blagosklonny, M. V., Toretsky, J., and Neckers, L. (1995) Geldanamycin selectively destabilizes and conformationally alters mutated p53. *Oncogene* **11**, 933–939.

84. Mimnaugh, E. G., Chavany, C., and Neckers, L. (1996) Polyubiquitination and proteasomal degradation of the p185c–erbB-2 receptor protein–tyrosine kinase induced by geldanamycin. *J. Biol. Chem.* **271**, 22,796–22,801.

85. Whitesell, L., Sutphin, P., An, W. G., Schulte, T., Blagosklonny, M. V., and Neckers, L. (1997) Geldanamycin–stimulated destabilization of mutated p53 is mediated by the proteasome in vivo. *Oncogene* **14**, 2809–2816.

86. Blagosklonny, M. V., Toretsky, J., Bohen, S., and Neckers, L. (1996) Mutant conformation of p53 translated in vitro or in vivo requires functional HSP90. *Proc. Natl. Acad. Sci. USA* **93**, 8379–8383.

87. Sasaki, K., Yasuda, H., and Onodera, K. (1979) Growth inhibition of virus transformed cells in vitro and antitumor activity in vivo of geldanamycin and its derivatives. *J. Antibiot. (Tokyo)*, **32**, 849–851.

88. Scheibel, T. and Buchner, J. (1998) The Hsp90 complex—a super–chaperone machine as a novel drug target. *Biochem. Pharmacol.* **56**, 675–682.

89. Supko, J. G., Hickman, R. L., Grever, M. R., and Malspeis, L. (1995) Preclinical pharmacologic evaluation of geldanamycin as an antitumor agent. *Cancer Chemother. Pharmacol.* **36**, 305–315.

90. Neshat, K., Sanchez, C. A., Galipeau, P. C., Cowan, D. S., Ramel, S., Levine, D. S., and Reid, B. J. (1994) Barrett's esophagus: a model of human neoplastic progression. *Cold Spring Harb. Symp. Quant. Biol.* **59**, 577–583.

91. Vogelstein, B. and Kinzler, K. W. (1994) Colorectal cancer and the intersection between basic and clinical research. *Cold Spring Harb. Symp. Quant. Biol.* **59**, 517–521.

92. Neshat, K., Sanchez, C. A., Galipeau, P. C., Blount, P. L., Levine, D. S., Joslyn, G., and Reid, B. J. (1994) p53 mutations in Barrett's adenocarcinoma and high–grade dysplasia. *Gastroenterology* **106**, 1589–1595.

93. Campomenosi, P., Conio, M., Bogliolo, M., Urbini, S., Assereto, P., Aprile, A., Monti, P., Aste, H., Lapertosa, G., Inga, A., Abbondandolo, A., and Fronza, G. (1996) p53 is frequently mutated in Barrett's metaplasia of the intestinal type. *Cancer Epidemiol. Biomarkers Prev*, **5**, 559–565.

94. Tobey, N. A. and Orlando, R. C. (1991) Mechanisms of acid injury to rabbit esophageal epithelium. Role of basolateral cell membrane acidification. *Gastroenterology* **101**, 1220–1228.

95. Seto, Y. and Kobori, O. (1993) Role of reflux oesophagitis and acid in the development of columnar epithelium in the rat oesophagus. *Br. J. Surg.* **80**, 467–470.

96. Jaskiewicz, K., Louw, J. and Anichkov, N. (1994) Barrett's oesophagus: mucin composition, neuroendocrine cells, p53 protein, cellular proliferation and differentiation. *Anticancer Res.* **14,** 1907–1912.
97. Ramel, S., Reid, B. J., Sanchez, C. A., Blount, P. L., Levine, D. S., Neshat, K., Haggitt, R. C., Dean, P. J., Thor, K., and Rabinovitch, P. S. (1992) Evaluation of p53 protein expression in Barrett's esophagus by two– parameter flow cytometry. *Gastroenterology* **102,** 1220–1228.
98. Reid, B. J., Barrett, M. T., Galipeau, P. C., Sanchez, C. A., Neshat, K., Cowan, D. S., and Levine, D. S. (1996) Barrett's esophagus: ordering the events that lead to cancer. *Eur. J. Cancer Prev* **5 Suppl 2,** 57–65.
99. Casson, A. G., Tammemagi, M., Eskandarian, S., Redston, M., McLaughlin, J., and Ozcelik, H. (1998) p53 alterations in oesophageal cancer: association with clinicopathological features, risk factors, and survival. *Mol. Pathol.* **51,** 71–79.
100. Hopwood, D., Moitra, S., Vojtesek, B., Johnston, D. A., Dillon, J. F., and Hupp, T. R. (1997) Biochemical analysis of the stress protein response in human oesophageal epithelium. *Gut* **41,** 156–163.

Index

A

Activator protein 1 (AP-1), 205, 347
AICA riboside as an activator of
    AMPK, 66, 70
Amino acid analysis, 30
AMP-activated protein kinase (AMPK),
    assay of 63, 65, 68
Analyis of SAGE tags, 301, 312
Anion exchange chromatography of
    eluted IKK signalsome, 119
Anticancer drugs, 472
Antiphosphotyrosine antibodies, 79
Apoptosis, 466, 469
Ataxia-telangiectasia (A-T), 99, 448
ATM, expression constructs, 102
ATM recombinant, assay of, 99, 105
ATPase assay (of Hsp70), 400

B

β-galactosidase assays, 291
Bilirubin and biliverdin, detection and
    assay of, 379–382
Biliverdin reductase, preparation and
    assay of, 382
Buthionine [S,R]-sulfoximine
    (BSO), 261

C

Calculation of carbonyl content, 20
Carbonyl groups, 15, 35
Carcinogenesis, 473
Cell cycle checkpoints, 99, 468
Chaperones, 393, 421, 472
Chaperone activity in vivo , 408–411
Chloramphenicol acetyl transferase
    (CAT) assays, 284, 287, 289

Cloning and sequencing SAGE
    concatamers, 311
Clonogenic cell survival assay, 411
Colorectal cancer progression
    model, 474
Competitive hybridisation, 322
Contaminating linkers in SAGE, 308

D

Data interpretation for ESI-MS, 8
Decrease in electrophoretic mobility as
    an assay for SAPKs, 137
Deuterium oxide (D_2O), 264
Dichlorodihydrofluorescein-diacetate
    (H_2DCF-DA), 36
Dichlorofluorescein (DCF)
    fluorescence as an assay of
    oxidative stress in
    lymphocytes, 35
Differential display PCR (DD-PCR),
    348, 353
2,4-Dinitrophenylhydrazine, 16, 17,
    19, 20
DNA damage and repair, 99, 447, 467
DNA-dependent protein kinase (DNA-
    PK), 85,87, 89, 90, 91, 448, 467
DNA double-strand break repair, 85

E

Electrophoretic mobility shift assays
    (EMSAs), 206, 207, 209, 229
Electrospray ionisation-mass
    spectrometry (ESI-MS), 5
Enzyme-linked immunosorbent assay
    (ELISA) kinase assay, 183
ERK5 (BMK-1) assay of, 134

F

Flow cytometry assay for DCF
    fluorescence, 40
Fluorogenic peptides as substrates for
    proteases, 52
Fluorescence *in situ* hybridisation, 238
Free radicals, 15, 25

G

GAL4 fusions as sensors for kinase
    activation, 147
Gel filtration analysis of proteins
    228, 401
Generating SABRE libraries, 332
Genome instability, 99
GF 109203X, inhibitor of protein
    kinase C, 164, 165, 171, 172
Glutathione depletion, 261, 369
Glycerol gradient fractionation of
    proteins, 227

H

Heat shock proteins (Hsp's) 217, 244,
    393, 469–472
Heat shock response, 217, 393, 447, 465
Heat shock transcription factors (HSFs)
    217, 470
Heme oxygenase-1(HO-1) activity
    assay, 374, 378
Heme oxygenase-1 (HO-1), induction
    of, 257, 264, 369, 469
High pressure liquid chromatgraphy
    (HPLC), 16, 29
HIV-1 LTR-promoter, 277, 279
HIV-1 LTR, reporter constructs, 284, 288
Human immunodeficiency virus type-1
    (HIV-1), 277
Hydrogen peroxide (H_2O_2), 35, 78, 267
Hydroxyl radical (•HO), 35, 266
4-Hydroxynonenal adducts, 25, 28, 29, 30

I

IκB, 110, 205, 279
IκB kinase, assay of, 115,

IKK signalsome, 109
Immune complex kinase assays, 80
Immunofluorescence analysis of HSF
    proteins, 225
Immunological detection and analysis
    of HSF proteins, 223
Immunological detection of hsp25, 438
Immunoprecipitation of the IKK
    signalsome, 118
Immunoprecipitation of p53
    protein, 453
Immunoprecipitation and detection of
    ATM, 103
Immunoprecipitation of tyrosine
    kinases, 78
"In gel" kinase assays, 136
In vitro footprinting, 234
In vivo genomic footprinting, 231
Instrument conditions for ESI-MS, 8
Interpretation of phosphopeptide maps,
    457–459
Ionising radiation, 466
Iron levels in cells, 269, 371

K

Kinase inhibitors, 161

L

Lipid peroxidation, 25
Luciferase activity assays, 155, 213
LY 294002, inhibitor of PI 3-kinase,
    164, 165, 170

M

MAPKAP kinase-1 (p90Rsk), 168, 172
MAP kinase activated protein kinases
    (MAPKAP kinases), assay of,
    130, 132, 133, 440
MAP kinase kinase (MKK) assays, 131,
    138, 139
Mass spectrometry,4
Matrix-assisted laser desorption
    ionisation-mass spectrometry
    (MALDI-MS), 4

Metabolic labelling of cellular
proteins, 57
Microsequencing , protein sample
preparation for, 120
Microtitre plate assay for DCF
fluorescence, 39
Mitogen-activated protein kinase
(MAPK), 127, 145, 161, 280
Mitogen and stress activated protein
kinase-1 (Msk-1), assay of
130, 133

N

Native gel electrophoresis assay, 404
Nitric oxide (NO), 3, 35, 369
Nitrosative stress, 3
Nitrosothiol (RNSO) formation of, 4
Nonradioactive kinase assays in vitro,
181, 184
Northern blot analysis, 357
Nuclear factor-κB (NFκB), 110, 205,
279, 347
Nuclear "run on" transcription, 236

O

Oxidative modification of proteins, 15
Oxidative stress, 35, 49, 57, 75, 145,
218, 277, 369, 393

P

P53, as "guardian of the genome," 466
P53 tumor suppressor, 100, 178, 184,
299, 447, 465, 468, 472–476
P53, phosphorylation of, 447
P70 S6 kinase, 169,
PD98059 inhibitor of MAP kinase
kinase 1(MEK1), 163, 165,
167–169
Phosphopeptides, immunization
with, 179
Phosphospecific antibodies, 177,
431, 432
Poly (ADP-ribose) polymerase
(PARP), 467

Polymerase chain reaction (PCR),
321, 334
Preparation of SAGE libraries, 305
Primer extension analysis of RNA, 242
Protein aggregation, 422, 426
Protein degradation (proteolysis),
measurement of, 49, 54, 56
Protein folding, 393
Protein kinase B (AKT), 170
Protein kinase C (PKC), 171
Protein kinase recognition motifs, 191
Protein phosphorylation, analysis
in vivo, 180
Protein purification (of Hsp's),
398–400
Protein refolding assays, 404, 405,
424–426
Protein solubility assay, 403
Pull down assay for DNA-PK, 88, 92

R

Radiolabeling of cells, 452
Random peptide libraries, 194, 199
Raney nickel catalysis, 26
Rapamycin, 164, 165, 169
Rational design of peptide variants,
194, 198
Reactive nitrogen species (RNS), 3
Reactive oxygen species (ROS), 35, 257
Ro 318220 inhibitor of protein kinase
C, 164, 165, 171, 172

S

S1 nuclease protection assay, 243
SABRE, selection of tester
homohybrids, 336
Sample preparation for ESI-MS, 7
SAP kinase kinase (SKK) assays, 131,
138, 139
SB203580 and SB202190, inhibitors of
SAPK2/p38, 163–167, 280, 441
Selective amplification via biotin- and
restriction- mediated enrichment
(SABRE), 321

Serial analysis of gene expression (SAGE), 297
Singlet oxygen, 264
Small heat shock proteins, 394
Small heat shock protein kinase activity, 438
Small heat shock proteins, phosphorylation of, 431
Snapshot assay of DCF fluorescence, 42
Src-family kinases, 76
Stress-activated protein kinases (SAPKs), 127, 130, 132, 134–136, 145, 148, 471
Stress-induced gene expression 277, 321, 347
Stress protein responses, 465
Superoxide anion (O_2^-), 35, 272
Synthetic peptide kinase assays, 193, 197

T

Thermotolerance, 393
Transcription factor activation domains, 147
Transcriptional reporter assays, 145, 206, 208, 211, 240

Transfection of transcription reporter constructs, 150, 153, 154, 212, 288, 289
Transforming growth factor-β (TGF-β), 468
Two-dimensional gel electrophoresis of proteins, 245, 436
Two-dimensional tryptic phosphopeptide mapping, 454
Tyrosine kinases, 75

U

Ubiquitination, 110, 205
Unit activities of protein kinases, calculation of, 139
Ultraviolet-A radiation and inducible gene expression, 257
Ultraviolet-B radiation and inducible gene expression, 347
Ultraviolet irradiation of cells, 78, 352

V

V(D)J recombination, 85

W

Western blot of DNA-PK, 94
Wortmannin, inhibitor of PI 3-kinase, 164, 165, 170